Delivering performance in food supply chains

Related titles:

Improving traceability in food processing and distribution
(ISBN 978-1-85573-959-8)
In the light of recent legislation and a number of food safety incidents, traceability of food products back from the consumer to the very beginning of the supply chain has never been so important. This book describes key components of traceability systems and how food manufacturers can manage them effectively. The first part of the book reviews the role of traceability systems not only in ensuring food safety but in optimising business performance. Part II looks at ways of building traceability systems whilst Part III reviews key traceability technologies such as DNA markers, electronic tagging of farm animals, ways of storing and transmitting traceability data and the range of data carrier technologies.

Food processing technology: principles and practice Third edition
(ISBN 978-1-84569-216-2)
The first edition of *Food processing technology* was quickly adopted as the standard text by many food science and technology courses. The publication of a completely revised and updated third edition consolidates the position of this text book as the best single-volume introduction to food manufacturing technologies available. The third edition has been updated and extended to include the many developments that have taken place since the second edition was published. In particular, advances in microprocessor control of equipment, 'minimal' processing technologies, functional foods, developments in 'active' or 'intelligent' packaging, and storage and distribution logistics are described. Technologies that relate to cost savings, environmental improvement or enhanced product quality are highlighted. Additionally, sections in each chapter on the impact of processing on foodborne micro-organisms are included for the first time.

Chilled foods: a comprehensive guide Third edition
(ISBN 978-1-84569-243-8)
The key requirements for chilled food products are good quality and microbiological safety at the point of consumption. The first edition of *Chilled foods* quickly established itself as the standard work on these issues for all those involved this important sector. The third edition has strengthened that reputation. The latest edition has been extensively revised, with entirely new chapters on raw materials, the operation of chilled food manufacturing plants, non-microbial hazards such as allergens, predictive microbiology, shelf-life, management of product safety and quality and packaging. Greater input has been sought from those in the chilled food industry to increase the publication's relevance to practitioners.

Details of these books and a complete list of Woodhead titles can be obtained by:

- visiting our web site at www.woodheadpublishing.com
- contacting Customer Services (e-mail: sales@woodheadpublishing.com; fax: +44 (0) 1223 893694; tel.: +44 (0) 1223 891358 ext. 130; address: Woodhead Publishing Limited, Abington Hall, Granta Park, Great Abington, Cambridge CB21 6AH, UK)

Woodhead Publishing Series in Food Science, Technology and Nutrition:
Number 185

Delivering performance in food supply chains

Edited by
Carlos Mena and Graham Stevens

CRC Press
Boca Raton Boston New York Washington, DC

WOODHEAD PUBLISHING LIMITED
Oxford   Cambridge   New Delhi

Published by Woodhead Publishing Limited, Abington Hall, Granta Park,
Great Abington, Cambridge CB21 6AH, UK
www.woodheadpublishing.com

Woodhead Publishing India Private Limited, G-2, Vardaan House, 7/28 Ansari Road,
Daryaganj, New Delhi – 110002, India
www.woodheadpublishingindia.com

Published in North America by CRC Press LLC, 6000 Broken Sound Parkway, NW,
Suite 300, Boca Raton, FL 33487, USA

First published 2010, Woodhead Publishing Limited and CRC Press LLC
© Woodhead Publishing Limited, 2010
The authors have asserted their moral rights.

British Library Cataloguing in Publication Data
A catalogue record for this book is available from the British Library.

Library of Congress Cataloging in Publication Data
A catalog record for this book is available from the Library of Congress.

Woodhead Publishing ISBN 978-1-84569-471-5 (book)
Woodhead Publishing ISBN 978-1-84569-777-8 (e-book)
CRC Press ISBN 978-1-4398-2726-0
CRC Press order number: N10144

The publishers' policy is to use permanent paper from mills that operate a sustainable
forestry policy, and which has been manufactured from pulp which is processed using
acid-free and elemental chlorine-free practices. Furthermore, the publishers ensure that the
text paper and cover board used have met acceptable environmental accreditation
standards.

Typeset by Toppan Best-set Premedia Limited, Hong Kong
Printed by TJ International Limited, Padstow, Cornwall, UK

Contents

Contributor contact details

(* = main contact)

Chapter 1

Dr Carlos Mena*
Senior Research Fellow
Centre for Logistics and Supply
   Chain Management
Cranfield School of Management
Cranfield University
Cranfield
Bedfordshire
MK43 0AL
UK

E-mail: carlos.mena@cranfield.
   ac.uk

Graham Stevens
Visiting Fellow
Cranfield University
Cranfield
Bedfordshire
MK43 0AL
UK

E-mail: graham@gcsconsulting.org.
   uk

Chapter 2

Dr Andrew S Humphries*
1 Castle Rose
Woughton Park
Milton Keynes
MK6 3BQ
UK

E-mail: andrew.humphries@
   sccindex.com

Linda McComie
42 Stone Hill
Two Mile Ash
Milton Keynes
MK8 8LR
UK

E-mail: linda.mccomie@sccindex.
   com

Chapter 3

Dr Martin Hingley*
School of Management
Harper Adams University College
Newport
Shropshire
TF10 8NB
UK

E-mail: mhingley@harper-adams.
ac.uk

Professor Adam Lindgreen
Hull University Business School
Cottingham Road
Hull
HU6 7RX
UK

E-mail: a.lindgreen@hull.ac.uk

Chapter 4

Dr Denyse Julien
Cranfield University
School of Applied Sciences
Bldg 50, College Road
Cranfield
Bedfordshire
MK43 0HT
UK

E-mail: d.m.julien@cranfield.ac.uk

Chapter 5

Dr Silvia Estrada-Flores
Principal Consultant
Food Chain Intelligence
PO Box 1789
North Sydney 2059
NSW
Australia

E-mail: silvia@food-chain.com.au

Chapter 6

Dr Öznur Yurt
Department of Logistics
  Management
Izmir University of Economics
Sakarya Cad. No:155 35330
Balcova/Izmir
Turkey

E-mail: oznur.yurt@ieu.edu.tr

Dr Carlos Mena*
Senior Research Fellow
Centre for Logistics and Supply
  Chain Management
School of Management
Cranfield University
Cranfield
Bedfordshire
MK43 0HT
UK

E-mail: carlos.mena@cranfield.
ac.uk

Graham Stevens
Visiting Fellow
Cranfield University
Cranfield
Bedfordshire
MK43 0AL
UK

E-mail: graham@gcsconsulting.org.
uk

Chapter 7

Dr Andrew Thomas*
Newport Business School
University of Wales Newport
Allt Yr Yn Campus
Newport
NP20 5DA
UK

E-mail: andrew.thomas@newport.ac.uk

Dr Yingli Wang and Dr Andrew Potter
Logistics Systems Dynamics Group
Cardiff Business School
Cardiff University
Colum Road
Cardiff
CF10 3EU
UK

E-mail: wangy14@cardiff.ac.uk;
  potterat@cardiff.ac.uk

Chapter 8

Daniel Chicksand*
Operations Management Group
Warwick Business School
Warwick University
Coventry
CV4 7AL
UK

E-mail: Daniel.Chicksand@wbs.ac.uk

Professor Andrew Cox, Chairman,
  Newpoint Consulting and visiting
  Professor at University of
  San Diego
Newpoint Consulting Limited
147 Luddington Road
Stratford-upon-Avon
Warwickshire
CV37 9SQ
UK

E-mail: a.w.cox@bham.ac.uk

Chapter 9

Dr Keivan Zokaei
Senior Research Associate and
  Director of MSc in Lean Ops
  (Service)
Lean Enterprise Research Centre
Cardiff Business School CBTC
Senghennydd Road
Cardiff
CF24 4AY
UK

E-mail: zokaeiak@Cardiff.ac.uk

Chapter 10

Professor Alan Harrison
Cranfield School of Management
Cranfield University
Cranfield
Bedfordshire
MK43 0AL
UK

E-mail: a.harrison@cranfield.ac.uk

Chapter 11

Dr Paul A. Chapman
Saïd Business School
Egrove Park
University of Oxford
Oxford
OX1 5NY
UK

E-mail: paul.chapman@sbs.ox.
    ac.uk

Chapter 12

Simon Templar
Teaching Fellow
Centre for Logistics and Supply
    Chain Management
Cranfield School of Management
Cranfield University
Cranfield
Bedfordshire
MK43 0AL
UK

E-mail. simon.templar@cranfield.
    ac.uk

Dr Carlos Mena*
Senior Research Fellow
Centre for Logistics and Supply
    Chain Management
Cranfield School of Management
Cranfield University
Cranfield
Bedfordshire
MK43 0AL
UK

E-mail. carlos.mena@cranfield.
    ac.uk

Chapter 13

Dr Charles Stephens
5 Croft Way
Everton
Doncaster
DN10 5DL
UK

E-mail: stephenseverton@
    btinternet.com

Chapter 14

Dr Marian Garcia Martinez
Senior Lecturer in Agri-Food
    Marketing
Kent Business School
The University of Kent
Canterbury
Kent CT2 7PE
UK

E-mail: m.garcia@kent.ac.uk

Chapter 15

Dr Richard Baines
Royal Agricultural College
Cirencester
Gloucestershire
GL7 6JS
UK

E-mail: Richard.baines@rac.ac.uk

Chapter 16

Professor Ludwig Theuvsen
Georg-August University of
    Goettingen
Platz der Göttingen Sieben 5
37073 Göttingen
Germany

E-mail: theuvsen@uni-goettingen.
    de

Chapter 17

Duncan Hobday
Cranfield School of Management
Cranfield University
Cranfield
Bedfordshire
MK43 0AL
UK

E-mail: d.hobday.s05@cranfield.
ac.uk

S. P. J. Higson
Cranfield Health
Cranfield University
Cranfield
Bedfordshire
MK43 0AL
UK

Dr Carlos Mena*
Senior Research Fellow
Centre for Logistics and Supply
Chain Management
Cranfield School of Management
Cranfield University
Cranfield
Bedfordshire
MK43 0AL
UK

Chapter 18

Stephen J. James* and Christian
James
Food Refrigeration & Process
Engineering Research Centre
(FRPERC)
The Grimsby Institute (GIFHE)
HSI Building
Origin Way
Europarc
Grimsby
North East Lincolnshire
DN37 9TZ
UK

E-mail: jamess@grimsby.ac.uk

Chapter 19

Professor Jack G. A. J. van der
Vorst*
Logistics, Decision and
Information Sciences
Wageningen University
P.O. Box 8130
6700 EW
Wageningen
The Netherlands

E-mail: Jack.vanderVorst@wur.nl

Dr Durk-Jouke van der Zee
Department of Operations
Faculty of Economics & Business
University of Groningen
P.O. Box 800
9700 AV Groningen
The Netherlands

E-mail: d.j.van.der.zee@rug.nl

Seth-Oscar Tromp
Agrotechnology and Food Science
Group
Research Center of Wageningen
P.O. Box 17
6700 AA Wageningen
The Netherlands

E-mail: Seth.Tromp@wur.nl

Chapter 20

Professor Maro Vlachopoulou
Department of Applied
  Informatics
University of Macedonia
156 Egnatia
54006
Thessaloniki
Greece

E-mail: mavla@uom.gr

A. Matopoulos*
Department of Marketing and
  Operations Management
University of Macedonia
156 Egnatia
54006
Thessaloniki
Greece

E-mail: arismat@uom.gr

Chapter 21

Katerina Pramatari*, Angeliki
  Karagiannaki, Cleopatra Bardaki
Athens University of Economics
  and Business
Department of Management
  Science and Technology
47A Evelpidon Str.
Athens, 11362
Greece

E-mail: k.pramatari@aueb.gr
akaragianaki@aueb.gr
cleobar@aueb.gr

Chapter 22

Dawn Fisher (MMath, MILT)
Senior Consultant, Transportation
AECOM
Lynnfield House
Church St
Altrincham
WA14 4DZ
UK

E-mail: dawn.fisher@aecom.com

Professor Alan McKinnon*
Logistics Research Centre
Heriot-Watt University
Edinburgh
EH14 4AS
Scotland
UK

E-mail: A.C.McKinnon@hw.ac.uk

Dr Andrew Palmer
Visiting Fellow
Centre for Logistics and Supply
  Chain Management
Cranfield School of Management
Cranfield University
Cranfield
Bedfordshire
MK43 0AL
UK

E-mail: andrew.palmer@cranfield.
  ac.uk

Chapter 23

Gabriela Alvarez
Director, Latitude
Executive DBA student
Cranfield School of Management
Ruelle des Halles 2
1095 Lutry
Switzerland

E-mail: alvarez@latitudeglobal.com

Chapter 24

M. Bourlakis*
Brunel Business School
Brunel University
Uxbridge
Middlesex
UB8 3PH
UK

E-mail: Michael.Bourlakis@brunel.
    ac.uk

A. Matopoulos
Department of Marketing and
    Operations Management
University of Macedonia
156 Egnatia
54006
Thessaloniki
Greece

E-mail: arismat@uom.gr

Woodhead Publishing Series in Food Science, Technology and Nutrition

1

Delivering performance in food supply chains: an introduction

C. Mena and G. Stevens, Cranfield University, UK

Abstract: The aim of this book is to provide a supply chain perspective of the food and drinks industry from farm to fork. The focus is on managing across functions and improving performance by addressing the challenges affecting the industry. The book is targeted at both practitioners and academics who are interested in understanding the key levers of performance in food supply chains. This introductory chapter provides an overview of the scale and structure of the food and drinks supply chain. This is followed by a discussion of the key trends affecting the industry, such as globalisation, price volatility, economic recession, product diversification, sustainability and corporate social responsibility. Six key challenges associated with managing food and drinks supply chains are introduced: managing relationships, aligning supply and demand, managing processes efficiently and effectively, maintaining quality and safety, leveraging technology and managing responsibly. Each of these challenges forms a section of the book and each section comprises three to five chapters. In this chapter we provide a roadmap for the book introducing the reader to the key challenges and discussing the contribution of each topic to supply chain performance.

Key words: challenges, food and drink supply chains, performance, trends.

1.1 The changing nature of the food supply chain

The earliest human settlements were located in the region between Mesopotamia and lower Egypt known as the fertile crescent. This region had the right conditions to support a stable food supply, such as access to water for agriculture and opportunities for hunting and fishing. Compared to nomadic tribes, these first settlers faced a different set of challenges: trying to balance supply and demand and making decisions about how much food to produce, store, transport and trade. They could be called the first supply chain managers, although the term was not in vogue at the time!

The term 'supply chain', emerged in the 1980s to describe 'a system whose constituent parts include material suppliers, production facilities, distribution services and customers linked together via a feed-forward flow of materials and a feedback flow of information' (Stevens, 1989). Initially the study of supply chains focused on industries that involve complex assembled products, like automotive, electronics and aerospace. However, food supply chains are different to those in these industries and blindly importing their concepts and tools without recognising the differences would be risky. Some of the main differences are:

- Seasonality: many industries are subject to seasonality, but food chains have seasonality of both demand and supply, and organisations need to structure their supply chains around these cycles.
- Health, nutrition and safety: food has an impact on the health of the consumers. As a result, issues of quality, traceability, safety and risk management are critical success factors.
- Short shelf life and volatile demand: products often have short shelf life and demand is sensitive to many factors such as weather changes, promotions and special events. Since holding stock to cover against unexpected demand is not an option, responsiveness and speed are more important.
- Impact on the environment: all industries have an impact on the environment; however, food has a disproportionate effect because of the extensive use of resources like water, energy and land, and the unintended outputs such as carbon dioxide (CO_2) emissions, pollution and waste.

Food supply chains have evolved over the years, influenced by an array of economic, social, environmental and technological factors. Early supply chains were short, local and relatively simple in terms of the number of activities, inputs and outputs. Modern chains, on the other hand, are global and complex, offering a vast array of products to cater for the ever-changing needs of consumers.

Changes in the food system have brought many benefits for the consumer: economies of scale have helped to keep costs down; quality and hygiene standards ensure food is safe to eat; convenience foods have made cooking easier and quicker; food preservation methods have reduced the need for frequent purchases and allowed the consumption of out-of-season foods; and online shopping has allowed customers to buy food from the convenience of their home (or office). Furthermore, the range of products on offer has expanded greatly, catering for diverse tastes, trends and health requirements. However, these changes have also brought with them new challenges for the industry. Issues such as sustainability, health, nutrition and corporate social responsibly are high on the agenda of most food producers and retailers.

The evolution of food supply chains has taken place at different rates in different parts of the world, depending on factors such as income levels, degree of urbanisation, natural resources, infrastructure and trade policies. However, trends appear to show convergence in food consumption and delivery systems, with developing nations following a similar pattern to that of developed ones (Frazão *et al.*, 2008). This book will take the perspective of food supply chains in developed countries, assuming that chains in other parts of the world are likely to follow a similar path.

In this initial chapter we will provide a top-level view of the global food supply chain, setting the context for the entire book. We will refer to the global food chain as being the complex network of individuals and organisations involved in producing, distributing and trading food and beverages across the world. First we will discuss the scale and structure of the food system, outlining the key players in the industry. We continue with a discussion of the trends and challenges that are shaping food supply chains and establish a structure for the book considering these challenges. Finally, we conclude with a section focusing on the key factors for delivering performance in food supply chains.

1.2   Scale and structure of the global food chain

The global food chain is extremely fragmented, with millions of participating organisations around the world, and this makes it difficult to assess its structure and scale. Figure 1.1 depicts the general structure of the food supply chain; at one end of the chain we have fishing and agriculture (including agricultural input such as seeds and fertilisers), in the middle there are processors, packaging suppliers, distributors, wholesalers, retailers and caterers, and finally there are the consumers. This is a supply chain in which everybody has a stake.

Estimates for global food retail spending for 2008 range between US$ 3.6 and 4 trillion – up by almost 20% from 2004 figures (USDA, 2008a; Datamonitor, 2009). This trend is expected to continue over the next five years and it has been estimated that by 2013, global sales of food will reach US$ 4,602 billion (Fig. 1.2 presents the trend and forecast of food retail sales over the period 2004–2013). Furthermore, the World Bank estimates that by 2030, worldwide demand for food will increase by 50% from 2009 (Evans, 2009).

The economic and social role that the food industry plays varies from country to country, but its impact is substantial even in the most developed countries. It has been estimated that the entire food industry is responsible for around 10% of the USA economy (USDA, 2008b) and around 7% in the case of the UK (McDiarmid *et al.*, 2008). Similarly, estimates indicate that in the USA, the food industry employs around 16.5 million people (10.6% of total employment) (USDA, 2008b) and around 3.7 million people in the UK (14% of total) (McDiarmid *et al.*, 2008). In developing countries, figures tend to be considerably higher and it is estimated that around 40%

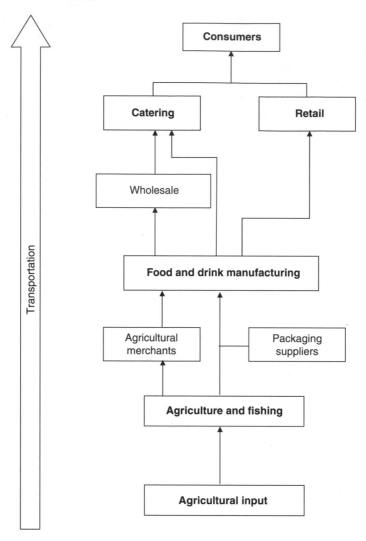

Fig. 1.1   Structure of the food supply chain.

of the world's labour force is employed in agriculture alone (CIA, 2009). These figures highlight the significance of the industry both economically and socially.

The value of global agricultural output has been growing at an average rate of 2.3% per year since 1961, outpacing the average population growth during the same period which was 1.7% per year (FAO, 2007). This is attributable to substantial increases in developing countries, but also to a shift towards higher value commodities such as livestock and horticulture. This reflects consumer trends which indicate an increasing demand for

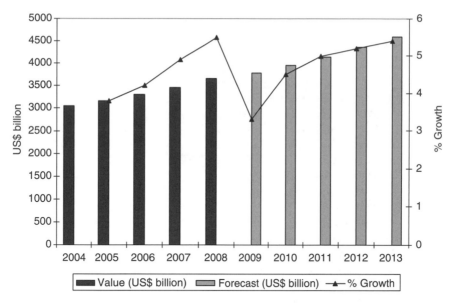

Fig. 1.2   Value of global food retail sales (Datamonitor, 2009).

items such as meat, fruit and vegetables, particularly in developing countries which have experienced substantial increases in income in recent
years. A study estimates that between 2000 and 2025 consumption of meat
in India will increase by 176% and by 70% in the case of milk and vegetables, but only 26% for grains (von Braun, 2007). Similar trends can be seen
in other large, developing countries such as China, Brazil and Nigeria
(FAO, 2007). As a result of these changes, the per capita consumption of
food around the world has increased by more than 20% in five decades,
from an average of 2280 kcal/person/day in the early 1960s to 2800 kcal/
person/day (FAO, 2007).

1.3   Trends affecting the food supply chain

1.3.1   Globalisation

Food supply chains have been globalising for centuries; however, the pace
of change has accelerated in recent years. According to figures from the
WTO, global exports of food products have increased from US$ 224,000
million in 1980 to US$ 913,000 million in 2007 (WTO, 2009). However,
although the agricultural exports have increased tenfold since the 1960s,
the share of agricultural trade compared to merchandise trade has fallen
from around 25% to less than 10% (FAO, 2007).

    According to the most recent agricultural outlook prepared jointly by
the Organisation for Economic Cooperation and Development (OECD)

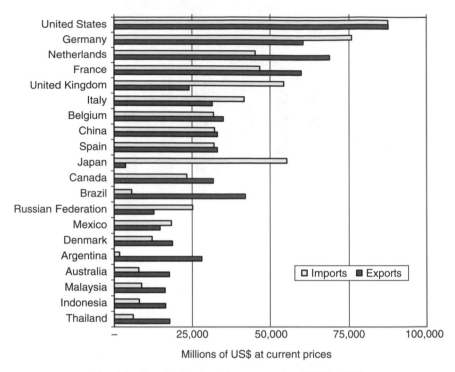

Fig. 1.3   Top 20 food trading countries (WTO, 2009).

and the Food and Agriculture Organisation of the United Nations (FAO) (OECD – FAO, 2008), growth is expected to continue, at least until 2016 and it is expected to be particularly strong for products such as beef, pork and vegetable oils.

Figure 1.3 shows the top 20 food trading countries. As expected, the largest and most developed economies tend to be the largest traders of food, with the USA being both the leading importer and exporter of food. However, some medium-sized countries, for example the Netherlands and Belgium, appear as some of the leading food exporters in the world.

From the perspective of management of a supply chain, globalisation has significant implications because it involves longer and more fragmented chains, which tend to be more difficult to manage. Longer chains require more transport, which has an impact on costs and lead times as well as having environmental repercussions. Moreover, globalisation creates a more interconnected food system which is more sensitive to disruptions.

1.3.2   Economic trends
Over the last couple of years we have experienced substantial volatility in food prices. Figure 1.4 presents the FAO's food price index, which consists

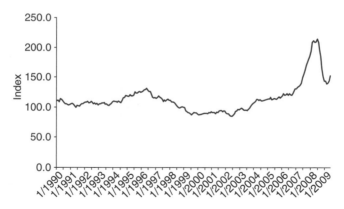

Fig. 1.4   FAO food price index 1990–2009 (FAO food price indices, June 2009).

of a weighted average of six commodity groups including, meat, dairy, sugar, cereals, oils and fats. The chart shows that food prices started to rise in late 2006 and reached a peak in mid-2008. The situation was so severe that a food crisis was declared and protests took place in many countries as people struggled to pay for staple foods. The index has decreased substantially over the last year, from 213.5 in June 2008 to 152.1 in May 2009, although current figures are still higher than pre-2007 data. The FAO has recently declared that 'international prices of most agricultural commodities have fallen in 2009 from their 2008 heights, an indication that many markets are slowly returning into balance' (FAO, 2009).

The causes of the food crisis in 2008 include a growing demand for food from Asian countries, higher energy costs, certain policies such as support for the use of biofuels, and temporary factors such as droughts leading to poor harvests (Cabinet Office, 2008). Since many of these factors are likely to continue affecting food supply and demand in the future, organisations and governments need to be ready to respond to future food crises.

The demand for food products has also been affected by other economic factors such as the credit crisis, a reduction in consumer confidence and a general economic slowdown. The food industry is less sensitive than other industries to variations in income and the share of food in total spending tends to increase during times of economic slowdown. However, it is possible that consumers will trade down to cheaper alternatives. This has been popularised by the 'Aldi effect', referring to the German discount retailer which has seen a substantial increase in sales, whilst other retailers have reacted by expanding their range of own brand products (Lyons, 2008). Another possible consequence of the economic downturn is that customers try to throw less food away and shop more smartly by avoiding unnecessary purchases.

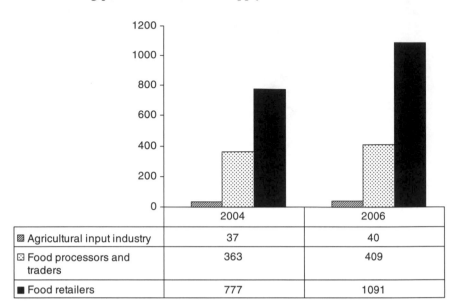

	2004	2006
⊠ Agricultural input industry	37	40
⊡ Food processors and traders	363	409
■ Food retailers	777	1091

Fig. 1.5   Sales of top ten companies by sector (US$ billion) (von Braun, 2008).

1.3.3   Shifting power structures

Another significant trend in the structure of food supply chains is the growing size and power of large corporations, particularly in retail. Figure 1.5 compares the sales of the top ten companies in the agricultural input, food processor and food retailer sectors. The chart indicates that the top firms have been increasing in size and this often represents more power.

In supply chain relationships, the term 'power' often has a negative connotation, but organisations can choose to use power in different ways. A dominant firm can choose to follow a cooperative approach, acting as a channel captain, helping to improve communication, coordination and performance across the chain. Similarly, they can choose to take an adversarial position, dictating terms and conditions, and using power to exploit their supply chain partners. The balance of power and the way in which companies choose to exercise that power have substantial implications for the structure and management of the chain.

1.3.4   Product diversification

Product diversification has also been dramatic in recent years; the average number of products on offer in USA supermarkets increased from 14,145 in 1980 to 49,225 in 1999 (Richards and Hamilton, 2006). These changes have been made possible by developments in food processing, storage, distribution and retailing.

Diversification is generally positive for the consumer, but from a supply chain management perspective it presents many challenges. First, it involves introducing new products, which are notoriously difficult to forecast. Poor forecasts can in turn produce overstocking and its associated costs, or under-stocking, leading to poor customer service, and could even cause demand amplification upstream of the supply chain, a phenomenon known as the bullwhip effect (Lee *et al.*, 1997; Geary *et al.*, 2006). Furthermore, increased product diversification can lead to increasingly complex warehousing and distribution operations.

1.3.5 Sustainability and corporate social responsibly (CSR)

In recent years companies in the food industry, particularly major retailers and manufacturers, have appeared to be more concerned with issues of corporate social responsibly and sustainability. Arguably, this reflects changing consumer attitudes such as the emergence of the so-called 'ethical consumer', concerned with issues such as fair trade, animal welfare, support for local farmers and impact on the environment including climate change, waste, pollution, pesticides and food miles (Cabinet Office, 2008).

Another consumer trend which is affecting the food supply chain is obesity and the increasing attention given to healthy eating. In many countries, consumers are becoming more health conscious, increasing the demand for products with lower fat, calories or salt, as well organic produce. In the UK this has been reflected in the increasing demand for fruit and vegetables (Cabinet Office, 2008). This trend affects the entire supply chain and companies have to address these issues in order to remain competitive.

1.4   Structure of the book

Improving supply chain performance involves providing solutions for the conflicts and challenges that are present in the industry. For this reason we decided to structure the book around six key challenges facing the food and drinks industry:

- managing supply chain relationships
- aligning supply and demand
- managing processes efficiently and effectively
- maintaining quality and safety
- using technology effectively
- delivering food and drinks responsibly.

Figure 1.6 depicts the structure of the entire book and presents the chapters included in each of the sections; below we will describe each of the sections and the role of each of the chapters.

	1 Introduction: Delivering performance in food supply chains
Part I Managing relationships in food supply chains	2 Performance measurement in the management of food supply chain relationships
	3 Living with power imbalance in the food supply chain
	4 Supplier safety assessment in the food supply chain and the role of standards
	5 Understanding innovation in food supply chains
Part II Aligning supply and demand in food supply chains	6 Sales and operations planning in food supply chains: case study
	7 Food supply chain planning, auditing and performance analysis
	8 Aligning marketing and sourcing strategies for competitive advantage in the food industry
Part III Managing processes efficiently and effectively in food supply chains	9 Value chain analysis of the UK food sector
	10 Improving responsiveness in food supply chains
	11 Reducing product losses across the food supply chain
	12 Methods for assessing time and cost in a food supply chain
	13 Improving food distribution performance through integration and collaboration in the supply chain
Part IV Maintaining quality and safety in food supply chains	14 Enhancing consumer confidence in food supply chains
	15 Quality and safety standards in food supply chains
	16 Developments in quality management systems for food production chains
Part V Using technology effectively in food supply chains	17 Role of diagnostic packaging in food supply chain management
	18 Advances in the cold chain to improve food safety, food quality and the food supply chain
	19 Simulation modelling for food supply chain redesign
	20 Adoption of e-business solutions in food supply chains
	21 Radio frequency identification (RFID) as a catalyst for improvements in food supply chain operations
Part VI Delivering food sustainably and responsibly	22 Reducing the external costs of food distribution in the UK
	23 Fair trade and beyond: voluntary standards and sustainable food supply chains
	24 Trends in food supply chain management

Fig. 1.6   Structure of *Delivering performance in food supply chains.*

1.4.1   Part I Managing relationships in food supply chains

Christopher (2005) defines supply chain management as the 'the management of upstream and downstream *relationships* with suppliers and customers in order to create enhanced value in the final market place at less cost to the supply chain as a whole'. Following this rationale, we argue that managing relationships is one of the fundamental challenges and one that is particularly relevant in the food and drinks industry.

Part I is comprised of four chapters. Chapter 2, by Humphries and McComie, discusses how the management of relationships can bring benefits to the entire supply chain and advocates the use of a relationship performance measurement approach to support continued improvement. Case studies from the confectionery and bakery sectors are used to highlight the benefits of the approach. In Chapter 3, Hingley and Lindgreen deal with the very important issue of power in supply chains and argue imbalance in power is not necessarily a barrier to entering collaborative relationships or to their success. They propose that accepting power imbalances is a first step in developing collaborative relationships that can be beneficial for all parties involved. In Chapter 4, Julien discusses supplier assessment as an approach to managing supply chain relationships and introduces the hazard analysis and critical control points (HACCP) methodology and the use of standards such as ISO 22000 and the global food safety initiative (GFSI) as tools to manage the relationship across partners in the supply chain. This section concludes with Chapter 5 in which Estrada-Flores discusses the advantages of using a collaborative approach to food innovation and presents a framework for co-innovation.

1.4.2   Part II Aligning supply and demand in food supply chains

Aligning supply and demand is an essential activity in any supply chain and failure to do so can lead either to dissatisfied customers or to excessive inventory. The food industry presents particular alignment challenges owing to the seasonal nature of both supply and demand, the high degree of uncertainty in demand and to the relatively short shelf life of many products, which complicates the use of inventory as a mechanism for buffering uncertainty.

Three chapters focus on the challenge of alignment. In Chapter 6, Yurt, Mena and Stevens present the sales and operations planning (S&OP) process as a solution to the alignment problem. In this chapter the authors describe the process and rationale behind S&OP and discuss the challenges in implementing it, using a case study to exemplify the key issues. In Chapter 7, Thomas, Wang and Potter discuss the four key sources of uncertainty in supply chains, namely, supply, demand, process and control systems, and propose an approach for dealing with uncertainty through planning, auditing and performance diagnostics. Finally, in Chapter 8, Chicksand discusses the need for alignment between the marketing and sourcing functions of an organisation and uses a case study from the meat sector to highlight the benefits of such alignment.

1.4.3   Part III Managing processes efficiently and effectively in food supply chains

The food and drinks industry is notorious for its low margins and recent price volatility has increased the pressure on producers and retailers alike to maintain efficient and effective processes. In Part III, five different approaches to improving efficiency and effectiveness are discussed.

Chapter 9 by Keivan Zokaei, introduces the value chain analysis (VCA) approach as applied to the food industry. This approach, often referred to as 'lean' because it aims to use fewer resources, has proven to be effective in many industries and this chapter shows how it can be applied to the food industry using a case study from the pork sector. In Chapter 10, Harrison presents the 'agile' approach to supply chain management which aims to master turbulence and improve responsiveness to end customers. Two cases from the fish and salads sectors are used to exemplify the approach. In Chapter 11, Chapman focuses on a major problem affecting supply chain efficiency: shrinkage. He introduces the problem of shrinkage in the food industry and presents structure to help companies address it. Templar and Mena, in Chapter 12, focus on two key indicators of supply chain performance: cost (as a measure of efficiency) and time (as a measure of effectiveness). They outline a methodology to gain visibility of time and cost across the supply chain and use a case study to demonstrate its use. Finally, in Chapter 13, Charles Stephens concentrates on the distribution stage of the food chain and discusses how efficiencies can be gained through integration and collaboration.

1.4.4   Part IV Maintaining quality and safety in food supply chains

In developed countries, food safety tends to be a given as people assume the appropriate systems are in place to eliminate safety risks. However, major incidents like salmonella in peanuts in the USA, melamine in milk produced in China and avian influenza in UK poultry have highlighted the vulnerabilities of food systems. Given the potential implications of food scares, companies tend to take this issue very seriously.

The three chapters in Part IV provide different perspectives of food quality and safety. In Chapter 14, Garcia Martinez takes the perspective of the customer and the perception of risk in the system. Baines, in Chapter 15, focuses on the use of standards, both pre-farm gate (SQF1000, GlobalGAP and ISO 22000) and post-farm gate (SQF2000, BRC Global, IFS and ISO 22000), to maintain quality and safety. Finally, in Chapter 16, Theuvsen discusses the use of a quality management system across the supply chain.

1.4.5 Part V Using technology effectively in food supply chains

Technology has had a major impact on the way supply chains are managed. Information and communication technologies, for instance, have made possible almost instant communication between customers and suppliers across the chain. This fast flow of information has allowed companies to be more responsive to changes in demand, making supply chains more effective in delivering the required products on time. Other technologies in packaging, processing and refrigeration have allowed food to be stored for longer, maintaining its quality and appearance. This has enabled many other supply chain trends such as global sourcing and the use of complex distribution networks.

Part V contains five chapters looking at different technologies that are having an impact on the management of food supply chains. Hobday, in Chapter 17, looks at intelligent packaging and its impact on issues such as traceability, inventory management and waste minimisation. Chapter 18, by James and James, discusses the role and impact of refrigeration technologies in the supply chain and how to design and improve a cold chain system. In Chapter 19, van der Vorst, van der Zee and Tromp present the use of simulation and modelling to improve supply chain management and uses a case study of fresh fruit to highlight the potential impact of this technology. In Chapter 20, Vlachopoulou and Matopoulos discuss the impact of information and communication technology (ICT) and e-business in food supply chains and analyse the factors that influence the adoption of such technologies. Finally, in Chapter 21, Pramatari, Karagiannaki and Bardaki discuss the use and impact of radio frequency identification (RFID) technologies in food supply chains.

1.4.6 Part VI Delivering food sustainably and responsibly

Sustainability and CSR issues have already been discussed as one of the key trends in the industry. It is no surprise that as customers become more concerned with issues such as global warming, pollution, deforestation and overfishing, companies in the food chain try to address their concerns by embarking on initiatives to reduce or eliminate the negative impact of their operations on the environment.

Part VI includes three chapters. First we have Fisher, McKinnon and Palmer in Chapter 22 discussing the issue of food transport, specifically looking at vehicle emissions and the impact on global warming. In Chapter 23, Alvarez discusses the development and use of environmental and fair trade standards and discusses how companies can become engaged in this field. Finally, Bourlakis and Matopoulos in Chapter 24 conclude the discussion by analysing future trends in technology and the environment.

1.5   Concluding remarks

In this chapter we have introduced the concept of food and drink supply chains as complex and continually changing systems which, at their core, involve farmers, processors/manufacturers, retailers and consumers. Around these core groups there are many other participants such as hauliers, wholesalers, packaging providers, government agencies and non-governmental organisations, all with different roles and different needs which have to be satisfied. To succeed in using a system with multiple stakeholders, organisations need to tick many boxes:

- They need to take a supply chain perspective and work together with other organisations to satisfy the needs of the consumers.
- They need to understand the needs and concerns of their customers and consumers and align their supply chains to satisfy these needs.
- They need to design and manage their processes to be efficient and effective in order to avoid wasting resources while maintaining responsiveness.
- They need to ensure quality and safety of the products across the chain.
- They need to use technology appropriately to leverage their capabilities and improve performance.
- They need to make their businesses environmentally, socially and economically sustainable.

These are the challenges managers in the food industry are facing and in this book we will provide different perspectives on how to deal with these challenges. There is no silver bullet in dealing with them, in some cases there is not even a clear path to resolving them. However, it is by understanding different approaches and exploring alternatives that we can continue developing solutions to these challenges. We hope this book can serve as a stepping stone in this process of exploration.

Ultimately we are all participants in the food and drink supply chain. We want this supply chain to deliver quality products to our table in a safe and efficient way, but we also need to recognise that these chains are a result of our individual decisions, the products we buy, where, when and how often we buy them. We all play a role in shaping the food supply chains of the future.

1.6   References

CABINET OFFICE (2008). *Food: An Analysis of the Issues*, Revision D – August 2008, Strategy Unit, Cabinet Office, London, UK (available from http://www.cabinetoffice.gov.uk/media/cabinetoffice/strategy/assets/food/food_matters1.pdf; last visited, 21/06/09).

CHRISTOPHER M (2005). *Logistics and Supply Chain Management: Creating Value-Adding Networks*, Financial Times – Prentice Hall, Harlow, UK.

CIA (2009). *The World Factbook*, World Statistics (available from https://www.cia.gov/library/publications/the-world-factbook/; last visited 05/06/09).

DATAMONITOR (2009). *Global Food Retail: Industry Profile*, Datamonitor, London, UK (reference code: 0199-2058; publication date: May 2009).

EVANS A (2009). *The Feeding of the Nine Billion: Global Food Security for the 21st Century*, Royal Institute of International Affairs, Chatham House, London, UK (www.chathamhouse.org.uk).

FAO (2007). *State of Food and Agriculture 2007*, Food and Agriculture Organization of the United Nations, Rome, Italy (available from: ftp://ftp.fao.org/docrep/fao/010/a1200e/a1200e00.pdf; last visited 21/06/09).

FAO (2009). *Food Outlook: Global Market Analysis*, Food and Agriculture Organization (FAO) (available from http://www.fao.org/docrep/011/ai482e/ai482e01.htm; last visited 15/06/09).

FRAZÃO E, MEADE B and REGMI A (2008). 'Converging patterns in global food consumptions and food delivery systems', *Amber Waves*, The Economics of Food, Farming, Natural Resources and Rural America, February 2008 (available from www.ers.usda.gov/AmberWaves/February08).

GEARY SM, DISNEY DR and TOWILL D (2006). 'On bullwhip in supply chains – historical review, present practice and expected future impact', *International Journal of Production Economics*, **101**, 2–18.

LEE H, PADMANABHAN S and WHANG S (1997). 'Information distortion in a supply chain: the bullwhip effect', *Management Science*, **43**(4), 546–58.

LYONS T (2008). 'Tesco's plot to counter Aldi effect'. *The Sunday Times*, 13 July.

MCDIARMID S, HOLDING J and CARR J (2008). *Food Statistics Pocketbook 2008*, DEFRA and National Statistics (available from https://statistics.defra.gov.uk/esg/publications/pocketstats/foodpocketstats/FoodPocketbook2008.pdf; last visited 05/06/09).

OECD–FAO (2008). *Agricultural Outlook 2008–2017*, Organisation for Economic Cooperation and Development (OECD) and Food and Agriculture Organization of the United Nations (FAO).

RICHARDS TJ and HAMILTON SF (2006). 'Rivalry in price and variety among supermarket retailers', *American Journal of Agricultural Economics*, **88**(3), 710–26.

STEVENS GC (1989). 'Integrating the supply chain', *International Journal of Physical Distribution & Materials Management*, **19**(8), 3–8.

USDA (2008a). *Global Food Markets, April 2008* (available from http://www.ers.usda.gov/Briefing/GlobalFoodMarkets/Industry.htm; last visited 21/06/09).

USDA (2008b). *Global Food Markets: Global Food Industry Structure*, United States Department of Agriculture, March 2008; (available from http://www.ers.usda.gov/Briefing/GlobalFoodMarkets/; last visited 21/06/09).

VON BRAUN J (2007). *The World's Food Situation: New driving forces and required actions*, International Food Policy Research Institute, Washington DC, USA (December 2007).

WTO (2009). *World Trade Organization, International Trade and Tariff Data* (available from http://www.wto.org/english/res_e/statis_e/Statis_e.htm; last visited 21/06/09).

Part I

Managing relationships in food supply chains

2

Performance measurement in the management of food supply chain relationships

A. S. Humphries and L. McComie, SCCI Ltd, UK

Abstract: The management of supply chain relationships is poorly understood. There is confusion over standpoints (customer, supplier, joint), manager's roles (professional, part-time, devolved) and methods (relationship management, supplier management, key account management). This chapter attempts to bring order to this chaos. It shows how formal relationship management methods can bring tangible benefits to collaborative supply chains including increased effectiveness and value for money and uses a number of food industry case studies to illustrate these points. In particular, it advocates the use of performance measurement and diagnosis to support continuous improvement and concludes by showing how these factors help food industry managers to meet future challenges.

Key words: supply chain management, supply chain relationships, relationship management, performance measurement, collaboration.

2.1 Introduction

Supply chain management (SCM) has come a long way over the last 25 years. It evolved from logistics, which concentrated on breaking down the barriers between departments to enable organisational output to be produced efficiently and effectively. SCM now aims to extend that ethos across the network of producers to the end customers. However, the management of process still tends to predominate: planning, scheduling, materials handling, IT systems, ordering and paying while supply chains fail to achieve their full potential. This is because the wider aspects of supply chain relationships such as contract, investment, problem and communication management are ignored or left to chance. As a result, the complete benefits of teamwork are missed. Objective, formal relationship management is thus essential for strategically important, collaborative partnerships and this, like every other aspect of management, depends on objective performance measurement.

This chapter describes how it is possible firmly to grasp a problem that is generally considered to be intangible – supply chain relationship performance measurement and management. Moreover it shows that if this is done correctly, significantly enhanced business effectiveness is possible. It first describes the rationale for supply chain relationship management and the problems that managers face amongst increasingly complex networks of customers and suppliers. Second, it examines the challenges of identifying the key dynamics that characterise supply chain relationships. Next, it describes an approach that utilises a unique and innovative perspective of behavioural feedback loops. It shows how the technique can be used practically to overcome the problems we have identified and three case studies are provided by way of illustration. It concludes with some thoughts on future trends.

2.2   Supply chain relationships

2.2.1   Supply chain management and collaboration

Supply chain aficionados continue to press managers to form closer, longer-term relationships between customers and fewer suppliers as the only response to increasing market sophistication and globalisation. This focus on relationships is underlined in Martin Christopher's (2005) definition of SCM: 'Supply chain management is the management of upstream and downstream *relationships*, with customers, suppliers, and key stakeholders in order to increase value and reduce cost for all members of the supply chain'. It is now expected that 'being good at managing relationships' is no longer the qualifying strategy but the winning one. Moreover, it is proposed that the fullest benefits of SCM will only be achieved from very close collaborative relationships. Collaboration in the supply chain means: 'working together to bring resources into a required relationship to achieve effective operations in harmony with the strategies and objectives of the parties involved thus resulting in mutual benefit' (Wilding and Humphries, 2006).

The UK food and drinks industry comprises a wide variety of organisations performing activities such as farming, manufacturing, catering and retailing. Together these organisations play an essential economic and social role with sales for over £148 billion, around 8% of the gross domestic product and employing at least 3.7 million people (12.4% of the UK work force) (IGD, 2006). All these organisations are linked together within supply chains of varying complexity and sophistication and thus finding ways to increase their effectiveness is of critical importance.

However, although it is widely agreed that cooperative supply chain relationships can achieve substantial benefits for the participants, it is also apparent that there are potential disadvantages. These include reduced flexibility and competition options, the risk of increased dependence, the risk of losing intellectual property rights (IPR) and, more complex, onerous organisation and management arrangements (Cooper *et al.*, 1997). This increased complexity is especially present when the number of potential partners expands in the ever more prevalent networks, alliances, joint ventures and consortia.

It is thus not surprising that full supply chain management implementation is seldom achieved in practice and this is a source of frustration, missed opportunities and cost. In addition to the lack of skills, tools and resources available to managers, a number of behavioural problems are evident. Partners still take a short-term view when faced by increasing market-place complexity and uncertainty and limit the extent of their collaborative focus. Adversarial practices such as power abuse, lack of transparency, poor/misleading communication and reluctance to adopt changes in attitude exist and reduce collaborative effectiveness (Kemppainen and Vepsalainen, 2003).

2.2.2   Nature of supply chain collaboration

Supply chain collaboration brings together capabilities and investment in time, money, infrastructure and intellectual capital to create value that neither party can achieve on its own (1 + 1 = 3 or more), not just in the present contract but in future ones too. The core of this value is a product or service that is unique and inherently difficult to copy. The essential ingredient that secures the superior revenues and competitive advantage for both partners is the creation and maintenance of a productive, harmonious relationship. In globalised competitive markets the ability to deliver value to shareholders and ensure that particular products or services are the first call of discerning customers depends on productive collaboration. This is characterised by:

- joint innovation
- customer focus
- high quality output
- world beating practices
- continuous improvement
- flexible commercial frameworks
- objective performance measurement
- improved business forecasting
- coordinated processes
- honest and open communication
- two-way information flow.

Investments in people, infrastructure and systems create a climate in which innovation and the free flow of ideas flourish. The result is the creation of reliable business systems, capable of delivering high quality goods and services which delight customers, and an overarching desire to be part of a 'win–win' relationship. In effect, a virtuous cycle is created to secure long-term profitability and market share. These relationships are valuable core assets which must be managed effectively.

2.2.3   Problems for supply chain relationship managers

Supply chain relationship managers (RMs) face a number of difficulties in doing their job. Good contracts can still conceal operational failures and poor teamwork which lead to a focus on the 'small print' rather than building long-term value for the customer. No matter how close the relationship, it will not tolerate underperformance. A traditional emphasis on the management of time, cost and quality, often called supplier or project management, usually ignores the organisation interaction effects. In a tightly coupled arrangement where little slack exists and communication is complicated by distance and number of organisations involved, small issues can 'snowball' and by the time they emerge, have the capability seriously to jeopardise the stability of the operation. These negative feedback situations usually occur because of a lack of attention to managing the wider relationship, that is supply chain relationship management.

'Shark effects' are those factors beneath the surface of the relationship that insidiously nibble away at the investments and undermine the value of the collaboration (see Fig. 2.1). They are particularly difficult to identify and eradicate.

Shark effects include:

- complacency (accepting average performance and normalisation of problems)
- distrust, i.e. 'they never do what they say they will do so why should we?'
- opportunism (seeking gain at the expense of the partnership)

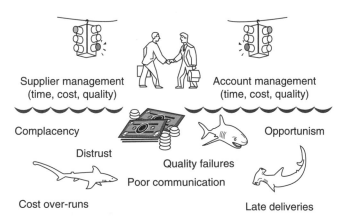

Fig. 2.1   Supply chain relationship management challenge.

- quality failures
- poor (untimely, incomplete and inaccurate) communication
- cost overruns/poor financial control
- late/incomplete deliveries.

As a result, the RM spends most of his or her time fire-fighting symptoms and is unable to see the source of the problems let alone fix them. Worse still, many problems do not reach the level of senior managers until they have become serious and when they do, the first reaction is often to pull out of the contract and consider the penalty clauses. This is likely to have a very damaging effect on the stability of the relationship.

Case study: bitter sweet
A global snackfood manufacturer (the customer) with operations characterised by few products and stable marketing programmes, decided several years ago to enhance its product range with complementary offerings from niche comanufacturers (the suppliers) rather than developing new ranges themselves. Although considerably smaller than their partner, the suppliers were world class operators offering high quality, cost-effective, flexible production which the larger company could not achieve economically in-house. Of these two relationships, one was of three years duration and one of over 20 years. The partners clearly understood what each brought to the relationship and saw it as a long-term collaboration:
   'Combining their flexibility and innovation with our know how ensures we have an enduring relationship.' [customer]
   As both relationships settled down they began to be managed by the customer as if they were extensions of its own production facilities using management styles and tools utilised for its in-house operations.
   'Their performance measures are focussed at low levels rather than the overall outcome. This does not match the high level service they buy from us.' [supplier]
   The customer firm aimed to free their partners from the need to manage packaging, components and logistics suppliers. However, this highly controlling and 'inflexible' approach proved to be inappropriate for interactions with dynamic entrepreneurial partners who want to innovate.
   Although the newer relationship was still in the 'honeymoon' period and there was still much optimism, operational issues were increasing and beginning to take on a life of their own. The older relationship had sunk to the point where contract penalties were being invoked on a regular basis and day-to-day communications were soured by acrimony and frustration.
   'When we charge exception bills I see no indication of any passion to fix the underlying causes. This is a fundamental difference in our attitudes.' [supplier]
   The unintentional disconnections between the needs of these partners resulted in poor communication, unreliable logistics, high penalty costs, frustrated staff, fragile cooperation and a low incentive to innovate by

either party. Moreover the 'punchy', attitude of supplier personnel owing to their entrepreneurial, small or medium enterprise (SME) culture and the 'resolute' processes imposed by the customer was not conducive to resolving these problems and getting the relationships out of a slowly worsening impasse. The 'sharks' were circling.

2.2.4   Supply chain relationship management

Supply chain RMs are in the front line of corporate governance. They are primarily responsible for ensuring that the complex mixture of hard and soft issues that represent the teamwork between organisations works at optimal performance and adapts to changing internal and external stimuli. But, how often are relationship management responsibilities vested in the account manager, a role that is split informally between sales, business development and commercial? More often than not, any effort to influence production or supply chain management to solve particular problems or make special arrangements are considered as interference.

The ability of a firm effectively to manage its key relationships is a strategic capability (Dyer et al., 2001). This not only affects its ability to create and sustain successful supply chain partnerships but also influences its reputation amongst stakeholders, such as financial institutions, and the market. It is therefore clear that flexible, continuously improving supply chain relationships must be managed positively. Professional supply chain RMs must be appointed with the authority to champion the needs of the relationship across the departments within his or her own firm and be empowered to agree collaboration tactics and solve relationship problems with his or her opposite number in partner companies. The supply chain RM and their team must become the firm's centre of expertise for all partnership knowledge and experience so the organisation continuously improves. The RM should also be the focus and coordinator for exploiting inter-firm learning opportunities. Dyer et al.'s (2001) research shows that firms that have a recognisable alliance management function consistently do better than those who do not.

A regular review meeting of the RMs and the appropriate senior managers of each firm must:

- review performance targets in the last period and issue statistics
- review work/orders in progress
- review forecasted sales and orders in the next period
- preempt and solve problems before they become serious
- actively seek out and initiate process improvements
- review future plans (including new products) and initiate preparation
- examine and discuss industry and technology updates
- identify policy issues to refer to senior management
- involve other supply chain partners

- prepare joint communications, information and team-building events for the collaborating firms.

Underpinning the success of the supply chain RM is an objective relationship performance measurement and diagnostic capability which allows his or her role to be carried out proactively.

Case study: the baker's tale

Two major UK companies, the customer being a baker and its supplier being a grain merchant, carried out business with each other for many years. The baker is a nationwide supplier to all the big supermarket chains and most of the smaller ones. The relationship was considered by the partners to be excellent and they could not be more positive about each other.

'Working with them is a joy!' [supplier]

Yet at the operational level little 'disturbances' (sharks) were present which undermined the efficiency of this near perfect collaboration. These showed up as aggravating behaviour.

'Sometimes they display a "we know best attitude" which rankles with us.' [supplier]

One of the main issues was communication, both within the organisations and between the partners. A number of key people did not always have the planning and quality information they needed and this caused some friction between the departments. The occasional serious failures in internal information flows such as un-communicated changes in production schedules had an effect on communications between the partners which led to some opportunistic behaviour such as firm orders being rejected on delivery without any explanation.

Nevertheless, despite these difficulties innovation was taking place and headway was being made in collaborative behaviour.

'One of the biggest gains has been the ability to move beyond the day-to-day issues and have a completely open dialogue about absolutely anything that might "add value" to our relationship.' [customer]

Another area of concern was operational reliability where there was a low but continuous level of dissatisfaction with service and product delivery, joint cost and risk management, and performance within the supply chain itself. One partner complained: 'They need to focus more resources on meeting our targets and converting issues into solutions'. [customer]

These problems had been recognised and piecemeal initiatives were taking place to rectify them. One particular review of the commercial arrangements had resulted in a significant improvement in relationship quality.

'By constructing the right contract we've reached a point where pricing is simple (and non-combative!). This has led to a huge rise in "value" in all other parts of the relationship.' [customer]

This was a valuable relationship with long-term potential but complacency had crept in and efforts to sustain performance were proving to be hard to manage and implement. Both partners had recognised that their inability to manage and solve issues proactively were tying-up staff in firefighting and reducing their focus on customers. There was also a drift away from recognising and using each other's knowledge, skills and experience for the benefit both parties. The overall results were costs that were higher than they needed to be and operations that were below par. These put additional pressure on the firms during a period when high market prices for wheat were putting pressure on the industry as a whole.

'A more dedicated resource should be made available to manage and develop the relationship. This would provide us with a focus for reducing the gap which has opened up between us and our competitors and allow us to deploy our strengths more effectively in the market.' [supplier]

2.3 Measuring relationship performance

2.3.1 Why measure?

When one asks managers to assess the performance of their supply chain relationship objectively, they usually look blank. Most believe that managing collaborative business relationships is an art rather than a science. Others consider that control through Gantt charts, project management techniques such as Prince and a close eye on budgets and balance sheets are the keys to success. Traditionalists, especially in commercial departments, believe that the letter of the contract is the only sound basis for achieving realistic relationship value. However, some recent prominent failures such as those in the UK defence and construction sectors have led to a realisation that key account managers have a crucial part to play in managing these relationships in their widest aspects. Although many companies have established supply chain RM posts, they have failed to provide simple and powerful tools to support them and the old adage: 'If you can't measure it you can't manage it' comes into play. The essential question is: Can complex relationship performance be measured, so that management effort can be targeted most effectively?

2.3.2 When to measure?

The common sense demands of good governance require that value-laden supply chain relationships are kept constantly under scrutiny. We have already mentioned that problems can 'bubble up' and become serious without warning. In addition there is a need to ensure that complacency does not set in and that pressure is kept up on joint relationship improving projects. Periodically, the relationship partners should also review performance and ensure that continuous improvement does not flag.

There are other *ad hoc* reasons for performance measurement and deeper diagnosis. The customer may have concerns over supplier performance and wish to carry out a joint review to get the relationship back on track. There may be a need to benchmark the position before a strategic change in direction occurs (e.g. partners jointly seeking new business, developing a new product or impressing an external stakeholder, such as a government agency or financial institution). A supplier might initiate the assessment as a way of getting closer to its main customer and achieving greater understanding of the relationship benefits than it had so far been able to communicate to its partner. There is a critical need to provide supply chain RMs with the data to allow them to manage these relationships more effectively.

2.3.3 Measurement requirements

We have researched the importance of proactively managing key collaborative supply chain relationships and asserted that this cannot be carried out effectively without adequate performance measurement and diagnostics. The requirements of a system that can do this are:

- To provide an objective performance benchmark which allows managers to determine easily where action is needed and against which future progress can be planned and measured.
- To offer both strategic and tactical perspectives.
- To expose precise details of the good areas, the poor areas, the inefficiencies and where wastage may be occurring in the functions and processes within both partner organisations that service the relationship.
- To make sense of a 'messy'/complex situation and offer recommendations for improvement and change.
- To allow bottom line and competitive edge improvements to be made by enhancing communication and enabling well managed, cost effective change to be implemented.
- To transform perceptions by promoting understanding of the way that the partners view each other, thus strengthening the relationship.

Given this clear specification, the challenge is to use supply chain and relationship management theory to develop an approach that is fit for use in operational settings. In the next section, we explore how this can be done.

2.4   Weighing up a concept

2.4.1 All-encompassing objectivity

Academics have long struggled with the problem of trying to capture all the key facets of interorganisational, operational and interpersonal dynamics when attempting to understand supply chain relationships. For example, Giannakis and Croom, (2004) proposed a '3S Model' containing the *synthesis* of business resources and networks, the *synergy* between network actors and the *synchronisation* of operational decisions. The International Market-ing and Purchasing Group considered the interactions between supply chain partners (Kern and Willcocks, 2002), Fawcett and Magnan (2002) examined supply chain integration and Harland *et al.* (2001) researched supply chain networks. The general consensus is that the task is highly problematical. What is needed is an objective means of measuring relation-ship performance which covers the full extent of the collaboration and encompasses both soft and hard aspects and provides useful diagnostics, that is, why things are happening as well as what.

2.4.2   Relationship quality

Skarmeas and Robson (2008) have defined perceived relationship quality as less conflict and higher trust and commitment with the aim of achieving greater partner satisfaction. Research by Wilding and Humphries (2006) into over 80 substantial-sized relationships in both the public and private sectors has shown that supply chain partnership value is indeed positively correlated with supply chain effectiveness (see Fig. 2.2). This suggests that there is some predictability in the performance of these important com-mercial structures.

However, relationship quality is clearly a nebulous idea containing many other elements including ethical behaviour, forbearance, customer orienta-tion, information sharing, learning, communication value, long term horizons, C3 behaviour (cooperation, coordination, collaboration) and will-ingness to invest. Numerous previous studies have attempted to character-ise the main features of business relationships and use them in the correct proportions in order to understand the performance drivers. But, they have generally failed to identify the measurable forces that provide impetus or motivation to what is essentially a dynamic activity.

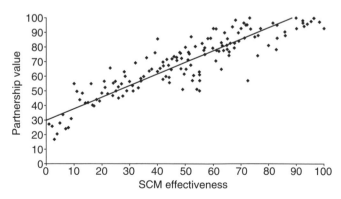

Fig. 2.2   Supply chain management effectiveness correlated with partnership value.

2.4.3 Relationship spirals

The perspective discussed above suggests that managing close proximity relationships seems to require new understanding of the dynamics involved. For instance, collaborative relationships are likely to be far more prone to positive feedback than an arms length relationship. In these circumstances minor problems can, if not recognised and managed, become personal and emotional, increasing the likelihood that new substantive conflicts will emerge and accelerate (Hanbrick *et al.*, 2001). Conversely, it is also possible for collaborative enterprise to bring operational advantages in the longer term as the partners become more effective as they develop through prior experience and active management of the learning process. Cooperation induces further cooperation over time and the emergence of trust and loyalty generates increasing benefits (Gulati, 1995; Lambert *et al.*, 1996, Luo and Park, 2004; Muthusamy and White, 2006). These dynamics accord with Williamson (1975) who proposed that within any organisation there would be a tendency for personnel to behave in a self-seeking opportunistic way. This could lead to a self-reinforcing cycle of poor performance. The increased (transaction) costs of managing (governance) in this situation thus became an important consideration in deciding to contract for a good or service or create it in-house (make or buy). These concepts are represented in the spirals in Fig. 2.3 and demonstrate that active management of supply chain relationships is thus essential to keep them in the success spiral and prevent them from falling back into the failure spiral.

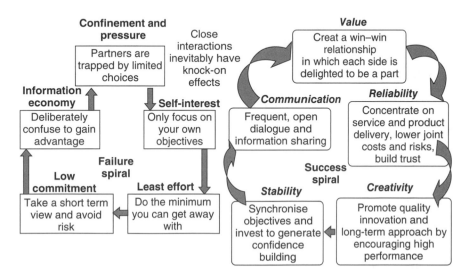

Fig. 2.3 Negative and positive relationship behaviour spirals (after Williamson's Organisations Failure Framework; Williamson, 1975).

Case study: end-to-end cereals
As part of an industry-wide study of the cereals industry, the linkages in a complete supply chain from raw materials (grain) to finished product (beer) were examined. The farmers who grew the grain were treated as a single supply source for the grain merchant. A single haulier, the most often used, was treated as the next link in the chain. The maltsters and brewers, who were technically two departments of the manufacturer, were treated as separate partners, one supplying the other.

There were a number of trends running through this supply chain. Operational reliability, including delivery timing, trust and management of cost reduction, was a concern at every level. All partners accepted that the industry was 'tight' and that there were not many opportunities for reducing cost. Moreover, there appeared to be no systematic approach across the chain for identifying and exploiting the potential for savings. For example, transportation was regarded as a transactional relationship despite the fact that it formed an essential part of the supply process. Trust was very much an issue too, even between the maltsters and brewers within the same company where attitudes were polarised between 'them' and 'us'. Between the farmers and the grain merchants this was particularly evident because of doubts over the partner's ability to deliver reliable quality products when required.

Communication was patchy, excellent in some links but needing considerable improvement in others. There was a lack of detailed knowledge of routine plans, schedules and orders necessary for the smooth functioning of the supply chain. There was also a lack of understanding of partners' contributions to the success of the relationships across the supply chain links, which suggested that training and routine information flows needed attention. This particular weakness was clearly a major contributor to many of the other issues identified. Because of the lack of information sharing, the supply chain had become compartmentalised which resulted in protective attitudes and a breakdown of trust which undermined effectiveness. There was a danger of slipping into a negative spiral, displaying all its classic symptoms including withholding information, focusing on the participant's own objectives, taking a short-term view and doing the minimum to get by. These partners only vaguely realised that their poor relationships were affecting the rest of the supply chain. The study made clear to all the supply chain partners that an approach to relationship management was needed that involved all the partners in a process of maintaining transparency and preventing isolation caused by negative feedback spiral behaviour.

2.4.4   Key performance drivers
The spirals in Fig. 2.3 provide a means of characterising either extreme in a spectrum of performance dynamics within supply chain relationships (Humphries and Wilding, 2003). The five positive dimensions described

below are derived from Williamson's (1975) Organisations Failure Framework (the failure spiral) using supply chain management and relationship marketing ideas. These are the supply chain relationship key performance measures which can be measured.

- Creativity – promoting quality, innovation and a long-term approach by encouraging high performance
- Stability – investment, synchronisation of objectives and confidence building
- Communication – frequent, open dialogue and information sharing
- Reliability – concentrating on service and product delivery, lowering joint costs and risks, building up trust
- Value – creating a win–win relationship in which each side is delighted to be a part.

In addition to these key drivers, a number of intrinsic characteristics of relationship performance have been confirmed. These provide further penetrating insights:

- Long-term orientation – encouraging stability, continuity, predictability and long-term joint gains
- Interdependence – loss of autonomy is compensated through the expected gains
- C3 Behaviour – collaboration, cooperation, coordination, jointly resourcing to achieve effective operations
- Trust – richer interaction between parties to create goodwill and the incentive to go the extra mile
- Commitment – the relationship is so important that it warrants maximum effort to maintain it
- Adaption – willingness to adapt products, procedures, inventory, management, attitudes, values and goals to the needs of the relationship
- Personal relationships – generating trust and openness through personal interaction.

2.4.5 Performance assessment process

The assessment should be simple, powerful and objective, sponsored by senior management and coordinated by the supply chain RMs. It should also be carried out quickly (within a few weeks) and in such a way that it does not disrupt day-to-day business within the organisations concerned. To this end it should use data collection methods such as scientifically designed on-line surveys combined with structured telephone interviews. These should only be targeted at knowledgeable staff. Confidentiality and impartiality should be central features in order to protect the interests of the companies and their staff. This guarantee of anonymity will ensure full and frank views are expressed.

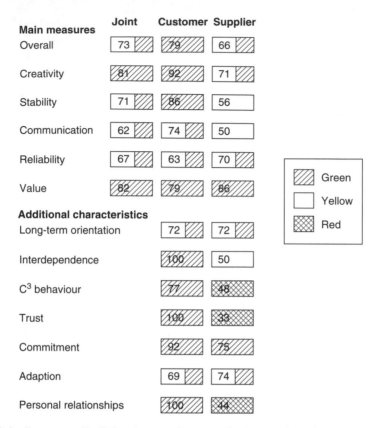

Fig. 2.4   Report traffic lights: key performance indicators (KPIs) at a glance. The numbers are the percentage satisfaction scores.

The findings of the assessment should be presented in clear, unambiguous, business friendly language so that the managers on both sides of a supply chain relationship can understand what is happening and why. The traffic lights in Fig. 2.4 show at a glance where the strengths and weaknesses of the relationship lie. The customer, supplier and joint percentage scores are shown and allow agreements and gaps to be clearly highlighted. Senior managers and supply chain RMs from both firms should have the opportunity to question the assessors jointly and discuss the underlying issues. This should be the beginning of the process of change or the continuation of one already in progress.

The details in the report should be sufficient to allow managers from both companies to hold discussions about relationship development. They could conduct joint workshops with staff, identify improved practices, develop implementation programmes and set continuous improvement objectives. This would be coordinated by the RMs who would use the

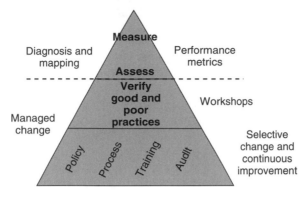

Fig. 2.5   Regular performance evaluation to sustain continuous improvement.

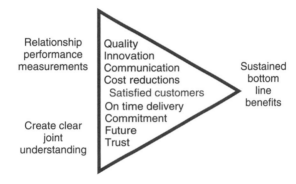

Fig. 2.6   Slaying the 'sharks' and regaining the competitive edge.

outcomes to improve the relationship continuously. This process is shown in Fig. 2.5.

Having used the performance assessment to set a benchmark, it should be repeated at regular intervals to measure and monitor performance in order to maintain the sharpness of the partner's competitive edge.

2.4.6   Supply chain performance outcomes

The view of the relationship revealed by a supply chain relationship performance assessment should create clear joint understanding that empowers supply chain RMs to fix the problems, monitor and sustain forward momentum and enter a virtuous circle of continuous improvement (see Fig. 2.6). The result must allow the joint teams to focus on quality, innovation, communication, cost reductions, on time delivery, commitment to the future and trust, to achieve satisfied customers and sustained bottom line benefits. The tangible results will be:

- opportunities to increase revenue and shareholder value (growth)
- increased customer satisfaction from better product/service quality and delivery (CRM)
- retained customers by differentiating the joint offering and locking out competitors
- reduced administration and production costs (margins)
- more integration and satisfaction from bridging the hidden gaps in teamwork (Continuity Risk)
- building joint capability to seize future business opportunities
- impressing stakeholders with the seriousness of partner intentions.

Bottom line benefits in the order of 15% should be achievable to both partners.

2.5   Future trends

The UK Institute of Grocery Distributors in their October 2008 Conference agenda listed the following questions as important concerns for the future:

- How you can best manage your business through a downturn?
- What will life after the credit crunch look like?
- What are the next steps on the path to a low carbon economy?
- What are the implications of the rapid growth in China, India and Russia?
- What role can ethical sourcing still play in these challenging times?
- How will trading relationships bear the strain of such rapid change?

When times are hard and there is increased uncertainty, building stronger relationships will bring extra security because collaboration with customers and suppliers demonstrates understanding, concern and the desire to work with them over the longer term. The vital challenge facing managers, regardless of economic conditions, is taking control of business relationships with the same level of attention that they pay to their financial accounts. Companies that have developed strong relationship management skills and implemented shared, continuous improvement techniques will be in a stronger position to weather storms such as the 'credit crunch'. They will manage expensive 'surprises' out of their supply chains, thus reducing their need for credit and increase their ability to work within their predicted cash flow. Collaboration between businesses in a supply chain can open up benefits not available to businesses working alone. They will realise that simple changes, such as better utilisation of transportation assets through better coordination and cooperation with logistics partners, are both 'green' and cost efficient. Rapid growth in emerging markets should be viewed as providing opportunities to form new customer and supplier collaborations in order to gain competitive advantage. Lessons learned in the home market about relationship management and inter-organisational teamwork will provide firms with a head start.

Ethical sourcing puts increased pressure on supply chain partners to put aside opportunistic behaviour. The penalties for getting this wrong include market-weakening bad publicity and legal penalties. The development of collaborative relationships, which depend strongly on high levels of trust, is thus critical to achieving a high reputation and maintaining long-term profitability in the growing market for ethically sourced products. Truly collaborating partners can thus create an environment which acts as a shelter from difficult and changeable market conditions. Within this setting, more effective teamwork will allow the supply chain to acquire greater flexibility, agility and increased predictability. The foundations to these future-proofing capabilities are professional relationship management supported by objective performance measurement.

2.6   Sources of further information and advice

The Institute of Grocery Distributors
The Institute of Grocery Distributors (IGD) works with consumers, companies and individuals throughout the UK food and grocery chain to provide information, research and leading edge best practice to help companies grow their business and develop their people. More information is available on the IGD website:

http://www.igd.com/

The Food Chain Centre
The Food Chain Centre is part of the national strategy to improve the competitiveness and profitability of UK farming. Reports from the UK Home Grown Cereals Authority (HGCA) sponsored joint project with the Cranfield School of Management Centre for Logistics and Supply Chain Management and Cardiff Business School Lean Enterprise Research Centre to improve the UK cereals supply chain are available on the FCC website:

http://www.foodchaincentre.com/

SCCI Limited
SCCI Ltd is a UK company that specialises in performance improvement within commercial relationships. At the core of its offering is care to ensure that all parties use objective, joint performance measurement as a means of promoting understanding between businesses to exploit the significant collaborative benefits that are potentially available. For more information visit:

http://www.sccindex.com/

2.7 References

CHRISTOPHER, M (2005). *Logistics and Supply Chain Management: Creating Value-added Networks*, 3rd edition, FT Prentice Hall, Harlow, UK.

COOPER, MC, LAMBERT, DM and PAGH, JD (1997). 'Supply chain management: more than a new name for logistics'. *International Journal of Logistics Management*, **8**(1), 1–14.

DYER, JH, KALE, P and SINGH, H (2001). 'How to make strategic alliances work', *MIT Sloan Management Review*, **42**(4), 37–43.

FAWCETT, SE and MAGNAN, GM (2002). 'The rhetoric and reality of supply chain integration', *Internal Journal of Physical Distribution & Logistics Management*, **32**(5), 339–61.

GULATI, R (1995). 'Does familiarity breed trust? The implications of repeated ties for contractual choice in alliances', *Academy of Management Journal*, **38**(1), 85–112.

GIANNAKIS, M and CROOM, SR (2004). 'Towards the development of a supply chain management paradigm: a conceptual framework', *Journal of Supply Chain Management*, **40**(2), 27–36.

HANBRICK, DC, LI, J, XIN, K and TSUI, AS (2001). 'Compositional gaps and downward spirals in international joint venture management groups', *Strategic Management Journal*, **22**, 1033–53.

HARLAND, CM, LAMMING, RC, ZHENG, J and JOHNSEN, TE (2001). 'A taxonomy of supply networks', *The Journal of Supply Chain Management*, 21–7.

HUMPHRIES, A and WILDING, R (2003). 'Sustained monopolistic business relationships: an interdisciplinary case', *British Journal of Management*, **14**(4), 323–38.

IGD – THE INSTITUTE OF GROCERY DISTRIBUTORS (2006). *UK Grocery Retailing, Fact-sheet*, Institute of Grocery Distribution, UK. Available 26/05/2005, http://www.igd.com/CIR.asp?menuid=51&cirid=114.

KEMPPAINEN, K and VEPSALAINEN, APJ (2003). 'Trends in industrial supply chains and networks', *International Journal of Physical Distribution & Logistics Management*, **33**(8), 701–19.

KERN, T and WILLCOCKS, L (2002). 'Exploring relationships in information technology outsourcing: the interaction approach', *European Journal of Information Systems*, **11**, 3–19.

LAMBERT, DM, EMMELHAINZ, MA and GARDNER, JT (1996). 'Developing and implementing supply chain partnerships', *International Journal of Logistics Management*, **7**(2), 1–17.

LUO, Y and PARK, SH (2004). 'Multi-party co-operation and performance in international equity joint ventures', *Journal of International Business Studies*, **35**, 140–60.

MUTHUSAMY, SK and WHITE, A (2006). 'Does power sharing matter? The role of power and influence in alliance performance', *Journal of Business Research*, **59**(7), 811–19.

SKARMEAS, D and ROBSON, MJ (2008). 'Determinants of relationship quality in importer–exporter relationships', *British Journal of Management*, **19**, 171–84.

WILDING, R and HUMPHRIES, AS (2006). 'Understanding collaborative supply chain relationships through the application of the Williamson organisational failure framework', *International Journal of Physical Distribution & Logistics Management*, **36**(4), 309–29.

WILLIAMSON, OE (1975). *Markets and Hierarchies: Analysis & Anti-trust Implications*, The Free Press, New York, USA, 39–40.

3

Living with power imbalance in the food supply chain

**M. Hingley, Harper Adams University College, UK, and
A. Lindgreen, University of Hull, UK**

Abstract: In the context of vertical agri-food industry business-to-business relationships in the United Kingdom, the majority of control lies with large multiple retailers. Predominant in agri-food channels is a reduced supplier sourcing model; category management and network supply coordination, through super middlemen, also are widely applied. Power-imbalanced business relationships appear to be important for understanding business exchanges and power should be a central consideration in business relationships. However, imbalance in power is no specific barrier to parties entering collaborative relationships or to their success. The acceptance of power imbalances is a key step to successful relationship building in agri-food channels and though collaborative chain activity may be beneficial, suppliers should recognise that such activity still means operating with imbalances in power and rewards.

Key words: business relationships, power, UK food supply.

3.1 Introduction

This chapter examines the nature of power in a business-to-business vertical supply chain context by studying the UK agri-food supply channel and its particular and vital interface with multiple food retailing organisations. Such an investigation is of particular interest because of the large-scale upheaval suffered in recent years in the UK agri-food industry and subsequent initiatives to address the ensuing difficulties, notably at the primary end of the food production system. The analysis of fresh food suppliers and retailers employs material from qualitative in-depth interviews with key proponents that clarify issues related to business relationships and power. The resultant discussion considers these issues of power and business relationships and their impact on evolving food industry structures (i.e. category management and preferred supplier arrangements in a network centred on super middlemen). Conclusions and recommendations pertain to the appropriateness of a relationship approach and theories of power for the development and operation of agri-food channels.

3.2   Role of power

Any understanding of the nature of business exchanges in vertical chains must address the nature of power and its influence on relationship formation. Power as a construct in business-to-business relationships has received irregular and contrasting treatment from analysts, including those who view the concept of power as alien to the effective workings of exchange relationships and determine success through principles of cooperation and trust. According to this view, power negates cooperation (Bretherton and Carswell, 2002; Doney and Cannon, 1997; Pole and Haskell, 2002). Furthermore, Gummesson (1999) considers a power imbalance to be detrimental to sustaining a business relationship and Naudé and Buttle (2000) express the common view that power has a negative influence which is not able to help build relationship quality. In their comparative study of power in UK and Australian business-to-business retail relationships, Dapiran and Hogarth-Scott (2003) note that other writers, including Kumar (1996) and Kumar *et al.* (1998), view power as the antithesis of trust and only in a negative light.

Yet a negative view of the role of power is by no means universal. A contrasting viewpoint (Blois, 1998; Campbell, 1997; Kalafatis, 2000; Svensson, 2001) emphasises that not all relationships result in mutual benefit; not all relationships are based on joint trust, nor do they need to be, and trust alone cannot be a sole source of dependence. For example, Earp *et al.* (1999) warn that viewing relationships as if they must involve commitment and trust means ignoring the rich diversity of business exchanges that not only exist but are appropriate in different contexts.

Perhaps, in researchers' enthusiasm for the ideal business exchange conditions, they have overlooked the role of power or dealt with it solely as a side issue, so that it rarely gets discussed in supply chains except to deny its importance (Williamson, 1995). Some authors argue power should be at the centre of any study of buyer–seller relationships (e.g. Cox *et al.*, 2003) and recent texts have revisited seminal works with this perspective. French and Raven (1959), for example, explore power and influence in organisational buying situations; Berthon *et al.* (2003), Farrell and Schroder (1999), Collins and Burt (2003) and Dapiran and Hogarth-Scott (2003) all cite this early source. So in the context of exploration of both human interaction and business relationships, power is seen as an ever-present influence, but this does not necessarily mean that it always has negative consequences.

Another significant contribution to this debate comes from the study of transaction cost economics or transaction cost analysis. Williamson's (1975, 1995) work receives multiple citations in studies of power (Loader, 1997), especially his treatment of surplus value (i.e. the supplier's cost of production versus the buyer's utility function) and how the share of surplus value is divided between parties (Cox et al., 2003). Williamson (1975) contends that each partner is motivated by its self-interest to retain as much of this share as possible for itself (Cox, 1999); therefore, a situation of power must be the ideal position. Furthermore, in buyer–seller relationships, there must be continual manoeuvring for power superiority to secure a greater share of the surplus value created through the possession and/or control of resources (Berthon et al., 2003; Stern and El-Ansary, 1996).

Kumar (1996) adds that the vast majority of manufacturer–retailer relationships are imbalanced and suggests a more extreme view, namely, that channel imbalance means relationships are inherently unstable and those that exist are in danger of becoming 'fiefdoms' that tie suppliers to powerful, dominant partners (Blois, 1997). Johnsen and Ford (2002) also cite the dangers of such asymmetrical relationships, claiming that the power of the dominant party disadvantages the weaker party, and Kumar et al. (1998) warn that powerful channel members can enforce punitive capabilities and actions that inflict negative consequences. Kumar (2005) deftly summarises existing definitions of power as dependence, punitive capability, non-coercive influence strategies and punitive actions; the last feature constitutes the antithesis of trust in relationships.

Yet power has no real regard to whether exchange relationships are balanced or unbalanced (i.e. symmetrical or asymmetrical). Therefore, it is not safe to assume that the natural state of exchange relationships should be one of symmetry and equilibrium, or even fairness. Rather, organisations may actively seek to unbalance their symmetrical relationships to gain a greater share of the benefits (Feldman, 1998). Narayandras and Rangan (2004), in their studies of pairs of relationships, suggest that asymmetric exchanges can thrive, which leads to ongoing trusting/committed relationships. Both outcomes are perfectly possible, depending on the context of the exchange partners, as well as the shifting interplay of power between buyers and sellers. Thus, relationship participants may experience buyer dominance, interdependence, independence, or supplier dominance (Cox, 2001). The degrees of power in different exchange circumstances may change or fluctuate, even within an ongoing relationship. A business organisation's response to the influence of power therefore must be appropriate to the particular and changing circumstances it faces (Cox, 2004). Ultimately, an imbalance of power should be regarded as a 'normal' phenomenon (Batt, 2004). Most organisations are calculating in their dealings with more or less powerful organisations, such that parties may accept imbalance in the pursuit of their business objectives (Geyskins et al., 1996; Newman et al., 2004).

In relation to fairness, relationships are seldom fair in the division of power or reward, nor are all parties equally committed to a relationship (Gummesson, 1996; Kumar, 1996). Fearne *et al.* (2004) consider issues surrounding fairness and justice in agri-food supply chains, gathering empirical data from supplier and retailer organisations. This information helps identify the applications of power imbalances and relational conflict. However, regardless of a desire to be fair, such partnership arrangements ultimately tend to offer the most benefit to the more powerful business partner (Christopher and Jüttner, 2000). It is not surprising that benefits are, or seem to be, unevenly shared, but this unfair distribution does not mean that power-imbalanced relationships are not workable or enduring. Davies (1996) offers a compromise as a means of living with power imbalance: admit that one channel member is normally in charge, so that channel members who wish to cooperate in order to attain mutual advantages must focus on the joint satisfaction of common objectives, regardless of the inevitable imbalance. Therefore, most business relationships exist in a state of what Bengtsson and Kock (2003) call 'co-opetition', in which competition and cooperation exist at the same time (Hingley *et al.*, 2006a).

3.3   Relationship approach and power issues in agri-food channel relationships

The scant and mostly negative treatment of the context of power in business relationships, as well as the predominance of investigations into 'positive' relationship factors (i.e. trust, commitment, cooperation and mutuality), leaves a gap in business relationship literature in relation to the role of power and the ability of organisations to manage power imbalances. Methods of building lasting, meaningful and workable relationships characterised by power imbalances and power dependency, remain highly pertinent to studies of agri-food industry supply chain relationships, in which power generally is skewed in favour of large retail buyers, whereas suppliers of (usually unbranded) products suffer power dependency.

Issues of power dependency, conflict, trust, commitment, cooperation and collaboration (Earp *et al.*, 1999; Johnson *et al.*, 1999) have been applied specifically to food supply chain and retailing industry contexts by, for example, Christopherson and Coath (2002), Hogg *et al.* (1996), Matanda *et al.* (2001), Siemieniuch *et al.* (1999), Egan (2000) and O'Keefe and Fearne (2002). In the United Kingdom, the agri-food industry has experienced concentration in most parts of the supply chain through backward vertical integration, initiated by powerful multiple retail buyers (Collins and Burt, 1999; Galizzi and Venturini, 1996; Howe, 1998; Robson and Rawnsley, 2001). As a result, the shift in power within food marketing channels favours multiple retailers (Bourlakis, 2001; Fiddis, 1997), which represent the main gateways to consumers and the gate keepers between producers and consumers (Lang, 2003).

Hughes (1994) describes food marketing and supply channels as including senior partners, channel captains, or channel leaders (Shaw and Ennis, 2000); O'Keefe and Fearne (2002) describe these category leaders as large processors or retailer buyers (Strak and Morgan, 1998). Retailers search for fewer and larger suppliers that can work with them in partnerships (Fearne and Hughes, 2000; Hingley, 2001; White, 2000; Rademakers and McKnight, 1998), inducing a trend toward multiple retailers that develop exclusive relationships with fewer, favoured, single-source or dedicated partnerships. In turn, suppliers are locked or tied into the relationship (Grunert *et al.*, 1997; Larson and Kulchitsky, 1998) in a type of vertical channel quasi-integration (Howe, 1998).

However, among those that advocate collaboration in the supply chain, a reduced source model may require consideration of the origins of partnering behaviour in a business context, as identified for the lean-thinking concept pioneered by Toyota in car manufacturing (Womack *et al.*, 1990). Cox (1999) calls this approach an 'operational innovation treadmill to oblivion', because the reduced supply base benefits from preferred supplier status but remains forever engaged in the vicious cycle of efficiency gains and cost-led competition.

In the medium-term, oligopolistic benefits accrue to suppliers that survive the consolidation process, but buying organisations then use their power to apply leverage aggressively to these supply chain survivors as a means of maximising value for them. Exclusivity may provide initial gains for the suppliers, but the longer-term result of the consolidation–rationalisation cycle and recycles is a power play that perpetuates the process. The outcome is not a true partnership and all it would imply in terms of mutuality, but rather a state of exclusivity derived from cycles of supply base reductions. Even then, the natural state of a supply chain may not feature stability; instead, stability may be just an interim phase in the continued power battle for share of surplus value. Furthermore, when a preferred or even single-source supplier exists, that supplier is much more likely to bear the burden of asset-specific investment. As a result, channel consolidation may not produce security at all; conversely, it might increase dependence on the buyer (Feldman, 1998).

Genuine two-way interactive partnerships are not fully developed in the UK food industry (Robson and Rawnsley, 2001). Collins and Burt (1999) consider the risk in vertical supply chains to be asymmetric; the width and depth of a retailer's business facilitates its survival if it loses a supplier, whereas the consequences for a supplier that loses a retailer can be much more serious. Further support for this view emerges from initiatives such as category management (CM), which may move more risk to the supplier and away from the retailer (Allen, 2001).

In CM approaches, a preferred supplier takes greater responsibility for the entire supply chain of a product category and aims to maximise sales and profitability through end-consumer orientation (Jarvis and Woolven, 1999). The Institute of Grocery Distribution (IGD) (2006a) for example, identifies and catalogues ongoing case illustrations of whole-chain benefits of consumer oriented CM in shared data, open communication channels and joint strategic planning. In CM, the retailer reduces the number of suppliers in order to guarantee consistency rather than relying on the varying qualities and specifications of different suppliers engaged in continual renegotiations of prices and terms. The CM process inevitably means devolved responsibilities, such that a preferred/nominated lead supplier becomes predominant for one or a group of products. The positive benefits of this for suppliers are greater engagement and devolved responsibility. The concept of channel 'captaincy' as identified in Fearne and Hughes (2000) highlights the positive benefits of 'partnered' marketing and mutual supply channel benefits despite overall imbalances in power. The implementation of modern business practices has helped improve efficiency in the UK fresh produce supply chain. This has allowed the chain to break out of the commodity trap and take the fresh produce category out of the commodity trading environment (Fearne and Hughes, 2000) by means of innovation and value creation (White, 2000). In fact both Hingley (2001) and White (2000) identify the positive empowerment of suppliers engaged in CM roles, such that there is actually some return of power and authority to channel captains who perform such an enhanced role in tandem with retailers.

By contrast, critics such as Dapiran and Hogarth-Scott (2003) contend that the development of CM has not necessarily increased cooperation in supply chains but that retailers instead may use it to reinforce their power and control. Duffy *et al.* (2003) similarly state that CM in food supply chains prompts retailers to prefer larger suppliers that dominate a specific product category. As Bevilacqua and Petroni (2002) argue, the larger buying organisations usually streamline the number of suppliers in order to gain a competitive advantage and, in the process, upset those that are excluded. In conclusion, there are divergent views concerning the role of power, with some seeing channel power as entirely negative and others who believe that devolved power can provide some kind of redress. But it is certain that the issue of power has, in the past, provoked some perhaps unwarranted negativity and clearly it is an issue that cannot be ignored and would benefit from being better understood. This certainly applies in a business and vertical channel context such as fresh food supply to multiple retailers.

3.4   Methodology

UK fresh food channels serve as the context for investigation because of their imbalanced business relationships, which function through the interface of powerful buyers and largely dependent suppliers. Fresh foods typically are unbranded in the UK and sector origin and growth relies mainly on the growth and channel predominance of supermarket chains. The empirical study therefore draws on qualitative and inductive in-depth interviews conducted across the dyadic interface between leading UK multiple food retailing organisations and fresh food supplier organisations. In-depth interviews were conducted with the seven leading multiple food retailing buying organisations and 15 fresh food supplier organisations in the UK. Given the concentrated nature of buying in the hands of so few large retailers, this gives a good and comprehensive view of buying policy. The interviews concerned primarily chains of supply, where suppliers were engaged in long-term vertical relationships, however, some suppliers dealt with more than one retailer customer; whereby, for example, they may have a predominant category leading relationship with one retailer and a subsidiary role with another for their produce. The retailer respondents are all senior managers (i.e. at least category manager level) and are responsible for all operational, as well as some wider strategic issues associated with a specific fresh food category.

The supplier businesses represent sub-sectors of the industry and provide supply from both domestic and global sources. All respondents were asked to express their perceptions about the nature, implementation and monitoring of business relationships, especially with regard to issues of power dependency and mutuality. For reasons of confidentiality, the identities of the case organisations remain anonymous.

Semi-structured personal interviews that allowed access to respondents' thoughts, opinions, attitudes and motivational ideas were used. In total, 21 interviews, each of one hour in length were conducted using a standard and consistent interview schedule. Semi-structured questioning concerned issues in general about supply chain relationships, power and so forth, but also involved specific questioning about particular dyadic exchange with named suppliers or retailers. So, typically relationships were explored between a retailer and two of their principal suppliers. Since cases were selected for their ability to contribute new insights, as well as in the expectation that these insights would be replicated, both theoretical breadth and category saturation was obtained.

A sample of suppliers with the desired characteristics was first located. Suppliers were selected because they provided typical examples of organisations (Miles and Huberman, 1994; Patton, 2002), in this case being fresh food category leading supply chain members. All of the respondents are typical in that they are preferred suppliers for at least one of the multiple retailers in a specific fresh produce category. None of the suppliers exceed £100 m in turnover and in this too they are typical. Additionally, each supplier is a second or third tier supplier to one or more retailers.

Interview questions were standardised around a number of topics and questions were kept deliberately broad to allow interviewees as much freedom in their answers as possible. The findings are taken from the words of the respondents themselves, thereby aiding the aim of the research, whilst gaining much more information than would have been available from alternative research methods. All interviews were taped first to increase accuracy of data presentation and later transcribed to allow detailed analysis. Each case analysis involved writing up a summary of each individual case in order to identify important case level phenomena. Following this process, a coding scheme was developed to assist with the cross-case analysis that involved searches for cross-case patterns. The case studies and interpretive reports were returned to the respondents for their comments, a step that helped enhance the validity of the research method further. Guided by considerations raised in the literature, the following research issues are explored: the nature and application of power in fresh food supplier–retailer vertical channel relationships and the impact of management trends on supply base concentration and category management (CM).

3.5   Findings

Table 3.1 summarises the key issues and findings of this study of power in fresh food industry relationships, as derived from the in-depth interviews and illustrated by quotations from study respondents. This table is designed to be read in tandem with the following discussion.

Power is notably imbalanced in fresh food industry relationships in favour of retailer buying organisations. However, the evidence from this study does not indicate any instability derived from these power-imbalanced structures; rather the opposite is true. Many long-standing vertical supply chain relationships exist, though retailers certainly remain in control of mini-fiefdoms. Furthermore, retailers maintain the potential to exert punitive action on suppliers who fail to conform to their wishes. Some suppliers worry about the expression of retailer power, enabled by the power imbalances, and fear abuses of that power (see issue 1.1 in Table 3.1).

Powerful UK retail buying organisations use collaborative relationship-based concepts to obtain fresh food supplies. Specifically, chains have been shortened, supplier numbers are rationalised, and partner/category leadership arrangements apply to dedicated and exclusive suppliers. However, true partnership is difficult to achieve amongst unequal members of vertical supply chains, which require a lead partner. As a result, ideals of symmetrical mutuality and equal trust within a relationship are largely unattainable. Suppliers broadly accept the state of asymmetrical power imbalance and all that goes with it, as long as they attain a reasonable proportion of the relationship value and/or this method is preferable to exchange channel structures with inherently higher transaction costs (see issue 1.2 in Table 3.1).

Fresh food vertical (and other similar) food industry relationships between suppliers and retailers employ what have been described as relationship-building tools, such as CM and exclusive/preferred supplier arrangements. Evidence from this study, however, appears to indicate a

Table 3.1 Findings from case interviews

Research issue	Findings	Illustrative quotation
1.1 Power balance in fresh food supply chains	Fresh food relationships are asymmetric and not mutual.	'In general the balance of power is in the retailer's favour.' (Retailer 3) '...risk is tipped in the direction of the supplier, reward is tipped in the direction of the multiple [supermarket chain retailer].' (Fresh Food Supplier 8) 'Suppliers are very important to us, but it is an imbanced partnership. It is a lot harder for them (suppliers) to find another customer, than it is for us to find another supplier.' (Retailer 5) 'We are in a reasonably positive relationship with our retailers but it stops short of being a partnership in that the shared objectives are relatively narrow.' (Fresh Food Supplier 4)
	Powerful channel members can exploit their position and even implement punitive action.	'...they (supermarket buyers) are almost like god in a lot of ways. That is the problem, buyers tend to be a bit egocentric and they know (it). One supermarket (named), kicks for the sake of kicking.' (Fresh Food Supplier 8) 'We deal with them (suppliers who do not meet standards) very forcefully (if they are at fault), but it is very difficult for them to deal with us if it is the other way (around).' (Retailer 3)
1.2 Reduction in supplier numbers/ supplier exclusivity	Supply base rationalisation puts the focus on fewer but more significant suppliers.	'We have a stronger relationship now with fewer suppliers. We are focused on delivering more with less suppliers.' (Retailer 3) '...Supplier 1 used to be one of 20 suppliers of salads.... [they] are now one of only two salad suppliers and quality, service and price have all improved.' (Retailer 1) '...we invested a lot of money in the business, so we increased our volume and became one of two main suppliers to (named retailer), through seven abattoirs for beef and lamb. I am only guessing but there had been (prior to this) in the region of 20–25 abattoirs supplying (the retailer).' (Fresh Food Supplier 11)
	Exclusivity delivers relatively short-term relationship gains for suppliers (as a result of the 'cycle–recycle' effect of supply base rationalisation).	'(Exclusivity can mean a) Poor negotiating position, [an] unstable business position.' (Fresh Food Supplier 8) 'When that (a new technological) process was put in it required a lot of collaboration, because everybody was learning and it required us to work very closely together (with the retail customer). But that is not the case now; it is all established technology, so the need isn't there so much.' (Fresh Food Supplier 11)

Table 3.1 _Cont'd_

Research issue	Findings	Illustrative quotation
1.3 Impact of category management (CM) approach	CM in fresh food channels has developed partial mutuality.	'We feel we are an important part of their business. We are much closer to the retailer and more important to the consumer.' (Fresh Food Supplier 1)
		'[There is] Much greater openness now moving business forward at a much improved pace than before. Previously we used to tell people what to do, we now consult.' (Retailer 3)
		'... we share data, plan jointly (which limits risk) and have an open forum in which to discuss concerns.' (Retailer 1)
		'Feedback is a two-way process, so we would judge the strength of our relationship by the fact that we share our research and our consumer information and the supplier shares their market information upwards with us ...' (Retailer 1)
		'...they (suppliers) don't sometimes realise how powerful they are, because without their goods, we'd have no business.' (Retailer 3)
		'As suppliers consolidate, their position will strengthen.' (Fresh Food Supplier 8)
	CM results in some (but under-exploited) countervailing power for suppliers.	'[There is a] Continued emphasis on less suppliers and [a] partnership approach. This will however get to a point where balance of power may switch to the supplier.' (Retailer 1)
		'I am sure that (named retailer) would weigh up that there are benefits in reducing the supply base, but is that outweighed by being dependent on a smaller number of suppliers?' (Fresh Food Supplier 12)
1.4 Living with power imbalance	Despite some power redress for suppliers, ultimate control still lies in the hands of the retailer.	'...this [information] flow is always to mainly benefit the retailer and plc shareholders.' (Fresh Food Supplier 7)
		'God help us if we drop below 99% (service level of delivery on time into the depot) ...' (Fresh Food Supplier 7)
		'...(supplier) prices are not rising. Costs are rising the whole time....' (Fresh Food Supplier 10)

degree of mutuality through the CM approach, which allows suppliers some inclusiveness and two-way exchange (e.g. of data). Although this study partially concurs that CM is a power tool used by retailers, it also can deliver rewards to preferred suppliers and may be a countervailing mechanism that works for the benefit of suppliers.

What suppliers would prefer, however, is a greater demonstration by retailers of the mutuality of the relationship. It is ironic, therefore, that some retailers state that preferred status suppliers do not realise the strength of their position and that retailers are indeed reliant on those suppliers. However, despite this finding, many fresh food suppliers are only short- to medium-term beneficiaries. Therefore, business-to-business exchanges in vertical fresh food channels have not reverted to transactional dealings and a reduced supply base model remains, such that key suppliers are more involved and regularly consulted in the business process. In addition, CM has brought to the fore the issue of power dependence, but the application of CM does not necessarily weaken the balance of mutuality in fresh food relationships (from the supplier perspective). Despite the imbalance, there have been some improvements in mutuality through CM in fresh food supply chain relationships, but again, retailers retain the ultimate say and may restrict or ration the free flow of exchange in key areas such as information. As a result, some suppliers believe that the imbalance in power, which favours the retailer, is counter productive in the longer term for an effective (mutual) working relationship, despite the efforts of suppliers to meet retailers' considerable quality and service demands. Suppliers may appreciate the short- to medium-term benefits to be derived from exclusive supply, but they also feel vulnerable with regard to the longer term cycle and recycle effects of rationalisation (see issue 1.3 in Table 3.1).

Even if asymmetry and power imbalance are not barriers to the formation of close and workable relationships, suppliers do not necessarily want to see a greater demonstration of the mutuality of relationships by retailers, with more emphasis on collaboration and reciprocity; in the interim, suppliers will live with asymmetry and power imbalances. Alternative routes to market are constricted by the process of channel rationalisation and, for many suppliers, the reduced source model is preferable and desirable. Fresh food channel relationships thus can be described as workable examples of co-opetition. The notion devised in relationship literature – that power is always a negative and divisive influence that precludes relationships – is clearly flawed.

Power is ever present in business-to-business exchange. Because a relationship approach does not replace the friction and continual power play between business exchange partners, power plays and relationship development coexist. Trust and mutuality appear to some degree in fresh food relationships, but their expression is conditional and often at the behest of the retailer. However, the experience revealed in fresh food relationships suggests that the situation remains workable, in line with Earp et al.'s (1999) view that commitment- and trust-based relationships are not the only effective way to conduct business exchanges. Organisations, even those engaged in partnering activity, seek to gain the upper hand. The ongoing power play in vertical supply chain relationships endures in a state of flux, although that state is not always overt, which does not mean that effective exchange relationships cannot endure (see issue 1.4 in Table 3.1).

3.6   Network of relationships

The dyadic relationship between large multiple retailers and their preferred suppliers or super middlemen (Hingley, 2005) are the axis around which the modern agri-food supply chain (and wider network) revolves. Figure 3.1 illustrates the central dyadic interface and wider network sets of interactions, linked by two-way exchanges (i.e. products and communications). This identification of super middlemen as intermediaries and performers of wider tasks places the structure and debate in the context of network concepts. Relationships exist in a network context (Gummesson, 1996; Healy et al., 2001), as demonstrated by Håkansson and Snehota (1990) and Anderson et al. (1994) with their contention that individual organisations and dyadic relationships both are part of a network of interrelationships (Johnsen and Ford, 2002).

Anderson et al. (1994) describe this type of corporate structure as the emergence of deconstructed firms, such that an organisation focuses on a sub-set of value-adding functions and relies on coordinated relationships within networks of other firms to provide the remainder of its offering. The application of a network approach creates further issues regarding the

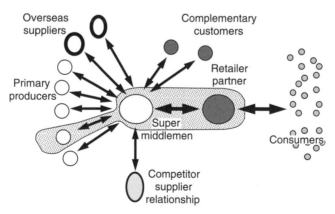

Fig. 3.1 Key network relationships in UK agri-food industry, featuring the central super middleman–retailer dyad and the extended triadic link with a primary producer.

nature of competition in business exchange. Wilson (1996), for example, compares the concept of competition under neo-classical theories of the firm (i.e. organisation competes with organisation) with a new paradigm, in which competition pits channel against channel rather than firm against firm. Low (1997) concurs and believes that industrial networks reject the notion of pure competition with faceless, unconnected firms; instead, business transactions are conducted within the framework of enduring business relationships characterised by mutual cooperation and adaptation. Furthermore, Spekman *et al.* (1998) believe that only close collaborative links throughout the supply chain can return the benefits of cost reduction and revenue-enhancing behaviour.

A network approach to supply chain relationships is borne out in practice in the agri-food industry, in which economic and trading circumstances result in changes to organisational structures. Competition in the UK agri-food industry, for example, occurs between the supply network led by Tesco and that of J. Sainsbury versus that of Asda. Each employs a hub of supply centred on its own super middlemen. Profitability therefore relies on the competitive success of one network compared with another. The status of super middlemen represents the culmination of the process of supply base rationalisation, resulting in a progressive reduction in the number of suppliers with which the major multiple retailers deal. Retailers do not want to deal directly with primary producers; similarly, farmers prefer to avoid interactions with participants further downstream (i.e. prefer a main interface with super middlemen). In fresh food supply, retailers adopt a portfolio approach to supplier relationships and develop a mix of close, ongoing network relationships and more distant ones with transactional suppliers via super middlemen.

In this network context, the manifestation of supermarket power through supply base reduction and concentration is not necessarily bad for agri-food industry suppliers. With preferred supply status, the suppliers can avoid the continual and fierce horizontal competitiveness inherent in securing retailer business and super middlemen develop a more significant role at the supply hub. As a result, retailers may control the networks in which they exist, but the hands-off approach and reliance on a new breed of intermediaries means that such control does not have to be destructive. This approach builds a higher level of reliance on the retailer, which is to its advantage, but it also benefits the lead supplier through the provision of devolved services.

Retailers do not want to be agricultural producers, importers, or food processors but instead prefer to focus on the business of retailing and the add-on services provided by their strong brands; they contract out most other activities. In response, responsibilities have devolved in specific ways. Super middlemen handle the bulk of contacts with primary agri-food producers and thereby sit at the hub of the triadic links between the primary producer, middleman and retailer (see Fig. 3.1). They shoulder the burden of securing supply; if a crop fails or a key product is in short supply, it is no longer just the buyers' problem to find an alternative source. Maintaining the continuity of supply also often falls on super middlemen, who may even procure from competitors to satisfy a supermarket's needs. Thus, procurement decision-making is delegated by the retailer to the category lead supplier.

As the hub for both domestic and overseas products, super middlemen may or may not be primary producers, yet they consistently manage the flow and mix of supply. It may make perfect sense to procure home-grown products, but these principal intermediaries will incorporate and rely on overseas supply if necessary. The supervision of quality assurance similarly has devolved as a requirement for creative product development. In such a system, super middlemen may be involved in food manufacture for some supermarket private-label products, from the semi-processed to the fully prepared.

Yet super middlemen may not necessarily provide increased product profitability; category leaders actually may suffer a decline in their direct profit margins. However, in the broader context, they also enjoy reduced transactional costs and reduced overheads, derived from channel consolidation efficiencies. Profitability also should be understood from the perspective of network market share. Market share for each product category becomes a far more important determinant of success than individual product or corporate profitability. Studies of agri-food supplier–retailer relationships undertaken by Hingley (2001), Hingley and Lindgreen (2002), Hingley et al. (2006b), and White (2000) reveal that suppliers broadly accept states of asymmetrical power imbalance, assuming, as previously noted, that they receive a reasonable proportion of the relationship value or recognise this method as preferable to alternative routes to market.

The tolerance of suppliers to imbalances in financial returns and continual contractions of supply chain profit margins remains difficult to gauge. It is equally difficult to assess precisely the level of profitability that is acceptable to suppliers, especially in the context of fierce global competitiveness in the food industry, which results in progressively tighter margins. However, when suppliers engage in preferred relationships with multiple retailers, price setting becomes less relevant and they often give up the right to negotiate prices in return for exclusivity. Retailers therefore determine the price, but preferred suppliers earn the reward of market share gains and the ability to lead a wider network, which adds value. Suppliers' profit margins thus are sacrificed to some degree for increased turnover, exclusivity and access to wider network arrangements, as well as the spin-off businesses associated with a retail customer (e.g. international markets). Some suppliers (i.e. producers of commodity and generic agrifood products) adopt the view that they will accept low prices and low margins when dealing with supermarkets, to the point that they become the 'last man standing' in a given sector or product category. In turn, they reap the rewards of enhanced market share and access to a retailer's business. However, UK multiple retailers generally will not reduce the supply base in a specific category to a sole organisation, because then they could not play off suppliers against one another. Sole supply does exist, but generally only in a sub-category of a larger commodity group of products.

These exclusivity arrangements with retailers may not be as binding as they first appear and suppliers may develop qualified exclusivity arrangements. Super middlemen maintain predominant relationships with their retailer customers, but they also may engage in secondary relationships with complementary businesses, such as supplying a food service customer. This secondary arrangement enables the supplier to sell in alternative outlets without upsetting the exclusivity that its retail customer requires.

Retailers may prefer larger suppliers for reasons of continuity, economy and market knowledge, although small suppliers can provide flexibility and all-important retail market differentiation. This benefit emerges in the relationships between small and medium enterprises that engage in direct or indirect, through super middlemen, interfaces with powerful partners. If the inherent imbalance of these relationships were a problem, no organisations would enter into them and they would not endure. This scenario clearly is not accurate and smaller suppliers frequently enter into such relationships and tolerate power imbalances (Blundel and Hingley, 2001). Larger partners perceive benefits in dealing with smaller suppliers, who offer advantages because of, rather than despite, their size.

Most recently, channel leading retailers have revised the concept of the role of super middlemen by considering matters of cost consciousness and control. Despite the benefits of CM, its infrastructure features significant central control cost and overheads. Some questions pertain to whether CM has run its course; perhaps circumstances again favour a leaner and more direct sourcing philosophy, in which stripped-down suppliers can gain direct routes to multiple retailers. The context for this premise derives from ongoing price pressures in the channel management of what are essentially low value fresh food commodities. Such cost and price pressures may prompt retailers to develop slimmed down and more direct supply routes for many fresh products. In addition, supplier criticism and discussion of the more negative aspects of the approach has led retailers to reappraise the future of CM. Perhaps retailers no longer require such a close relationship with their suppliers, because the supplier quality and integrity issues and protocols that initiated this approach have been well established and could be addressed by freelance, outsourced agents.

Despite the predominance of channel and network systems, an alternative to the CM approach with a super middleman hub, proposed by Hingley (2008) and shown in Fig. 3.2, provides options for retailers to deal directly with suppliers, including preferential direct routes for new and innovative suppliers. The UK retail market, as in much of the industrialised world, is

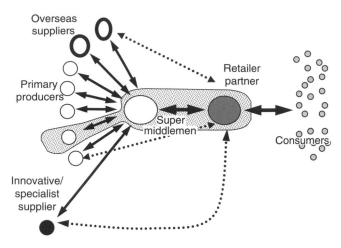

Fig. 3.2 Key network relationships in UK agri-food industry, showing the direct supply alternative (dotted lines), by-passing the super middleman.

determined by fierce competition between fewer and bigger multiple retailer chains. Pressures on expected returns to shareholders mean an ever-tightening squeeze on suppliers, which may also mean that some products could be 'value-added' at the country of origin (e.g. where labour costs are cheaper), by-passing CM-type intermediaries.

However, retailers will find it difficult to return to purely transaction-based direct supply in the fresh produce industry. First, retailers have down-sized and outsourced much of their operations and managing a large supply network requires more hands-on effort and more direct decision makers. Second, super middlemen provide a valuable service role that enables retailers to remain relatively remote. These intermediaries add value, ensure quality and consolidate products from many sources. Even if some activities, such as quality inspections, could be outsourced to freelance agents, the process would still require retailers to devote management time and attention. However, the temptation for retailers to cut costs by removing the intermediary infrastructure, which they perceive as costly, may be strong. In such circumstances, the products still must be imported, consolidated in the retailer's home or host country, and delivered to the store. The market for fresh food and its complex network of interrelationships entails a costly process and stripping it down to a basic transactional exchange can work only for the simplest commodities. For most products, a high service level is still necessary, especially in this age of corporate accountability and product traceability. Therefore, retailers would be unwise to sacrifice the value-adding input of their partner suppliers, which often have accumulated years of experience in dealing with supply chain problems that could not be easily solved at arm's length.

3.7   Conclusions

Literature on business relationships, distilled from diverse antecedents, combines the common ingredients of profitable ongoing interactions, collaborative value creation and the underpinning themes of trust, commitment and mutuality. However, it also undervalues the significance of power in the formation and operation of business relationships. For example, agri-food industry relationships are power dependent, but in contrast to some views, these relationships can exist and thrive despite power imbalances.

Although power is notably imbalanced in agri-food relationships, in favour of the retailer buying organisations, a state of instability is not a guaranteed result, rather, many long-standing vertical supply chain relationships exist (Hingley, 2001). Despite recent rhetoric, these exchanges cannot be considered (by suppliers at least) to be partnerships. Retailers remain in control of mini-fiefdoms, as envisaged in Blois (1997) and, as Kumar (1996) suggests, the potential remains for punitive action by retailers if suppliers do not accede to their wishes. The situation is not fair and it may not even be just, but that does not mean the system is not workable or even beneficial for agri-food suppliers. Some suppliers may worry about the expression of retailer power allowed by the imbalance, as identified by Fearne *et al.* (2004) and Duffy *et al.* (2003), as well as the potential for abuses of power. However, asymmetry and unfairness do not mean that organisations are unwilling to enter into and continue relationships with major multiple retail chains, which remain the largest and most consistent market outlet for UK agri-food.

Furthermore, as UK retailers apply collaborative, relationship-based constructs to agri-food supply, chains have shortened, supplier numbers have been rationalised and partner/category leadership arrangements have begun using dedicated, exclusive super middlemen. This process benefits the remaining category suppliers (at least in the medium-term) by providing them with more business and greater access to retailer customers. However, true partnerships are difficult to achieve in vertical supply chains, which demand a lead partner. Therefore, the relational ideals of symmetrical mutuality and equal trust are largely unattainable.

Mutuality also cannot be achieved through the CM approach, unless some inclusiveness and two-way exchange is established. However, business-to-business exchanges in agri-food channels have not reverted to transactional dealings. Despite retailers' renewed interest in direct sourcing, a reduced source model persists in which key suppliers are more involved and regularly consulted. Asymmetry and power imbalance do not bar the formation of close and workable relationships, although suppliers still would prefer a greater demonstration of the mutuality of relationships by retailers, with more emphasis on collaboration (Fearne *et al.,* 2004; Hingley, 2005). In the interim, they accept asymmetry and power imbalances, contrary to the view purported by some literature that implies asymmetrical exchange is not functional and only has a negative influence.

Ironically, some retailers surveyed in studies of agri-food supplier–buyer exchanges state that their preferred status suppliers do not realise the strength of their own position (Hingley, 2005; White, 2000). For suppliers, the danger lies in the short- to medium-term gains (largely as a result of consolidation), which may create a treadmill effect once those gains have been absorbed. To avoid such an outcome, suppliers, even those important and pivotal super middlemen, must stay ahead of the game and anticipate future retailer service requirements, which enable them to become essentially invaluable to their key customers.

The notion that power is always a negative and divisive influence that precludes relationship forming is clearly flawed. Power is ever present in business-to-business exchanges, even if a relationship approach exists. Trust and mutuality characterise agri-food relationships, but their expression is conditional and often at the behest of the retailer, which implies that methods other than commitment- and trust-based relationships provide effective ways of conducting business exchanges. Organisations, even those engaged in partnering activity, commonly seek to gain the upper hand, so ongoing power plays in vertical supply chain relationships create a state of flux (Cox *et al.*, 2001; Howe, 1998; Ogbonna and Wilkinson, 1996), which does not mean that effective exchange relationships cannot endure.

The abuse of power certainly is a destructive force, but the exercise of power in asymmetric relationships (whether punitive action or functional conflict) is more typical than perpetual cooperation or power symmetry. Although all organisations hope to gain advantages and thus disrupt symmetry, striving for self-interest does not preclude cooperative actions and cooperative and competitive business strategies can coexist. Weaker parties exhibit a degree of tolerance to imbalances of power, which means that such relationships are not necessarily unstable or short in duration (Blundel and Hingley, 2001; Narayandras and Rangan, 2004). A key element appears to be the existence of channel captains. Asymmetry is no barrier to entry, especially if suppliers have something a buyer wants or believe they may profit from the situation, despite the power imbalance.

Only by admitting the presence of channel leadership and acknowledging where power lies can suppliers move forward and strive for what is important – market survival and an acceptable level of organisational profit. Thus, they must accept power imbalances and the inherent nature of inequity and unfairness in supply chains. To this end, Davies (1996) provides a compelling argument: rather than fruitlessly fighting against imbalance, it may be more rewarding to concede control. This first step may enable agri-food suppliers to benefit from the efficiencies of a truly integrated (but retailer-controlled) network.

In the agri-food channel context, expressions of power by retailers also confirm the first three features described by Kumar (1996): dependence, punitive capability and non-coercive influence strategies. However, trust appears to exist by degree and its expression may progress or recede according to circumstances. Such shifts do not necessarily depend on the absence or presence of the types of power outlined previously.

A final point pertains to when a dominated party might decide that its domination has become unacceptable. For Kumar (2005), punitive action provides the line in the sand. Although this chapter identifies weaker partners' tolerance to displays of power dominance and asymmetry of trust and their acceptance of the inequity of outputs, further studies should investigate the boundaries of this tolerance from a stakeholder perspective, including the identification of the specific circumstances that might create disincentives to forming relationships or force their dissolution. For example, just how much income loss will a weaker party tolerate to retain an important customer's business? This question remains unanswered in empirical work, largely because such studies require an investigation of failed business relationships, which participants are notoriously reticent to discuss. However, suppliers will continue to trade off retailer power play and their resultant outcome imbalances for the prospect of business continuity, increased market share and acceptable profitability.

3.8   Future trends

The overall trend in the UK is for retail-driven vertical food chains, most notably in areas where retailers are very strong (fresh, perishable and retailer-branded sectors). The food industry is being dominated by a few large retail corporations on a national level, with some even operating on a European or global scale. In the UK the takeover of one of the largest food retailers (Safeway by Wm. Morrison) has resulted in four major supermarket chains (Tesco, Sainsbury, Wal-Mart Asda and Morrisons) accounting for three-quarters of retail grocery sales (IGD, 2006b). Tesco takes a third of the value of UK grocery sales alone. This concentration has put further pressure on fresh produce suppliers who have been channelled down the CM route. That is not to say that CM is not beneficial for suppliers, as IGD (2006a) and others highlight the mutual success of the approach. However, is this phenomenon peculiar to the highly centralised arena of UK food retailing? Certainly the UK has some specific features concerning the nature of fresh foods, for example, there is a low level of supplier branding within these sectors in the UK and a correspondingly strong retailer own label heritage (Hingley, 2001).

Given that there is much that is similar in the European food industry, there are comparable levels of retailer concentration in most industrialised western European nations (Bourlakis, 2001), resulting in similar rationalisation and supplier arrangements across Europe. Notably, both IGD (2006a) and ECR Europe (2008) talk of European-wide initiatives identifying consumer benefits of supply chain collaboration; however, such initiatives do not tend to dwell on the power implications. In the UK, the national and trade media continues to monitor power-based issues in supply chains, *The Guardian* in Anon (2008), for example, provides a historical timeline of power-based contention in food supply in the UK and there are regular accounts of the negative consequences of the result of power inequity on supply channels (Grant Thornton, 2007). Despite this, other nations look at the UK as being ahead in the mechanics and operations of supply chain relationships including issues of category management and power.

There is evidence of application of these issues in other global regions, for example, Parker and Byrom (2009) chart power issues concerning supplier–retailer relationships in an Australian context, as does Dunne (2008). The latter example is published in a recent special edition of *British Food Journal* ('Relationships, networks and interactions in food and agriculture business-to-business marketing and purchasing' – see below) which includes papers from authors who consider issues of supplier–retailer relationships, trust, power and much more from a New Zealand fresh produce perspective (Clements *et al.*, 2008), as well as contributions from Danish, Finnish, Brazilian and Chinese contexts. As the march of food industry globalisation continues, these issues of retail driven supplier concentration, power and supplier chain relationships will follow in new countries and contexts. However, different countries have different interpretations of the issues, which may be culture bound. A good illustration of this is the *guanxi* relationship approach, based on personal connections and allegiances, particular to Chinese culture and this is ideally explained and put into context of Western learning by Lu *et al.*, (2008) and is again to be found in the *British Food Journal* special edition identified above and referenced below.

3.9   Sources of further information and advice

British Food Journal (2008) special edition, Guest editors: Dr. Adam Lindgreen, Dr. Martin Hingley and Dr. Jacques Trienekins, 'Relationships, networks and interactions in food and agriculture business-to-business marketing and purchasing', vol. **110** (4/5), Emerald Group Publishing.

Industrial Marketing Management, 2005, **34** (8), includes the paper cited here by Hingley (2005) but also three further papers, by eminent Professors Blois, Kumar and Naude, respectively, who debate the power issues raised in the paper. The special treatment is concluded by a further response by Hingley.

Useful web site links:
English Farming and Food Partnerships: http://www.effp.com/
Institute of Grocery Distribution: http://www.igd.com/

3.10 References

ALLEN, S. (2001), 'Changes in supply chain structure: the impact of expanding consumer choice', in *Food Supply Chain Management: Issues for the Hospitality and Retail Sectors*, Eastham, J.F., Sharples, L. and Ball, S.D. (eds), Butterworth-Heinemann, Oxford, 314–23.

ANDERSON, J.C., HÅKANSSON, H. and JOHANSSON, J. (1994), 'Dyadic business relationships within a business network context', *Journal of Marketing*, **58**, 1–15.

ANON (2008), 'The 10-year journey to curb supermarket power', *The Guardian*, 15th February, http://www.guardian.co.uk/business/2008/feb/15/retail.supermarkets (accessed 15/12/08).

BATT, P.J. (2004), 'Power-dependence in agricultural supply chains: Fact or fallacy?' in *Proceedings of 20th Annual Conference of the Industrial Marketing and Purchasing Group*, Copenhagen, 2–4 September.

BENGTSSON, M. and KOCK, S. (2003), 'Tension in co-opetition', *Developments in Marketing Science*, **26**, 38–42.

BERTHON, P., PITT, LEYLAND, F., EWING, M.T. and BAKKELAND, G. (2003), 'Norms and power in marketing relationships: Alternative theories and empirical evidence', *Journal of Business Research*, **56** (9), 699–709.

BEVILACQUA, M. and PETRONI, A. (2002), 'From traditional purchasing to supplier management: A fuzzy approach to supplier selection', *International Journal of Logistics*, **5** (3), 236–51.

BLOIS, K.J. (1997), 'Are business-to-business relationships inherently unstable?', *Journal of Marketing Management*, **13**, 367–82.

BLOIS, K.J. (1998), 'Don't all firms have relationships?' *Journal of Business and Industrial Marketing*, **13** (3), 256–70.

BLUNDEL, R.K. and HINGLEY, M.K. (2001), 'Exploring growth in vertical inter-firm relationships: Small-medium firms supplying multiple food retailers', *Journal of Small Business and Enterprise Development*, **8** (3), 245–65.

BOURLAKIS, M.A. (2001), 'Future issues in European supply chain management', in *Food Supply Chain Management: Issues for the Hospitality and Retail Sectors*, Eastham, J.F., Sharples, L. and Ball, S.D. (eds), Butterworth-Heinemann, Oxford, 297–303.

BRETHERTON, P. and CARSWELL, P. (2002), 'Trust me–I'm a marketing academic! A cross-disciplinary look at trust', in *Proceedings of: Academy of Marketing Annual Conference*, July 2–5, 2002, Nottingham University Business School.

CAMPBELL, A.J. (1997), 'Buyer–supplier partnerships: Flip sides of the same coin?', *Journal of Business and Industrial Marketing*, **12** (6), 417–34.

CHRISTOPHER, M. and JÜTTNER, U. (2000), 'Supply chain relationships: Making the transition to closer integration', *International Journal of Logistics: Research and Applications*, **3** (1), 6–23.

CHRISTOPHERSON, G. and COATH, E. (2002), 'Collaboration or control in food supply chains: Who ultimately pays the price?' in *Proceedings of the Fifth International Conference on Chain and Network Management in Agribusiness and the Food Industry*, Noordwijk, Wageningen University, 6–8 June 2002.

CLEMENTS, M.D., LAZO, R.M. and MARTIN, S.K. (2008), 'Relationship connectors in NZ fresh produce supply chains', *British Food Journal*, **110** (4/5), 346–60.

COLLINS, A. and BURT, S. (1999), 'Dependency in manufacturer- retailer relationships: The potential implications of retail internationalisation for indigenous food manufacturers', *Journal of Marketing Management*, **15** (1), 673–93.

COLLINS, A. and BURT, S. (2003), 'Market sanctions, monitoring and vertical coordination within retailer-manufacturer relationships: The case of retail brand suppliers', *European Journal of Marketing*, **37** (5/6), 668–89.

COX, A. (1999), 'Power, value and supply chain management', *Supply Chain Management: An International Journal*, **4** (4), 167–75.

COX, A. (2001), 'The power perspective in procurement and supply management', *The Journal of Supply Chain Management*, **37** (2), 4–7.

COX, A. (2004), 'The art of the possible: Relationship management in power regimes and supply chains', *Supply Chain Management*, **9** (5), 346–56.

COX, A., SANDERSON, J., WATSON, G. and LONSDALE, C. (2001), 'Power regimes: A strategic perspective on the management of business-to-business relationships in supply networks', in *Proceedings of the 17th Annual IMP Conference: Interactions, Relationships and Networks: Strategic Dimensions*, Håkansson, H., Solberg, C.A., Huemer, L. and Steigum, L. (eds), Norwegian School of Management BI, Oslo, 9–11 September.

COX, A., LONSDALE, C. and WATSON, G. (2003), 'The role of incentives in buyer-supplier relationships: Industrial cases from a UK study', in *Proceedings from the 19th Annual IMP Conference*, Lugano, Switzerland, 4–6 September.

DAPIRAN, G.P. and HOGARTH-SCOTT, S. (2003), 'Are co-operation and trust being confused with power? An analysis of food retailing in Australia and the UK', *International Journal of Retail and Distribution Management*, **31** (5), 256–67.

DAVIES, G. (1996), 'Supply-chain relationships', in *Relationship Marketing: Theory and Practice*, Buttle, F. (ed.), Paul Chapman Publishing, London, 17–28.

DONEY, P.M. and CANNON, J.P. (1997), 'An examination of the nature of trust in buyer-seller relationships', *Journal of Marketing*, **61**, 35–51.

DUFFY, R., FEARNE, A. and HORNIBROOK, S. (2003), 'Measuring distributive justice and procedural justice: an exploratory investigation of the fairness of retailer-supplier relationships in the UK food industry', *British Food Journal*, **105** (10), 682–94.

DUNNE, A.J. (2008), 'The impact of an organization's capacity on its ability to engage its supply chain partners', *British Food Journal*, **110** (4/5), 361–75.

EARP, S., HARRISON, T. and HUNTER, A. (1999), 'Relationship marketing: Myth or reality?', in *Proceedings of the 15th Annual IMP Conference*, University College, Dublin, 1999, 1–21.

ECR EUROPE (2008), Working together to fulfill consumer wishes better, faster and at less cost. http://www.ecrnet.org/ (accessed 15/12/08).

EGAN, J. (2000), 'Drivers to relational strategies in retailing', *International Journal of Retail and Distribution Management*, **28** (8), 379–86.

FARRELL, M. and SCHRODER, W. (1999), 'Power and influence in the buying centre', *European Journal of Marketing*, **33** (11/12), 1161–70.

FEARNE, A. and HUGHES, D. (2000), 'Success factors in the fresh produce supply chain: Insights from the UK', *British Food Journal*, **102** (10), 760–76.

FEARNE, A., DUFFY, R. and HORNIBROOK, S. (2004), 'Measuring distributive and procedural justice in buyer/supplier relationships: an empirical study of UK supermarket supply chains', in *Proceedings of 88th Seminar, European Association of Agricultural Economics*, Paris, France, 5–6 May.

FELDMAN, L.J. (1998), 'Industry viewpoint: relational interdependency and punctuated equilibrium', *Journal of Business and Industrial Marketing*, **13** (3), 288–93.

FIDDIS, C. (1997), *Manufacturer Retailer Relationships in the Food and Drink Industry: Strategies and Tactics in the Battle for Power*, FT Retail and Consumer Publishing/Pearson Professional, London.

FRENCH, J.R.P. and RAVEN, B.H. (1959), 'The bases of social power', in *Studies in Social Power*, Cartwright, D. (ed.), Institute for Social Research, The University of Michigan, Ann Arbor, MI, 150–67.

GALIZZI, G. and VENTURINI, L. (eds) (1996), *Economics of Innovation: The Case of the Food Industry*, Physica-Verlag, Heidelberg.

GEYSKENS, I., STEENKAMP, J-B.E.M., SCHEER, L.K. and KUMAR, N. (1996), 'The effects of trust and interdependence on relationship commitment: A transatlantic study', *International Journal of Research in Marketing*, **13**, 303–17.

GRANT THORNTON (2007), *Supermarket Code of Practice Offers no Protection According to Three Quarters of Food Suppliers*, 25 August 2007, http://www.grant-thornton.co.uk/press_room/supermarket_code_of_practice_o.aspx [accessed 15/12/08]

GRUNERT, K.G., LARSEN, H.H., MADSEN, T.K. and BAADSGAARD, A. (1997), *Market Orientation in Food and Agriculture*. Kluwer Academic, Boston, MA.

GUMMESSON, E. (1996), 'Relationship marketing and the imaginary organisation: A synthesis', *European Journal of Marketing*, **30** (2), 31–44.

GUMMESSON, E. (1999), *Total Relationship Marketing: From the 4 Ps – Product, Price, Promotion, Place – of Traditional Marketing Management to the 30 Rs – the Thirty Relationships – of the New Marketing Paradigm*, Butterworth-Heinemann, Oxford.

HÅKANSSON, H. and SNEHOTA, I. (1990), 'No business is an island: The network concept of business strategy', in *Understanding Business Markets: Interactions, Relationships and Networks*, Ford, D. (ed.), Academic Press, San Diego.

HEALY, M., HASTINGS, K., BROWN, L. and GARDINER, M. (2001), 'The old, the new and the complicated: A trilogy of marketing relationships', *European Journal of Marketing*, **35** (1), 182–93.

HINGLEY, M.K. (2001), 'Relationship management in the supply chain', *International Journal of Logistics Management*, **12** (2), 57–71.

HINGLEY, M. (2005), 'Power to all our friends? Learning to live with imbalance in UK supplier-retailer relationships', *Industrial Marketing Management*, **34** (8), 848–58.

HINGLEY, M. (2008), 'Evolution of category management in UK supermarket fresh produce networks: A return to direct supply channels?' in *Proceedings of 2nd International European Forum on Innovation and System Dynamics in Food Networks*, Innsbruck-Igls, Austria, 18–22 February.

HINGLEY, M.K. and LINDGREEN, A. (2002), 'Marketing of agricultural products: case findings', *British Food Journal*, **104** (10), 806–27.

HINGLEY, M., CUSTANCE, P. and WALLEY, K. (2006a), 'Coopetition within the UK agri-food chain', in *7th International Conference on Management in AgriFood Chains and Networks*, Bijman, J., Omta, O., Trienekens, J., Wubben, E. and Wijnands, J. (eds), 31 May–2 June, Wageningen, Wageningen Academic Publishers, Department of Business Administration, Wageningen University.

HINGLEY, M, LINDGREEN, A. and CASSWELL, B. (2006b), 'Supplier–retailer relationships in the UK fresh produce supply chain', *Journal of International Food Products and Agribusiness Marketing*, **18** (1/2), 49–86.

HOGG, A., KALAFATIS, S.P. and BLANKSON, C. (1996), 'Customer–supplier relationships in the UK trade of rice', *British Food Journal*, **98** (2), 29–35.

HOWE, W.S. (1998), 'Vertical market relations in the UK grocery trade: analysis and Government Policy', *International Journal of Retail and Distribution Management*, **26** (6), 212–24.

HUGHES, D. (ed.) (1994), *Breaking with Tradition: Building Partnerships and Alliances in the European Food Industry*, Wye College Press, Wye.

INSTITUTE OF GROCERY DISTRIBUTION, IGD (2006a), *Category Management: A Global Perspective*. IGD, Watford.

INSTITUTE OF GROCERY DISTRIBUTION, IGD (2006b), *UK Grocery Retailing*. IGD, Watford.

JARVIS, M and WOOLVEN, J. (1999), *Category Management in Action*. IGD Business Publications, Watford.

JOHNSEN, R.E. and FORD, D. (2002), 'Developing the concept of asymmetrical and symmetrical relationships: Linking relationship characteristics and firms' capabilities and strategies', in *Proceedings from the 18th Annual IMP Conference*, Spencer, R., Pons, J-F. and Gasiglia, H. (eds), Graduate School of Business and Management, 5–7 September, Dijon, France.

JOHNSON, W.C., CHINUNTDEJ, N. and WEINSTEIN, A. (1999), 'Creating value through customer and supplier relationships', in *Proceedings of the 15th Annual IMP Conference*, 2–4 September, McLoughlin, D. and Horan, C. (eds), University College Dublin.

KALAFATIS, S.P. (2000), 'Buyer–seller relationships along channels of distribution', *Industrial Marketing Management*, **31**, 215–28.

KUMAR, N. (1996), 'The power of trust in manufacturer-retailer relationships', *Harvard Business Review*, (November–December), 92–106.

KUMAR (2005), 'The power of power in supplier–retailer relationships', *Industrial Marketing Management*, **34** (8), 863–66.

KUMAR, N., SCHEER, L.K. and STEENKAMP, J-B.E.M. (1998), 'Interdependence, punitive capability, and the reciprocation of punitive actions in channel relationships', *Journal of Marketing Research*, **35** (May), 225–35.

LANG, T. (2003), 'Food industrialization and food power: Implications for food governance', *Development Policy Review*, **21** (5–6), 555–68.

LARSON, P.D. and KULCHITSKY, J.D. (1998), 'Single sourcing and supplier certification: Performance and relationship implications', *Industrial Marketing Management*, **27** (1), 73–81.

LOADER, R. (1997), 'Assessing transaction costs to describe supply chain relationships in agri-food systems', *Supply Chain Management*, **2** (1), 23–35.

LOW, B.K.H. (1997), 'Managing business relationships and positions in industrial networks', *Industrial Marketing Management*, **26**, 189–202.

LU, H., TRIENEKINS, J., OMTA, S.W.F. and FENG, S. (2008), 'The value of *guanxi* for small vegetable farmers in China', *British Food Journal*, **110** (4/5), 412–29.

MATANDA, M., MAVONDA, F. and SCHRODER, B. (2001), 'Impact of relational constructs on the specific supply chain performance dimensions', in *Proceedings of the Academy of Marketing Annual Conference*, 4–7 July 2001, Cardiff.

MILES, B. and HUBERMAN, A.M. (1994), *Qualitative Data Analysis: An Expanded Sourcebook*, 2nd edn, Sage Publications, Thousand Oaks, CA.

NARAYANDRAS, D. and RANGAN, V.K. (2004), 'Building and sustaining buyer-seller relationships in mature industrial markets', *Journal of Marketing*, **68** (3), 63–7.

NAUDÉ, P. and BUTTLE, F. (2000), 'Assessing relationship quality', *Industrial Marketing Management*, **29**, 351–61.

NEWMAN, A., LINGS, I. and LEE, N. (2004), 'What's in a handshake? Exploring business-to-business relational exchange', in *Proceedings of Academy of Marketing Annual Conference*, July 2004, University of Gloucestershire, Cheltenham.

OGBONNA, E. and WILKINSON, B. (1996), 'Inter-organisational power relations in the UK grocery industry: Contradictions and developments', *International Review of Retail Distribution and Consumer Research*, **6** (4), 395–414.

O'KEEFE, M. and FEARNE, A. (2002), 'From commodity marketing to category management: Insights from the Waitrose category leadership programme in fresh produce', *Supply Chain Management*, **7** (5), 296–301.

PARKER, M. and BYROM, J. (2009), 'The elusive written contract: dependence, power, conflict, and opportunism within the Australian food industry', in *Controversies in Food and Agricultural Marketing*, Lindgreen, A., Hingley, M. and Vanhamme, J. (eds), Gower Publishing, Aldershot, forthcoming.

PATTON, M.Q. (2002), *Qualitative Research and Evaluation Methods*, 3rd edn, Sage Publications, Thousand Oaks, CA.

POLE, K.L. and HASKELL, J. (2002), 'Managing a modern relationship: Critical factors for business to business markets', in *Proceedings of: Academy of Marketing Annual Conference*, 2–5 July, Nottingham University Business School.

RADEMAKERS, M.F.L. and MCKNIGHT, P.J. (1998), 'Concentration and inter-firm co-operation within the Dutch potato supply chain', *Supply Chain Management*, **3** (4), 203–13.

ROBSON, I. and RAWNSLEY, V. (2001), 'Co-operation or coercion? Supplier networks and relationships in the UK food industry', *Supply Chain Management: An International Journal*, **6** (1), 39–47.

SHAW, S. and ENNIS, S. (2000), 'Marketing-channel management', in *The Oxford Textbook of Marketing*, Blois, K. (ed.), Oxford University Press, Oxford, 245–70.

SIEMIENIUCH, C.E., WADDELL, F.N. and SINCLAIR, M.A. (1999), 'The role of 'partnership' in supply chain management for fast-moving consumer goods: A case study', *International Journal of Logistics: Research and Applications*, **2** (1), 87–101.

SPEKMAN, R.E., KAMAUFF JR, J.W. and MYHR, N. (1998), 'An empirical investigation into supply chain management: a perspective on partnerships', *Supply Chain Management*, **3** (2), 53–67.

STERN, L.W. and EL-ANSARY, A.I. (1996), *Marketing Channels*, 5th edn, Prentice-Hall, New Jersey.

STRAK, J. and MORGAN, W. (1998), *The UK Food and Drink Industry: A Sector by Sector Economic and Statistical Analysis*, Euro PA and Associates, Northborough.

SVENSSON, G. (2001), 'Extending trust and mutual trust in business relationships towards a synchronised trust chain in marketing channels', *Management Decision*, **39** (6), 431–40.

WHITE, H.M.F. (2000), 'Buyer–supplier relationships in the UK fresh produce industry', *British Food Journal*, **102** (1), 6–17.

WILLIAMSON, O.E. (1975), *Markets and Hierarchies: Analysis and Antitrust Implications*, The Free Press, New York/London.

WILLIAMSON, O.E. (1995), 'Hierarchies, markets and power in the economy: an economic perspective', *Industrial and Corporate Change*, **4** (1), 21–49.

WILSON, N. (1996), 'The supply chains of perishable products in Northern Europe', *British Food Journal*, **98** (6), 9–15.

WOMACK, J.P., JONES, D. and ROOS, D. (1990), *The Machine that Changed the World: The Story of Lean Production*, Harper-Collins, New York.

4

Supplier safety assessment in the food supply chain and the role of standards

D. M. Julien, Cranfield University, UK

Abstract: Food supply chains of today are increasingly global, with organisations having to source materials from outside traditional boundaries in order to remain competitive. Additionally, the interconnectivity of these global supply networks can mean that a problem in one country often results in a global crisis. These and other trends bring with them many challenges that need to be managed to safeguard the end consumer. The safety and quality of the finished product is dependent on the integrity of the entire chain from the farm to the fork, which requires systems and approaches to be in place to ensure that there are no breaks or deviations that will result in adverse effects further downstream. This chapter will review an approach to evaluating and assessing suppliers in the food sector, advances in the sector in order to harmonise standards globally with the introduction of ISO 22000 and the Global Food Safety Initiative (GFSI). Also included are examples from industry of a number of approaches followed.

Key words: supplier evaluation, supplier assessment, supplier audits, food and beverage sector, ISO 22000, GFSI, standards, quality management, food safety, HACCP.

4.1 Introduction

Supplier selection, assessment and ongoing evaluation is of paramount importance to the food and beverage sector. In particular, the control of issues relating to food safety is key at all stages of the supply chain, as bad practices on the farm can result in contamination of the finished consumer product if the problem is not detected. Food safety is always at the heart of any supplier evaluation process in the sector as it is the minimum requirement that must be met at all stages of the process. Non-compliance issues in this area will result in large public recalls which are extremely damaging to the brand and potentially to the business.

The global marketplace of today, where food processors and retailers are having to source materials from outside traditional boundaries in order to remain competitive, brings with it many challenges that need to be managed to safeguard the end consumer. This shift away from small traditional/local manufacturers with local supply chains to a dispersed network of companies with a global manufacturing and supplier base has been the result of a diverse set of business drivers such as reduced costs, growth potential, access to new markets, competitive pressures and access to qualified personnel.

Understanding local cultures and regulations is fundamental to avoiding the pitfalls that come from assuming that product and process requirements are interpreted consistently and will operate in the same way. Careful translation of specifications and the transfer of intrinsic knowledge to new partners is vital. Given the interconnectivity of supply networks today, a problem in one country often results in a global crisis, as evidenced by the recent melamine in milk scandal (BBC, 2008a).

> The problem shows how big food companies can struggle to impose food safety standards on suppliers in the developing markets they increasingly rely on for sales growth. (Patrick *et al.*, 2008)

One lesson learned from this and other examples is the importance of understanding your supply base, knowing where the risks are, and having confidence in the suppliers' practices and monitoring their ongoing performance. Midler (2007) discusses what he calls quality fade on products from some Chinese subcontractors where the quality is deliberately reduced to increase the profit margins on the products. One of the reasons for this is the lack of effective government controls and the overwhelming short-term view held by many manufacturers where their future existence is always in question. Despite these problems, organisations cannot turn their backs on China or other developing economies as their long-term growth and competitiveness will involve understanding how to operate and manage their supply chains in these countries.

> Whether a company views China as a manufacturing base, an attractive market or both, world-class execution will be necessary to succeed, and success in China will be needed to survive not only there but around the globe. (Hexter and Woetzel, 2007)

The current trend of tiering the supply base in parallel with the reduction in the number of core suppliers who interact with the focal company helps to reduce the management burden for the focal company. In parallel supply networks have become more complex in the last few years partly owing to the increased levels of outsourcing and offshoring of key stages in the manufacture of products to low cost countries around the world (Christopher, 2005; Harland *et al.*, 2003). So whilst the focal company may be interacting with fewer suppliers they tend to be located globally and may also be managing critical materials flows into the company that were previously managed by the focal company.

This transfer of responsibility into the supply network depends on the suppliers at all levels of the network acting in an ethical manner and accepting responsibility for their part of the chain, it also assumes that legislative bodies in the new economies have the expertise and capability to regulate the sector to the required level. Recent crises in the sector have illustrated that this is not always the case and there is room for improvement in the management and development of the supply base to the appropriate level.

> Despite a nationwide campaign to raise food safety standards and reassure consumers, China's broken-down food safety inspectorate is still failing to catch and report lapses in standards when they happen. (BBC, 2008b)

The safety and quality of the finished product is dependent on the integrity of the entire chain from the farm to the fork requiring systems and approaches to be in place to ensure that there are no breaks or deviations that will result in adverse effects further downstream.

Various approaches to evaluating and assessing suppliers in the food sector will be reviewed. Additionally, advances in the sector to harmonise standards globally by the introduction of ISO 22000 and the Global Food Safety Initiative (GFSI) and their importance in reducing waste and improving food safety will be considered.

4.2   Material risk assessment

> Understanding food ingredients and the variances among them is a must in order to ensure food quality and safety. (Stier, 2006)

An important precursor to the sourcing decision for any material used in the manufacture of food and beverages is to complete a risk assessment of the material. The risk assessment should include the inherent food safety risk of the material, the planned use of the material and the nature of the operation. The planned use of the material is important, as further processing may eliminate certain hazards that may be present and so it is less of a concern, compared to a material which will not be processed any further and used in the assembly of the final product. Good examples of this can be found in the chill chain where materials are purchased in and used directly in the assembly of fresh sandwiches and salads (CFA, 2006).

Alternatively, the same material used in two different products can have a very different level of risk associated with it because of the target group of consumers, for example milk powder for infant formula is assessed as high risk compared to milk powder blended and processed into a yoghurt.

The risk classification of the material is independent of the supplier. Many companies use three different classifications for materials: high, medium and low risk. Table 4.1 illustrates the various risk categories for a

Table 4.1 Examples of risk categories for different raw materials

	High risk	Medium risk	Low risk
For ingredient suppliers	• Liquid egg • Milk powder for a dry mix infant formula • Spices	• Dried vegetables • Dried herbs • Nuts	• Sugar for a wet mix • Salt
For packaging suppliers	Food contact packaging with inherent risk: • Glass jars • Sterilised tin cans and aseptically filled containers • Premiums/gifts for children under 3 years old	Food contact packaging with minimal inherent risk or low speed line: • Pre-formed trays for confectionery • Closures for plastic bottles	No contact with food, low speed line: • Shipping cartons • Display cartons • External shrink wrapping

range of raw and packaging materials based on examples from both large and medium sized food manufacturers.

The Chilled Food Association (CFA) (2006) in the UK has developed a decision tree to help its members to target their supplier quality assurance resources at the riskiest raw materials. In many of the larger food companies a network of material experts are being developed as an important internal competency; these material experts provide key support for the regional purchasing groups. The material experts are responsible for scanning the environment to stay up-to-date on the latest developments related to their particular incoming material group. In addition, they codify their knowledge in the form of material-specific lists that will be shared with assessors and other concerned groups. The material experts typically participate in the assessment of the key suppliers for their material whenever possible as this helps keep them in touch with current practices in the supply chain and they may also be more likely to identify possible innovations at the supplier site. On the other hand they are also better able to spot possible deviations from the norm that need to be managed. The selection of these individuals should take into consideration both their technical expertise and their communication skills.

4.3 Supplier assessment and management

Food processors need a well organised and rigid vendor quality program, which includes a vendor selection and approval process. (Stier, 2006)

The management of the inbound quality of materials is concerned with the sourcing, evaluation and selection of suppliers, provision of education and training, monitoring of supplier performance and supplier certification. The supplier management process involves individuals with a range of functions who are in contact with suppliers. Stier (2006) emphasises the importance of purchasing, quality and technical staff being involved as their skills complement each other and provide for a more thorough evaluation of the suppliers. The following functions tend to be core of any supplier management team:

- quality
- purchasing
- manufacturing.

Suppliers must be very carefully selected. The purchasing function typically evaluates the business-related factors such as:

- management structure and competence
- financial situation
- ownership of the company and
- business reputation of the company.

Purchasing will also determine if the supplier's offer is competitive and makes good business sense for the company. Depending on the material that the potential supplier will be providing, the technical know-how of the supplier also needs to be evaluated. At this point, the quality and manufacturing functions take the lead role in the evaluation of the supplier's:

- manufacturing capabilities
- quality assurance system
- technical capabilities
- HACCP (hazard analysis and critical control points) study and
- openness and acceptance of assessments and inspections.

In addition, individuals from the research and development, regulatory affairs, agricultural services and new product introduction departments are often required to interact with suppliers, depending on the business requirements.

The total system cost approach implies that when negotiating with a supplier, the focus is not just on the price of the material being purchased (Mangan et al., 2008). Other aspects that add value should also be taken into consideration. Examples of these value-adding aspects, which are harder to quantify are:

- performance of the material on the line
- high quality in terms of food safety
- consistency of the material delivered
- flexibility in meeting delivery requirements
- potential to grow with demand.

The purchasing department is also involved in the selective development of partnerships with certain suppliers in order to develop unique solutions that provide a competitive advantage and mutual benefit for both parties. As in any sector, companies will need to have a process defined for the assessment of new and existing suppliers in their supply network. The most common approach currently taken for new suppliers is to send the prospective supplier a self-assessment questionnaire for completion and return to the focal company; see Fig. 4.1 for an outline of the process.

Whilst the questionnaire differs between companies, it tends to cover similar topics and the type of information that is requested from the supplier is similar, see Fig. 4.2. Typically at the core is a strong focus on issues relating to food safety with additional details about the business and commercial aspects. Where the focal company has no first hand experience with new suppliers, it is important at this point to ascertain what other companies they supply and what certifications they may already hold to give a sense of the level of their operations. For existing suppliers it tends to be more of an updating exercise in case of changes, either in the products that they supply or to the process.

Based on the supplier's track record and review of the completed questionnaire, a confidence level is assigned to the supplier. Different companies have different naming conventions, be it A, B, or C grade suppliers, or high, medium or low level suppliers. One definition used for the latter of these conventions is as follows:

- High confidence supplier: a supplier previously assessed and formally approved, supplying materials corresponding to the agreed specifications and general conditions, with reliable deliveries and rapid positive response in case of deviations.
- Medium confidence supplier: a supplier previously assessed and formally approved, where deviations from agreed specifications or the general conditions have occurred but the response to complaints has been positively dealt with, or suppliers for whom a previous assessment revealed requests for corrective action(s).
- Low confidence supplier: suppliers that do not totally meet our requirements or have lost our confidence, but which, owing to a lack of alternatives, we are forced to use.

The importance of assessing the suppliers is emphasised by the CFA (2006) as the way to safeguard the safety and quality of the products. It is also recognised that this involves the effective combination of different approaches from on-site audits either using internal auditors or third party auditors and self-assessment questionnaires and certificates of analysis (CFA, 2006).

Depending on the outcome of the review of both the risk level for the material sourced and the supplier ranking, the sourcing company will decide if an on-site audit is required and, for existing suppliers, the

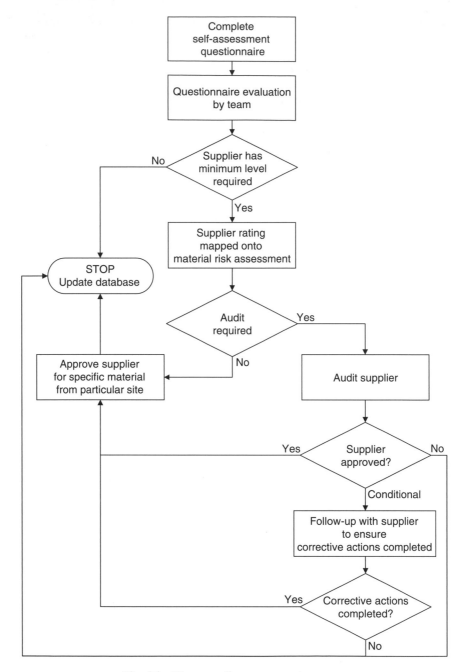

Fig. 4.1   New supplier assessment process.

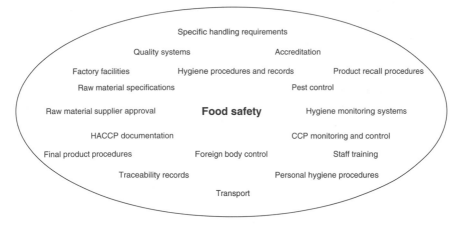

Fig. 4.2 Focus areas for self assessment questionnaire.

		Raw material risk level	
		Low	**Medium**	**High**
Supplier confidence level	**High**	Non-significant hazard controlled by **CP** at the supplier's site. Certificate of analysis. Random sampling of each lot. Minimal inspection of samples. Simplified audit of supplier.		Significant hazard controlled by **CCP** at the supplier's site. Certificate of analysis. Random sampling of each lot. Normal inspection of samples. Simplified audit of supplier.
	Medium	The following decisions are made on a case-by-case basis: Location of the CCP/CP Type of sampling (random, statistical) Frequency of analysis Type of supplier audit		
	Low	Non-significant hazard controlled by **CP** at focal company's factory. Random sampling of each lot. Normal inspection of samples. Simplified audit of supplier.		Significant hazard controlled by **CCP** at the focal company's factory. Statistical sampling of each lot. Defined acceptance quality levels. Reinforced inspection of samples to ensure the conformance of the lot. Tightened audit of supplier.

Fig. 4.3 Material risk versus supplier confidence.

frequency of the re-audits. The decision whether or not to visit the supplier for an on-site assessment should also take into consideration the time lapse since the last visit made to the site, whether this is a new supplier, whether the supplier has already been approved by another large food manufacturer and, in the case of existing suppliers, the track record of their performance.

One example from a leading manufacturer is provided in Fig. 4.3, which maps the material risk onto the supplier confidence level to determine the most appropriate approach to their control and management.

At this point if an audit is required the focal company will need to decide who will carry out the audit. Similar to many other internal functions, numerous companies have opted to use third party auditors instead of maintaining and developing the resource internally. However, most of the larger organisations will maintain a small in-house auditing function for certain types of audit and outsource the less critical audits to the external auditor.

> A Kraft Foods spokesperson says that the company's auditing program is integral to its food-safety program. All of Kraft's suppliers must meet its 'Supplier Quality Expectations' and are audited before the company buys the ingredients. Across global operations, the company uses internal and external auditors, depending on the circumstances. (ASQ, 2008)

4.4   Hazard Analysis and Critical Control Points (HACCP) overview

The HACCP concept is central to most supplier assessment and certification schemes. HACCP is an international approach followed by the food sector. It is a system of process control that was developed by the National Aeronautic and Space Administration (NASA) in preparation for space flight and has been adopted in many industries. HACCP is designed to ensure product safety and is a preventative approach to food safety that addresses physical, chemical and biological hazards that would result in the product being unsafe for human consumption; it is seen as a means of prevention rather than finished product inspection.

HACCP is built around seven key principles (HACCP, 2008):

- Principle 1: Conduct a hazard analysis.
  Processing plants need to identify any food safety hazards related to their operations that must be prevented, eliminated, or reduced to an acceptable level.
- Principle 2: Identify critical control points.
  A critical control point (CCP) is a point, step, or procedure in a food process at which control is essential to prevent or eliminate a hazard or to reduce it to acceptable levels.

- Principle 3: Establish critical limits for each critical control point.
  A critical limit is the maximum or minimum value to which a physical, biological or chemical hazard must be controlled which separates acceptability from unacceptability.
- Principle 4: Establish critical control point monitoring requirements.
  Effective monitoring activities are necessary to ensure that the process is under control at each critical control point. It is normally a requirement that each monitoring procedure and its frequency be listed in the HACCP plan.
- Principle 5: Establish corrective actions.
  These are actions to be taken when monitoring indicates that a critical control point is not under control.
- Principle 6: Establish record keeping procedures.
  Food regulations require that all processing plants maintain documents which demonstrate the effective application of the measures outlined in the HACCP plan. In addition to the HACCP plan, records documenting the monitoring of critical control points, critical limits, verification activities and the handling of processing deviations should also be maintained.
- Principle 7: Validation of the HACCP system:
  Validation of the HACCP plans ensures that the processing plants do what they were designed to do and are able to ensure the production of safe product.

The HACCP plan aims to map out the entire process and for every stage or step of the process to identify if there are any hazards associated with it. If there is a hazard, it is either classified as a critical control point (CCP) or as just a control point (CP). Once they are all classified, the team will try to understand what are the necessary controls to put in place and the limits for each one. When this has been completed, monitoring and corrective action plans can be put in place. Manning and Baines (2004) have highlighted that globalisation of food supply chains has led to an increase in food safety risks, which increases the likelihood of pandemics of food-borne disease and that HACCP is an essential approach for assessing and managing these inherent risks. In a special report, the FAO/WHO (2002) also reinforce the importance of HACCP and other risk assessment tools in particular for developing nations.

> The application of HACCP and risk assessment concepts in recent years are leading to fundamental changes in the approach to food safety. (FAO/WHO, 2002).

4.5 Supplier audits and performance management

Manufacturers have a responsibility to manage their suppliers, to collaborate with the key suppliers to look for possible improvements and also to work with weaker suppliers to eliminate any non-conformities (Humphreys *et al.*, 2001; Chin *et al.*, 2006). The performance of the company's suppliers will influence customer satisfaction and ultimately long-term market success. Most practitioners view supplier performance as contributing to enhancing the competitive advantage of a firm and it has become key to achieving good quality leading to a world-class success (Lemke *et al.*, 2003).

> We want our relationships with suppliers to be mutually beneficial, and to buy from companies that have high standards. We apply very stringent food safety and quality assurance processes across all of our product range. All of our suppliers must meet our requirements to ensure our customers get fresh, wholesome food they can trust from our stores. (Tesco, 2008)

The frequency of re-evaluation of the supplier is dependent on both the supplier and the material classification (see Table 4.2 for an example from one small company). Most companies will express a target frequency for auditing suppliers in their quality manual.

However, owing to the thinning out of internal audit capabilities, companies recognise that many of their subsidiaries are unable to perform audits of all their suppliers in a timely and regular fashion. This will result in the purchase of raw materials from suppliers who have not been audited on a regular basis. The focal companies in general consider this situation to be undesirable and extremely risky for the safety of their products and, as such, their business and brand. To solve this problem, many organisations are engaging the services of an external auditing body to perform supplier audits on their behalf. The increasing use of external auditors to perform the crucial activity of supplier audits needs to be managed and monitored to ensure that the company's supplier management is more robust than before.

In one large global food manufacturer the approach taken was to select two third party auditors to partner globally and to work with the companies to agree on the content and style of the audits performed on the company's behalf. In addition, it was recommended that all suppliers of high risk materials and suppliers who are key, owing to the volume of business, should be audited by internal company auditors and not the third party auditor.

Table 4.2 Audit planning

Supplier ranking	Material risk		
	High	Medium	Low
A	Full audit every 12 months	Documentation audit every 18 months	Documentation audit every 2 years
B	Full audit every 12 months	Documentation audit every 18 months	Documentation audit every 2 years
C	Full audit every 12 months	Full audit every 12 months	Documentation audit every 18 months

Once the audit is complete the report needs to be reviewed and action taken where necessary. The findings from the assessment are classified into the following categories:

- Compliance: no non-conformances reported
- Minor: a sub-section of the audit document has not been fully met but this would not affect the quality or safety of the product being supplied
- Major: a failure to comply with any sub-section of the audit document or a situation that would raise significant doubt about the safety or quality of the product being supplied
- Critical: is a failure to comply with a food safety or legal issue or a fundamental requirement within the audit document.

Depending on the list of findings, their classification and their quantity, the focal company will adapt the way of working with the supplier to accommodate these differences. An amalgamation of approaches from several food companies is presented in Table 4.3.

Once the supplier has been approved and supply has started the focal company puts in place an appropriate monitoring plan for the supplier. This includes decisions about sampling frequency, where the sample is taken (at the supplier or on receipt), use of external laboratories for producing 'certificates of analyses' (COAs), what is being tested for and the method of communication with the supplier about their performance. The majority of companies have performance metrics which track quality (including microbiological issues, foreign bodies, plus other defined quality criteria for the material), on-time delivery and delivery in-full type measures. These three measures are core and apply to any supplier, however, there could be other measures that are linked more to the type of relationship with the supplier, for example, if there is any joint new product development.

4.6   Impact of globalisation and need for a global standard

Increasingly competitive global marketplaces are forcing manufacturers constantly to look for ways to increase and sustain their competitive position. The pressures of globalisation emphasise the need for businesses to overcome cost pressures and improve efficiency throughout their supply chains. Cost reduction is the most quoted benefit of global sourcing (Fagan, 1991; Kohn, 1993; Rajagopal and Bernard, 1993). However, there are many other benefits that can be attained through global sourcing including access to new markets, higher quality goods, access to worldwide technology, better delivery service and better customer service (Birou and Fawcett, 1993; Scully and Fawcett, 1994). The successful implementation of a global sourcing strategy is therefore paramount for manufacturing supply chains in order for businesses to achieve their full potential. The removal of many of the trade barriers between countries has also contributed to this

Table 4.3 Audit actions

Findings		Action
Compliance	No action required	
Minor	Corrective actions fixed according to a defined action plan, for completion within 4 weeks Evidence of corrective action required (photos, copies of documents, etc) Follow-up in 1 year	No additional or special controls need to be implemented before sourcing from them
Major	Undertake corrective action within 4 weeks Corrective actions fixed and should be completed as soon as possible, but in not more than 6 months Evidence of corrective action required (photos, copies of documents etc) Follow-up in 3–6 months	Detailed list of the deviations and the potential hazards and financial impacts to control or compensate for each one should be made Proactive joint improvement plan, or search for alternative sources or development of the supplier base with the goal to replace current supplier Supplier not approved until corrective action evidence received (new suppliers)
Critical	Food safety problems Undertake corrective action immediately	Supplier should not be used until non-conformance eliminated Supplier not approved (new suppliers) Re-audit required before supply can start

phenomenon by allowing the free movement of goods and services across borders. The EU, EEA and NAFTA are good examples of the types of economic regions that are driving the optimisation of supply chains. The removal of trade barriers has allowed large global organisations to rationalise and optimise their manufacturing and supplier base globally, leverage their purchasing power and streamline their logistics networks.

A study carried out by Lewin and Peeters (2006) reinforces the reasons provided for offshoring and highlights the opportunities for growth as one of the key reasons. Certainly the benefits for companies following this route are many but it is important to understand that it brings with it an increased risk to the operating companies. Harland *et al.* (2003) make the point that risk in supply chains today is far greater than it has been in the past and that there is a need for companies to address these risks. One difficulty is that the risk is shifting around the supply networks and may lie outside the company's direct control. Understanding the challenges faced by food manufacturers to manage the quality of their inbound materials, because of today's highly outsourced and offshore supply chains, cannot be underestimated. Companies need to consider the impact on their business of moving production offshore to low cost economies due to different cultures and their interpretation of requirements, language barriers and the increased physical separation of customer and supplier. Once the challenges are better understood, it should be possible to make recommendations about how to manage them better.

The food sector has seen a huge growth in the number of certification and company standards which organisations along the supply chain need to be audited against if they plan to grow their business. Figure 4.4 represents some of the plethora of standards for which organisations operating or supplying to the UK would need to be accredited, depending on their particular customer requirements.

This type of proliferation of standards is not unique to the UK market and similar patterns are observed around the world. A number of countries have developed their own national standards for the supply of safe food. In addition different companies and groups within the industry have devel-

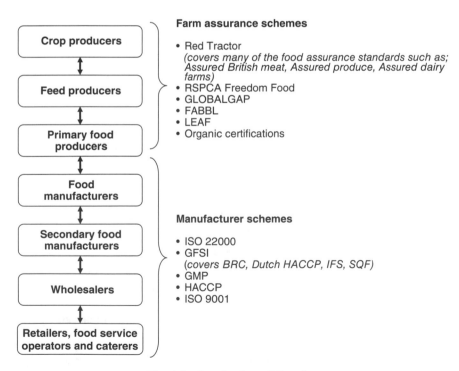

Fig. 4.4   Standards proliferation.

oped their own standards for auditing their suppliers. The end result is a proliferation of standards in the food supply chain internationally. The problem with this lack of a single truly internationally accredited and recognised standard is that each of these individual schemes is perceived as being superior by its sponsor country or organisation.

This all leads to lots of confusion in the marketplace over the exact requirements for the suppliers, uneven levels of food safety globally, increased cost and complexity for the suppliers who find themselves obliged to conform to different standards and programmes and disruption to the business activities owing to audits several times a year by different groups. Two groups responded to the need in the marketplace for harmonisation across all the standards and introduced two global standards; the Global Food Safety Initiative (GFSI) launched in May 2000 and ISO 22000 which was ratified in September 2005.

4.7   Global Food Safety Initiative (GFSI)

The Global Food Safety Initiative (GFSI) is coordinated by CIES – The Food Business Forum. It has a retailer-driven board that includes some advisory members from manufacturers (CIES, 2008a).

> The GFSI vision of 'once certified, accepted everywhere' has now become a reality. Carrefour, Tesco, Metro, Migros, Ahold, Wal-Mart and Delhaize have agreed to reduce duplication in the supply chain through the common acceptance of any of the four GFSI benchmarked schemes. (CIES, 2008a)

The GFSI mission is to work on continuous improvement in food safety management systems to ensure confidence in the delivery of food to consumers (CIES, 2007a). The GFSI objectives (CIES, 2007a) are to:

- maintain a benchmarking process for food safety management schemes to work towards convergence between food safety standards;
- improve cost efficiency throughout the food supply chain through the common acceptance of GFSI recognised standards by retailers around the world;
- provide a unique international stakeholder platform for networking, knowledge exchange and sharing of best food safety practice and information.

The GFSI approved schemes consist of three key elements which cover the range of food safety management criteria: food safety management systems, good practices and HACCP (CIES, 2007a), see Fig. 4.5.

The four currently recognised manufacturing schemes included in the GFSI are (CIES, 2008a):

- BRC – British Retail Consortium Global Food Standard (Version 5)
- Dutch HACCP (Option B)

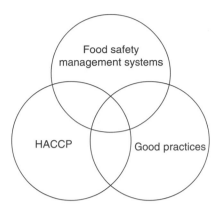

Fig. 4.5   GFSI key elements.

- IFS – International Food Standard (Version 5)
- SQF 2000 – Safe Quality Food Program (Level 3).

The growth of the GFSI standard was in response to increases in the numbers of private label products, which brought about collaboration between retailers and manufacturers on the development of shared standards.

> Wal-Mart has announced that suppliers of its private label and other food items, like produce, meat and fish, must comply with Global Food Safety Initiative (GFSI) recognized standards. (CIES, 2008b)

4.8   ISO 22000 – food safety management systems

The international food safety management standard, ISO 22000, was developed in response to a need for a worldwide standard supported by an independent, international organisation, which would encourage harmonisation of national and private standards for food safety management (CIES, 2007b). The development of the standard was undertaken by a working group whose members represented mirror groups of 23 national standard bodies and other organisations with liaison status. The participating countries included Japan, United States, Australia, Canada, Korea, Malta, Belgium, Greece, United Kingdom, Germany, France, The Netherlands, Thailand, Italy, Sweden, Denmark, Switzerland and Ireland. The participating organisations included the Codex Alimentarius Commission, the Confederation of the Food and Drink Industries of the European Union (CIAA), the International Hotel and Restaurant Association and the World Food Safety Organization (WFSO) (CIES 2007b). Unlike the GFSI standard, which was retailer driven, the development of the ISO 22000 standard gathered input from a much broader representation across global food chains. An international generally accepted ISO standard that integrated the necessary food safety principles with accepted quality management principles would also promote consensus in deciding the necessary elements of food safety systems for businesses in the food chain. Last, it would also be aligned with ISO 9001 in order to enhance compatibility with existing overall management approaches in the food businesses concerned (Frost, 2005).

> The goal of ... ISO 22000 ... is to harmonize at a global level the requirements for food safety management systems throughout the food chain. (Pillay and Muliyil, 2005)

The standard defines food safety as the concept that foodstuffs should not be harmful to the consumer and recognises that food safety hazards can be introduced at any stage of the food chain. The standard would be applicable at all stages of the supply chain. Recognising that food safety problems can originate at any point in the supply chain, the standard requires that an organisation in the food chain takes into account the safety hazards to the consumer of the final food product and, if necessary, take measures to control those hazards. Companies that adhere to the requirements of the standard should have in place systems to support the effective communication and information exchange between all parties in the chain, so as to promote understanding of the risks at any particular stage in the chain as a whole.

Ultimately, ISO 22000 considers food safety as a joint responsibility that is principally assured by the combined efforts of all parties participating in the food chain and encourages effective communication of food safety issues to suppliers, customers and other relevant interested parties in the chain (Frost, 2005). The key elements of the standard (Færgemand and Jespersen, 2005) are presented in Fig. 4.6 and outlined below.

- Interactive communication implies the communication of the needs of the focal company to organisations both upstream and downstream in the food chain.
- System management understands that the most effective food safety systems are designed, operated and updated within the framework of a structured management system and incorporated into the overall management activities of the organisation.
- Hazard control combines the Codex Alimentarius HACCP principles and application steps with prerequisite programmes which enhance and maintain operational conditions to enable more effective control of food safety hazards.

Numerous benefits have been linked to the introduction of ISO 22000. The most fundamental being that ISO 22000 has been designed to allow all types of organisations across the food chain to implement a food safety

Fig. 4.6   ISO 22000 key elements.

management system. These range from feed producers, primary producers, food manufacturers, transport and storage operators and subcontractors to retail and food service outlets – together with related organisations such as producers of equipment, packaging material, cleaning agents, additives and ingredients (Frost, 2005; Pillay and Muliyil, 2005). Figure 4.7 presents an example of the communication links supported between the focal food producer in the centre and the entire supply chain, and also with the shadow supply chain that provides key services and inputs to the core supply network and the linkages with the regulatory bodies and the end consumer.

Other benefits include reduction in duplication and cost through the existence of one single internationally accepted standard. ISO 22000 extends the successful management system approach of the ISO 9001:2000 quality management system standard which is widely implemented in all sectors but does not itself specifically address food safety. The two standards are fully compatible and companies already certified to ISO 9001 will find it easy to extend this certification to include ISO 22000. Some additional cited benefits are:

- Traceability – identification of an organisation's impact on food safety within the supply chain
- Control/reduction of food safety hazards
- Legal compliance
- Smooth conversion from existing food safety certifications
- Continuous improved business performance in line with the ISO 22000 food safety policy and objectives.

Another important consideration is the compatibility of the standard with existing schemes worldwide based on Codex guidelines for good manufacturing practices and HACCP. This includes many national standards from

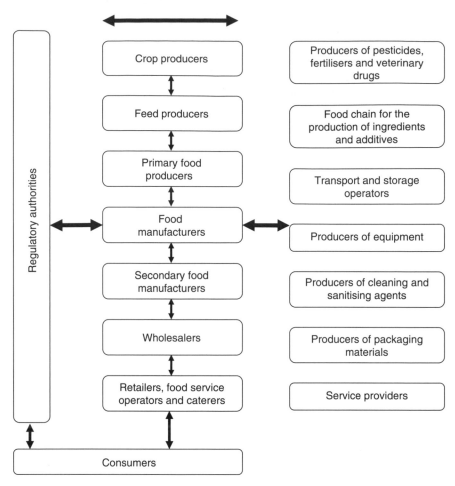

Fig. 4.7   ISO 22000 communication links (adapted from Frost, 2005).

North American, European and Asian countries. An organisation with an existing food safety programme can incorporate the elements of ISO 22000 into their existing system by using a stepwise approach. All this makes ISO 22000 more than a set of standards as it helps organisations to develop by extending their reach, providing a more logical and structured approach to food safety management, to gain easier access to global markets.

Both GFSI and ISO 22000 have at their core the desire to reduce duplication of activities across supply chains and in so doing improve the efficiency of the chains and reduce costs. However, the difference in ownership will result in difference in the acceptability and the responsiveness of the two standards. Owing to the internationally owned nature of the ISO structure, any changes proposed to the board could be difficult and time consuming to implement, whilst the GFSI board consider that retailer-driven GFSI schemes will have a specific reactivity which helps meet market demand in a timely and efficient manner. That aside, the ISO 22000 standard has the support of governments and food authorities globally and will probably have a wider acceptance across the entire food chain.

4.9   Conclusion and future trends

There are numerous other trends and issues concerning food supply chains that may ultimately have an impact on the supplier base and how they are assessed. Numerous certification schemes exist and are emerging in response to increasing consumer and government sensibilities about environment and animal welfare.

The FAIRTRADE mark is an example of a niche that looks set to continue to grow in the coming years and requires suppliers of Fairtrade branded products to be certified in accordance with Fairtrade standards. Estimated retail sales of Fairtrade products globally during 2007 reached in excess of €2.3 billion, which represented a growth of 47% (FLO, 2008).

Organic food is another significant trend which is claimed to have become mainstream across Europe (CONDOR, 2005). The Condor report (2005) also reports that organic food represented less than 10% of the total food spend in Europe, but was forecast to continue to grow. Organic food and drink sales in the UK nudged the £2 billion mark for the first time in 2006, with a sustained market growth rate of 22% throughout the year (Soil Association, 2008). Based on research from eight EU member states, the motivations and perceived benefits for consumers of organic foods are very similar: improved taste, contribution to a healthier life and a beneficial impact on the environment and society are seen as the major benefits by consumers (CONDOR, 2005).

Many of these schemes have an impact on the farm and whilst we have seen some consolidation in the manufacturing end of the food supply chain, there is still an explosion of different schemes being introduced for primary producers upstream of the manufacturers. In addition, the focus of many of these schemes is not food safety-related but more focussed on organoleptic and sensory aspects of the food, that is better taste and social consciousness and animal welfare.

Changes in global food supply chains during the last 50 years have resulted in a large increase in the distance food travels from the farm to consumer, known as 'food miles', with its associated environmental, social and economic impact. Growing concern over this impact has led to a debate about whether to try to measure and ultimately reduce food miles (Smith *et al.*, 2005). The question is will the global debate about social responsibility change behaviour on the ground or do the consumers now expect to have a year-round supply of produce and products, in so doing reinforcing the requirement on organisations to keep sourcing globally to meet the demand? This ultimately results in ever more complex supply networks that need to be managed.

In addition, despite all the progress made to-date, many developing countries are still at an early stage of evolution with respect to strengthening their food safety control programmes. National food safety policies may be limited in scope and food control systems may involve fragmentation and duplication. This results in them experiencing difficulties in prioritising and resourcing their work in relation to food safety. Global organisations need to understand these limitations when sourcing from countries that are less sophisticated in terms of the technical and scientific infrastructure needed to support the sector (FAO/WHO 2002).

Responsibility for protecting the consumer through the provision of safe food will always lie with the brand owner of the product. These organisations need to protect their brand identity through the careful selection and management of their supplier base wherever those suppliers are located. The benefits of global standards like both GFSI and ISO 22000 cannot be underestimated as they play a key role in reducing waste and inefficiencies in the supply networks and also provide a benchmark for suppliers to work towards which will result in better food safety.

4.10 References

ASQ (2008). 'Using Third-Party Audits to Help Ensure Food Safety'. *National Provisioner*. October 16, 2008 (accessed 23rd October 2008). http://www.asq.org/qualitynews/qnt/execute/displaySetup?newsID=4880

BBC NEWS (2008a). *Why China's milk industry went sour* (accessed on 01/10/2008). http://news.bbc.co.uk/1/hi/world/asia-pacific/7635466.stm

BBC NEWS (2008b). *Chinese melamine scandal widens* (accessed on 12/11/2008). http://news.bbc.co.uk/1/hi/world/asia-pacific/7701477.stm

BIROU, L.M. and FAWCETT, S.E. (1993). 'International purchasing: benefits, requirements and challenges', *International Journal of Purchasing and Materials Management*, **29**(2), 27–37.

CFA (2006). *Best Practice Guidelines for the Production of Chilled Food*. 4th edn, Chilled Food Association, UK.

CHIN, K.S., YEUNG, I.K. and PUN, K.F. (2006). 'Development of an assessment system for supplier quality management', *International Journal of Quality & Reliability Management*, **23**(7), 743–65.

CHRISTOPHER, M. (2005). *Logistics and Supply chain Management. Creating Value-Adding Networks*. 3rd edn, Prentice Hall, UK.

CIES (2007a). *The Global Food Safety Initiative – GFSI Guidance Document*, 5th edn, CIES September 2007.

CIES (2007b). *What is ISO 22000?* GFSI Technical Committee, September 2007.

CIES (2008a). *Global Food Safety Initiative* (accessed 23/10/2008). http://www.ciesnet.com/2-wwedo/2.2-programmes/2.2.foodsafety.gfsi.asp

CIES (2008b). *Wal-Mart suppliers must meet GFSI-approved standards*. CIES Press Release. February 2008 (accessed 20th October 2008). http://www.ciesnet.com/4-press/4.2-press-release/index.asp

CONDOR (2005). Consumer decision making on organic products. Final brochure (accessed 20/11/2008). http://www.condor-organic.org/

FÆRGEMAND, J. and JESPERSEN, D. (2005). 'Key elements and benefits of ISO 22000', *ISO Management Systems – November–December*, 18.

FAGAN, M.L. (1991). 'A guide to global sourcing', *The Journal of Business Strategy*, March/April, 21–5.

FAO/WHO (2002). *Principles and Guidelines for Incorporating Microbiological Risk Assessment in the Development of Food Safety Standards, Guidelines and Related Texts*. Kiel, Germany.

FLO (2008). *An Inspiration for Change. Fairtrade Labelling Organizations International (FLO) – Annual Report 2007* (accessed 20/11/2008). http://www.fairtrade.net/

FROST, R. (2005). 'ISO 22000 is first in family of food safety management system standards', *ISO Management Systems – November–December* 2005, 16–19.

HACCP (2008). *Hazard Analysis and Critical Control Points* (accessed 23/10/2008). http://en.wikipedia.org/wiki/HACCP

HARLAND, C., BRENCHLEY, R. and WALKER, H. (2003). 'Risk in supply networks', *Journal of Purchasing and Supply Management*, **9**, 51–62.

HEXTER, J. and WOETZEL, J. (2007). *Bringing Best Practice to China*, Operations, No 4. McKinsey & Co, http://www.mckinseyquarterly.com/Operations/Bringing_best_practice_to_china_2044

HUMPHREYS, P.K., SHIU, W.K. and CHAN, F.T.S. (2001). 'Collaborative buyer-supplier relationships in Hong Kong manufacturing firms', *Supply Chain Management: An International Journal*, **6**(4), 152–62.

KOHN, F.L. (1993). *Global sourcing: broadening your supply horizons*, Business Forum, Winter/Spring, 17–20.

LEMKE, F., GOFFIN, K. and SZWEJCZEWSKI, M. (2003). 'Investigating the meaning of supplier-manufacturer partnerships: an exploratory study', *International Journal of Physical Distribution & Logistics Management*, **33**(1), 12–35.

LEWIN, A. and PETERS, C. (2006). 'The top-line allure of offshoring', *Harvard Business Review*, 22–4.

MANGAN, J., LALWANI, C. and BUTCHER, T. (2008). *Global Logistics and Supply Chain Management*, John Wiley & Sons.

MANNING, L. and BAINES, R.N. (2004). 'Effective management of food safety and quality', *British Food Journal*, **106**(8), 598–606.

MIDLER, P. (2007) 'Quality fade: china's great business challenge', *Knowledge@Wharton* (accessed 29/07/2007). http://knowledge.wharton.upenn.edu/article.cfm?articleid=1776

PATRICK, A.O., JARGON, J., CANAVES, J. and DEAN, J. (2008). 'Food giants scrutinize chinese suppliers', *Wall Street Journal*, (eastern edition), New York, Sep 30, 2008.

PILLAY, V. and MULIYIL, V. (2005). *ISO 22000. Food Safety Management Systems – the one universal food safety management standard that works across all others*, SGS Systems & Certifications Services. October 2005.

RAJAGOPAL, S. and BERNARD, K.N. (1993). 'Globalization of the procurement process', *Marketing Intelligence and Planning*, **11**(7), 44–56.

SCULLY, J.I. and FAWCETT, S.E. (1994). 'International procurement strategies: challenges and opportunities for the small firm', *Production and Inventory Management Journal*, **35**(2), 39–46.

SMITH, A., WATKISS, P., TWEDDLE, G., MCKINNON, A., BROWNE, M., HUNT, A., TRELEVEN, C., NASH, C. and CROSS, S. (2005). *The Validity of Food Miles as an Indicator of Sustainable Development: Final report*, AEA Technology, July 2005.

SOIL ASSOCIATION (2008). http://www.soilassociation.org/

STIER, R.F. (2006). 'As strong as your weakest link', *Food Engineering*, **78**(10), 22.

TESCO (2008). *Our suppliers and ethical trading* (accessed 29/10/2008). http://www.tescoreports.com/crreview08/suppliers-ethical.html

5

Understanding innovation in food chains

S. Estrada-Flores, Food Chain Intelligence, Australia

Abstract: Co-innovation involves cooperation and integration of existing knowledge from different organisations across food supply chains. This chapter discusses the advantages of using a concerted approach to food co-innovation. Such an approach should recognise the highly dynamic nature of the food manufacturing industry, the benefits of market driven innovation and the usefulness of policy as an instrument to encourage innovation. It is proposed that a sustainable co-innovation framework could create a paradigm shift in the way food chain participants innovate. The transformation of supply chain relationships and innovation processes could make the food industry more resilient to recent financial and environmental challenges.

Key words: food chains, co-innovation, supply chain, national innovation model.

5.1   Introduction

The aim of the food industry is to transform agricultural raw materials into safe, convenient, good tasting and nutritious products for consumers, in a profitable and sustainable manner. The food supply chain is defined as a set of interdependent companies that work closely together to manage the flow of goods and services along the value-added chain of agricultural and food products, in order to realize superior customer value at the lowest possible costs (Folkerts and Koehorst, 1997). Value addition in the food supply chain is generated by activities linked to primary and secondary processing, packaging, distribution and retail, as illustrated in Fig. 5.1.

The industry is strongly driven by the following trends:

- Horizontal issues: global trends in ageing and health awareness, consumer trust and consumer satisfaction (convenience), food safety and traceability, financial sustainability related to costs of production/packaging/transport, innovation as an engine for growth and sustainability in production systems.

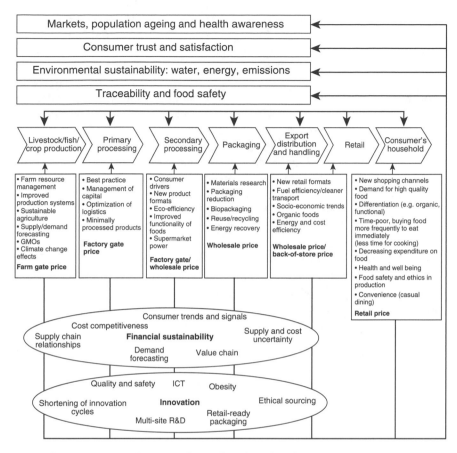

Fig. 5.1   Interaction of horizontal and vertical issues in the food chain.

- Vertical (specific) issues: each link in the chain has particular concerns and drivers. However, some of these have a cumulative effect (e.g. acceptance of genetically modified organisms affects both growers and manufacturers).

The question of how innovation happens in food chains is relatively unexplored. Some authors attribute this to the prevalent deterministic view of food production systems in economic theories, in that agriculture and food production are determined by nature and the biological cycles of crops and livestock (Morgan and Murdoch, 2000). Others view food-related innovation as a low capital, low risk activity, with low requirements of technological novelty or specialized skills when compared to other industrial sectors such as electronics or aircraft manufacturing. Furthermore, the number of patented inventions with direct application in the agri-food sector have been historically low (Arundel *et al.*, 1995). Yet, industries that participate in the food supply chain have successfully applied technologies developed in other sectors, such as the pharmaceutical industry, biotechnology and electronics (Beckeman, 2007).

Although the aspects mentioned above contribute to the scarcity of published research on innovation in food supply chain systems, the definition of innovation itself has been a subject of much controversy. Historically, innovation and research and development (R&D) have been used as synonyms. However, they are different in that research activities seek to increase knowledge in the areas investigated, without any expectations of the project's outcomes. In this sense, research projects are always successful as they normally deliver new knowledge. In innovation projects, the research team has commercial expectations from the onset and therefore, these projects can fail (Environmental Innovations Advisory Group, 2006).

In the context of this chapter, innovation is a process that goes beyond theoretical conception, through technical invention to commercial exploitation (Department of Agriculture Fisheries and Forestry Australia, 2001). A similar description was used in a recent review of the public support for science and innovation in Australia, where innovation was defined as a deliberative process undertaken 'by firms, governments and others that add value to the economy or society by generating or recognizing potentially beneficial knowledge and using such knowledge to improve products, services, processes or organizational forms' (Productivity Commission, 2006). The challenge of adopting this wider concept of innovation is distinguishing true innovation in services, businesses and marketing activities from simple organizational changes.

5.2   Innovation as a driver of global competitiveness

The competitiveness of a firm or a country depends on its ability to generate sustainable and relatively higher revenues from the factors of production and high employment as a result of exposure to international competition (Hatzichronoglou, 1996). Therefore, firms must develop strategic advantages that allow them to maximize the capabilities that distinguish them from competitors (Porter, 1980).

A study (Grunert *et al.*, 1995) investigated the link between innovation activity and business performance in the European food industry. The authors found a lack of conclusive evidence of the relationship between the intensity of R&D activities and business performance. These inconclusive results may be attributed to the concept of innovation itself, which extends beyond R&D to encompass a wide range of technological and business activities. Adding to this complexity, food chains involve various players, including farmers, manufacturers, retailers and suppliers of various goods and services such as information technology, ingredients, transport, storage and packaging. Further, the flow of benefits of innovations onto the supply chain partners are difficult to quantify. For example, increasing the availability of raw materials or improving the processing or sensory characteristics of plants and animals through innovation in primary production also benefits the manufacturing and retail partners by improving availability, product quality and sales.

A second European study (Wijnands *et al.*, 2007) compared the competitiveness of the European food industry with their counterparts in Canada, Australia, New Zealand, the United States and Brazil. The study highlighted four characteristics of innovation in the food industry:

- Consumers are relatively conservative about the foods they select and many 'new products' are, in fact, variations of products already accepted in the market. Although these variations may not be true innovations (Moskowitz and Hartmann, 2008), they do lead to considerable market value and therefore increase the competitive advantage of food firms.
- The most significant innovations in the food industry are market driven, with consumers putting emphasis on quality and convenience. Examples include packaging, logistics, integration with domestic appliance manufacturers (e.g. coffee and domestic appliances for coffee making) and changes oriented to improve convenience at point-of-sale (e.g. the development of vending machines for schools and hospitals offering prepacked fresh fruit salads).
- The food industry frequently innovates with a focus on management and organizational processes. The integration of precision agriculture and global positioning systems to improve planting and harvesting efficiencies is an example of organizational innovation (Cutler *et al.*, 2008).
- Agriculture and food manufacturing absorb innovations from other industries (e.g. ICT, logistics and marketing) at a fast rate. Through early adoption of potentially profitable innovations already proven successful in other industries, food companies can still achieve a 'first mover' advantage.

Innovation indexes used by the OECD (Organisation for Economic Cooperation and Development) and other international organizations do not capture these subtleties. For example, innovation and R&D activities in packaging, logistics, storage and food processing machinery, which are central to efficient food chains, are normally accounted for in sectors other than food manufacturing (e.g. services, chemical manufacturing, plastics products, fabricated metal products). Further, non-technological R&D expenditure is a significant component of the total business expenditure in innovation (Cutler *et al.*, 2008).

Although the inadequacy of using R&D indicators as direct measures of innovation activity is recognized, R&D remains an important component of innovation. Therefore, trends in R&D expenditure for OECD countries are analysed in the following section.

5.2.1   Historical trends in innovation in the food industry worldwide

Figure 5.2 shows the historical business enterprise expenditure on R&D (BERD) in food, beverages and tobacco for nine OECD countries. This figure suggests significant differences in the prioritization of strategic R&D funding for the food sector, whereby most countries lag behind the investment observed in Japan and the USA.

Figure 5.3 presents a slightly different landscape of R&D activity by using two more indicators: (a) R&D intensity using production (RDIP), which measures BERD expenses as percentages of production (gross output) in current prices and (b) R&D intensity using value added (RDIV), which measures BERD expenses as percentages of the contribution of the industry to total gross domestic product (GDP) in current prices. Figure 5.3 suggests that Denmark has the highest R&D intensity of the countries included in this comparison, both in terms of production and value addition. Finland, the Netherlands and Japan present high RDIP values. In contrast, the United States presents a lower R&D intensity than countries which ranked lower using BERD expenditure only as an indicator (Fig. 5.2).

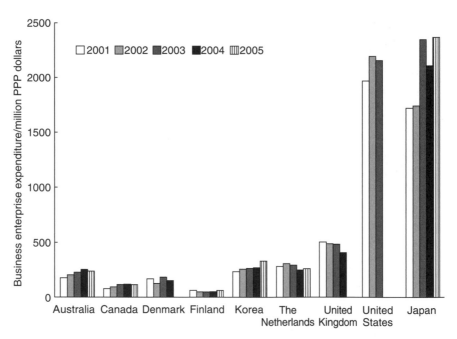

Fig. 5.2 Business enterprise expenditure on R&D (BERD) in the food, beverages and tobacco sectors for nine OECD countries in purchasing power parity (PPP) dollars at current prices.

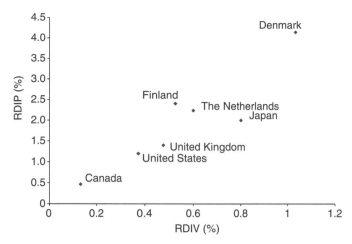

Fig. 5.3   Intensity of R&D expenditure as a percentage of gross production output (RDIP) and as a percentage of value addition to the gross domestic product (RDIV) in foods, beverages and tobacco. Values represent an average for the period of 2000–2003.

5.2.2   Government-led initiatives focused on food innovation

The New Zealand Ministry of Agriculture and Forestry recently released a report (MAF, 2008) that outlines the basis of collaboration between government, industry and research providers to lift the long-term science base, capability, environmental performance and competitiveness of the country's pastoral and food industries. The report highlights the similarities between the economies of Finland, Sweden, Norway, Denmark, Iceland, The Netherlands, Canada, Australia and New Zealand, which are all heavily reliant on natural resources such as agriculture, forestry and fishing.

Interestingly, the European countries mentioned have managed to develop knowledge-based industries and advanced technology sectors in parallel with their resource-based industries. These countries seem to be successful in transferring the knowledge and innovations created in the advanced industries to the rural industries.

Government-led policies on innovation have a strategic role in achieving the right balance between investment dedicated to creating and developing new knowledge from science and research and investment dedicated to increasing the capacity of enterprises to receive, absorb and commercialize such knowledge (Cutler *et al.*, 2008). Government policies on innovation also define national priorities, remove structural impediments to innovation and create the conditions necessary to introduce new technologies and services to the market. The following paragraphs discuss some government-led initiatives for developing food innovation in Europe, Australia and the United States. A summary of the food innovation systems in these three regions is presented in Table 5.1.

Table 5.1 Summary of food innovation systems in Europe, USA and Australia

Region	Food and drinks retail turnover[a] ($US billion), 2007	SMEs (%)	Largest current government funding programme (food R&D)	Current topical areas	Public funding for agricultural production and technology in 2007 ($US million)	Sources
Europe	1168.5	99	Food, Agriculture and Fisheries, and Biotechnology (FP7)	Food safety and quality; health; sustainability and renewable bioresources; aquaculture and fisheries production; animal welfare; dietary needs of consumers	3167.7	Wijnands *et al.*, 2007; European Commission, 2007b
USA	901.8	99	National Institute for Food and Agriculture (NIFA); Agriculture and Food Research Initiative (AFRI);	Renewable energy and sustainability; food safety, nutrition and health; plant health and production & plant products; animal health and production & animal products; agricultural systems and technology; rural economics and social aspects	2471.3	US Dept. of Commerce, 2008; OECD, 2008
Australia	75.3	>99	Rural R&D corporations (RDCs); Cooperative Research Centres (CRCs)	Sustainability and renewable bioresources; water conservation; biosecurity; dietary needs of consumers	288.2	DAFF, 2008; OECD, 2008; Estrada-Flores, 2008b

[a] Includes supermarket and foodservice sales.

Europe

The European food and drink industry encompasses about 280,000 companies and provides jobs for 4 million people. The estimated annual turnover exceeds €800 billion and is based on exports, despite extremely competitive domestic and international markets (European Commission, 2008a).

The European agro-food industry is dominated by small and medium size enterprises (SMEs), which represent 99% of the industry. The production is highly diverse and SMEs often lack resources and personnel to invest in research and innovation. Further, these companies traditionally perceive that they have low returns on investment and low profit margins. As a result, little expenditure is dedicated to cover innovation costs such as patenting (Confederation of Danish Industry, 2008).

Europe's Framework Programme for Research and Development is the main financial tool through which the European Union supports research and development activities, covering almost all scientific disciplines (European Commission, 2008b). The programme is traditionally run in 5-year periods, but in the present FP7 programme the budget implementation period has been extended to seven years.

The current 'Food, Agriculture and Fisheries, and Biotechnology' research programme has been allocated more than €1.9 billion funding for the duration of FP7. The objective of the research programme for 2007–2013 is to build a European knowledge-based bio-economy (KBBE), addressing the following needs:

- growing demand for safer, healthier, higher quality food;
- sustainable use and production of renewable bioresources;
- increasing risk of epizootic and zoonotic diseases and food related disorders;
- sustainability and security of agricultural, aquaculture and fisheries production;
- increasing demand for high quality food, taking into account animal welfare and rural and coastal contexts and response to specific dietary needs of consumers.

Under a KBBE framework, life sciences and biotechnology will undoubtedly play a significant role. Areas such as the development of eco-efficient products, processes and technologies and the development of healthier, wholesome foods that respond to specific nutritional needs of consumers are particularly encouraged.

The FP7 programme aims to promote innovation in companies with low to medium technological capabilities and with little or no research capability. The programme also aims to enable the access of these companies to research institutes and universities to outsource their R&D. However, SMEs in Europe still have difficulties in accessing the various financial instruments and capital for their development, mainly owing to the lack of financial and human resources to create direct collaboration with universities and other research centres (European Commission, 2008c).

The long timeframes needed to comply with the current legislation for new product introduction into the marketplace are also considered to be a hindrance to innovation. This issue is not exclusive to Europe: similar concerns have been raised in Canada (Standing Committee on Industry Science and Technology, 2007), the Unites States (Food Directorate, 2007) and in Australia (Annison, 2008). Innovations particularly affected are those with nutritional and health claims, pre-market requirements for GMOs and novel functional foods.

Australia

The Australian government has historically placed more emphasis in pre-production and primary production research than in manufacturing and post-manufacturing research. For example, the national R&D expenditure during 2005 was AUD$1 billion in the primary production of foods (including plant and animal production and primary products), while R&D investment in food manufacturing was just below AUD$300 million (Australian Bureau of Statistics, 2005). Interestingly, the value of primary production goods trebled by the time the food reached the shelves in supermarkets in 2005, with the largest increment in value occurring at the manufacturing stage. The low levels of R&D in the food manufacturing stage are at odds with the knowledge-intensive character of the industry (Howard Partners, 2005).

The Australian Government currently supports 16 rural R&D corporations (RDCs), which are the most significant instrument of engagement between government and private primary industry. Government funding for RDCs is matched by private funding through an industry levy. The combined total annual budget for RDCs is about AUD$500 million.

The Cooperative Research Centres (CRC) programme was established in 1990 to bring together researchers in the public and private sectors with the end users. The programme has resulted in the establishment of 168 CRCs during its lifetime, operating across the manufacturing, information technology, mining and energy, agriculture and rural-based manufacturing, environment and medical science and technology sectors. About 15 CRCs have tackled particular aspects of the food chain, with more than half of these centres focusing on primary production issues.

While RDCs are strongly driven by industries through a heavy involvement in priority setting and contribution of levy funding, CRCs can be characterized as research organisations pooling their financial, human and infrastructure resources, involving industry partners as advisors and funding partners. Other past government funding initiatives for food innovation closely linked to market demands include:

- The supermarket to Asia (STA) Strategy from 1996 to 2002, which had as a main objective the expansion of Australia's agri-food exports to Asia. The Technical Market Access Programme was created through STA and it is still used by the Australian Department of Agriculture, Fisheries and Forestry as a mechanism for enhancing market access of Australian foods exports.
- The National Food Industry Strategy (NFIS) Ltd (2002–2007) was an industry-led company funded by the Australian Government to implement most recommendations emerging from the National Food Industry Council (NFIC). NFIC was a leadership team where key food industry companies and Commonwealth Ministries were represented. Food industry grants were the most representative instrument for R&D funding from NFIS, providing funding for projects involving technical and scientific R&D with strong prospects for commercialization.

Future directions of public funding for Australian research in food chains are uncertain. A recent report discussing national research priorities for the period 2010–2020 highlighted agriculture, food security, nutraceuticals, food safety and certification and biological testing as priority areas for investment (Cutler *et al.*, 2008). However, concerted efforts between Government, industry and research organisations to set a national agenda for food innovation have not been developed as yet.

United States

In the American system for public funding of R&D, federal funding normally flows through agencies that support research activities in universities, national laboratories and, to some extent, the industry (Lane, 2008). The United States Department of Agriculture (USDA) is the main federal agency supporting food innovation. Examples of project areas include advanced computer-aided design and manufacturing technologies for food products, processes, and equipment to enhance food safety, quality and value (Interagency Working Group on Manufacturing R&D, 2008).

There seems to be a scarcity of US government programmes that directly support food innovation developed in private industry. The US government strategy for innovation has been historically based on enacting regulations and tax reductions that ease the way for innovative companies, rather than directing federal funding to encourage innovation. This strategy has led to a flattening of government funding for food and agricultural research since the late 1980s. In parallel, private investment in food and agricultural research – especially in new technologies – has increased (Patrick, 2006). However, the private food sector in the USA has not invested enough in 'blue horizon' R&D, where payoffs are over long timeframes, the potential market is highly speculative, the need for investment is the highest and where the benefits are widely diffused.

In view of these facts, a new National Institute for Food and Agriculture (NIFA) was recently opened to enhance existing food research programmes. The objectives of NIFA include streamlining agricultural research into six programmes: (1) renewable energy, natural resources and environment; (2) food safety, nutrition and health; (3) plant health and production and plant products; (4) animal health and production and animal products; (5) agriculture systems and technology; and (6) agriculture economics and rural communities.

Further, an Agriculture and Food Research Initiative (AFRI) was created in 2008, with a budgeted funding of US$700 million per year. Sixty percent of appropriated funds are earmarked to fund basic research and the remaining budget will be allocated to fund applied research programmes. The creation of both AFRI and NIFA are mandates of the recently approved Food, Conservation and Energy Act (US Department of Agriculture, 2008).

5.3   Changing face of innovation in food production and manufacturing

The sophistication of the innovation supply chain has increased considerably from the 'economic miracle' days after the Second World War to present times. The Second World War itself prompted the industry to develop new food preservation methods (e.g. canning), packaging and logistics to supply food for the army (Goldblith, 1989).

Until the 1950s research was focused on production and the predominant innovation process followed the linear science push model (Fig. 5.4) whereby new ideas, rather than consumer's needs, were the trigger to innovate. In the 1950s and 1960s, many OECD countries built on basic science projects, dedicating funds to support research in universities and dedicated research institutes (Arnold, 2007).

5.3.1   Producer-led innovation

In the post-war years, the key driver behind innovation in food production was the need to improve on-farm yield and productivity. The political and economic environment encouraged farmers to invest in research and development, utilizing government subsidies and the available public research institutions that were oriented to agriculture.

Fig. 5.4 Traditional linear science-push innovation model.

Farmers in the United States were particularly successful at capitalizing the opportunities for innovation between 1949 and 1999. The high levels of agricultural productivity in the USA have been attributed to four decades of innovation in seed crops and in livestock-raising practices. Table 5.2 presents some crucial innovations developed in the history of farming.

Morgan and Murdoch (2000) provide an excellent review on the innovations developed in the agrochemical industry, triggered by the need to combat weeds and pests in the 1940s and 1950s. Concerns about the use of chemical control methods in turn led to the development of integrated pest management in later years (Hall and Moffit, 2002).

5.3.2 Manufacturer-led innovation

After the collapse of communism and the cold war years, the attention of science and technology policy makers turned inwards, focusing on developing solutions to pressing domestic issues. National priorities such as health care, education and social justice took the front seat and access to R&D funding increasingly required a close alignment of research objectives with these priorities (Lane, 2008).

In the 1950s, innovation projects initiated at the manufacturer's level rarely involved food retailers and there was even less involvement of suppliers. This strategy has changed little over the past years. As a

Table 5.2 Producer-led innovations in USA agricultural history

Innovation	Decade of introduction	Outcome targeted
Antibiotics in animal production	1950s	Enabled farmers to raise large numbers of chickens in confinement
Fully automated feeding	1940s	Advances allowed medication to be distributed with the feed
Selective breeding	1940s	Development of higher meat yields in birds, cattle and hog production
Vertical integration	1980s	Linking of breeders and growers to processors, usually through production contracts rather than by ownership. Concentration of market power in packers and processors
Hybrid seeds	1940s	Improved yields and resistance to pests and infections.
Genetic modification	1980s	Reduced levels of pesticide use and tillage. Businesses based on a life sciences model combining agriculture, food and pharmaceuticals around a genomics science base

consequence, new product development (NPD) projects initiated by processors have an extensive consumer research phase but often fail to engage
suppliers. This engagement is important to ensure that the raw materials
have the required quality and grading expectations for the new product.

An *ad hoc* approach to NPD may also lead to confusion about critical
supply chain conditions (e.g. time and temperature) that should be maintained throughout the distribution of the new product. Failure to maintain
these conditions may lead to quality losses or can even trigger serious food
safety issues. The lack of a coordinated innovation framework between
supply chain partners partly explains why NPD remains a high risk activity
in the food sector, where over 80% of NPD ventures fail (van der Valk and
Wynstra, 2005). Table 5.3 presents some manufacturing-led innovation initiatives in the past years.

In Table 5.3, the bold typeface highlights the most common areas of
collaboration between manufacturers and retailers. Innovation projects in
supply chain areas create significant synergies between these two parties.

5.3.3   Retailer-led innovation

From the three major organizational forces in food supply chains (i.e.
farmers, manufacturers and retailers) retail-led innovations have focused
the most on improving supply chain and logistics through initiatives including Efficient Consumer Response (ECR), Electronic Data Interchange
(EDI) and traceability, among others. In fact, ECR, Collaborative Planning, Forecasting and Replenishment (CPFR) and category management
rely on the joint development of strategic category plans and a collaborative framework between retailers and their suppliers (Dapiran and Hogarth-
Scott, 2003). Table 5.4 shows some of the retail-led innovations in the
past years.

5.4   Contemporary food innovation models

As consumers became a political force and their demands of manufacturers
became more complex, a shift in power in food chains occurred. In the
1980s and 1990s, the increase in disposable income, decrease in population
growth and development of global supply chain systems resulted in a decel-
eration in the demand for food (Costa and Jongen, 2006). Shorter product
development cycles and a buyer's market rendered the traditional
technology-push approach to food innovation obsolete.

5.4.1   Influence of supply chain relationships in innovation

Table 5.5 presents the relationships that currently exist between buyers and
sellers in the supply chain.

Table 5.3 Variation in manufacturer-led innovation. Bold typeface indicates innovations that overlap with retail drivers (adapted from Linnemann *et al.*, 2006 and Keh, 1998)

Innovation	Outcome targeted
New product development (NPD)	• *'Me-too' products*: a product that replicates characteristics of existing successful products in the market, thus avoiding some NPD risks. The objective is to erode the market of a competitor • *Line extensions*: variations of a well-known product (e.g. flavours, colours, etc). The aim is to increase market share and improve product positioning with relatively little effort and development time, making small changes in manufacturing processes, marketing strategy and storage and/or handling operations. • *Repositioning of products*: changing the promotion strategy of current products in the market, to reposition these as products responding to current consumer's demands. The major effort is, thus, in marketing. For example, repositioning of products as 'healthy' or functional products. The aim is to capitalize opportunities in niche markets. • *New form/formulations for existing products*: these encompass products that have been altered to another form (e.g. dried, granulated, concentrated, spreadable or frozen) or products that have been reformulated. In the former category, extensive R&D and development time may be required, together with changes in supply chain operations. Formulation changes can have various impacts on the supply chain, according to the degree of variation in the product. The outcomes sought refer to convenience, value addition, cost reduction, unreliable supply of some raw materials, or the availability of better/less costly ingredients. • *Evolutionary innovative products:* substantial changes in an existing product, other than described above. The changes must add value/functionality in a significant manner to the original version. R&D times, costs and risks are generally greater than for other modifications. Marketing can also be costly. • *Radically innovative products:* a 'never seen before' product. These require extensive product development, have high R&D, marketing and capital (new equipment) costs and have the highest failure chance of all categories. Having said this, these products potentially offer greater rewards than others. Some products can be potentially disruptive.
New packaging development	Added functionality, better preservation of foods, variety in volumes/portions, more attractive designs for targeted consumer segments, labelling, convenience, **retail ready**.
New processes	Cost reduction (e.g. less labour, energy efficient), occupational health and safety (OH&S) compliance, reduction of environmental impact, requirement for manufacturing new product
New supply chains	**Response to changes in client's (e.g. retail, foodservice, etc) business formats, supply chain initiatives, traceability (e.g. RFID)**

Table 5.4 Past and current retail-led innovation projects (adapted from Dapiran and Hogarth-Scott, 2003 and Keh, 1998)

Innovation	Outcome targeted
Air conditioning control	Comfort in supermarkets
Scanner systems	Efficient inventory
Bar codes	Traceability, supply chain management
Electronic cash register	Efficiency, shopping experience
Electronic Data Interchange (EDI)	Paperless management of supply chain, reduced order lead time, fewer out-of-stock situations, lower inventory costs, reducing errors in ordering, shipping and receiving, reduction in labour costs, higher service levels
Category management	Vertical integration, matching of consumer's preferences by sellers offerings and growth of categories
Cross-docking	Cost efficiency in distribution
Efficient consumer response (ECR)	Efficiency gains in store assortment, promotion, new product introduction and replenishment, through constant flow of product and information between suppliers and retailers
Collaborative planning, forecasting and replenishment (CPFR)	Coordination of supply–demand

Table 5.5 Modern types of relationship between buyers and sellers in the supply chain (adapted from Fearne *et al.*, 2001)

Type of relationship	Characteristics
Quasi-vertical integration	Both parties enter into a long-term contractual obligation, where the costs, risks, profits and losses are shared in a venture arrangement. Examples: joint venture, franchises, licenses
Tapered vertical integration	A firm obtains a proportion of its inputs through backward vertical integration with a supplier. Examples include the acquisition of supply sources or equity in the supplier's business
Full vertical integration	One firm carries out two or more sequential stages of the food chain. Integration can occur backwards (upstream) or downwards (downstream)
Horizontally integrated networks	Relationship between businesses serving similar markets or locations; the combination of forces can occur in activities such as accessing new markets, new product development and shared purchasing functions

The past decade has seen a transformation in supply chain relationships from an adversarial, transactional-based operation into a cooperative approach. This shift has strengthened the position of the food industry in the face of disruptive external events, such as the oil crisis in the mid-1970s (Fearne *et al.*, 2001). Collaborative approaches have also led to faster and more efficient responses to consumer's demands.

As supply chain relationships evolve, new vertical or horizontal cooperation models can stimulate innovation by pooling existing knowledge from different organizations throughout the chain. For example, collaboration between retailers and manufacturers could lead to new retail formats that could cross over the convenience sector (Earle, 1997). As mentioned previously, modern retail-end processes rely on the joint development of strategic category plans and a collaborative framework between retailers and their suppliers.

There are three concepts that are related to innovation models in evolved supply chain relationships: market-led innovation, co-innovation and forward commitment procurement. These concepts are described next.

5.4.2   Market-led innovation

There are two distinctive types of market-led innovation in the food industry: consumer-led and user-led innovation.

Consumer-led product development was introduced in the early 1990s as a market-oriented innovation concept concerning the use of consumers' current and future needs (Urban and Hauser, 1993). The aim of consumer-led product development is to create product differentiation, leading to higher consumer satisfaction, increased levels of consumption of specific products, or increased overall value of the given level of consumption (Grunert and Valli, 2001).

User-led innovation refers to the phenomenon observed in the late 1970s, whereby customers proceeded to modify or adapt existing products according to their own needs of their own accord (Grunert *et al.*, 2008). Customers may be intermediate users of the product or consumers. For example, the equipment and process for preparing free-flowing ice cream in the form of deep frozen beads was developed by Dippin' Dots founder Curt Jones (Jones, 1992; Jones *et al.*, 2001). In fact, only about 50% of the patents registered between 1991 and 2002 in the field of cryogenic freezing for food manufacturing were granted to equipment manufacturers (Estrada-Flores, 2002; Estrada-Flores, 2008a).

In a consumer-orientated approach, the development of new products, processes or services begins with consumer and market research to identify the specific characteristics required by consumers. In a user-led approach, a deep understanding of the users' wants, needs and preferences plays a key role (Grunert *et al.*, 2008). Therefore, market-led innovation requires

an understanding of the complex social, technological and ecological connections in agri-food systems (Lowe *et al.*, 2008). The role of social scientists
and consumer researchers becomes crucial, as the 'translators' of descriptive and qualitative terminology in which consumers express themselves
using accurate technological specifications that all collaborators can follow
(Linnemann *et al.*, 2006).

User-led innovation arguably offers the best platform for collaborative
projects between suppliers, retailers and manufacturers. Consider the
highly perishable fruit and vegetable category, which has a shelf life of
sometimes days only. For these products, some retailers plan their resourcing more than once a day, because the potential wastage cost exceeds the
savings through economies of scale in transportation and warehousing
activities. In this case, the flow benefits of co-innovation can be captured
through information sharing and forecasting collaboration. For less perishable items (e.g. dried or canned foods) a highly efficient supply chain
depends on low inventory levels and high capacity utilization (Holweg
et al., 2005).

5.4.3   Collaborative innovation

Collaborative innovation, also known as 'open innovation' (Sarkar and
Costa, 2008) or 'co-innovation' (Fearne, 2007) can be particularly advanta-
geous to small and medium-sized enterprises (SMEs), which can tap into
skills and knowledge beyond their own capabilities (Grunert *et al.*, 1995).
Large manufacturers can achieve a more systematic approach to NPD, an
increased emphasis on market-oriented NPD and a stronger network for
product market intelligence.

Fearne (2007) proposed the model presented in Fig. 5.5, which is based
on the partnership of supply chain players in three platforms: innovation,
service and operations. He also pointed out that there are three drivers for
undertaking co-innovation:

• It allows the development of new (value added) products/services for
  distinct customers and targeted consumer segments
• Process improvement occurs for existing products/services beyond
  organizational boundaries
• Innovation at the interfaces in the value chain are much more difficult
  for others to copy, thus increasing the competitive advantage.

Co-innovation often requires the development of global innovation networks, which add further advantages, such as:

• supporting collaboration between geographically dispersed teams of
  suppliers, manufacturers and retailers
• speeding time-to-market and reducing costs for all the parties involved
  in the NPD process through resource pooling and reduction of the learning curve times

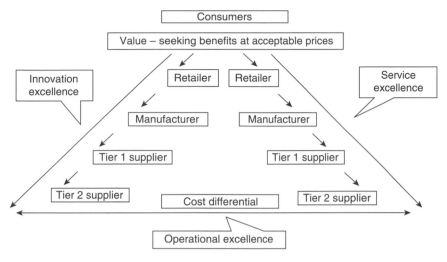

Fig. 5.5 Co-innovation framework (source: A. Fearne, 2007, with permission).

- meeting customer expectations in a more accurate manner, owing to the direct input of retailers into the main purchase drivers of products
- enforcing consistency and quality of brands and innovations
- creating a compliance audit trail through sharing of quality documentation
- most importantly, creating a repository of protected know how and intellectual property, only available to the chain participants.

The challenges of co-innovating are better understood in the context of recent public hearings on the balance of market power between food suppliers, manufacturers and retailers: in Australia and in the UK, large food retailers are perceived as the parties with the greatest influence and control in the food chain. This influence received significant attention in the UK, which led to the introduction of a Supermarket Code of Practice in 2002. The Code explicitly mentions that all supply chain participants would benefit if they work together to expand the market for their products and develop a profitable and sustainable business (Fearne, 2005). The Australian Competition and Consumer Commission (ACCC) investigated the relationship between suppliers and retailers in a similar manner (Australian Competition and Consumer Commission, 2008). Although the ACCC report did not identify 'anything that is fundamentally wrong with the grocery supply chain', the report did not appease the concerns of grocery suppliers and the overall sentiment of a lack of transparency in the Australian grocery chain and the dominance of retailers remains (West, 2008). Therefore, creating the required level of communication, trust, commitment and interdependence required for co-innovation calls for significant political and organizational efforts.

Supermarkets are the 'last frontier' between the commercial food chain and the consumer. Given this fact and that retailers are often seen as the dominant power in the food chain, it would be fitting for large retailers to adopt the role of innovation leaders in the industry. Indeed, this leadership role may hold the key to improving trust and communication throughout the entire food supply chain.

5.4.4   Forward commitment procurement (FCP)

Forward commitment procurement (FCP) is another type of demand-side driven innovation, specifically led by the public sector. It is defined as 'a commitment to purchase, at a point in the future, a product or service that does not yet exist commercially, against a specification that current products do not meet, at a sufficient scale to enable the investment needed to tool up and manufacture products that meet the cost and performance targets in a specification' (St John's Innovation Centre, 2006).

In an FCP model, a public sector organization commits itself to purchase a predefined quantity of a product, service or technology currently under development but not yet commercially available (Environmental Innovations Advisory Group, 2006). The commitment is for a future date and is based on a specified product performance being achieved. When the product has been developed that meets this performance specification within the agreed timeframes and framework, the organization purchases the product at a specified volume and cost, at levels that encourage supplier investment to ensure economies-of-scale. The private sector reacts by freeing investment to search for innovations that respond to those specifications. Once the product/service has entered the market, normal market conditions determine competition and price (Environmental Innovations Advisory Group, 2006).

The FCP model is one of the most promising models that can encourage environmental innovation, where the government itself acts as an early adopter. The procurement process is also supported by regulations that enhance market conditions in order to create a demand for the new products and services. Key elements of FCP processes are its focus on market needs and outcomes and a clear translation of market needs into specifications for the tendering process. A recent report (European Commission, 2005) noted the following advantages of FCP innovation models:

- Firms are given the incentive to spend money on research in the knowledge that an informed customer is waiting for the resulting innovations and thus the risk of investing in R&D is reduced.
- Competition is shifted from a sole focus on price to the provision of solutions which offer the greatest advantage to users over the whole life use of the purchase.

- FCP opens up opportunities to improve the quality and productivity of public services offered to the citizens through the deployment of innovative goods and services.
- Technologies launched in this way may then move on to further deployment in private sector markets. Other policy objectives (e.g. sustainability, food safety) may also be achieved by procurement of innovative solutions.

The European Commission has examined precommercial procurement processes for R&D services, thus suggesting its preference for this approach in Europe's strategic sourcing of innovation (European Commission, 2007a).

5.5   A new model: sustainable co-innovation

Arnold and Kuhlman (2001) established that there are three major pillars of a national innovation system:

1   The industrial system, encompassing large, medium and small manufacturing enterprises and technology-based firms. These systems supply the demands of intermediate and final consumers.
2   The education and research system, encompassing professional education and training organizations, higher education and research organizations and public sector research. These systems supply solutions for the industrial systems sector, but also for innovation intermediaries (e.g. innovation brokers) and for final and intermediate consumers.
3   The political system, which encompasses government, governance and policies for research and technology development. The 'clients' of the political system are the education and research system and indirectly, the industrial system. The latter is served through framework conditions encompassing the financial environment and incentives (e.g. taxes) needed to support innovation.

These three systems are financed through banking, venture capital and public innovation funds. Although it is widely acknowledged that national innovation systems should integrate these three pillars and their financing sources to streamline the innovation supply chain in the manufacturing industries, the complexity of the task may seem daunting.

This section presents an example of how these elements could be integrated into a single innovation system. The model proposed, which has been named the sustainable co-innovation (SCOI) model, uses the concepts of consumer-led innovation, collaborative innovation and FCP to integrate the industrial, education and political structures discussed above.

5.5.1   Central innovation 'brain'

In the SCOI model, a central overseeing organization takes on the role of innovation coordinator. The organization, established as a partnership between private and public partners (e.g. a joint venture) could be modelled after previously tested organizational models in Australia (i.e. National Food Industry Strategy Ltd) and New Zealand (i.e. New Zealand Fast Forward Ltd). In the context of this chapter, the central organization will be called NFISC. NFISC would:

- Provide a strategic framework for national food innovation, from a market-led, supply chain perspective.
- Coordinate the activities required to introduce new technology in the marketplace, following an FCP approach, by:
  - engaging with government departments and food supply chain players to develop the buyer's needs into specifications for tenders;
  - acting as a technology broker, bringing venture/equity capital and innovator companies together;
  - coordinating legislative activities that provide adequate market conditions for adopting innovations in the wider industry; and,
  - promoting technology demonstrations and case studies to show the business case.
- Coordinate government-led food innovation activities (e.g. grants and strategic directions for public R&D), from a supply and value chain perspective. This would avoid a lack of supply chain focus and would increase critical mass in the initiatives undertaken.
- Provide timely competitive and technical intelligence for all stakeholders about the particular FCP projects targeted, bringing to the stakeholder's attention current and emerging technologies in the targeted technology markets.

Further, the SCOI model is based on the development of three types of consortia: the *buyer* consortia, the *supplier* consortia and the *legislative* consortia.

5.5.2   Buyer consortia

A buyer consortium is formed by two or more supply chain partners that establish an alliance to contract/purchase the new process/ product/ service developed by a supplier consortium. Examples of alliances may be (a) a retailer and a cooperative that enjoy mutual benefits in purchasing a new retail-ready format for fresh produce that is also environmentally friendly; (b) a fast-food chain, a third party logistics provider and a food manufacturer that all mutually benefit from the use of new logistics systems that allow direct dispatch of home delivery orders from the manufacturer to the consumer. This type of consortium becomes effectively a supply chain innovation network as illustrated in Fig. 5.6.

Fig. 5.6   Buyer consortium.

5.5.3   Supplier consortia

A supplier consortium is formed by two or more organizations that seek to deliver innovation at the specifications (e.g. cost, time, performance) set by legislative consortia. Public and private R&D organizations supply solutions and showcase their R&D to innovation entrepreneurs who are hoping to develop the new product/ technology/ service, either as a start-up company or as a new product development in an established company. Universities and RTOs contribute with gap analyses of the skills required to deliver the innovation to the marketplace and build training and education programmes to address these gaps. If needed, venture capital is sought to continue R&D activities in collaboration with the public/private R&D organizations and to prepare case studies highlighting the business case. Figure 5.7 illustrates this type of consortium. A crucial aspect of the operation of these consortia is technology transfer from public R&D organizations to the new venture. Fair intellectual property arrangements should be negotiated to ensure that all parties in the innovation supply chain are rewarded.

5.5.4   Legislative consortia

A legislative consortium aims to increase the receptiveness of the market to new technologies/ processes/ products by introducing standards, regulations and laws that increase performance targets in certain areas. Examples may include:

- new regulations on water usage encouraging innovation in water utilization;
- improved regulations on the limits of food poisoning cases per territory or state encouraging innovation in food safety;

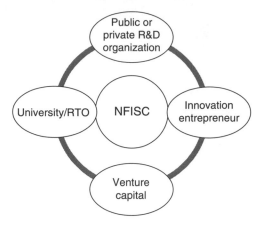

Fig. 5.7   Supplier consortium.

Fig. 5.8   Legislative consortium.

- new energy regulations for commercial and domestic refrigeration encouraging innovation in refrigeration systems.

Figure 5.8 illustrates a potential legislative consortium.

5.5.5   Innovation process

In the SCOI model, a buyer consortium (which may or may not include a public sector organization) establishes a commitment to purchase a pre-defined quantity of a product/technology/service, currently under development. NFISC and one or more buyer consortia agree about the performance sought for the innovation. A supplier consortium is formed and solutions are sought, based on past innovations applied to different industry fields or completely new concepts. Meanwhile, a legislative consortium develops standards, regulations and certification processes that enable fair competition and enhance the chances of the uptake of new solutions at the agreed performance specification. When the innovation has been developed, meeting all performance criteria, the buyer consortium purchases the product at a specified volume and cost, at levels that encourage other supplier consortia to enter the market. The private sector would react by freeing investment to search for innovations that respond to those specifications. The SCOI model encompasses the steps and processes shown in Fig. 5.9.

The SCOI model does not necessarily advocate a government role on becoming the early market buyer that executes the forward commitment options. This strategy may be effective in some areas with obvious links to public good, such as food safety. However, innovations with a clear commercial or industrial impact would benefit from an early intervention of food supply chain players that commit to buy the new product / service/ technology when this is developed. The drivers for such commitment should be based on (a) a superior value proposition, which may include financial, environmental and social performance parameters; (b) new regulations encouraging the uptake of the innovation; and (c) a demonstrated increase in competitiveness in the marketplace if the innovation is adopted.

5.6   Platforms for sustainable co-innovation in the food sector

In the view of this author, there are three areas where co-innovation can achieve its greatest impact:

1   Environmental co-innovation: For economies that largely depend on agricultural exports, it will become critical to demonstrate that their food supply chains are aligned with good environmental practices. A recent example of environmental co-innovation is the Sustainable Distribution Strategy, which involves 37 well-known food and consumer goods companies in the UK. It is expected that this initiative will lead to savings of 23 million litres of diesel fuel per year, through sharing of vehicles and optimising the use of warehouses (Anon, 2008).

2   New product development (NPD): Consumer-led NPD may offer the best platform for co-innovative projects between retailers and manufacturers (Grunert and Valli, 2001). In highly perishable new products, the flow benefits of co-innovation can be captured simply by information sharing and forecasting collaboration. For less perishable foods, synchronization of inventories can be an attractive proposition for the supply chain partners (Holweg et al., 2005). In a co-innovative environment, forecasting and synchronization have more probability of success than in traditional, manufacturer-led NPD efforts.

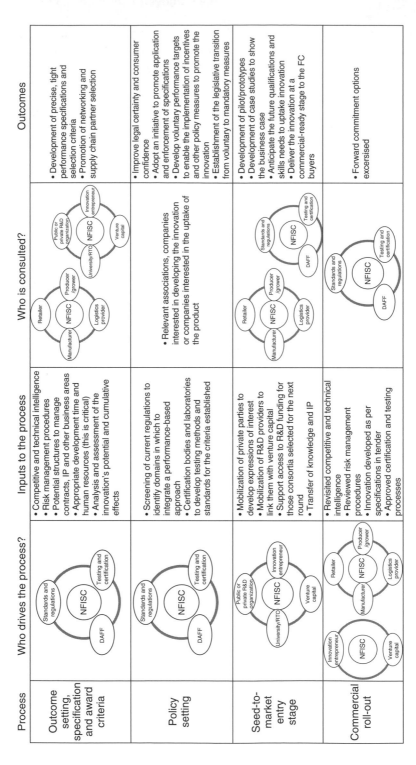

Process	Who drives the process?	Inputs to the process	Who is consulted?	Outcomes
Outcome setting, specification and award criteria		• Competitive and technical intelligence • Risk management procedures • Potential structures to manage contracts, IP and other business areas • Appropriate development time and human resources (this is critical) • Analysis and assessment of the innovation's potential and cumulative effects		• Development of precise, tight performance specifications and selection criteria • Promotion of networking and supply chain partner selection
Policy setting		• Screening of current regulations to identify domains in which to integrate a performance-based approach • Certification bodies and laboratories to develop testing methods and standards for the criteria established	• Relevant associations, companies interested in developing the innovation or companies interested in the uptake of the product	• Improve legal certainty and consumer confidence • Adopt an initiative to promote application and enforcement of specifications • Develop voluntary performance targets to enable the implementation of incentives and other policy measures to promote the innovation • Establishment of the legislative transition from voluntary to mandatory measures
Seed-to-market entry stage		• Mobilization of private parties to develop expressions of interest • Mobilization of R&D providers to link them with venture capital • Support access to R&D funding for those consortia selected for the next round • Transfer of knowledge and IP		• Development of pilot/prototypes • Development of case studies to show the business case • Anticipate the future qualifications and skills needs to uptake innovation • Deliver the innovation at a commercial-ready stage to the FC buyers
Commercial roll-out		• Revisited competitive and technical intelligence • Reviewed risk management procedures • Innovation developed as per specifications in tender • Approved certification and testing processes		• Forward commitment options excersised

Fig. 5.9 The roles and processes of the SCOI model.

3   New services and processes: Innovation in the services sector draws elements from engineering, science and management theory. New developments in this emerging area have the potential to tackle complex problems in which coordination and collaboration between professionals in different fields is essential (Paton and MacLaughlin, 2008). Service innovations include category management, coordination of supply/demand and track and trace technologies.

Examples of specific food supply chain projects where a national system could improve the success rate of co-innovations in the market include:

• information and communication technologies in modern food supply chains (e.g. traceability systems);
• food logistics and packaging systems (e.g. retail-ready, convenience markets, ageing population);
• functional foods and nutraceutical value chains;
• emerging technologies for improvement of food safety and quality;
• sustainability of food supply chains, with emphasis on water, energy and food waste (e.g. carbon-neutral supply chains);
• food security, including planning of pandemic scenarios and contingencies.

5.7   Conclusions and future trends

It has been long recognized that co-innovation, in the form of cooperation and integration of existing knowledge from different organizations across supply chains, can lead to exciting new products and services. A concerted approach in the agricultural and food industries would go a long way towards reaching the levels of openness, communication and commitment required for modern innovation models. Such an approach should recognize the highly dynamic nature of the food industry, the benefits of market driven innovation and the usefulness of policy as an instrument for encouraging innovation in the marketplace.

A sustainable co-innovation framework can help to achieve a paradigm shift in the way food chain participants create novel products, processes and services. Furthermore, transforming supply chain relationships from adversarial, transactional-based operations, to cooperative approaches would increase the resilience of food chains to the many challenges that have plagued the industry of late. Resilient food chains can only add value to companies, their partners and the society in general.

Current disruptive events such as the turmoil of global financial markets, growing concerns about the relationship between food production and environmental footprints and the effect of increasing production costs on the affordability and security of food can lead to a profound transformation of food production systems. A fundamental question raised by these recent events is 'Under this challenging scenario, How can the global agri-food sector become more competitive?'. The answer to this question is likely to require a holistic vision of food chains, whereby competitiveness is conceptualized as a function of the cumulative efficiency of the supply chain partners. The key role of innovation in lifting the competitiveness of the industry needs to be recognized and addressed in the strategic plans of industry and governments worldwide.

Increased urban population coupled with an increased awareness of the impact of food chains on the environment will lead to the development of new food distribution models. Examples of these trends include:

1 The rise of urban, local and regional chains. Issues such as food security, power imbalances in food chains, the environmental impact of food transport, obesity and other health issues attributed to the strategies of multinational food companies have led to some disillusion among consumers in current food systems and the growth of companies that embrace ethical sourcing and environmental awareness. Interest in local and regional distribution models that connect growers to consumers has ramped up globally since 2007 (Estrada-Flores, 2009) with examples such as the FoodConnect enterprise in Australia, the eFarm company in India and the 'Von hier' brand promoted in Germany.

2 Sharing distribution networks and infrastructure: Companies are now re-evaluating the way they store and move goods in the context of supporting continued economic growth, while protecting the environment. Some current examples demonstrate that enterprises are now willing to cooperate with other firms (even with competitors) to improve economies of scale and decrease logistics carbon footprints. Examples include:
    • The ECR Sustainable Distribution Group initiative (UK), which aims to save food and grocery industry mileage by sharing vehicles and improving the efficiency of warehousing networks.
    • The SmartWay^{SM} Transport (USA), an innovative collaboration between the Environmental Protection Agency and the freight sector designed to improve energy efficiency, reduce greenhouse gas and air pollutant emissions and improve energy security (EPA, 2009).
    • The Clean Cargo Working Group (USA), which is a multi-sector, business-to-business collaboration between ocean carriers, freight forwarders and shippers of cargo. Members of this group include Coca-Cola, Wal-Mart, Chiquita Brands and Starbuck's, among others. Tools used to enhance communication between participants are annual environmental surveys, intermodal emission calculators and CSR performance surveys (BSR, 2009).

3 The growth of supermarket-led initiatives to optimize supply chains. Examples include:
    • Woolworth's Limited Environmental Sustainability plan (Australia), which aims to achieve a 25% reduction in CO2-e per carton delivered

by 2012 through reduction in distance travelled, the introduction of new vehicle designs, the use of alternative fuels and the use of hybrid trucks (Woolworth's, 2008).

- The Woolworths Holdings Limited initiative (South Africa) to reduce relative transport emissions by 20%, by restricting airfreight of food products and sourcing food regionally wherever possible reducing reliance on long distance road transport (Woolworths Holding Ltd, 2009).
- The Wal-Mart's Sustainability 360 initiative (USA-global), which aims to reduce the number of trucks by redesigning the supply chain network, changing the presentation and size of food products and using auxiliary power units in their truck fleet (Anderson, 2008).
- The redesign of the Tesco distribution network (UK), which aims to reduce 50% emissions per case by 2012. Tesco's supply chain infrastructure includes 29 warehouses and over 2000 vehicles travelling 659 million km throughout the primary and secondary transport operations. Tesco is measuring the carbon footprint of three of its major food categories (tomatoes, potatoes and orange juice). Tesco has also committed itself to reduce packaging by 25% over the next three years (Watkins, 2008).

Other strategies that can also contribute to the sustainability of food distribution systems are:

- use of real-time telematics and computer routing and scheduling (Robson et al., 2007)
- implementation of alternative transport modes (e.g. rail and water ways)
- use of hybrid trucks, which increase fuel efficiency by 30–50% (Environmental Defense Fund, 2009)
- use of alternative energy sources for powering cold chain infrastructure, for example, the combination of grid electricity and power derived from wind (eolic) energy (TNO, 2009)
- packaging optimization and removal of excess packaging (Robson et al., 2007)
- use of hydrogen cells to power forklifts in distribution centres (DOE, 2009)
- use of supply chain network modelling as an aid to reducing food shipping carbon footprints (Robson et al., 2007).

Novel developments on these areas are likely to continue in the next decade. This author believes that innovation can successfully become the driver in transforming food supply chains. Innovation will enable the world to supply safe, wholesome and affordable food in an environmentally challenged future.

5.8   Sources of further information and advice

Readers wishing to understand innovation in the context of the Australian food industry in more detail may want to refer to recent submissions presented in response to the following reviews:

- The Garnaut Climate Change Review (2008) (www.garnautreview.org. au) which discusses carbon trading schemes for Australia to be implemented by 2010.
- The Cutler Review of the National Innovation System (2008) (http:// www.innovation.gov.au/innovationreview) which discusses a potential map from the government's innovation investment for 2010–2020.
- The Australian Competition and Consumer Commission inquiry into the competitiveness of retail prices for standard groceries (http://www. accc.gov.au/content/index.phtml?itemId=809228).
- The Senate's inquiry about food production in Australia (http:// www.aph.gov.au/senate/Committee/agric_ctte/food_production/tor. htm).

In all these reviews, the submissions presented by the food industry are relevant. Some comments as to how these reviews will affect the Australian food industry can be found in the Food Chain Intelligence website (http:// www.food-chain.com.au/events).

Also of particular interest are:

- Fearne *et al.*, 2001: This chapter examines the process of globalization, the emergence of food retailers as the dominant force in the food supply chain and the relationships that are emerging from the current balance in the market. This is key reading material to understand how retailers have a unique opportunity to drive innovation in new product developent and services throughout the chain.
- Grunert *et al.*, 1995: This is a very detailed paper that reviews the literature on the determinants of innovation and their impact on business performance. The authors propose a theoretical framework for the analysis of innovation in the food industry.
- Grunert *et al.*, 2008: The paper provides an overview of relevant streams of research that can form a basis for research on user-oriented innovation in the food sector. The authors show the relevance of research on: (a) the formation of user preferences; (b) innovation management; and (c) interactive innovation.
- Wijnands *et al.*, 2007: In this study funded by the European Commission, the competitiveness of the European food industry is compared to other countries (i.e. USA, Canada, Australia and Brazil). Innovation is highlighted as a key industry driver and as a potential source of solutions for increasing competitiveness. This study is one of the few which included all subsectors of the food industry and benchmarked these with important non-EU countries.

5.9   References

ANDERSON, M. A. (2008), *Supply Chain Sustainability.* Presentation for the SIF International Working Group. Accessed at http://www.socialinvest.org/projects/iwg/documents/Anderson_Presentation_10-08_v2.pdf on 23 Jan 2009.

ANNISON, G. (2008), 'AFGC submission to Department of Innovation, Industry, Science and Research', *Submission to the Review of the National Innovation System*, Australian Food and Grocery Council, p. 26.

ANON (2008), *'An idea whose time has come?'*, Logistics Manager. Accessed at http://www.logisticsmanager.com/Articles/10082/An+idea+whose+time+has+come.html on 25 November 2008.

ARNOLD, E. (2007), 'Governing the knowledge infrastructure in an innovation systems world', *1st Hemispheric Meeting of the Science, Technology and Innovation Network*, p. 19.

ARNOLD, E. and KUHLMAN, S. (2001), *RNC in the Norwegian Research and Innovation System*, Background Report No. 12 in the evaluation of the Resarch Council of Norway, Brighton, Technopolis, 47 pp.

ARUNDEL, A., VAN DER PAAL, G. and SOETE, L. (1995), 'Innovation strategies of Europe's largest industrial firms', *Results of the PACE Survey for Information Sources, Public Research, Protection of Innovations, and Government Programme*, University of Limbourg, Maastricht, The Netherlands.

AUSTRALIAN BUREAU OF STATISTICS (2005), *Research and Experimental Development Summary 2004/05.*

AUSTRALIAN COMPETITION AND CONSUMER COMMISSION (2008), *Report of the ACCC Inquiry into the Competitiveness of Retail Prices for Standard Groceries.* Commonwealth of Australia. Canberra. 457 pp. Accessed at http://www.accc.gov.au/content/index.phtml?itemId=809228 on 23 January 2009.

BECKEMAN, M. S. C. (2007), 'Clusters/networks promote food innovations', *Journal of Food Engineering*, **79**, 1418–25.

BUSINESS FOR SOCIAL RESPONSIBILITY (BSR) (2009), *Working Groups: Clean Cargo.* Accessed at http://www.bsr.org/membership/working-groups/clean-cargo.cfm on 12 Jan, 2009.

CONFEDERATION OF DANISH INDUSTRY (2008), 'Issues paper', *Working Group – Research and Innovation*, p. 2.

COSTA, A. I. A. and JONGEN, W. M. F. (2006), 'New insights into consumer-led food product development', *Trends in Food Science and Technology*, **17**(8), 457–65.

CUTLER, T., GRUEN, N., O'KANE, M., DOWRICK, S., KENNEDY, N., DAVIS, G., LIVINGSTONE, C., CLARK, M., FOSTER, J., PEACOCK, J., KELLY, P., HUGHES, A., LESTER, R., METCALFE, S. and SMITH, K. (2008), 'Venturous Australia: Building Strength in Innovation', in Cutler, T. (Ed.), *Review of the National Innovation System*, p. 228.

DAPIRAN, P. and HOGARTH-SCOTT, S. (2003), 'Are co-operation and trust confused with power? An analysis of food retailing in Australia and the UK', *International Journal of Retail and Distribution*, **31**(5), 256–67.

DEPARTMENT OF AGRICULTURE, FISHERIES AND FORESTRY (DAFF) (2008), *Australian Food Statistics 2007*, Australian Government Department of Agriculture, Fisheries and Forestry, Canberra, ACT, p. 177.

DEPARTMENT OF AGRICULTURE FISHERIES AND FORESTRY AUSTRALIA (2001), *Recipes for Success: case studies illustrating successful innovations by food businesses*, p. 51.

DEPARTMENT OF ENERGY (DOE) (2009), *Early Markets: Fuel Cells for Material Handling Equipment.* Accessed at http://www1.eere.energy.gov/hydrogenandfuelcells/education/pdfs/early_markets_forklifts.pdf on 23 January 2009.

EARLE, M. D. (1997), 'Innovation in the food industry', *Trends in Food Science and Technology*, **8**, 166–75.

ENVIRONMENTAL DEFENSE FUND (2009), *FEDEX partnership.* Accessed at http://www.edf.org/page.cfm?tagID=1453 on 14 January 2009.

ENVIRONMENTAL INNOVATIONS ADVISORY GROUP (2006), 'Bridging the gap between environmental necessity and economic opportunity', *1st report of the Environmental Innovations Advisory Group*, Department of Trade and Industry & Department for Environment Food and Rural Affairs, 10 pp.

ENVIRONMENTAL PROTECTION AGENCY (EPA) (2009), *SmartWay^{SM} Basic Information.* Accessed at http://www.epa.gov/smartway/basic-information/index.htm on January 02, 2009.

ESTRADA-FLORES, S. (2002), 'Cryogenic technologies for the freezing and transport of food product', *Ecolibrium: The Official Journal of AIRAH*, **6**(1), 16–21.

ESTRADA-FLORES, S. (2008a), *Freezing and Chilling by Cryogenic Gases and Liquids (static and continuous equipment)*, unpublished.

ESTRADA-FLORES, S. (2008b), *Sustainable Co-innovation: the food supply chain as a case study*, Submission to the Review Panel of the National Innovation System, Sydney, Australia, p. 39.

ESTRADA-FLORES, S. (2009), *Opportunities and Challenges Faced with Emerging Technologies in the Australian Vegetable Industry*, Technology platform 1: supply chain and logistics. Report for Horticulture Australia Ltd, 82 pp.

EUROPEAN COMMISSION (2007a), *Pre-commercial Procurement: Driving innovation to ensure sustainable high quality public services in Europe*, Communication from the Commission to the European Parliament, the Council, the European Economic and Social Committee and the Committee of the Regions, Brussels, Belgium, p. 11.

EUROPEAN COMMISSION (2007b), 'Work Programme 2008. Cooperation Theme 2. Food, Agriculture and Fisheries, and Biotechnology', p. 48.

EUROPEAN COMMISSION (2008a), 'Food Industry'. Accessed at http://ec.europa.eu/enterprise/food/index_en.htm on 24 September 2008.

EUROPEAN COMMISSION (2008b), 'What is FP7?'. Accessed at http://cordis.europa.eu/fp7/faq_en.html on 24 September 2008.

EUROPEAN COMMISSION (2008c), 'High level group on the competitiveness of the agro-food industry', in Industry, E. a. (Ed.), *Progress Report.*, Brussels, Belgium, p. 20.

FEARNE, A. (2005), 'Justice in UK supermarket buyer-supplier relationships: an empirical analysis.', *International Journal of Retail and Distribution Management*, **23**(7), 7–16.

FEARNE, A. (2007), *Co-innovation for Sustainable Competitive Advantage*, Presentation in the abare Outlook 2007 conference, abare Economics.

FEARNE, A., HUGHES, D. and DUFFY, R. (2001), 'Concepts of collaboration: supply chain management in a global food industry', in *Food Supply Chain Management. Issues in the hospitality and retail sectors*, Eastham, J.F., Sharples, L. and Ball, S.D. (eds), Butterworth Heinemann, Oxford, UK, 55–89.

FOLKERTS, H. and KOEHORST, H. (1997), 'Challenges in international food supply chains: vertical co-ordination in the European agribusiness and food industries', *Supply Chain Management: An International Journal*, **2**(1), 11–14.

FOOD DIRECTORATE (2007), *Managing Health Claims for Foods in Canada*, Discussion Paper, Nutrition Evaluation Division, Health Canada, 124 pp.

GARNAUT CLIMATE CHANGE REVIEW WEBSITE (2008), Accessed at www.garnautreview.org.au on 23 January 2009.

GEORGHIOU, L., CAVE, J., CANTALLOPS, C. B., CALOGHIROU, Y., CORVERS, S. DALPÉ, R., EDLER, J., HORNBANGER, K., MABILE, M., NILSSON, H., O'LEARY, R., PIGA, G., TRONSLIN, P. and WARD, E. (2005), *Public Procurement for Research and Innovation – Developing Procurement Practices Favourable to R&D And Innovation*, Expert Group Report. Office for Official Publications of the EU, Luxembourg, 46 pp.

GOLDBLITH, S. A. (1989), 'Fifty years of progress in food science and technology: from art based on experience to technology based on science', *Food Technology*, **43**(9), 88.

GRUNERT, K. and VALLI, C. (2001), 'Designer-made meat and dairy products: consumer-led product development', *Livestock Production Science*, **72**, 83–98.

GRUNERT, K., HARMSEN, H., MEULENBERG, M., KUIPER, E., OTTOWITZ, T., DECLERK, F., TRAILL, B. and GÖRANSSON, G. (1995), *A framework for Analysing Innovation in the Food Sector*, Centre for Market Surveillance, Research and Strategy in the Food Sector, MAPP Working Paper, 33 pp.

GRUNERT, K. G., JENSEN, B. B., SONNE, A., BRUNSØ, K., BYRNE, D. V., CLAUSEN, C., FRIIS, A., HOLM, L., HYLDIG, G., KRISTENSEN, N. H., LETTL, C. and SCHOLDERER, J. (2008), 'User-oriented innovation in the food sector: Relevant streams of research and an agenda for future work', *Trends in Food Science and Technology*, 1–13.

HALL, D. C. and MOFFITT, L. J. (2002), 'Adoption and diffusion of sustainable food technology and policy', in *Economics of Pesticides, Sustainable Food Production and Organic Food Markets*, Hall, D. C. and Moffitt, L. J. (eds), JAI – An Imprint of Elsevier Science, Amsterdam, **4**, 3–18.

HATZICHRONOGLOU, T. (1996), *Globalisation and Competitiveness: Relevant Indicators*, STI Working Papers No 16, OECD, Paris p. 62.

HOLWEG, M., DISNEY, S., HOLMSTRÖM, J. and SMÅROS, J. (2005), 'Supply chain collaboration: making sense of the strategy continuum', *European Management Journal*, **23**(2), 170–81.

HOWARD PARTNERS (2005), *The Emerging Business of Knowledge Transfer*, Report for the Department of Education Science and Training Canberra, Australia, 141 pp.

INTERAGENCY WORKING GROUP ON MANUFACTURING R&D (2008), *Manufacturing the Future: Federal priorities for manufacturing R&D*, National Science And Technology Council, Committee On Technology (CT), 86 pp.

JONES, C. (1992), *Method of Preparing and Storing a Free Flowing, Frozen Alimentary Dairy Product*, USPT 5,126,156, United States.

JONES, M., JONES, C. and JONES, S. (2001), *Cryogenic Processor for Liquid Feed Preparation of a Free-Flowing Frozen Product and Method for Freezing Liquid Composition*, US patent application number 6223542, United States.

KEH, H. T. (1998) 'Technological innovations in grocery retailing : retrospect and prospect', *Technology in Society*, **20**, 195–209.

LANE, N. (2008), 'US science and technology: An uncoordinated system that seems to work', *Technology in Society*, **30**, 248–63.

LINNEMANN, A. R., BENNER, M., VERKERK, R. and VAN BOEKEL, M. A. J. S. (2006), 'Consumer-driven food product development', *Trends in Food Science and Technology*, **17**, 184–90.

LOWE, P., PHILLIPSON, J. and LEE, R. P. (2008), 'Socio-technical innovation for sustainable food chains: roles for social science', *Trends in Food Science and Technology*, **19**, 226–33.

MAF (2008), *New Zealand Fast Forward: Science, Food, Farms*, Establishment Group, Report on Proposed Governance Structure, p. 68.

MORGAN, K. and MURDOCH, J. (2000), 'Organic vs. conventional agriculture: knowledge, power and innovation in the food chain', *Geoforum*, **31**(15), 159–73.

MOSKOWITZ, H. and HARTMANN, J. (2008), 'Consumer research: creating a solid base for innovative strategies', *Trends in Food Science and Technology*, 1–9.

OECD (2008), *Dataset: Government Budget Appropriations or Outlays for R&D*, OECD Stat, Organisation for Economic Co-operation and Development.

PATON, R. A. and MACLAUGHLIN, S. (2008), 'Services innovation: Knowledge transfer and the supply chain.', *European Management Journal*, **26**, 77–83.

PATRICK, S. M. (2006), 'Federal support for new food and nutrition research: possibilities for the next farm bill', *Journal of the American Dietetic Association*, **106**(12), 1951–3.

PORTER, M. E. (1980), *Competitive Strategy*, The Free Press, New York, USA, 396 pp.

PRODUCTIVITY COMMISSION (2006), 'Public Support for Science and Innovation, chapter 1: Introduction. Research Report.', Canberra, Australia. p. 9.

ROBSON, S., FISHER, D., PALMER, A. and MCKINNON, A. (2007), 'Reducing the External Costs of the Domestic Transportation of Food by the Food Industry'. Report for the Department for Environment, Food and Rural Affairs. Faber Maunsell. 73 pp.

SARKAR, S. and COSTA, A. I. A. (2008), 'Dynamics of open innovation in the food industry', *Trends in Food Science and Technology*.

STANDING COMMITTEE ON INDUSTRY SCIENCE AND TECHNOLOGY (2007), *Manufacturing: Moving Forward – rising to the challenge*, Report of the Standing Committee on Industry, Science and Technology, House of Commons, Canada, p. 55.

ST JOHN'S INNOVATION CENTRE (2006), 'Innovative procurement – opportunity for new solutions'. Accesed at http://www.cambridgenetwork.co.uk/news/article/default.aspx?objid=16135 on 30 Sept 2008.

THE NETHERLANDS ORGANISATION FOR APPLIED SCIENTIFIC RESEARCH (TNO), 'Nightwind: storage of energy in coldstores'. Accessed at http://www.tno.nl/content.cfm?context=markten&content=case&laag1=176&item_id=247&Taal=2 on 20 Aug 2009.

URBAN, G. L. and HAUSER, J. R. (1993), *Design and Marketing of New Products*, Prentice-Hall, Englewood Cliffs, NJ.

US DEPARTMENT OF AGRICULTURE (2008), *Federal Register Notices*, pp. 50926–8.

US DEPARTMENT OF COMMERCE (2008), *Industry Report: Food Manufacturing NAICS 311*, 13 pp.

VAN DER VALK, W. and WYNSTRA, F. (2005), 'Supplier involvement in new product development in the food industry', *Industrial Marketing Management*, **34**(7), 681–94.

WATKINS, D. (2008), *Tesco: Network Re-design for Sustainable Growth*. Accessed at http://www.igd.com/index.asp?id=1&fid=1&sid=3&tid=41&folid=0&cid=40   on 29 June 2009.

WEST, M. (2008), 'Milking the consumer', *Business Day*, Fairfax Digital. Accessed at http://business.theage.com.au/business/milking-the-consumer-20080811–3thn.html on 12 Jan 2009.

WIJNANDS, J. H. M., VAN DER MEULEN, B. M. J. and POPPE, K. J. (2007), 'Competitiveness of the European Food Industry: An economic and legal assessment', European Commission, The Hague, The Netherlands, 320 pp.

WOOLWORTHS HOLDING LIMITED (WOOLWORTHS) (2009), *Sustainability*. Accessed at http://www.woolworthsholdings.co.za/sustainability/environment.asp   on   13 January 2009.

WOOLWORTH'S LIMITED (2008), *Corporate Responsibility Report*. Accessed at http://crreport08.woolworthslimited.com.au/summary.php on 14 January 2009.

Part II

Aligning supply and demand in food supply chains

6

Sales and operations planning for the food supply chain: case study

Ö. Yurt, Izmir University of Economics, Turkey, C. Mena and
G. Stevens, Cranfield University, UK

Abstract: The aim of this chapter is to examine sales and operations planning (S&OP) in the food and drink industry. The chapter highlights the impact on performance of the planning process and specifically S&OP, for food and drink companies. First an overview of S&OP and its historical evolution is presented. Then aims, trends, best practices and implementation processes of S&OP, together with its challenges are examined. The section that follows provides a roadmap for implementing S&OP in the food and drinks industry and discusses the critical aspects that should be evaluated during implementation. Finally, different aspects of S&OP in the food industry are discussed, based on the distinctive characteristics of the sector. The case study of a global leading company in the beverage sector is used throughout the chapter to exemplify key points.

Key words: sales and operations planning, food and drink supply chains, planning process.

6.1 What is sales and operations planning (S&OP)?

The aim of sales and operations planning (S&OP) is simply to maintain a balance between supply and demand. However, defining the entire S&OP process is more complicated since its use has evolved and a number of new tools and techniques have been included in this decision making process.

First of all, S&OP is not a discrete activity and it is not an ultimate destination. It is mainly the management of an ongoing process. The Aberdeen Group defined S&OP (2005, p.i) as: 'a set of business processes and technologies that enable an enterprise to effectively respond to demand and supply variability with timely determinations of the right market and supply chain mix, all through the S&OP horizon'. The S&OP process is affected

Fig. 6.1   S&OP: orchestrating the dynamic marketplace (adapted from Aberdeen
Group, 2005, p 3).

by a number of factors as illustrated in Fig. 6.1. These factors are the
elements of a dynamic marketplace which are orchestrated by the S&OP
process.

S&OP is an ongoing and dynamic process (Aberdeen Group, 2005;
Wallace and Stahl 2008) which gives companies the ability to make changes
very quickly and to determine the potential changes that may arise from
new market conditions. It provides a logic and structure for managing
change (Wallace and Stahl, 2008) and requires effective harmonisation and
synchronisation of different functions of the business. Accordingly, coordination of a number of different plans is needed. Ling and Goddard (1988)
refer to S&OP as the process for 'orchestrating the success' since such a
decision making process aims to provide congruence of several operations
and sub-processes. Most significantly, S&OP should be coordinated and
aligned with financial planning activity to provide a continuum.

S&OP requires continuous updates and both operating plans and strategic plans need to be checked and revised regularly. Therefore, periodic
meetings and management of the required changes are critical in S&OP.
The responsibilities of the key players in each department are to compare
the actual results of the plan, assess their performance and prepare an
updated plan for the current period. This process enables companies to
develop realistic goals for the following period which are the basis for business success (Ling and Goddard 1988). In this context, S&OP is a valuable
strategic management tool for executives.

Although the processes for managing demand and supply have the
common goal of providing business success and increasing business performance, lack of coordination between these processes is common. To solve

this problem, S&OP aligns demand and supply processes by adjusting and harmonising demand and supply plans.

Businesses usually face performance problems owing to their misaligned operations. Although they have totally integrated their supply chain and utilised supply chain management practices, they may still have considerable problems with the integration and alignment of functions and departments in their organisation. One of the most significant misalignments is between sales and marketing which deals with demand management and production, and logistics and purchasing which deals with supply management (Mentzer and Moon, 2004; Crum and Palmatier, 2004). Again S&OP is the potential solution to this problem.

Before discussing the evolution of S&OP, it should be noted that some confusion about its definition and boundaries still exists. Balancing demand and supply at the aggregate level is the basis of S&OP and this activity does not deal with specific details about products, customers and orders. However, the boundaries of the approach have broadened to include some of these details (Wallace and Stahl, 2008). This evolution has created much confusion about the definition and scope of S&OP. Wallace and Stahl (2008, p. xvii) proposed two different terms to reduce the confusion: '... Executive S&OP to refer the executive activity' and S&OP '... refer to the larger set of processes, which include forecasting and planning at the detail level as well as Executive S&OP'. In this chapter we adopt the executive S&OP concept which has also been called dynamic S&OP (e.g. Aberdeen Group, 2005).

6.1.1   Evolution of sales and operations planning

Through its evolution, S&OP has been given different names including Integrated Business Planning, Integrated Business Management, Integrated Performance Management, Rolling Business Planning and Regional Business Management (Coldrick *et al.*, 2003). The term S&OP process was first introduced in the late 1980s when it mainly focused on the operations function, the main variable was product volume and there was no link between the metrics used for S&OP and financial plans. In the 1990s, the benefits of S&OP started to gain visibility. The majority of companies that implemented S&OP in the 1990s have successfully reduced inventory levels, increased customer service levels and boosted profits as a result (Coldrick *et al.*, 2003). In recent years the focus has expanded to provide an operational ability to evaluate demand and ensure that the required resources were in place. It turned into an ongoing process which aligns demand and supply processes and harmonises different functions of the organisation. It has become an executive tool which works in coordination with financial plans and is operated in coordination with master production scheduling (MPS) and the operational plans of the company (Coldrick *et al.*, 2003; Wallace and Stahl, 2008).

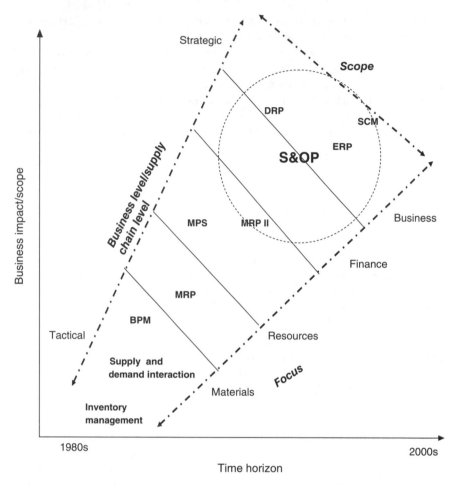

Fig. 6.2 Historical evolution of S&OP. More information on the concepts detailed here is provided in Russell and Taylor (2005).

S&OP has become a consolidation process of the multiple functions in a company as well as the synchronisation of that company with other supply chain members. Figure 6.2 depicts the evolution of sales and operations planning. In the 1980s, companies were managing their demand and supply processes by focusing on inventory. At that time, inventory management efforts aimed to determine the optimum inventory level, replenishment times and volumes, all of which are the basic elements of demand and supply management processes. In that period, the main focus was on materials and operations and these were being managed at a tactical level. Over time, inventory planning and management efforts, as well as supply and demand processes, have been embedded into business process management

(BPM) and materials requirements planning (MRP). However, these tools were only partially fulfilling the roles of S&OP. Through these approaches, the focus started to turn onto the whole resources from the particular materials while the business impact was becoming more operational than tactical.

In the 1990s, manufacturing resources planning (MRP II) and MPS processes started to be accepted as effective tools for planning production and inventory processes by the vast majority of companies. The business impact of those tools was approaching the strategic level. In the late 1990s, S&OP became an approach which is not far from our current understanding. It became an approach, a tool and an activity which affects the entire business and even other members of the supply chain. Accordingly, its focus has become mostly financial and covers the entire company. Today the scope of S&OP is still broadening.

6.1.2 Aims and objectives

Although the 'official' aim of S&OP is to balance supply and demand, companies embarking on S&OP usually have a number of specific objectives in mind. According to Wallace and Stahl (2008) the role of S&OP is to eliminate the disconnect between the strategic and business planning processes and the detailed plans. The linkage role of S&OP is illustrated in Fig. 6.3.

Managers engage in planning processes like S&OP with the aim of gaining benefits for their organisations. Ling and Goddard (1988) suggest that the following benefits can be obtained from a successful S&OP implementation:

- It helps departments work better together and it links company functions.

Fig. 6.3   Linkage role of S&OP (adapted from Wallace and Stahl, 2008, p 12).

- It provides a connection between the business plan with the operations of each department.
- It produces realistic plans.
- It reduces surprises and hidden decisions.

In order to exemplify the objectives of S&OP, we will use the case of the large multi-national food and drink company: The company, operating in over 25 countries with a turnover in excess of €6 billion in 2007, performs in a very dynamic market and over the past five years has been growing annually at double-digit rates. This has partly been achieved through geographical expansion, particularly in countries with high business growth rates such as Russia, but also through an increasingly complex product range with over 75 brands and several hundred stock keeping units (SKUs). The company has used S&OP over a number of years and when asked about the reasons for using the system, they provided a summary of the aims of their S&OP. Their views are summarised below:

- To have one set of numbers: 'We used to have multiple forecasts and plans and this was a recipe for disaster, creating conflicts and undermining performance' said one of the executives. The implementation of S&OP helped to ensure that everybody plans and organises activities based on the same set of numbers and that these numbers are agreed by everybody.
- To increase forecast accuracy: Forecasts are the centre of the planning process as they influence many decisions such as inventory levels, lot sizes and target service levels. It is acknowledged that some products are more difficult to forecast than others, but it is important to know where the problems lie and act on them as soon as possible. The S&OP process provides a framework for reviewing forecasts on a regular basis and taking targeted action.
- To work around constraints: One of the executives stated: 'Our facilities and infrastructure have constraints, and S&OP help us to consider all constraints before rolling out plans. This essentially produces more realistic plans'.
- To manage promotions effectively: Promotions are commonplace in the food and drinks industry, and achieving effective management of promotions is one of the central reasons for engaging in S&OP. The S&OP process ensures promotional activities are planned holistically, taking into consideration capacity and other constraints.
- To manage customer service: 'We aim for perfection in customer service, but we accept this is not always possible when you have a wide range of products like we do' said one of the managers. There is a perception that S&OP has helped to improve delivery and availability measures, but it also helped when making difficult decisions to manage service

levels in accordance with the reliability of forecasts and the impact on the business.

- To improve life cycle management: product introductions and withdrawals present some of the biggest challenges for supply chain managers. On the one hand, demand for new products is notoriously difficult to predict, since there is no history on which to base the initial forecast and this is coupled with other sources of uncertainty in production and supply. Withdrawals, on the other hand, can also leave the company with obsolete goods and raw materials if they are not properly managed and communicated. One of the managers said: 'S&OP has provided a forum to make life cycle decisions, making sure they are properly agreed and communicated and helping us to keep check on the complexity of our portfolio'.

- To manage cash: 'Managing cash is an important issue and we need to consider it when making supply and demand decisions. That is why our finance team is also involved in the S&OP process', stated an executive of the company. This helps to resolve cost/benefit trade-offs such as cost versus the benefit of certain fixed assets.

- To manage profitability and margin growth: 'In the end it all goes down to money. The ultimate goal of S&OP is to improve the profitability and market growth of the company', explained a manager of the company.

6.2   Trends in sales and operations planning and best practices

In today's marketplace, S&OP implementations and business practices are aimed at more effective demand and supply balancing, which is the fundamental goal of traditional S&OP. Thus, the characteristics of contemporary S&OP and traditional S&OP are not the same. Many of the characteristics of traditional S&OP have been replaced with new ones in today's dynamic S&OP. Table 6.1 illustrates these changes based on S&OP best practices.

Lessons learned from the best practices of S&OP have enabled other companies to benchmark their businesses and to manage their processes proactively. Aberdeen Group (2005, p. ii) determines the critical elements of best practices as:

'– Explicitly linking supply and inventories to demand dynamics,
– Contingency planning to shape demand and harmonize supply,
– Tightly managing the demand process and not just the numbers.'

Moreover, the following recommendations, which are based on the results mentioned in the studies of Aberdeen Group (2004 and 2005) and

Table 6.1 Changing characteristics of traditional and executive S&OP

Characteristics	Traditional	Executive S&OP
Timing	Quarterly	'Right time': weekly/daily Total time: approximately 18 months
Change	Discrete and incremental	Continuous (change management philosophy)
Objective	Volumetric Balance	Profitability
Approach	Single iterations	Multiple 'what-ifs'
Ideal capability	Responsiveness	Shaping
Organisational scope	Sales and operations	Enterprise + network partners (entire supply chain)
Predominant activity	Data gathering and cleaning	Dynamic decision making

Source: Adapted from Aberdeen Group, June 2005, p ii.

Muzumdar and Fontanella (2006), provide a roadmap for S&OP implementers based on best practice:

- Focus on your target market and positioning strategies. Do not try to dominate all segments.
- Focus on demand management processes first. Effective demand management provides a better management of other processes like supply management.
- Involve your supply chain members who play the role of business partner in the design of the S&OP processes.
- What-if analysis and contingency planning are two critical tools for a more effective S&OP.
- Responsiveness will be provided by the alignment of supply and demand processes.
- S&OP requires more than just 'planning'. Companies should test the validity of their planned assumptions and ask 'why' questions rather than 'what'.
- Generally, companies need less data than managers think they do during the S&OP process. Research results have shown that less than 10–15% of available data is vital to achieve the required results.
- Developing an 'outside-in' sequence of S&OP initiatives is necessary because uncontrollable factors have negative effects on the sales and operations planning processes.
- Managers need to focus on critical information, not just more data, since it has been indicated that less than 10–15% of all available data has a significant influence on business results.

6.3   Implementing sales and operations planning

Although the benefits of S&OP might be clear for many companies, effective implementation is a major challenge. Proper implementation of S&OP provides a linkage between tactical and strategic planning processes and helps to orchestrate all departments through horizontal and vertical communication. Accordingly, companies are able to formulate realistic plans, thus achieving their objectives. Furthermore, S&OP enables companies to coordinate their operations and financial processes.

There are still some companies which assume that implementing S&OP is only a series of meetings with the aim of balancing supply and demand. Actually the managers of such companies are implementing traditional S&OP, which offers an organised process which is not a revolutionary development since it only focuses on customer service and inventory levels. This can bring some benefits to organisations, but falls short of a full implementation of executive S&OP and could lead to poor overall performance of the process. Some symptoms of poor planning processes, such as S&OP, include lost sales, over-optimistic sales forecasts and low forecast accuracy, mismatch of available production and actual demand, limited flexibility to respond to changing market segment priorities, lack of collaboration with trading partners, poor on-time delivery owing to low level of inventory, and a lack of sustained focus on best market segment and supply partners (Aberdeen Group, 2005; IGD, 2007).

Companies which use S&OP usually have an adaptation process when they first start to implement it. Companies need to protect themselves from the above-mentioned symptoms by adapting and continuously improving S&OP practices. The evolution of S&OP is critical at a single company level as well, since it is an ongoing change management process. The vast majority of companies are not at the mature stage of S&OP implementation and they usually have a long way to go before they reach it. Insights and learning at the different stages of the maturity model of S&OP may provide sustainable benefits for the managers of these companies. A roadmap for S&OP implementation is depicted in Table 6.2 based on Lapide's (Lapide, 2005, p 14) 'S&OP maturity model', which presents the development of S&OP at the single business level. In this model, moving gradually from one stage to another is recommended for a successful S&OP process.

Different stages of the model are evaluated according to a number of key elements including people, technology, process and scope of focus which are critical to business success. A similar evolution process is also defined by Oliver Wight (2000) which provides a checklist for operational excellence. In this approach, companies are classified from A to D. Stage 4, a mature process of the model, is illustrated in Table 6.2 and represents a company which fits into group A of Oliver Wight's (2000) classification in terms of the S&OP process. To benefit completely from

Table 6.2 S&OP maturity model

Keys for business success	Stage 1 marginal process	Stage 2 rudimentary process	Stage 3 classic process	Stage 4 mature process
People	*Informal meetings:* • Sporadic scheduling	*Formal meetings:* • Routine schedule • Spotty attendance and participation	*Formal meetings:* • 100% attendance and participation	*Event-driven meetings:* • Scheduled when someone wants to consider a change or when a supply–demand imbalance is detected
Process	*Disjointed processes:* • Separate, disjointed demand plans • Supply plans not aligned to demand plans	*Interfaced processes:* • Demand plan reconciled • Supply plans aligned to demand plans	*Integrated processes:* • Demand and supply plans jointly aligned • External collaboration with limited number of suppliers	*Extended processes:* • Demand and supply plans aligned internally and eternally • External collaboration with most suppliers and customers
Technology	*Minimal technology-enablement:* • Multitude of spreadsheets	*Stand alone applications interfaced:* • Stand alone demand planning system • Stand alone multifacility advanced planning and scheduling system • Systems interfaced on a one-way basis	*Applications integrated:* • Demand planning packages and supply planning applications integrated • External information manually brought into the process	*Full set of integrated technologies:* • An advanced S&OP workbench • External-facing collaborative software integrated to internal demand–supply planning systems
Scope of focus	Tactical	Tactical–operational	Operational–strategic	Strategic

Source: Adapted from Lapide, 2005, p 14.

S&OP, companies should make an effort to reach group A of Oliver Wight's (2000) classification and/or stage 4 of Lapide's model (2005).

This is not a particularly easy process owing to the numerous challenges which are discussed in the following section. The basic requirement is having the insight of change management during S&OP implementation. However, many fundamental factors and infrastructure affect the success of change management including appropriate organisational culture, alignment of agenda, political environment, industrial environment, clarity of objectives, clarity of duties and responsibilities of each department during S&OP, and implementation and integrity of data.

The critical question for companies is 'How do you equip the organisation to implement S&OP?' The answer to this question is more important than the implementation process but the above-mentioned factors help companies to answer it. Actually Oliver Wight (2000) proposed a proven path, an ABCD checklist for companies, to help them to be aware of the requirements of being competitive and achieving world-class levels of performance. Regular use of the checklist can help companies evaluate their progress and their journey from D to A.

Ling and Goddard (1988, p 18–19) highlight five critical factors in deciding how to equip organisations before implementing S&OP:

'1. Each department must gain an understanding of the S&OP process
2. The company must commit the time and resources to the process
3. The company must define product groupings or families
4. The company must establish an adequate planning horizon
5. The company must establish and manage time fences'

One of the confusions about S&OP implementation is whether it should be implemented in individual products or at an aggregate level. By aggregate we mean product groupings or families. S&OP generally focuses on the entire business and it usually does not aim to achieve results at an individual product level. Actually it aims to maximise business profitability. Although it may change according to the industry, S&OP should be carried out at the aggregate level if forecasts based on product families provide sufficient information for S&OP (Lapide, 2002). Since planning at the SKU level would cause problems by the increasing volume and complexity of data as well as decreasing data accuracy, implementation of S&OP at an aggregate level is more practical and it provides more control by managers. However, grouping products into appropriate families is a critical task in the S&OP process and managers of different departments should achieve a consensus on the product families in terms of size and meaningfulness (Ling and Goddard, 1988).

The entire planning process includes different planning stages including operational planning, materials planning and S&OP, as illustrated in Fig. 6.4. These planning processes, which are significant for businesses, are interdependent and should be managed in a coordinated manner. It should

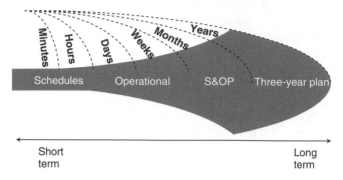

Fig. 6.4   Typical planning process.

be noted that the timescales of different stages in the planning process vary in different industries.

The best time for the implementation of the process is another source of confusion in S&OP implementation. The S&OP process generally takes 18–24 months or even longer, although longer and shorter planning periods are common (Lapide, 2002). Ultimately this period will depend on the size and complexity of the organisation.

Once implemented, the S&OP process repeats itself on a monthly basis as depicted in Fig. 6.5. The process starts by gathering the necessary sales and operations data and finishes with the executive meeting. Attendees of the meeting should include the president (general manager), directors of sales, marketing, supply chain, operations, product development, finance, logistics and human resources (Wallace and Stahl, 2008).

Many issues can affect the S&OP monthly process. Thus, coordinating and adjusting the different plans holistically will ensure the proper implementation of the S&OP process. To this extent, Lapide (2004) proposes to accept marketing and sales plans as rough-cut plans and to adjust them to the supply plans. The S&OP process does not recognise demand plans as fixed. Supply and demand management should be aligned along the entire process. Therefore demand planners and supply planners should first accept that they must work as a team with a common purpose. They must also accept that their plans can be changed and/or adjusted according to the other party for the purpose of a more effective S&OP process. Hence developing a basis for the consensus after eliminating the discrepancies between demand and supply plans is the main aim of S&OP meetings.

Accurate forecasting is the main input of S&OP. Although past data is a great source of forecasting, Wallace and Stahl (2008) proposed some factors that show how history is not the best predictor:

> . . . field input regarding large customers, potential new customers, new products, promotion plans, open bids, price changes, level of inventory in

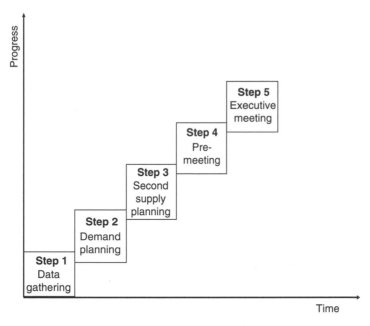

Fig. 6.5   Monthly S&OP process (adapted from Wallace and Stahl, 2005, p 54).

distribution, point of sales volume, competitive activity, industry and market dynamics, economic conditions, intra-company demand (from other business units within the company), and . . . a review of the forecast errors from the prior month.

Therefore, many factors including those mentioned above should be evaluated in an integrated manner during the forecasting process.

Another confusion within S&OP implementation is whether to use spreadsheets or sophisticated software to implement it. Accurate and timely information is essential for the S&OP process. Although a number of software packages support the S&OP process, many companies rely heavily on spreadsheets in their planning decisions. This is a great advantage for companies since S&OP does not require sophisticated, complex and expensive software. Also, in some cases, graphs are used to make spreadsheet data easier to understand. Using graphs is a good way to highlight the critical issues along the process and/or significant results of the process. On the other hand, relying on spreadsheets has some disadvantages such as the difficulty of integration and synchronisation of various spreadsheets (Supply Chain Digest, 2004).

When we focussed on the S&OP implementation process of our case study company, we were faced with a similar definition of S&OP implementation:

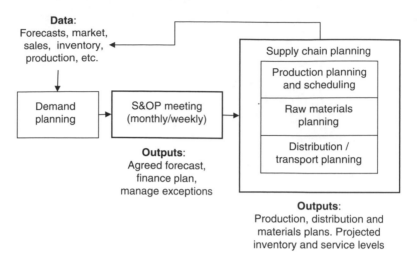

Fig. 6.6   The S&OP process in context.

A continuous process that follows a monthly cycle. The starting point is the collection of all the necessary data to create an initial demand plan. This includes forecasts, orders, market information, inventory, and some production information. This plan, prepared by the demand planning team, is circulated to all the attendees of the S&OP monthly meeting which includes the regional director as well as managers from sales, supply chain, finance and demand planning. The main outputs of the meeting are an agreed forecast and a financial plan and most of the discussion will focus on managing exceptions such as promotions, new product introductions and interruptions in demand, as well as assessing their financial implications. Once a forecast has been agreed, it is used to produce more specific plans for production, raw materials and distribution.

Figure 6.6 depicts the main flows of S&OP.

Managers of the company also stated: 'In addition to the monthly meeting, which is dedicated to tactical and strategic issues such as profitability and long-term investment, the company has weekly meetings involving smaller groups of people and are targeted at resolving day-to-day issues and addressing short-term changes in the plan'. Together with the managers of the company, we developed Fig. 6.7 which illustrates the sequence of events and the main milestones over a four-week cycle.

6.4   Challenges of sales and operations planning

Although there are many companies implementing or trying to implement S&OP, it is difficult to say that it is being done correctly. There are several challenges to implementation which are discussed below:

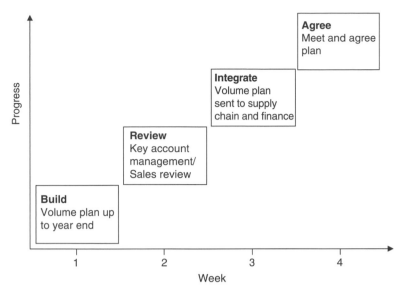

Fig. 6.7   S&OP milestones.

- Implementers' misunderstanding of S&OP: This is the most important challenge of S&OP. The manager who can understand the aims, processes, benefits and challenges of S&OP will be able to implement it and to gain the most from it. On the other hand, the S&OP process will not succeed if the implementers do not understand the process.

  As depicted in Fig. 6.4, planning processes should be managed in a coordinated manner. If a specific problem about a particular product or a supplier cannot be resolved in the first stages of the planning process, i.e. the operational planning and materials planning stages, then extra effort is needed during the S&OP stages to solve these problems, but such an implementation will obviously slow down the S&OP process.

- S&OP should be embedded in the entire business processes: If it is not, then it will be difficult to gain benefits from it. Most frequently, managers complain about not achieving positive results although they have had S&OP training.

  People need to have S&OP insight during implementation. Although it is possible to manage the S&OP process in a particular department, this is not sufficient. S&OP requires effective change management skills. However, if S&OP is not adopted by all parts of the company and if the process is not embedded into other business processes, then implementing will be more challenging.

- Mentally non-aligned staff: In many cases it is difficult to align the views of the marketing and operations functions since they are responsible for different parts of S&OP (i.e. supply management and demand management) and they tend to have different measures of performance. If their different perspectives cannot be aligned before starting to implement S&OP, it would obviously be difficult to benefit from S&OP.

- Management's bias toward a certain department: As mentioned before, S&OP deals with several functions in the company and the involvement of key members of these functions is essential. Accordingly, general management's balanced view toward each function is critical but providing that view is not an easy task (Ling and Goddard, 1988).
- Fear of detail (Ling and Goddard, 1988): Although S&OP requires aggregate data, in some cases the planning process needs some degree of back-up detail. Therefore, finding the balance between the volume of product family level data and individual SKU level data is necessary and sometimes a challenge to make the planning process work.
- Failure to involve all stakeholders (Lee, 2005): If management does not involve all stakeholders, including first line manufacturing supervisors, schedulers, salespeople, sales and marketing managers, engineers, plant managers and new product coordinators in the process, it will be difficult to achieve the expected results at the end of the process.
- The perception of S&OP as simply a software implementation project (Lee, 2005): Although it utilises software as well as spreadsheets, S&OP is an executive management and capability improvement tool.
- Recognition of S&OP only as a tool to improve planning and scheduling: S&OP plays a role as a significant decision support tool which provides critical insights for the future.
- Recognition of S&OP as a very costly implementation: It actually is not as expensive as managers may think. The main cost items regarding S&OP are training costs and consulting costs – if indeed a consultancy is required. So, although it is not free, it does not require a high investment. All it actually requires is people's time (Wallace and Stahl, 2008).
- Poor training: Another challenge proposed by Wallace and Stahl (2008) is insufficient education of users, insufficient discipline and conflict aversion in the company.
- The need for change in organisational culture: The S&OP process brings a new approach into the decision making process of companies. It needs a transformation and change process within the corporate culture (Lee, 2005). Systems approach and change management philosophy are the main prerequisites of S&OP implementation. Therefore, any lack of these prerequisites is a great challenge to S&OP.

During interviews in the case study a manager stated: 'Implementing and sustaining S&OP presents substantial challenges because it requires cross-functional collaboration and involves the continuous evaluation of trade-offs which can generate disagreement and possibly conflict'. Based on interview results we present some additional challenges for successful implementation of S&OP:

- Achieving cultural change: The effective use of S&OP requires a cultural change that can break the silo mentality in the organisation and allow people to work together for a common cause. The main problem of a silo mentality is that it leads to local optimisation; however, S&OP provides clarity about the global optimum and helps to resolve trade-offs. This requires a cultural change which involves a robust process, with structured regular meetings (such as monthly and weekly), a common set of key performance indicators (KPIs) as part of a balanced scorecard and, finally, appropriate training and information dissemination.
- Developing corporate capability: In many organisations, S&OP does not operate at a corporate level but only at a regional level, with each unit managing the process independently. Inevitably, practices vary from one country to the next and, while some countries exhibit very sophisticated practices, others are still developing. However, the case study company has used the recent implementation of an ERP system to homogenise practices and disseminate leading practices. This will also lead to more consistent reporting across all countries, based on common KPIs.
- Effective use of tools: S&OP requires the filtering and analysing of vast amounts of data and for this we need the appropriate analytical and information tools and trained users. The most important set of tools relates to forecasting, particularly for a company with a wide diversity of SKUs and very different characteristics, which require different approaches to forecasting. To decide the most appropriate approach to forecasting, products are classified according to their financial impact and demand variability. This classification framework is presented in Table 6.3.

6.5   Sales and operations planning in the food and drink industry

The food and drink supply chain has certain characteristics that make it different from any other industry. These characteristics need to be taken into consideration throughout the planning process and particularly in S&OP. In this section we will review the key characteristics of the industry and discuss their implications for S&OP.

- Volatility of demand: In the food and drink industry, companies face a major problem in consolidating and normalising demand. This is difficult since demand tends to be volatile in this sector caused by a variety of factors including promotions, weather fluctuations and rapidly changing customer preferences. This also makes forecasting difficult. Any

Table 6.3   Forecasting approaches for different products

high	• Reliable demand forecasting and high impact on business • Aim for low safety stock and high service levels • Use of statistical tools • Keep an eye on them	• Unreliable forecasting and high impact on business • Aim for high safety stock and modest service levels • Use statistical tools and work on improving forecast accuracy • High priority in S&OP review
Financial impact	• Reliable forecasting and low impact on business • Aim for low safety stock and high service levels • Use replenishment models • Plan with low risk, higher frequency and	• Unreliable forecasting and low financial impact • Aim for high stocks and modest service levels • Increase frequency of planning • Use replenishment models and statistical tools
low	large lot sizes	• Consider delisting
	low Demand variability *high*	

problem in demand management affects customer service negatively. To solve this problem companies increase the number of their back-orders. Usually, the cost of purchasing and transportation increases because of backorders. Companies tend to solve these problems through improved financial planning but a better solution is implementing an effective sales and operations plan (Wallace and Stahl, 2008).

• Supply volatility: Equally, supply volatility also affects planning processes in the food and drink industry. Unlike other industries, supply cycles change in short periods. This mostly affects product mix rather than volume and requires a more careful S&OP. In classical S&OP examples, demand is the driver and it is independent, whereas supply is dependent (Wallace and Stahl, 2008). However, in the vast majority of cases in the food and drink industry, supply is the key driver. In highly variable supply cases, supply determines what is going to be sold.

In one example of a food processing company, the size and amount of the crop to be harvested determines the sales volume of the processed food product. In such cases, the five steps of the S&OP process should be adopted (Fig. 6.5). In the new version of executive S&OP, the process begins with the initial supply planning phase (Fig. 6.8). Initial supply planning is usually the responsibility of the purchasing function. Predicted supply is passed to the sales and marketing department for the demand planning process. Then the new forecast, determined in the demand planning stage, is given to the

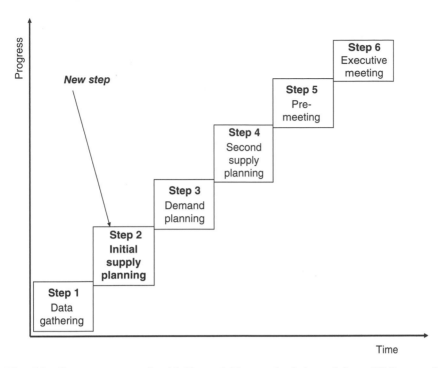

Fig. 6.8 Six-step process for highly variable supply (adapted from Wallace and Stahl, 2008, p 209).

second supply process. Following this, premeeting and executive meeting stages take place (Wallace and Stahl, 2008).

- Seasonality: This is a barrier in the balance of supply and demand and plays a critical role in the food and drink industry from both a demand and supply management aspect. Seasonal demand requires careful inventory management whereas seasonal supply necessitates multiple sourcing locations – generally in different regions. For products which are subject to seasonal supply such as olives or products that are subject to seasonal demands, such as ice cream, annual planning is required to have a perspective of the whole year. However, annual S&OP should be updated according to the periodically held cross-sectional meetings. It should be noted here that seasonality will not be a big risk if an effective S&OP process is implemented. In such cases, accurate forecasting will enable companies to manage seasonal fluctuations.
- Natural cycles: Although S&OP is a great tool for making fundamental changes and improvements in business processes, it is not easy to realise in the food and drink industry. For instance, in the poultry industry it is inevitable that short-term based supply and demand alignments will be

made. Producers must manage their processes according to predetermined rules. However, after the first stages of the production, adjustment of volumes of different types of end products such as fresh, frozen and chilled can be possible. This implementation is basically related to product mix rather than product volume.

- Short lead times: Companies need to react quickly to changes in demand in the food and drink industry. Therefore, lead time tends to be short, usually between 24 and 48 hours, which is a critical input of S&OP.
- Large number of SKUs: In the food and drink industry, there are high numbers of SKUs and, accordingly, product families. This characteristic of the sector requires different approaches to forecasting. The appropriate classification of SKUs according to their financial impact and demand variability is essential for effective forecasting and S&OP.
- Products with short shelf life: Food and drink products tend to have short shelf life. Although there are different types of products in this industry, a large proportion of food and drink products are perishable. This characteristic of the sector increases the importance of the balance between demand and supply which can only be realised by successful S&OP.
- Promotion intensive: Food and drink companies frequently face customer service challenges and high levels of forecast errors. This is partly caused by the promotion intensive nature of the industry. Therefore S&OP processes are even more significant for these companies. Companies need a more accurate demand forecasting system, especially during promotion periods. In such cases, S&OP terms should be reduced to weekly periods although the longer term S&OP period continues. For this reason, horizons for planning updates have to be shorter in this sector. Therefore, S&OP should be managed in at tactical as well as at a strategic level.

In the light of these characteristics, if the food and drink industry is compared to other industries such as automotive, electronics and chemicals, a number of differences in S&OP are visible. The planning process is more stable in sectors such as chemicals and pharmaceuticals which require long horizon stability. Therefore, companies in these industries will not require the same levels of agility that food and drink companies need. As a result, in the food and drink industry the planning horizon tends to be shorter when compared to other industries and MPS tends to be even more critical than S&OP. It should be noted that MPS is more important for short shelf life products, although S&OP is more critical for long shelf life products.

One of the best practice examples was identified by the Aberdeen Group (2005) as Campbell Soups – a global manufacturer and marketer in the food and drink industry. Their main products of Campbell Soups include soups, sauces, beverages, biscuits, confectionery and prepared foods.

Campbells operates in an environment in which promotion activities are usual. During these promotion periods, managing customer service and optimising inventory levels have been the great challenges for Campbells. Promotions aim to decrease prices and to increase sales. This environment requires a higher level of inventory turnover and immediate replenishment. The target forecast accuracy level was 75%+; however, it was found to be impossible for the company to achieve such a target. Campbells solved this problem by using a new technology solution which allowed them to reduce weekly forecast error to less than 25%. The Aberdeen Group (2005) reported the Campbells case as follows:

> ...Campbell learned that the technologies necessary to enable real-time forecasting for immediate replenishment as necessarily those that are effective for medium term planning purposes and that it is critical not to take a 'one forecasting technology does all' approach. They also determined that it was possible and very important to accomplish this while seamlessly integrating with their existing powerful suite of planning applications. Finally, the company proved that individual best intuitive 'guesses' were no match for a formal decision making process, enabled by planning and forecasting models capable of generating multiple 'what-if' scenarios and simulated outcomes.

6.6 Summary and conclusions

In this chapter we have highlighted the significance of S&OP in the food and drink industry. Better planning processes, as well as balancing supply and demand with the coordination and integration of all related functions in a company, can lead to favourable outputs such as lower costs, lower stock, better service, increased sales and improved forecast accuracy. We propose a roadmap for good practice of S&OP implementation in the food and drink industry by placing emphasis on the main characteristics, challenges and trends in S&OP in this sector. We aim to capture the attention of current and future implementers of S&OP in this industry and to provide them with a framework to facilitate the difficult process of adopting S&OP.

6.7 References

ABERDEEN GROUP (2004), *The Sales and Operations Planning Benchmark Report: Leveraging S&OP for Competitive Advantage*, Boston, MA.

ABERDEEN GROUP (2005), *Best Practices in S&OP–A Benchmark Report*, Boston, MA.

COLDRICK A, LING D and TURNER C (2003), *Evolution of Sales & Operations Planning–from Production Planning to Integrated Decision Making*, Gloucester, Strata Bridge Report (http://www.stratabridge.com/news/sept03_s_op_evolution.pdf, accessed 19th May, 2009).

CRUM C and PALMATIER G E (2004), 'Demand collaboration: What's holding us back', *Supply Chain Management Review*, **8**(1), 54–61.

IGD SUPPLY CHAIN ANALYSIS RESEARCH (2007), *Demand Planning*, Institute of Grocery Distribution, Watford, UK.

LAPIDE L (2002), 'New developments in business forecasting', *Journal of Business Forecasting Methods and Systems*, **21**(2), 11–14.

LAPIDE L (2004), 'Sales and operations planning Part 1: the process', *The Journal of Business Forecasting*, **23**(3), 17–19.

LAPIDE L (2005), 'Sales and operations planning part III: A diagnostic model', *The Journal of Business Forecasting*, **25**(2), 14–16.

LEE J (2005), *S&OP: It's Not A Meeting–It's a Culture Change*, Supply Chain Consultants, Wilmington, DE (http://www.supplychain.com/Downloads/SOPCulture Change.pdf, accessed 2nd May, 2009).

LING C and GODDARD W (1988), *Orchestrating Success: Improve control of the business with sales and operations planning*, John Wiley & Sons, New York.

MENTZER J and MOON M (2004), 'Understanding demand', *Supply Chain Management Review*, **8**(4), 38–45.

MUZUMDAR M and FONTANELLA J (2006), 'The secrets to S&OP success', *Supply Chain Management Review*, **10**(3), 34–41.

RUSSELL R S and TAYLOR B W (2005), *Operations Management: Quality and competitiveness in a global environment*, 5th revised edition, John Wiley and Sons, New York.

SUPPLY CHAIN DIGEST JUNE (2004), *Sales and Operations in Complex Discrete Manufacturers – A Research Report on Challenges and Opportunities*, Springboro, OH (http://www.scdigest.com/assets/reps/SCDigest_Demand_Mgmt_Discrete_Mfg_ Report.pdf, accessed 21st May 2009)

THE OLIVER WIGHT (2000), *The Oliver Wight ABCD Checklist for Operational Excellence*, 5th edition, John Wiley & Sons, New York.

WALLACE T F and STAHL B (2008), *Sales & Operations Planning: The How-to Book*, 3rd edition, TF Wallace and Company, USA.

7

Food supply chain planning, auditing and performance analysis

A. J. Thomas, University of Wales Newport, UK, Y. Wang and A. Potter, Cardiff University, UK

Abstract: The food supply chain has faced significant uncertainties over recent years leading to a heightened interest in effective supply chain management within the industry. By considering the four key systems inputs integral to a supply chain (those of demand, supply, process and control) a company can take action first to stabilise supply chain performance and second to improve its effectiveness and operational capabilities by reducing uncertainty within the chain. This chapter outlines the key drivers and uncertainties that exist within a food supply chain and describes a structured supply chain planning, auditing and performance diagnostic approach (quick scan audit methodology) that can be used to measure supply chain effectiveness. By measuring and assessing the symptoms of complex information and material flow it is possible to identify the root causes of supply chain uncertainty. The chapter describes how these uncertainties can be reduced and systems performance improved through effective business systems re-engineering methods. The chapter also describes the understand, document, simplify and optimise (UDSO) concept. It shows how UDSO enables the supply chain to be simplified and optimised before reviewing a number of approaches to supply chain planning used to improve supply chain systems effectiveness. The approaches described are efficient consumer response (ECR), collaborative forecasting, planning and replenishment (CPFR) and vendor managed inventory (VMI). These approaches encourage collaboration along the supply chain leading to improved supply chain performance.

Key words: uncertainty, supply chain auditing, business systems re-engineering, ECR, CPFR, VMI.

7.1 Introduction

The food supply chain has faced significant uncertainties over recent years and this has heightened the interest in effective supply chain management within the industry. Major events such as the BSE crisis in the UK, foot and mouth in Europe and the spread of swine fever in Western Europe have had a significant affect on the red meat industry over the last ten years or so. It was the effective control and management of its supply chain in ensuring tracking and tracing of animal movement during this time that played an important role in ensuring that the spread of these diseases was contained and finally destroyed (Van Der Vorst and Beulens, 2002).

From a business and political perspective, the introduction of the Common Agricultural Policy has served in some cases to discourage production efficiencies and the development of a competitive edge amongst suppliers (Corbett, 1992). This, together with increased competition amongst retailers, especially supermarkets, have increased the level of supply and demand uncertainty within traditional food supply chains. Supermarket retailers have been the major driver of global sourcing of food products, resulting in an increase in the number and diversity of food supply channels. This, in turn, has forced companies to address their supply chain strategies in order to survive in an increasingly complex environment.

From an economic viewpoint, the ability of a company to survive during times of uncertainty depends upon its ability to work as part of an efficient, integrated, collaborative supply chain system, rather than as an independent entity. Using the strength and power of a collaborative supply chain to compete, rather than work in isolation, was highlighted by Christopher (1998). This is particularly relevant for food supply chains because of the shelf life constraints of products, increased consumer demands for safe and animal welfare-based processing methods, the pressure to reduce food miles and a general industry requirement to deliver food products on a just-in-time basis.

The benefits of working within a collaborative supply chain are many; however, managers operating within such systems need to be aware that there is a risk that actions taken to improve the performance of individual companies can have a detrimental effect on other companies within the chain. Therefore, an integrated and holistic approach to the control and management of the supply chain is needed and effective collaboration between the players within the supply chain is essential.

It is clear that the management of collaborative food supply chains is becoming increasingly more complex and when working as part of a collaborative supply chain system the strength of such a supply chain is only as good as its weakest link. The overall performance of each company within the supply chain needs to be ascertained so that poorly performing companies within the chain are identified and assisted in order to strengthen overall supply chain capabilities. A robust and rigorous method of supply chain auditing is required to review individual company performance and assess how effectively it contributes to the overall supply chain system.

The aim of this chapter, therefore, is to present a method for assessing uncertainty within food supply chains and measuring individual company performance. To assist in this process, a structured supply chain planning, auditing and performance diagnostic approach, quick scan auditing methodology (QSAM), is presented. Later the chapter reviews a number of approaches to supply chain planning which enables companies to integrate and perform at their optimum levels of performance.

7.2 Food supply chain uncertainty

The uncertainty circle model is a convenient way of categorising the disturbances that may be encountered by a company in a food supply chain. For example, a company may find uncertainties associated with:

- erratic, frequent and problematic downtime of its machinery and equipment thus affecting processing performance
- changing customer schedules leading to increased demand disturbance
- poor supplier delivery performance leading to unreliable production plans
- unstable process control features which inhibit an optimum way of working.

Understanding which of the four areas causes the greatest uncertainty enables the company to prioritise its resources and focus on improving the areas which provide the greatest leverage. These four areas of uncertainty, identified as process, supply, demand and control are shown in Fig. 7.1 in the form of an 'uncertainty circle. The elements of the circle are directly taken from a traditional manufacturing system but can be applied effectively to look at food supply chains with the same effect.

The uncertainty circle method is based on the control of a company's internal processes in responding to the effects of customer demand and in turn the ability of the company's internal processes to place orders with their suppliers (Towill, 2006). The uncertainty circle relating to a standard red meat supply chain is shown in Fig. 7.2.

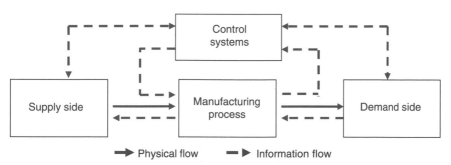

Fig. 7.1   The uncertainty circle (from Mason-Jones and Towill, 1998).

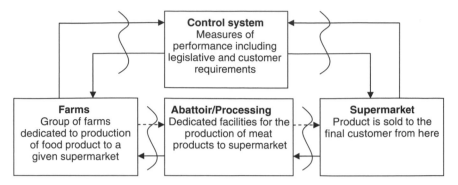

Fig. 7.2 Typical elements of a meat supply chain (adapted from Taylor, 2006).

Here the typical uncertainty drivers can be defined for each area of the circle. For instance, process uncertainty can be affected because the company's own processes are not yielding the required product on time. This could come from poorly performing slaughtering equipment leading to excessive downtime and maintenance, poorly performing or inadequate refrigeration systems causing product to be discarded and so on.

Supplier interface uncertainty results from the supply chain's inability to cope with the requested demand patterns as well as the effects caused by poorly performing suppliers within the chain. Demand pattern uncertainty can come from a significant increase (or drop) in supermarket demand for a product (which is outside the planned demand increases experienced caused by seasonal demand patterns). Disease outbreaks or product recalls from supermarkets are typical uncertainty drivers, as are issues surrounding large scale price and product promotions, especially if certain products are thought to improve health and vitality. Supplier performance can also play a major role in creating uncertainty in the supply system. This can be through poorly performing machinery and equipment at farms and abattoirs which have an adverse impact upon the quality and delivery of products, poor adherence to hygiene and quality procedures rendering product unsuitable for further processing and so on.

Demand interface uncertainty is compounded by a lack of accurate tracking of customer requirements by volumes and frequency of delivery or players in the system generating 'rogue' orders for price leverage. The uncertainty drivers here are very similar to the supply chain uncertainty where the outbreak of disease and the subsequent containment controls result in a complete removal of a particular product from a supermarket's product portfolio until such time as the product is considered suitable for human consumption. Finally, further uncertainty is induced by poor system controls based, for example, on the use of wrong decision rules regarding anticipation of customer orders or the lack of logistics control which can render products out of date and no longer saleable. As various business improvement programmes are successfully implemented, uncertainty will be reduced resulting in a more effective and focused supply chain capable of moving towards a seamless supply performance (Childerhouse *et al.*, 2004).

In seeking to establish the degree of uncertainty in an individual value stream, complex material flow is a primary lead indicator (Towill, 1999). It is possible to produce checklists to analyse supply chain behaviour based around four groups of complex material flow symptoms. These groups are termed:

- dynamic behaviour
- physical situation
- operational characteristics and
- organisational characteristics.

Table 7.1 shows the elements contained in each group. The presence of uncertainty will result in the occurrence of a high frequency of these complex material flow symptoms. This grouping allows the checklists to be used to identify the symptoms clearly, leading to the identification of short, medium and long-term improvement opportunities. Codification via Likert Scales allows the statistical relationship between these four groups of uncertainty symptoms and the four uncertainty circle segments to be evaluated (Childerhouse and Towill, 2003).

A smooth well controlled material flow system lies at the heart of best practice supply chain management. By identifying shortfalls in material flow it is possible to highlight those areas most in need of re-engineering. A set of good practice, simplicity rules for improving material flow has been devised. These rules are shown in Table 7.2.

These 12 rules point the way forward to smoothing material flow throughout the chain. They are an amalgam of the principles of material planning and control. The philosophy associated with the twelve rules is simple yet effective, reflecting the need to design out problems at source rather than applying complex controls to an already complex system.

The main features of a number of key business concepts such as just in time (JIT), lean flow systems and supply chain management have been extracted, analysed and distilled to create the 12. The rules in turn provide a simple set of guidelines which can be used for auditing the supply chain dynamics of a given business system. If material flow is overly complex numerous symptoms become clearly visible, resulting in ineffective product delivery process performance. Towill (1999) identified 24 symptoms that can be categorised into dynamic, physical, organisational and process characteristics. All can be observed either physically or by analysing numerical data and/or written communication within the chain. The 'digital nature' of the results obtained allows consistency to be built up between different parties auditing the same value stream. The 12 simplicity rules provide a complete set of guidelines that can be used to simplify material flows.

Table 7.1   Four classes of symptoms observed in complex material flow
(from Towill, 1999)

Class of symptoms	Symptoms observed in complex material flow
Dynamic behaviour	Systems-induced behaviour observed in demand patterns System behaviour often unexpected and counterintuitive Causal relationships often separated geographically Excessive demand amplifications as orders are passed upstream Rogue orders induced by system 'players' Poor and variable customer service levels
Physical situation	Large and increasing number of products per pound turnover High labour content Multiple production and distribution points Large pools of inventory throughout the system Complicated material flow patterns Poor stores control
Operational characteristics	Shop floor decisions based on batch-and-queue 'Interference' between competing value streams Causal relationships often well separated in time Failure to synchronise all orders and acquisitions Failure to compress lead times Variable performance in response to similar order patterns
Organisational characteristics	Decision making by functional groups Excessive quality inspection Multiple independent information systems Overheads and indirect costs allocated across product groups and not by activity Excessive layers of management between the CEO and the shop floor Bureaucratic and lengthy decision-making process

7.3   Understand, document, simplify and optimise (UDSO)

In reducing uncertainty within food supply chains, the business systems re-engineering approach (UDSO) suggested by Watson (1994) provides a suitable framework for managers today. A number of generic methodologies exist to ensure a consistent approach is adopted from problem definition through to the implementation and operation of a solution. The UDSO method allows supply chain performance to be assessed using a clear and systematic method of analysis. It combines both qualitative and quantitative forms of information and data to describe and subsequently solve a given systems-based problem.

Table 7.2   The 12 simplicity rules (from Towill, 1999)

Rule 1	Only process and supply products that can be quickly despatched and invoiced to customers
Rule 2	Only process in one time bucket those products needed for processing in the next
Rule 3	Streamline material flow and minimise throughput time
Rule 4	Reduce information lead times via the use of the shortest planning periods
Rule 5	Only take deliveries from suppliers in small batches as and when needed for processing and assembly
Rule 6	Synchronise time buckets throughout the chain
Rule 7	Form natural clusters of products and design processes appropriate to each value stream so that diverse customer requirements can be best served
Rule 8	Eliminate all uncertainties in all processes
Rule 9	Develop a structured approach to change. Understand, document, simplify, optimise (UDSO)
Rule 10	Have highly visible and streamlined information flows
Rule 11	Use proven and robust decision support systems in the management of the supply chain
Rule 12	The operational target of the seamless supply chain needs to be commonly accepted and shared by all members of the change team

The UDSO stages are:

Understand – define the problem, system boundaries and performance metrics for the supply chain under review.

Document – model the flows through a system and highlight opportunities for solutions.

Simplify – eliminate waste from companies within the supply chain and the interconnections between the companies, focusing on opportunities to create seamless interfaces across functional boundaries (Towill, 1997).

Optimise – only once the processes have been identified and streamlined should 'sophisticated' methods of control be applied to ensure their consistency and reliability (Mason-Jones and Towill, 1998). Methods include computer simulation, experimental design, network analysis and statistical process control. Without the preceding three stages of re-engineering, there is a risk of developing complex management systems.

An auditing approach can be followed to understand and document the supply chain. Supply chain auditing enables an objective third party assessment to be made of the performance of a supply chain. In addition, it highlights best practice within the chain enabling measures to be implemented. There is a range of different supply chain auditing tools available;

see Foggin *et al.* (2004) for a review of some of these. Quick scan audit methodology is outlined below.

7.4   Quick scan audit methodology (QSAM)

The quick scan audit methodology (QSAM) is a diagnostic tool used to analyse the performance of a company's supply chain. The term itself was first coined by the Eindhoven University of Technology for a technique used by their postgraduate students (Lewis *et al.*, 1998). However, it has since been developed into a robust diagnostic tool through the work of the Logistics Systems Dynamics Group at Cardiff University. The aim is to 'understand and document the supply chain…and to identify quick hit (not quick fix) improvements and longer-term strategic action plans' (Lewis *et al.*, 1998). This definition highlights three elements: recording processes within the supply chain, analysing current behaviour and providing ideas for improvements. Although initially developed for the automotive sector, QSAM has been applied to food supply chains within the UK, Thailand and New Zealand.

The first stage in QSAM is to identify a suitable supply chain. Securing 'buy in' from the people who will be involved is essential in order to facilitate the data collection and ownership of the solution. At this stage the scope of the study is refined, particularly through the identification of which products to monitor through the data analysis phase. A detailed 'walk through' of company operations should also be undertaken in order to contextualise the questionnaire information and interviews.

Techniques that are used for data collection (Naim *et al.*, 2002) include:

- Process mapping – This technique allows the supply chain to be documented clearly, in a language understandable by all. There are a wide range of techniques that can be used. Value stream mapping is a simple way of illustrating a supply chain on a single sheet of paper (Rother and Shook, 1998).
- Data analysis – Having documented the supply chain, historic data is collected and analysed to provide a greater understanding of its behaviour, as well as allowing other tools such as time series analysis to be used.
- Questionnaires – These provide a starting point for the QSAM and are used to gain an overview of the business, information on the staff profile and relationships both within the organisation and with customers and suppliers (Lewis *et al.*, 1998). Ideally, the questionnaires should be issued at the launch meeting so that they can be completed by the time the data collection starts.
- Semi-structured interviews – This final technique should provide more depth to the study and cover similar areas to the questionnaires (Naim

et al., 2002). Interviews are also used to answer any queries that arise during data analysis.

The data collection phase normally takes two days. However, this is flexible and will change depending upon the scope of the study and the resources available. Once completed, the next step is to analyse the data further, to identify causal relationships and potential areas for improvement.

Whilst QSAM focuses on the main elements of the uncertainty circle, the interactions that exist between the elements are also of major concern. QSAM includes the study of the logistical systems including transportation of raw materials between the supplier and processing plant as well as the transportation and logistics strategies employed to deliver the product to the customer.

From the QSAM findings, a targeted list of improvement opportunities can be produced. These will help to simplify and optimise the supply chain. In simplifying the supply chain, there has been much activity within the UK relating to the application of lean thinking and other business process re-engineering approaches throughout the food supply chain (Simons and Taylor, 2007; Simons *et al.*, 2008).

7.5   Application of quick scan audit methodology in the food industry

The following section provides a comparative analysis of the performance of three supply chain companies operating within the food processing industry. Two of the companies were involved in red meat processing (companies A and B) whilst the other was a dairy company (company C).

Figure 7.3 shows the QSAM uncertainty scores obtained for each company. It can be seen that the dairy company had a lower overall systems uncertainty score whereas the red meat processing companies performed less well. Upon analysing the uncertainty scores further, it is possible to see that the process uncertainty is better for the red meat companies compared with the dairy company (this is shown in Fig. 7.3 where the process uncertainty element of the red meat companies is smaller compared to the uncertainty element for the dairy company). Likewise, the supply, demand and control scores for the red meat companies were not as good as the dairy company.

Through closer and more detailed analysis of the processing capabilities of each company it is possible to put the QSAM scores into context. For instance, the dairy company operates using state of the art automated processing equipment whereas the red meat companies are more labour intensive in nature using relatively basic processing equipment. As a result, the process uncertainty scores for the red meat processors are lower since their basic machinery and tooling is very reliable, requiring little maintenance.

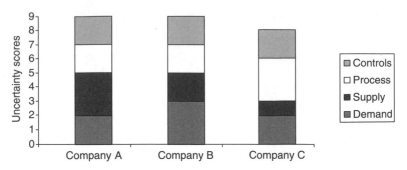

Fig. 7.3 QSAM uncertainty scores for three food processing companies.

Being more labour intensive in nature, the red meat processors are able to adjust labour levels quickly to cater for fluctuations in customer demand patterns. In this situation the red meat processors are essentially more agile and responsive to demand disturbances.

This section shows how QSAM provides a basis for comparing the performance of companies against a series of best practice production norms. It enables companies to identify quickly where the greatest levels of uncertainty lie, allowing the company to mobilise resouces to attack areas of poor performance.

7.6   Optimisation to reduce supply chain uncertainty

There are a number of different initiatives that have been taken by companies within the food supply chain to optimise operations. Many of these are focused upon the information flows and how shared information can be used to improve supply chain performance.

7.6.1   Efficient consumer response (ECR)

Efficient consumer response (ECR) was developed in the grocery industry in the early 1990s. It was believed that companies can serve consumers better, faster and at less cost by working together with trading partners. The core objective of ECR is to increase consumer value while optimising the supply chain and by doing so, generate profitable growth for the trading partners involved. It is estimated that from 1995 to 2005, 3.6% of consumer sales value was saved through successful ECR implementation across the Western Europe through reductions in operating cost (Hofstetter and Jones, 2005). The study also suggested that today's top tier ECR adopters enjoy 6% better service levels, 5% higher on-shelf availability and ten days

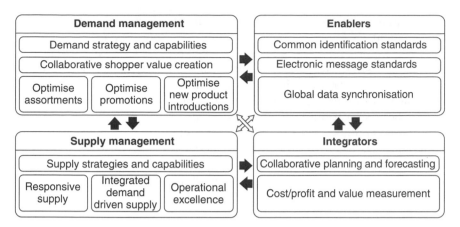

Fig. 7.4 Four key areas of ECR (from www.ecrnet.org, 2008).

fewer finished goods inventories than low or non-adopters of ECR practices.

The focus of ECR is to integrate supply chain management with demand management. There are four focus areas under ECR. These areas are broken down further into core and advanced improvement concepts as shown in Fig. 7.4.

Demand management is at the heart of ECR. Careful management of product category and promotional efficiency for both existing and new products helps retailers to maintain a high level of product availability for consumers and to improve their relationship with suppliers. Supply management, on the other hand, aims to fulfil customers/consumers' demand through the configuration of effective supply systems. It needs to be demand driven and responsive in order to cope with the ever-changing needs of customers. At the same time, there are constant pressures for both retailers and suppliers to pursue 'operational excellence' so that the cost of supply is under control and a reasonable profit margin can be retained while they pass on increased value to the customers.

A collaborative approach is often encouraged between retailers and suppliers to implement various initiatives for joint benefit. For example, during 2006, Mills and ICA Norway worked collaboratively on a Shelf-Ready-Package solution for Soft Flora Light in order to reduce replenishment costs and increase sales through improved product visibility. This collaborative initiative was implemented in all Norwegian stores and led to significant productivity savings of approximately €375 000 (ECR Europe and Accenture, 2006). Many retailers have benefited by deploying sophisticated technology such as extranets or portals to share their point of sales (POS) data with manufacturers. For example, Nestlé France uses POS data provided by its retail partners to analyse the performance of its products in the area of efficient assortment, leading to more effective category management.

A key enabler for ECR is the use of technology, such as electronic data interchange (EDI) for effective data transfer and information sharing. Having accurate and consistent product master data is fundamental to allowing many other ECR practices to take place and allows effective integration between various business partners along the value chain. Capturing true cost and profits is the 'bridge between supply and demand, enabling the underlying strategies to be optimally linked and the necessary resources to be coherently managed' (globalscorecard.net, 2008). Having a uniform value measurement system across companies will also help to assess any benefits or costs resulting from the introduction of various ECR initiatives.

Implementing ECR invariably requires the business processes to be redesigned if retailers and suppliers are to work in collaboration. However, a number of factors act as barriers to effective collaboration. For instance, intensifying competition between trading partners results in a fear of information sharing. There are also concerns regarding the balance of costs and benefits between participants, as retailers are normally in a more powerful position and may not pass on the benefits obtained through ECR practices.

7.6.2   Collaborative planning, forecasting and replenishment (CPFR)

Collaborative planning, forecasting and replenishment (CPFR) originated in 1995, when a pilot project between Wal-Mart, Warner-Lambert, Benchmarking partners SAP and Manugistics took place, to develop a business model aimed at forecasting and replenishing inventory collaboratively (Danese, 2007). The initiative quickly became popular as organisations saw the potential benefits of eliminating demand and supply uncertainties through improved communications between supply chain partners. Although the methodology is applicable to any industry, CPFR applications to date have largely focused on the food, apparel and general merchandise sectors.

ECR Europe (2008) defines CPFR, as a cross-industry initiative designed to improve the relationship between retailers, manufacturers and their suppliers through comanaged planning processes and information sharing. By improving the collaboration between trade partners, customer service is expected to be improved while inventory management is made more efficient. A more technology-oriented definition is provided by Voluntary Inter-industry Commerce Standards (VICS, 2002): 'CPFR is a collection of new business practices that leverage the internet and electronic data interchange in order to radically reduce inventories and expenses while improving customer service'.

CPFR, defined by VICS (2002), is based on a nine-step process model consisting of:

(1) developing a collaboration agreement
(2) creating a joint business plan
(3) creating a sales forecast
(4) identifying exceptions for a sales forecast
(5) resolving/collaborating on exception items
(6) creating an order forecast
(7) identifying exceptions for an order forecast
(8) resolving/collaborating on exception items and
(9) generating orders.

Skjoett-Larsen *et al.* (2003) further classified CPFR into three types: basic, developed and advanced, according to the various degrees of information sharing and process integration between organisations. Their main characteristics have been summarised in Table 7.3.

Basic CPFR is appropriate where the supply chain members want to have a transaction cost approach (TCA) to collaboration and only a few business processes need to be integrated. Developed CPFR is usually adopted when two organisations decide to share more information and have more areas to integrate.

Advanced CPFR expands to coordinate processes within forecasting, replenishment and planning. For instance, the planning processes involve collaboration on production planning, product development, transport planning and marketing activities. It may also be that CPFR only occurs at certain times of the year, or for certain products. For example, in planning

Table 7.3 Three types of CPFR

Dimensions	Basic CPFR	Developed CPFR	Advanced CPFR
Shared Information	Sales orders and confirmation Inventory data	Demand data Order planning data Promotion data Production data	Demand data Order planning data Promotion data Production data
Degree of discussion	No	Some	Frequently
Coordination / synchronisation	No	Some	All activities
Competence development	No	No	Knowledge
Evaluation	No	No	Experiences
Type of relationship	Transactional	Information sharing	Mutual learning
Theoretical explanation	TCA	Network	Resource and competence-based

Source: Skjoett-Larsen *et al.* 2003.

and managing promotions, Tesco and Nestlé share significant amounts of information to ensure that stock is available in the retail outlets for the whole promotion period. For products not under promotion, additional real time information is provided to Nestlé from Tesco (through an intranet), but there is no collaborative planning.

According to Hill (1999), when applied to Listerine products, CPFR has led to improved in-stock availability from 87% to 98% and a reduced lead time from 21 to 11 days. Wal-Mart, the pioneer of CPRF, together with Sara Lee reported an 18% reduction in inventory levels, up to a 20% reduction in replenishment cycle and a 40% more accurate forecast with a 32% increase in sales (Attaran and Attaran, 2007). Another popular example given by Attaran and Attaran is Heineken USA. The company implemented CPFR in late 1995, with the aim of reducing delivery time from 10–12 weeks to between 4–6 weeks. Using commercial CPFR-compliant software and the internet, Heineken established a web-based private network connecting its customers and/or suppliers. The company delivers customised forecasting data to its distributors through individual web pages. Distributors can login and view their sales forecast, and modify and submit their order online. The system has helped the company to reduce order cycle times from three months to four weeks. Other benefits include lower procurement costs, reduced inventory and fresher products for consumers.

Although CPFR can potentially bring significant benefits to organisations which adopt the practice, implementing it is a challenging task. It requires intensive organisational resources as well as mutual trust between multiple trading partners. For example, Procter & Gamble is doing collaborative planning with several hundreds of retailers, but is only performing full CPFR with a handful of retailers (Sliwa, 2002). A recent report by ECR Europe and Accenture (2006) also shows that many companies are still at the early stage of adopting this initiative.

7.6.3 Vendor managed inventory (VMI)

VMI is an approach in which the vendor controls the buyer's inventory level, so as to ensure that predetermined customer service levels are maintained (Waller et al., 1999). The main difference from CPFR is that control of the retailer's inventory passes to the supplier rather than the supply chain, retaining decision points within both organisations. VMI has been introduced by a number of retailers to their suppliers, including Tesco and Londis (Brenchley and New-Fielding, 2001).

Figure 7.5 presents a simple process map of a VMI supply chain based on research in the UK grocery sector. In total, there are three cycles: restocking the store from the distribution centre (DC), replenishing the DC from the supplier and the production process itself. The VMI relationship particularly influences the second of these cycles, between the supplier and

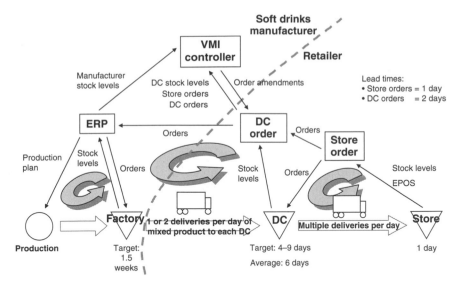

Fig. 7.5   Process map of the VMI supply chain.

the DC. The aim of the VMI controller is to maintain stock levels at each DC between four and nine days. Initial orders are generated by the DC replenishment system (DC order), in the same manner as if the supplier did not have a VMI relationship with the retailer. The VMI controller then carries out a number of checks, focusing in particular on production issues, deviation from forecast store demand and DC stock levels. Should any of these checks highlight a potential problem, the orders on the DC order system can be adjusted up or down. Potential also exists to substitute products to maintain vehicle fill levels. Once these checks have been carried out, the orders are confirmed on the supplier's ERP system.

It has been shown that VMI brings benefits to all parties in the relationship. In the case of the above supply chain, the visibility provided by VMI leads to increased flexibility in production planning, enabling customer service objectives for all customers to be maintained (Potter *et al.*, 2009). This is particularly useful when the production process is under pressure. In the above case, there were limited inventory benefits, although stock-holding for the VMI products was less variable. However, this is not always the case and inventory reductions of 10–20% can be achieved at both the retailer and supplier (Riley, 2003). Finally, VMI can enable an improvement in transport performance to be made, both in terms of reduced cost and increased vehicle fill levels. Returning to the above example, it was found that VMI effectively reduced transport demand by 4% compared to a supply chain operating under normal operating conditions (Potter *et al.*, 2007).

7.7   Concluding remarks

Uncertainty remains a challenge for the food industry. Only by considering all four internal sources in the supply chain – demand, supply, process and control systems – can a company start to take action to improve supply chain performance. Even so, this may not protect against uncertain external sources over which organisations have little control.

This chapter has highlighted a diagnostic approach that can be used to examine uncertainty in more detail. QSAM provides an insight into a company's operations over a relatively short timescale. By looking for the symptoms of complex material flow, it is possible to identify the root causes of uncertainty. Simplification of the supply chain can then be addressed using the simplicity rules.

Once a supply chain has been simplified, the next stage is to optimise the supply chain through effective planning. In the food industry, there have been a number of initiatives to support this, particularly at the retail end of the supply chain. Efficient consumer response, collaborative forecasting, planning and replenishment and vendor managed inventory (VMI) encourage collaboration along the supply chain leading to improved supply chain performance.

7.8   References

ATTARAN, M. and ATTARAN, S. (2007), 'Collaborative supply chain management', *Business Process Management Journal*, **13**(3), 390–404.

BRENCHLEY, D. and NEW-FIELDING, S. (2001), 'Co-managed inventory in a retail community', *Logistics and Transport Focus*, **3**(6), 27–31.

CHILDERHOUSE, P. and TOWILL, D.R. (2003), 'Simplified material flow holds the key to supply chain integration'. *OMEGA, The International Journal of Management Science*, **31**, 17–27.

CHILDERHOUSE, P., DISNEY, S.M. and TOWILL, D.R. (2004), 'Tailored toolkit to enable seamless supply chains', *International Journal of Production Research*, **42**(17), 3627–46.

CHRISTOPHER, M.G. (1998), *Logistics and Supply Chain Management: Strategies for Reducing Costs and Improving Services*, Pitman Publishing, London.

CORBETT, D. (1992), 'Milk quotas – benefit or constraint? Why a common agricultural policy?' *British Food Journal*, **94**(5), 38–40.

DANESE, P. (2007), 'Designing CPFR collaborations: insights from seven case studies', *International Journal of Operations & Production Management*, **27**(2), 181–204.

ECR EUROPE AND ACCENTURE (2006), *Shelf-ready Packaging, Addressing the Challenge: a comprehensive guide for a collaborative approach*, available from http://www.ecrnet.org/04-publications/blue_books/srp/ECR%20Europe%20SRP%20Blue%20Book_final.pdf.

FOGGIN, J.H., MENTZER, J.T. and MONROE, C.L. (2004), 'A supply chain diagnostic tool', *International Journal of Physical Distribution and Logistics Management*, **34**(10), 827–55.

GLOBALSCORECARD.NET (2008), *Guide to ECR concepts*, available from http://www.globalscorecard.net/guide_to_ECR/d_guide.asp, accessed 12 November 2008.

HILL, S. (1999), 'CPFR builds the unites partnership of apparel', *Apparel Industry Magazine*, **60**, 54–60.

HOFSTETTER, J.S. and JONES, C.C. (2005), *The Case for ECR. A review and outlook of continuous ECR adoption in Western Europe*, available from www.ecrnet.org/04-publications/blue_books/pub_2005_case_for_ecr_exe_summary.pdf, accessed 12 November 2008.

LEWIS, J.C., NAIM, M.M., WARDLE, S. and WILLIAMS, E. (1998), 'Quick Scan your way to supply chain improvement', *Control*, **24**(5), 14–16.

MASON-JONES, R. and TOWILL, D.R. (1998), 'Shrinking the supply chain uncertainty circle', *IOM Control*, September, 17–22.

NAIM, M.M., CHILDERHOUSE, P., DISNEY, S. and TOWILL, D. (2002), 'A supply chain diagnostic methodology: determining the vector of change', *Computers and Industrial Engineering*, **42**, 135–47.

POTTER, A., DISNEY, S.M. and TOWILL, D.R. (2007), 'Integrating transport into supply chains: Vendor Managed Inventory (VMI)', in *Trends in Supply Chain Design and Management: Technologies and Methodologies*, Jung, H., Chen F.F. and Jeong, B. (eds.), Springer-Verlag, London, 331–44.

POTTER, A., TOWILL, D.R., BOEHME, T. and DISNEY, S.M. (2009), 'The influence of multi-product production strategy on factory induced bullwhip', *International Journal of Production Research*, **146**(20), 5739–59.

RILEY, H. (2003), 'Stock response (with a twist)', *Supply Management*, 17th July, 30–1.

ROTHER, M. and SHOOK, J. (1998), *Learning to See: Value Stream Mapping to Add Value and Eliminate Muda*, Lean Enterprise Institute, Brookline, MA.

SIMONS, D. and TAYLOR, D.H. (2007), 'Lean thinking in the UK red meat industry: A systems and contingency approach', *International Journal of Production Economics*, **106**(1), 70–81.

SIMONS, D., ZOKAEI, A.K. and WHITEHEAD, P. (2008), *Applying Lean Thinking to the Cereals Industry*, Food Chain Centre Completion Report, available from http://www.foodchaincentre.com.

SKJOETT-LARSEN, T., THERNØE, C. and ANDRESEN, C. (2003), 'Supply chain collaboration: Theoretical perspectives and empirical evidence', *International Journal of Physical Distribution & Logistics Management,* **33**(6), 531–49.

SLIWA, C. (2002), 'CPFR clamor persists, but adoption remains slow', *Computerworld*, 1 July, 10.

TAYLOR, D. (2006), 'Strategic decisions in the development of lean agri-food supply chains: a case study of the UK pork sector', *Supply Chain Management: An International Journal*, **11**(3), 270–81.

TOWILL, D.R. (1997), 'The seamless supply chain – the predator's strategic advantage', *International Journal of Technology Management*, **13**, 37–56.

TOWILL, D.R. (1999), 'Simplicity wins: twelve rules for designing effective supply chains', *Control the Journal of the Institute of Operations Management*, **25**, 9–13.

TOWILL, D.R. (2006), 'Fadotomy – anatomy of the transformation of a fad into a management paradigm', *Journal of Management History*, **12**(3), 319–38.

VAN DER VORST and BEULENS (2002), 'Identifying sources of uncertainty to generate supply chain redesign strategies', *International Journal of Physical Distributions and Logistics*, **32**(6), 409–30.

VOLUNTARY INTERINDUSTRY COMMERCE STANDARDS (VICS) (2002), *Collaborative Planning, Forecasting and Replenishment*. Voluntary Guidelines, http://www.vics.org/docs/committees/CPFR_Whitepaper_Spring_2008-VICS.pdf

WALLER, M., JOHNSON, M.E. and DAVIS, T. (1999), 'Vendor-managed inventory in the retail supply chain', *Journal of Business Logistics*, **20**(1), 183–203.

WATSON, G.H. (1994), *Business Systems Engineering: Managing Breakthrough Changes for Productivity and Profit*, John Wiley and Sons, New York.

8

Aligning marketing and sourcing strategies for competitive advantage in the food industry

D. Chicksand, Warwick University, UK and A. Cox, Newpoint
Consulting, UK and University of San Diego, USA

Abstract: In a period of considerable change and uncertainty in food supply
chains and markets this chapter demonstrates how the alignment of marketing
and sourcing strategies can provide far-sighted firms with an effective way of
coping with increased levels of competition. The experience of Pioneer
Foodservice (a meat wholesaler in the UK) highlights the potential rewards for
companies wishing to develop branded products in the food industry. This
chapter also shows the way in which sustainable competitive advantages can be
achieved through the alignment of marketing and sourcing strategies.

Key words: competitive advantage, marketing, branding, sourcing, food service.

8.1 Introduction

The world economy has seen a remarkable transformation as a result of
increased globalisation. Although the global distribution of food is not a
new phenomenon (Hirst and Thompson, 1996) the recent pace of change
is unprecedented. Owing to technological advances in transporting, preserving and storing products, combined with the effective flow of information, production chains can now span long distance and are increasingly
controlled by a few large-scale transnational corporations. Apart from
technological advances, a number of other factors have contributed towards
the acceleration of global trade: trade and financial market liberalisation,
encouragement of foreign direct investment (FDI) and improved intellectual property and consumer protection laws (Farina, 2001; Murdoch

et al., 2000; Ramsey, 2003). These factors have created conditions that have increased global competition in many sectors, including the food industry.

The last two decades have also seen a period of considerable change and upheaval in domestic markets and supply chains. The UK farming and food industry, in particular, has had to cope with a number of unprecedented structural changes that have an enormous impact. These include:

- Common Agriculture Policy (CAP) reforms
- consumer concerns over the quality and safety of UK beef products as a result of outbreaks of livestock diseases (including foot and mouth disease (FMD) and bovine spongiform encephalopathy (BSE))
- a long-term decline in consumer demand for red meat, with a preference for alternative proteins (chicken) and healthier foot options
- a radical change in consumer preferences, both in the way we eat (in favour of convenience food choices) and what we eat (an increasing interest in GM free foods and organic products)
- the concentration of market power in the hands of a small number of multiple food retailers (there is a significantly greater concentration of market power in multiple retailers in the UK than in many other counties)
- increased foreign competition and finally
- embargoes on British beef exports owing to restrictions put in place after the outbreak of BSE and FMD (Hingley and Lindgreen, 2002; Taylor and Simons, 2004; Hingley, 2005; Cox *et al.*, 2006a; 2006b).

These challenges have contributed towards a general decline in the numbers of primary producers and in many cases production levels, in particular, within the red meat, pig and dairy industries. For example, the number of pig farm holdings fell from 17,100 in 2000 to 10,000 by 2007 (Competition Commission, 2008, pp. A9 (5)-1–A9 (5)-2), whilst the number of dairy farms has also declined considerably from 35,000 in 1995 to less than 20,000 in 2006 (Competition Commission, 2008, pp. A9 (3)-1). Furthermore, there is evidence that for many producers it is not possible to compete internationally. According to BPEX (2005), the rapid growth in imports has contributed to the decline in UK pig production. British producers cannot compete with cheaper imports (Competition Commission, 2008, pp. A9 (5)-1). This is also evident when we consider that imports of beef and veal, for example, grew by 69.2% between 1996 and 2005, whilst exports fell by 82.5% over the same period (Competition Commission, 2008, pp. A9 (4)-9).

Faced with these unparalleled challenges it is now widely accepted that the UK industry needs to develop strategies to address these acute problems. The next section focuses on the response from UK Government, industry and academics.

8.2   Literature review and discussion of marketing and sourcing strategies

In response to the crisis facing the UK food industry there have been a number of government policy documents, including: the *England Rural Development Programme: 2000–2006* (MAFF, 2001); the Curry Report (Curry, 2002); *The Strategy for Sustainable Farming and Food* (DEFRA, 2002) and *The Sustainable Farming and Food Strategy: Forward Look 2006* (DEFRA, 2006). Although these policy documents focus on broad issues relating to UK agriculture as a whole, there has been a clear message: farmers need to implement diversification strategies to appropriate greater value from the supply chain as a whole, add value to food and collaborate to eliminate waste and inefficiency.

There has also been considerable advice from industry bodies and academics suggesting how food chain participants might adapt to the structural changes occurring within their industries. Advice for UK beef industry participants has largely been focused upon how to improve sustainability, competitiveness and profitability. The literature focusing on improvement strategies for food supply chains in the UK is extensive, with the following key texts:

- horizontal collaboration (Hind, 1997; Hendrickson *et al.*, 2001; Curry, 2002; DEFRA, 2002; EFFP, 2004; Cox *et al.*, 2005),
- value chain analysis (lean thinking) (Curry, 2002; FCC and RMIF, 2003; Simons *et al.*, 2003; Francis, 2004; Cox and Chicksand, 2005c; Simons and Zokaei, 2005);
- vertical integration, collaboration and coordination (Shaw and Gibbs, 1995; Palmer, 1996; Fearne, 1998; Van Der Vorst *et al.*, 1998; Katz and Boland, 2000; Curry, 2002; Hornibrook and Fearne, 2003; Fearne *et al.*, 2004; Cox *et al.*, 2006a); and,
- establishing alternative supply chains including farm shop, export channels and food service supply chains (Curry, 2002; Cox *et al.*, 2006b).

Overall, there has been a considerable emphasis on the need for collaborative lean techniques for waste elimination and a focus on the importance of adding value to food (which can be achieved through the development of brands). Each of these opportunities for UK farming and food is discussed here.

8.3   Creating sustainable business success: the development of effective marketing strategies

The realisation that businesses need to find ways of engineering sustainable competitive advantages has been central to the debate about what strategies UK food chain participants should adopt. There are several factors in the creation of sustainable business success, but choosing the right product and/or service and marketing it (supported by a strong brand) to a target market is a key component of business success (Cox and Chicksand, 2007b). In part, sustainable business success requires companies to align their branding, marketing and sales, and pricing strategies to differentiate themselves from their competitors.

8.3.1   How can brands help develop sustainable business success?

To understand why differentiating products and/or services from others is increasingly important in an ever more competitive global environment, it is first necessary to define a brand. The following definition provides an insight into the concept: '... brand names convey the image of a product or service and refer to a name, term, symbol, sign, or design used by a firm to differentiate its offering from those of its competitors' (Czinkota and Ronkainen, 1995). A branding strategy may also be the combination of a product, brand name, packaging, symbols, themes and images (Vrontis, 1998).

Many writers believe that brands have now become an integral part of both the consumer's choice and organisational strategies and that competition is no longer at a core-product level, because consumers buy brands rather than products. This means that competition is no longer based upon the tangible quality or service attributes of a product, but on less tangible added attributes that the brand represents (Doyle, 1994; Simoes and Dibb, 2001; Vrontis, 1998). A consumer's choice of preferred brand is becoming a reflection of who they are. So much so, that according to some '... people choose brands in the same way as they choose friends' (Levitt, 1986).

Brands are becoming increasingly important in all consumer markets. Faced with a growing choice of products and with added concerns over animal welfare and food safety, customers are increasingly relying upon the 'quality signalling' of brands in food supply chains and markets (Shocker *et al.*, 1994; Simoes and Dibb, 2001; Cox *et al.*, 2007a). To adapt to these changing dynamics, organisations have created their own distinctive features to differentiate themselves positively in the eyes of the consumer (Simoes and Dibb, 2001). There are a number of important commercial benefits that can arise for businesses from the development of brands (see Table 8.1 below).

8.3.2   Branding in the UK beef industry

In the UK beef industry in particular, there has been considerable interest in developing product development and marketing strategies to defend British-grown produce. The Curry Commission (2002) called for the establishment and promotion of branded premium beef products arguing that: '... with rising real incomes and consumers' interest in variety and choice

Table 8.1 The importance of brand definition

- It differentiates product/service from those of competitors.
- It signifies a given quality or safety standard for a product/service.
- It enhances a product's/service's competitiveness.
- It influences the price-elasticity of demand.
- It eases the introduction of new products and services.
- It creates customer recognition and loyalty.
- It enhances leverage over supply chain partners.
- It creates long-term shareholder value.

increasing, there will be opportunities here that English farmers can seize'
(Curry, 2002).

Ever since Smith's (1956) pilot study into product differentiation as an
alternative marketing strategy, there has been considerable debate about
the development of branding strategies in the food industry (Koehn, 1999;
Loughlin, 1999; Beverland, 2001; Hingley and Lindgreen, 2002; Ramsey,
2003; USDA, 2003; Zylbersztajn and Filho, 2003). There has, however,
been far less academic debate about the need to develop branded beef
products in the UK beef supply chain (Cox and Chicksand, 2005a, 2007a;
Cox et al., 2006a, 2007a,b). Although considerable effort has been made by
EBLEX (English Beef and Lamb Executive, 2005) to promote English beef
and lamb, the pace of implementation of branding strategies in the UK beef
industry is far behind that of international competitors. For example, in the
USA, seven of the top 10 breeds of cattle have considered a branded programme (USDA, 2003).

A recent UK 'Beef Industry Summit' highlighted that although many in
the British beef industry recognise the need for change, this will not happen
overnight. This is because the CAP subsidy system has meant that British
beef has tended to be a commodity, with very few brands being developed
and little differentiation in the sector (Porter, 2006). The Pioneer Foodservice 'Lakeland Premium Branded Beef Case', which is discussed in
some detail here shows, however, that UK consumers value the attributes
(quality, safety and traceability) associated with branded products and are
willing to pay more for this.

8.3.3 Developing a successful differentiation strategy in the global beef industry

The benefits that arise from branding occur because of concerns about
food safety are leading consumers in developed countries to exercise
more caution when buying and/or demanding 'quality' products (Murdoch
et al., 2000). The quality-signalling role of brands, or country of origin
labelling (or a combination of both), is being increasingly exploited by
retailers, processors and producers, as consumers are encouraged to pay

a premium in return for a guarantee of perceived quality (Sans *et al.*, 2005).

The use of brand names is well developed in other protein sectors, such as poultry (for example, Bernard Matthews in the UK and Perdue, Tyson, Hudson and Butterball in the USA). Although beef brands have started to make advances in the USA beef industry (Katz and Boland, 2000), this has not typically been the case in the UK beef industry. With increased competition in the USA retail market, Safeway's and King Sooper's beef brands are increasingly important in the fight for customers. In 2001, King Sooper's introduced its 'Cattleman's Collection' private labels, to be followed by Safeway's 'Ranchers Reserve Angus Beef' and Albertsons 'AngusPride', a national brand. Boulder-based Wild Oats Markets recently added to the competition by introducing its new branded line of natural angus steaks. The 'Ranchers' and 'Cattlemans', brands are based upon a promise of eating quality and are both USDA 'select' and the next grade 'choice' lines (Denver Post, 2003).

Argentinean beef producers have long understood the need to develop brands or trademarks to differentiate their products. Prinex was created in the 1990s by a group of Buenos Aires farmers to signal their production of the highest quality beef. The Prinex business strategy was based upon market differentiation and segmentation using the 'Novillo Pampeano' trademark, basing the brand upon known origin and quality. Today the company exports over 1000 tonnes per annum, especially to Europe, Chile, Brazil and ex-Soviet countries. The entrance of the brand into the most select customer market segments (ABC1) in developed countries was key to its success. This has enabled them to obtain higher prices, with a 20% premium in final prices over the average selling price. In Spain, Prinex has been able to sell at 45% above the price of their other Argentinean competitors. Two other examples are the 'Carne Angus Certificada' (a trademark of the American Angus Association, with offices in Argentina) set up to supply traceable and quality beef to the USA, and the setting up of the Consorcio Pampas del Salado, an association of farmers in the provinces of Argentina, which has developed an origin and quality assurance protocol (Ordonez *et al.*, 2004).

Another example is the recent success of Brazilian beef as a brand. The brand has helped Brazil to increase exports and to become the world's leading export country (in terms of volume), despite recent foot & mouth disease outbreaks. The aggressive marketing effort of ABIEC (Association of Brazilian Processors and Exporters) is central to its success. Through a comprehensive promotion programme, approved by the National Export Promotion Agency (APEX), it has developed a successful brand – 'Brazilian Beef'. The brand is targeted worldwide, but with a heavy emphasis on the EU, where 60% of Brazilian beef is destined. The brand focuses on the natural 'healthy' grass-fed environment in which Brazilian beef is reared, as opposed to grain-fed beef. Brazil recognised the need, not only to increase quantity of sales, but also to increase the quality and value of sales. Speciality and niche markets are seen as future target markets for Brazilian beef and could pose a real threat to many indigenous beef producers in the UK and elsewhere (Steiger, 2006).

8.4   Creating sustainable business success: the development of effective sourcing strategies

The need to develop an effective branding and marketing strategy to differentiate food products is self-evident. What may not be as well understood, however, is the need to align this with an effective supply chain sourcing strategy.

There have been a number of strategies suggested to create a robust and sustainable farming and food industry. At the heart of much of the work in this field has been the concept of lean thinking. This needs to be considered in some detail, as the development of 'lean' integrated value chains is viewed by many people within the farming and food industry as the best way forward (Curry, 2002; BPEX, 2002). Lean and collaborative thinking has also been the theoretical driver underpinning much of the work of UK bodies, such as the Food Chain Centre (FCC), the English Food and Farm Partnership (EFFP) and various industry forums, such as the Red Meat Industry Forum (RMIF), set up as a result of the Curry Report (2002). Strategies that focus upon increased supply chain collaboration as a means of delivering waste reduction and improved efficiency are, therefore, viewed as one of the primary means of defending producers from cheap meat suppliers (Zokaei and Simons, 2006).

Lean thinking has received considerable attention as an approach for companies to adopt to achieve sustainable competitive advantage in the beef supply chain. This approach, by eradicating waste and inefficiency throughout the supply chain, seeks to find ways to deliver exceptional value to end customers (Womack and Jones, 1996; Hines et al., 2000). The FCC piloted the concept of lean thinking in the food industry and examined 33 chains from farm to fork. In partnership with the Cardiff Business School and using a value chain analysis (VCA) tool, the analysis involved engaging businesses within the food chain to encourage collaboration and to identify where cost and value are added. The FCC concluded that on average 20% of the cost in the food chain added no value (DEFRA, 2007a). In cooperation with others, the RMIF has also applied the lean thinking approach to complete nine VCAs of the 33 that FCC delivered, covering a mix of species (beef, lamb and pork), different distribution channels (retail and catering) and sizes of business (FCC and RMIF, 2003; DEFRA, 2007a). The VCA work was successful in highlighting significant opportunities for improving both operational and strategic efficiency in agricultural supply chains (Taylor, 2006; Zokaei and Simons, 2006). However, advocates of 'lean thinking' have also acknowledged that attempts to establish collabora-tive intra-company teams to generate 'win–win' integrated supply chain improvement have been less successful (Simons *et al.*, 2003; Fearne, 2005; DEFRA, 2007b).

One key reason, it can be argued, for the partial success of adopting of lean principles and vertical collaboration in many agri-supply chains has been the difficulty of achieving the desired levels of trust between partici-pants in the chain (Fearne, 2005). The traditional way of life and thinking of producers is often a powerful barrier to achieving effective supply chain management in sectors of the beef industry (Simons *et al.*, 2003; Fearne, 2005). Recent research has, therefore, advocated the need to address issues of mistrust in the red meat industry, through fundamentally changing the 'trading mentality' by adopting contractual commitments to source specific volumes, establishing agreements as to price / cost policies and adopting agreements for benefit sharing (Taylor, 2006).

Although this is sound advice, as some writers have started to acknowl-edge, the historical lack of trust between supply chain participants has much to do with the imbalance of power between the multiple retailers and the processors and farmers, a power imbalance which 'lean thinking' does not sufficiently acknowledge. The abuse of power by multiple retailers has led to reoccurring pressures on supplier prices and profits, with the continual threat of switching sources if suppliers do not comply (Hingley, 2005; Taylor, 2006; Competition Commission, 2008). It comes as little surprise, therefore, that this lack of trust discourages investment and acts as a real barrier to supply chain cooperation (Fearne, 2005).

Furthermore, although there has been some analysis of the commercial benefits (reportedly 2–3% potential savings at each stage of the red meat value chain), there has been insufficient focus on the potential operational pitfalls and difficulties that occur when implementing a lean approach (Zokaei and Simons, 2006). It has been argued that the adoption of a sourcing strategy will not improve profitability for those putting it into operation unless the strategy can generate power resources that improve the upstream and/or downstream power and leverage position of a firm within a supply chain (Cox *et al.*, 2003, 2004a,b). It can also be argued that in the UK beef supply chain, the adoption of lean strategies can result in a high level of dependency on buyers and to low or declining levels of profitability.

A further possible explanation for the partial success of introducing lean thinking, in particular within the UK beef industry, is the unique nature of supply and demand that characterises this industry. The nature of supply and demand within the UK beef industry is not always commensurate with the structural properties that are required to allow long-term lean collab-orative approaches to operate effectively. It is, therefore, essential to emphasise that the characteristics (production processes, supply and demand, etc) of the beef industry are quite different from process-based industries (such as the automotive), in which lean thinking was pioneered (Cox and Chicksand, 2004, 2005b).

It can be argued, therefore, given the different markets and supply chain circumstances operating in beef supply chains, that 'agilean' or a more responsive/agile approach may be more appropriate than lean. Recent research, based on a more robust understanding of the supply and demand and power and leverage characteristics of the UK beef industry, shows that there is a need for firms to understand how to select the most appropriate upstream and downstream supply chain management strategies, rather than copying a lean approach pioneered in a very different supply and demand and power and leverage environment (Cox and Chicksand, 2005d).

8.5   Creating sustainable business success: aligning brand and marketing strategies with sourcing strategies

Until recently, there has been very limited academic research focusing on the need to align marketing and sourcing strategies. There has also been very little attention given to the downstream risks to a brand from an inadequate assessment of the appropriateness of an organisation's upstream sourcing strategy. The case of Nike clearly demonstrates this issue. The development of the Nike brand and its phenomenal growth in the last 20 years is one of the most successful cases of product differentiation in modern business. But in the late 1990s, this growth came to an abrupt end. In order to support its differentiation strategy, Nike invested heavily in marketing and new product development. As a result, Nike became the most famous sports brand and in 1997 controlled over one-third of the global athlete footwear market, greater than the sum of its four major competitors – Reebok, Adidas, Fila and Converse – put together. Yet, in 1999, Nike, the largest and most famous sporting goods producer in the world, reported an 8% decline in revenue across the USA, Asia Pacific and Latin American markets. Several reasons were given for this poor performance, including the retirement of Michael Jordan and a shortened regular NBA season. Nike attempted to solve these problems by heavily investing in innovative designs, by aggressive marketing and by reducing operational costs. Despite its aggressive approach, revenue did not grow and by the end of 2000 Nike had to rethink its strategy (Locke, 2003).

Historically Nike had taken advantage of global sourcing opportunities to reduce costs by outsourcing production and relocating plants to Korea and Taiwan in the 1980s, and then to China, Indonesia and Vietnam in the 1990s. By 2006, Nike's products were manufactured by over half a million workers in 700 offshore plants in 51 countries, although the company has only 22,658 staff on its own pay roll. Nike's sourcing strategy was initially successful in obtaining high quality products at continuously reduced cost and this supported growing profitability as market share increased (Nike Inc, 2005).

This pressure to reduce costs was continuously passed onto suppliers but Nike failed to understand that this might result in the exploitation of the employees of its suppliers and the long-term implications for its own brand and market share when this exploitation was realised. The first Anti-Nike organisation – 'Boycott Nike' – was established in 1996 and since then Nike has been challenged by many human rights groups for its use of sweatshops in southeast Asian countries. Stories of Nike's underpaid workers in Indonesia, child labour in Cambodia and Pakistan, poor working conditions in China and labour abuse in Vietnam have been reported globally (BBC News, 2000, 2001, 2002). Nike's worldwide image was tarnished and forced them to ensure that their suppliers did not abuse and exploit their employees.

There is no doubt that the decline in sales in the late 1990s was a direct result of a misaligned sourcing strategy in relation to the company's avowed brand image. While Nike has certainly cleaned up its sourcing act since then and has continued to retain a major share of the global market for its products, this case clearly demonstrates the risks that a misaligned sourcing strategy can create for reputation, revenue and profitability (Locke, 2003; Cox and Chicksand, 2006).

With this in mind, the remainder of this chapter explains, through case material, how a food producer created a differentiated branded premium beef product and aligned it with its sourcing strategy. The findings from the case study suggest a need for firms in the food industry to align their upstream sourcing strategies, so as to support often more developed downstream marketing strategies.

8.6   Pioneer Foodservice case: developing a premium branded beef product

8.6.1   Methodology

The case material in this chapter reports on some of the findings from a study sponsored by the Engineering and Physical Science Research Council, North West Development Agency, North West Food Alliance and the Red Meat Industry Forum in the UK. Previous work within the industry has suggested several options for restructuring the industry, but it can be argued that previous work has lacked a robust methodology for understanding the unique supply and demand, and power and leverage, characteristics of the industry. As a result it had been difficult to provide clear guidance on the appropriateness of alternative marketing (banded and non-branded) and sourcing (proactive and reactive) strategies.

To rectify this gap a power regimes methodology was used to analyse UK beef supply chains and, through understanding supply and demand characteristics, gain a better understanding of the power and leverage dynamics within them. Extensive interviews were carried out with participants at key stages of the supply chain, reported in this chapter, using a power-positioning tool (Cox, 2004a, 2004b, 2004c). After constructing a standard questionnaire and collecting responses, content analysis was used to analyse the data collected. The power matrix is the analytical tool used to understand the appropriateness of particular upstream and downstream supply chain strategies. The matrix is constructed based on three primary variables (with sub-variables behind these as indicated in Fig. 8.1), the presence or absence of which, for the buyer or supplier, forms the basis for the standard questionnaire discussed earlier.

The three primary variables analysed are the relative utility and the relative scarcity of the resources that are exchanged between the two parties and the information advantages that arise in exchange transactions for buyers and suppliers (Cox *et al.*, 2000; Cox, 2001; Cox *et al.*, 2002). Each party within a transactional exchange can be located in one of four basic power positions: buyer dominance (>), interdependence (=), independence (0) and supplier dominance (<). What is important to understand is that

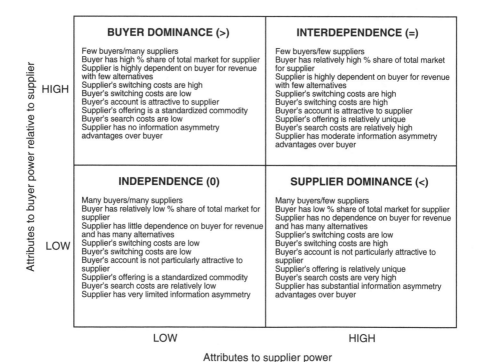

Fig. 8.1   The power matrix. © Robertson Cox Ltd, 2000 all rights reserved (from Cox, 2001, p 14).

buyers and suppliers should not only understand and manage the current power circumstance, but also use relationships in the future to create new power circumstances that provide for a more congenial leverage position for them to maximise their often divergent economic objectives (Cox *et al.*, 2004a,b).

Knowing the current power and leverage circumstances in dyadic (buyer and supplier) relationships is the first step. It is also important to understand that buyers have four major operational sourcing options that they can choose from when they work with suppliers to achieve continuous improvements in value for money. Buyers can act reactively or proactively, when they develop relationships with suppliers. Buyers must also decide whether they have the capability and resources to undertake first-tier relationship management or work throughout the supply chain with suppliers from the first tier through all tiers down to raw-material suppliers (Cox *et al.*, 2003, 2004a,b).

As Fig. 8.2 demonstrates, there are four sourcing options available for buyers to manage their suppliers and supply chains: supplier selection, supplier development, supply chain sourcing and supply chain management (Cox *et al.*, 2003). Buyers and suppliers must also think carefully about their internal capabilities and external power circumstances when they decide which sourcing option is the most appropriate for them to adopt (Cox *et al.*, 2003, 2004a,b). Knowing these variables then allows the analyst to understand the appropriateness of particular sourcing approaches and power circumstances, as indicated in Table 8.2. This methodology was used to assist in the selection of upstream and downstream strategies for the focal firm in the specific red meat supply chain analysed here.

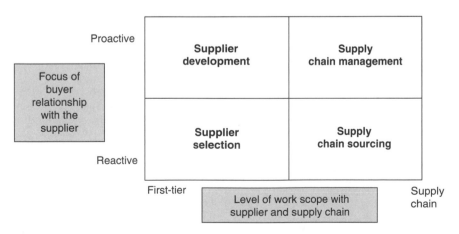

Fig. 8.2   Four sourcing options for buyers (from Cox *et al.*, 2003, p 5).

Table 8.2 Appropriateness in sourcing strategies, power circumstances and relationship management

Sourcing approach	Power and leverage circumstance	Appropriate relationship management styles
Supplier selection	BUYER DOMINANCE (>)	Buyer Adversarial Arm's-Length / Supplier Non-Adversarial Arm's-Length
	INDEPENDENCE (0)	Buyer and Supplier Adversarial Arm's-Length
	INTERDEPENDENCE (=)	Buyer and Supplier Non-Adversarial Arm's-Length
	SUPPLIER DOMINANCE (<)	Buyer Non-Adversarial Arm's-Length / Supplier Adversarial Arm's-Length
Supply chain sourcing	BUYER DOMINANCE (>)	Buyer Adversarial Arm's-Length / Supplier Non-Adversarial Arm's-Length
	INDEPENDENCE (0)	Buyer and Supplier Adversarial Arm's-Length
	INTERDEPENDENCE (=)	Buyer and Supplier Non-Adversarial Arm's-Length
	SUPPLIER DOMINANCE (<)	Buyer Non-Adversarial Arm's-Length / Supplier Adversarial Arm's-Length
Supplier development	BUYER DOMINANCE (>)	Buyer Adversarial Collaboration / Supplier Non-Adversarial Collaboration
	INDEPENDENCE (0)	Not Applicable
	INTERDEPENDENCE (=)	Buyer and Supplier Non-Adversarial Collaboration
	SUPPLIER DOMINANCE (<)	Buyer Non-Adversarial Collaboration / Supplier Adversarial Collaboration
Supply chain management	BUYER DOMINANCE (>)	Buyer Adversarial Collaboration / Supplier Non-Adversarial Collaboration
	INDEPENDENCE (0)	Not Applicable
	INTERDEPENDENCE (=)	Buyer and Supplier Non-Adversarial Collaboration
	SUPPLIER DOMINANCE (<)	Buyer Non-Adversarial Collaboration / Supplier Adversarial Collaboration

Source: © Robertson Cox Ltd, 2003 all rights reserved (from Cox, 2004c, p 355.)

8.6.2   The case: developing the brand and an aligned sourcing strategy

Pioneer Foodservice is a medium-sized beef processor and catering butcher, based in the Lakeland area of the North of England. Prior to 2000, Pioneer was just one of a large number of catering butchers in the Lakeland area producing undifferentiated raw and semi-prepared beef products for both catering and retail sales. In 2000 Pioneer created an alliance with a livestock auctioneer (Harrison & Hetherington, H&H) and an abattoir (Bowland Food), to consider differentiating their products, so as to increase their share of the catering service market and to make higher commercial returns.

As Barry Garret (2006) from Pioneer Foodservice stated: '...we [the three companies] believed that there must be a better way to manage our beef supply chain. We needed to differentiate our products to achieve a higher share of the catering service market and to gain better returns for all involved'. The initial idea was to improve service and quality levels and brand premium beef products. They soon discovered that this required not only considerable managerial effort and financial investment, but also very different expertise than they currently possessed.

The firms involved realised that they did not possess the internal capability to cope with all aspects of the restructuring of the supply chain. They decided, therefore, that the best way to progress was through vertical collaboration. In order to achieve this they needed to work together and share the risks and responsibilities generated from the collaborative approach. Within this new supply chain framework, Harrison & Hetherington was responsible for sourcing beef of a superior 'standard' specification from farm-assured producers in CA (Cumbria) and LA (Lancashire) postcode areas, either through the auction ring or direct from the farm. Bowland Food took ownership of the animals, provided a slaughtering and primary processing service, delivering primal cuts to Pioneer. Pioneer Foodservice would then further process these primal cuts to customer specifications and deliver the final branded beef product to the customer. Pioneer was to also play a central role in developing and selling the brand within the regional catering service market by improving service, product quality and creating brand recognition.

Pioneer was immediately faced with several dilemmas; should they develop a national brand or a local brand and what should be the basis of differentiation for their beef brand? A further question was whether to build the brand through existing sales channels (sales to restaurants, pubs and public services) or to develop and target the retail route? After much deliberation, it was decided that the costs and risks associated with trying to launch a national brand were too great and that as a regional company it was better to build on their local and regional presence. Therefore, it seemed that the most effective branding strategy would be to develop a brand associated with the region, which prided itself on full traceability of high quality, extensively reared beef, originating in the CA and LA postcode areas. Therein the 'Lakeland Beef' brand was born.

It was also decided that Pioneer would initially heavily promote the brand with existing customers, as it was felt that returns would be more favourable if restaurant and other catering customers were targeted in preference to developing relationships in the more competitive retail sector. A further decision was also made to pursue a single branding strategy over sub-branding or multi-branding for all their beef products, targeted at their diverse portfolio of customers. This was a potential risk as they would be targeting very different customers with the same brand, from up-market restaurants at one end of the spectrum to price-sensitive school meals contracts with the local LEAs at the other. Nevertheless, it was felt that different quality and priced cuts of beef products could be successfully targeted at different customers under a single 'Lakeland Beef' brand. The 'Lakeland Beef' brand was officially launched in 2002.

Having made decisions about the branding approach, Pioneer and its partners also had to consider the most appropriate sourcing strategy. They initially selected a reactive supply chain sourcing approach (see Fig. 8.2 and Table 8.2). The sourcing approach had moved beyond simple supplier selection, in that the focal company 'Pioneer' was concerned with relationships beyond its first-tier supplier. This contrasts from the way they previously managed their non-branded beef supply chain (supplier selection). However, it is important to highlight that none of the organisations made specific dedicated investments and all parties could easily switch from the newly developed 'Lakeland' supply chain, as the brand was not at the time a major contributor to these businesses. As Fig. 8.3 shows, the 'Lakeland' power regime was best described as independent, with the power resources slightly favouring the supplier, moving the relationship towards supplier dominance.

A further two years of promotion significantly increased the recognition of the 'Lakeland Beef' brand within North West of England. This enabled Pioneer to differentiate its products from other catering butchers in the region. The brand also gave Pioneer's existing business clients (a majority of whom were restaurants and independent contracted caterers) a way of differentiating and enhancing their own businesses (through directly promoting the 'Lakeland Beef' brand at the point of sale). New routes to

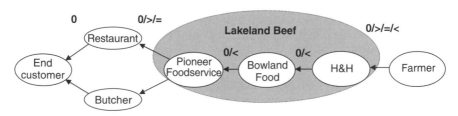

Fig. 8.3 Original 'Lakeland Beef' supply chain.

market, such as schools, hospitals and Pioneer's own restaurant/steakhouse also developed. The 'Lakeland' business continued to progress and as many as 100 carcasses were bought and sold each week.

The successful development of the brand enabled Pioneer to pass some extra value back to its upstream partners, Bowland, H&H and the farmers (a 1–2p per kilo premium). However, this growth in demand led Pioneer to become more concerned with securing a sustainable source of high quality CA/LA animal supply. In order to achieve this, Pioneer was eager to acquire H&H's know-how in sourcing the right quality beef. Although the premium was a gesture of goodwill to its partners, it was also a means of securing a quality beef supply (where there are inherent shifts in the power resources (0/>/=/<) between the farmers and their customers, primarily owing to periods of supply scarcity, see Fig. 8.3).

However, from late 2003 onwards the relationship between Pioneer and Bowland faltered. A number of factors contributed towards the deterioration of trust in the relationship. One key issue was the traceability and quality of the products, key determinants for the success of the 'Lakeland' brand. Given the existing supply chain structure, Bowland owned the carcass and sold primal cuts to Pioneer to be further portioned. Without owning the carcass, Pioneer was unable to ensure that they were receiving the desired levels of quality and traceability.

In 2004, Pioneer ended its relationship with Bowland Foods and changed the way they worked with their partners. The decision was made by Pioneer to work even more closely with its supply chain partners and, therefore, what was needed were partners who would be keen to invest in and be part of the long-term development of the brand. The idea was for Pioneer to source quality locally produced animals either directly from farmers or through the auctions, with the help of their procurement partners, H&H. Rose County became the new slaughter and primary processor for the brand.

Rose County is one of the largest and most technically advanced abattoirs in the UK and Ireland and has the capability to source livestock directly from local beef farmers. Rose County was also supplying beef products to leading multiple retailers (MR) in the UK and an internal customer (its sister company Dungannon Meats, a large Irish secondary meat processor and packer). Pioneer also made the decision to take ownership of the cattle and move the relationship to one of contract kill for the 'Lakeland' brand (based on fixed charges of £55 per head, plus 25p per kilo for carcass weight, plus removal costs of offal, less the value of by-products, i.e. skin, bone etc). Part of the reason for choosing a large abattoir with MR contracts was the fact that Rose County would be able to solve Pioneer's potential carcass balancing problem (because they owned the carcass) by buying back unwanted cuts. The relationship with Pioneer was seen as desirable to Rose County as the 'Lakeland' brand could potentially become a more significant account in the future.

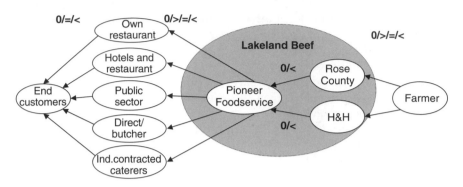

Fig. 8.4   Evolved 'Lakeland Beef' supply chain.

To reduce their reliance upon H&H, and to reduce the risk of supply shortages, Pioneer is now able to draw upon Rose County's direct sourcing capability (see Fig. 8.4) for the 'Lakeland' brand.

After a long process of supply chain restructuring by using a supplier development approach (see Fig. 8.2), Pioneer became the central party in the 'Lakeland' power regime. However, it is important to emphasise that this supply chain restructuring process did not alter the power position within the evolved supply chain in favour of Pioneer. 'Lakeland' beef still did not account for a significant part of Rose County's, or H&H's, business. As Fig. 8.4 shows, the 'Lakeland' power regime was still characterised as independent, with the power resources slightly favouring the suppliers, moving the relationship towards supplier dominance.

8.6.3   Need for a proactive sourcing strategy

By 2006, the brand had taken a much larger share of the catering service market in the North West of England and this enabled Pioneer to earn much higher returns than had been achieved from selling undifferentiated processed and raw meat products. Pioneer was now in a position to take the brand national and make much higher returns from sales with a differentiated product offering, delivering to their customers a superior product and service. By targeting high value end customers (i.e. discerning restaurateurs) who are willing to pay a premium for the high quality 'Lakeland Beef' brand, Pioneer have been able to 'grow the pie' for all involved. Farmers who supply into this chain receive a premium – that is higher farmgate prices for supplying what the end customer wants. By working together more closely, the end consumer gets the right quality meat, as information about their needs is fed through the whole supply chain back to the farmers.

With the brand's continuing success, there was now a need for Pioneer and its partners to develop a more proactive sourcing strategy, so as to align with Pioneer's proactive marketing strategy. With increased success, Pioneer had become focused on securing quality animals at the right price, however, as Fig. 8.4 emphasises, they were at the behest of their suppliers (Rose County and H&H) with very little control over the primary producers (characterised by shifting power circumstances). It is important to highlight that although UK primary cattle supplies, especially highly quality beast production, continue to fall, market demand for high quality beef has continued to grow (Harvey, 2004; Sinclair, 2004; Cox and Chicksand, 2005b, 2005c; Cox et al., 2006b).

Therefore, securing high quality primary beef supply was uncertain for Pioneer. It was crucial for Pioneer to understand that under the new Pioneer-led 'Lakeland' supply chain structure (see Fig. 8.4) current procurement partners have potentially less incentive for developing the brand and market expansion. It is, therefore, possible that in the event of a supply shortage, the suppliers would either not prioritise 'Lakeland' demands (in favour of larger customers), or secure supply at the best available price (owing to the commission structure). Therefore, it can be argued that both H&H and Rose County could potentially be in a supplier dominant position in the future.

Furthermore, if 'Lakeland' sales continued to grow, there might be a possibility that neither H&H nor Rose County would be capable of sourcing the right quality and quantity of CA/LA animals for Pioneer, owing to Common Agricultural Policy (CAP) reforms which may encourage more producers to leave the market or (in an increasingly competitive market) reduce the incentive to produce beef. Therefore, the only action that can reduce the inherent uncertainty in supply is to have direct contractual relationships with producers, thereby also reducing Pioneers' reliance upon the procurement role of H&H and Rose County (see Fig. 8.5). This can be achieved either by insourcing the procurement role of H&H/Rose County, thereby cutting out the currently outsourced procurement function, or by

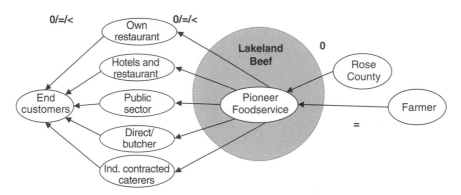

Fig. 8.5   Proactively restructuring the 'Lakeland Beef' supply chain.

changing H&H/Rose County's procurement role to 'management' of Pioneers' direct producer relationships.

Whichever management structure Pioneer opted for, the adoption of a supply chain management approach to establish direct interdependent, non-adversarial collaborative relationships (see Fig. 8.5) with primary animal suppliers would reduce Pioneer's inherent supply risk. Having direct contracts gives beef producers some security of demand at 'fair' prices, in return for a guarantee of a consistent primary cattle supply for Pioneer. Direct relationships also reduce Rose County's power resources within the 'Lakeland' supply chain and this moves the relationship towards independence (see Fig. 8.5). Direct relationships with producers also supports a truly proactive sourcing approach and enables Pioneer to influence primary producers' practices (i.e. breeding programmes, feeding and finishing regimes), to ensure that the animals meet Pioneer's high standards for the brand.

Direct contractual relationships also provide Pioneer with savings, through the elimination of sourcing commission, reductions in transaction costs and potential savings in logistics costs. These savings could be shared, although not necessarily equally, between Pioneer and the contracted beef producers. Consequently, Pioneer could secure cheaper beef prices by direct relationships. What's more, direct relationships would also be a great marketing story to sell, thereby backing up the brand value.

It is, however, important to understand the sunk and switching costs and operational risks associated with direct relationships. It is clear that the issue of supply security affects governance structures and ultimately power positions between buyers and sellers. Therefore, Pioneer may have to consider guaranteeing a considerable premium to entice beef producers initially to enter into a long-term relationship, as the full impact of CAP reforms still remains unknown. Pioneer will also have to take the risk of guaranteeing to buy a specified number of animals and will need to have the ability to solve the fluctuations in balancing demand and supply.

8.7   Conclusion and discussion

This case raises several issues. The first is that of the sustainability of the 'Lakeland Brand' and its ability to move from a successful regional to a nationally, or even internationally, recognised brand. The problem with brands is, of course, the potential ease with which they can be replicated. Once a brand has been replicated it loses its ability to gain a premium price for the product/service. If everyone has it, or can have it, then it becomes of less value to the purchaser. Thus, the ability to defend the brand against replication becomes critical. This means there are things that must be done if a brand-led strategy is pursued.

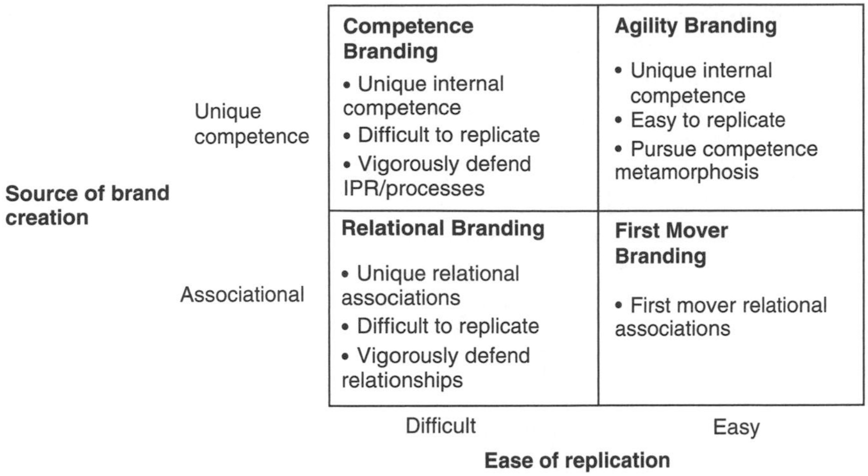

	Unique competence	Competence Branding • Unique internal competence • Difficult to replicate • Vigorously defend IPR/processes	Agility Branding • Unique internal competence • Easy to replicate • Pursue competence metamorphosis
Source of brand creation	Associational	Relational Branding • Unique relational associations • Difficult to replicate • Vigorously defend relationships	First Mover Branding • First mover relational associations
		Difficult	Easy

Ease of replication

Fig. 8.6 Brand vulnerability and risk analysis. © Robertson Cox Ltd, 2000, all rights reserved.

As Fig. 8.6 highlights brand creation can come from two sources – unique competence or associational. Unique competence can be in the form of unique location, quality or delivery; or a combination of these (i.e. Ritz Hotel (unique locations and quality) / Rolls Royce (unique quality) / Amazon (unique delivery), etc). Brand can also arise from association with the success of others, because people want to associate themselves with success or value the endorsement of a high profile individual they esteem (Nike and Michael Jordan/Tiger Woods or Ralph Lauren shirts/or Gucci handbags/Jamie Oliver steaks). 'I am successful because I can afford expensive things.'

The four boxes in Fig. 8.6 explain how this might work. If you have a unique competence that is not easy to replicate you must be careful to defend unique IPR and processes and systems – competence branding. In food this might be a unique recipe for a food product.

Relational branding means the need to maintain the association/ relationship and deny it to others. This is like the Lakeland case with their limited number of suppliers, but would also apply to having, for instance, Nigella Lawson or Jamie Oliver under a long-term contract that others could not easily replicate.

Agility branding means that since others can easily replicate your competence you must constantly strive to find a new competence. This could mean constantly striving for new food products, like the sweets and breakfast cereal manufacturers who constantly seek to bring new ideas/recipes to market.

First mover branding is similar but applies to the first mover opportunities that might come from spotting trends and new fads (i.e. understanding how endorsement from celebrities such as David Beckham might give short-term kudos to a particular product or service). In food, for David Beckham, it could be chocolate footballs.

The Pioneer Foodservice case highlighted the potential to add value and ring fence an acceptable return for all supply chain partners (including farmers) through brand differentiation. Whether this 'local brand' can be successful at a national or international level, or maintain its position in the North West with a plethora of 'local' and national beef brands entering the market (The 'Well Hung Beef Company', based in South Devon, 'Cumbrian Fellbred', based in Milnthorpe, Cumbria and the 'Certified Angus Beef' brand, to name a few) remains, however, to be seen. When developing a successful brand there are a number of factors that need to be considered. The brand risk and vulnerability analysis tool described here can be used to help organisations to think strategically about brand development and protect the considerable investment made in developing and promoting a brand.

The second important issue raised by this case is the importance of linking a firms' marketing strategy with its sourcing strategy. Understanding the appropriateness of a sourcing strategy can only be achieved by following a systematic approach. This requires an understanding of the power positions throughout the supply chain, an understanding of all of the strategic and operational options available in the chain and then the selection of the most appropriate option available to improve the power and leverage position. In the 'Lakeland Beef' brand case, Pioneer selected different sourcing options at different stages of its marketing campaign (see Fig. 8.7). This is because an organisation's internal capability, external power circum-

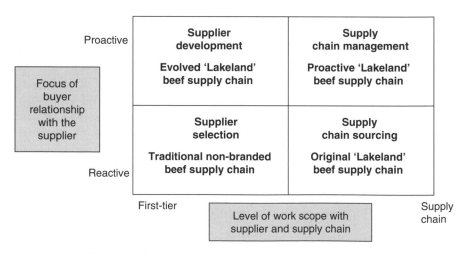

Fig. 8.7 Pioneer's appropriate sourcing option selection.

stances and market performance are subject to change. Appropriateness is about having an iterative mind-set, adapting to change and fully understanding the shifting circumstances in which a business operates.

It is also important that business managers have a clear understanding of the strategies and theories that have originated and evolved in very different industries before adopting them into their own business practice. Current supply chain theories rarely consider security of primary material supply as a major risk to a firm's marketing activities and overall competitive strategy. This is mainly because most of these theories were developed from studies of successful cases in manufacturing assembly supply chains, such as the automobile, computing and the garment industries (Ohno, 1988; Bovel and Martha, 2000; Christopher, 2000; Christopher and Towill, 2000; Mason-Jones *et al.*, 2000; Power, 2005). Sub-components used in these industries are typically industrial commodities and their production can usually be standardised in terms of quality, quantity and fabrication time. Manufacturing in these industries can also be more readily adjusted by using substitutes, or by changing machines and processing methods to avoid temporary supply shortages. The lack of raw material and natural resources such as metal, wood and oil is not often a significant issue in these industries over the long term.

However, unlike manufacturing industries, dispersed primary beef production requires a long fabrication time (minimum of 24 months) and complex disassembling and reassembling processing. Structural changes within the UK beef industry also reduce the long-term sustainability of high quality beef farming businesses. Therefore, without considering a proactive sourcing strategy, firms in the UK beef industry, like Pioneer, will be exposed to unacceptably high levels of business risk in the future. With this in mind, marketing strategies must go hand-in-hand with the establishment of appropriate sourcing strategies in industries with a potentially unsustainable and unsecured upstream supply chain.

The need for effective alignment of marketing and sales, and procurement and supply chain strategies, is clear if brand strategies are to be sustained and competitive advantage achieved. The key learning point for companies, therefore, is the need to understand how to link strategy and operational delivery across product and/or service brand development, as well as within internal and external supply chain execution. Our research has shown that successful companies adopt a 5-step approach to alignment:

1. mandated cross-functional involvement of strategy, marketing and sales, R&D, operations and procurement and logistics functions in business strategy, brand and product/service development and execution;
2. company-wide strategic source planning to link customer and brand differentiation strategies with internal operations and external supply chain execution strategies over time;

3. rigorous and robust segmentation of all internal and external sourcing requirements that have an impact on brand differentiation, with brand differentiation opportunity analysis across all categories of spending;
4. rigorous and robust segmentation of all internal and external sourcing requirements, with brand dilution risk and vulnerability analysis across all categories of spending;
5. development of an internal and external opportunity and supply chain vulnerability early warning system to capture brand differentiation opportunities and to forestall brand dilution risks.

These five steps are necessary because success requires, not only that the strategy and marketing functions work with R&D, operations, procurement and logistics functions to create a brand identity, but also their delivery. In a world of increasing risk and uncertainty, monitoring the external and internal opportunities and threats is clearly a continuous rather than one-off process. Only those organisations that can institutionalise continuous cross-functional implementation of a linked brand and internal and external sourcing strategy are likely to be successful in the future.

8.8   References

BBC NEWS, KENYON, P. (2000), 'Gap and Nike: No Sweat?' BBC News, October 15th.
BBC NEWS, (2001), 'Nike Admits Abuse in Indonesian Plants', BBC News, February 22nd.
BBC NEWS, GALPHIN, R. (2002), Spotlight on Indonesian 'Sweat Shops', BBC News, March 7th.
BEVERLAND, M. (2001), 'Creating value through brands: the ZESPRI™ kiwi fruit case', British Food Journal, **103**(6), 383–99.
BOVEL, D. and MARTHA, J. (2000), 'From supply chain to value net', Journal of Strategy Management, July/August, 24–28.
BPEX (2002), The Road to Recovery – A Strategy for the British Pig Industry, Meat and Livestock Commission, Milton Keynes.
BPEX (2005), The Road to Recovery 2006–2009, Meat and Livestock Commission, Milton Keynes.
CHRISTOPHER, M. (2000), 'The agile supply chain: competing in volatile markets', Industrial Marketing Management, **29**(1), 37–44.
CHRISTOPHER, M. and TOWILL, D. (2000), 'Supply chain migration from lean and functional to agile and customised', Supply Chain Management: An International Journal, **5**(4), 206–13.
COMPETITION COMMISSION (2008), The Supply of Groceries in the UK: Market Investigation, Competition Commission, London.
COX, A. (2001), 'Understanding Buyer and Supplier Power: A Framework for Procurement and Supply Competence', The Journal of Supply Chain Management, **37**(2), 8–15.
COX, A. (2004a), Win-Win? The Paradox of Value and Interests in Business Relationships, Earlsgate Press, Stratford-upon-Avon.
COX, A. (2004b), 'Business relationship alignment: on the commensurability of value capture and mutuality in buyer and supplier exchange', Supply Chain Management: An International Journal, **9**(5), 410–20.

COX, A. (2004c), 'The art of the possible: relationship management in power regimes and supply chains', *Supply Chain Management: An International Journal,* **9**(5), 346–56.

COX, A. and CHICKSAND, D. (2004), 'The exploding cow: the problem of implementing lean production and supply in beef supply chains', *British Academy of Management (BAM) Conference,* 31 Oct–2 Nov, 2004, St. Andrews.

COX, A. and CHICKSAND, D. (2005a), 'Developing local brands in the red meat industry: in the age of the "super supermarkets" can localised branding bring about sustained returns?' *British Academy of Management (BAM) Conference,* 13–15 September, 2005, Oxford.

COX, A. and CHICKSAND, D. (2005b), 'Understanding the impact of change in the red meat industry: a focus on demand and supply complexities in beef and lamb supply chains', *4th Global Conference on Business and Economics (GCBE),* 26–28 June, 2005, Oxford.

COX, A. and CHICKSAND, D. (2005c), 'Sustaining competitive advantage in the UK pig industry: implementing lean production and supply in red meat supply chains', *International Purchasing and Supply Education and Research Association (IPSERA) Conference,* Apr, Archamps, France.

COX, A. and CHICKSAND, D. (2005d), 'The limits of lean management thinking: multiple retailers and food and farming supply chains', *European Management Journal,* **23**(6), 648–62.

COX, A. and CHICKSAND, D. (2006), 'Backing the brand', *CPO Agenda,* **2**(3), Autumn, 46–9.

COX, A. and CHICKSAND, D. (2007a), 'Branding is a strong link to the food chain', *Farmers Weekly,* 10 August, **147**(6), 24–5.

COX, A. and CHICKSAND, D. (2007b), 'The power regimes perspective: power and business choices in food and farming supply chains', *Management Online Review,* morexpertise.com, February, 1–9.

COX, A., SANDERSON, J. and WATSON, G. (2000), *Power Regimes,* Earlsgate Press, Stratford-upon-Avon.

COX, A., IRELAND, P., LONSDALE, C., SANDERSON, J. and WATSON, G. (2002), *Supply Chains, Markets and Power: Mapping Buyer and Supplier Power Regimes,* Routledge, London.

COX, A., IRELAND, P., LONSDALE, C., SANDERSON, J. and WATSON, G. (2003), *Supply Chain Management: A Guide to Best Practice.* Financial Times/Prentice Hall, London.

COX, A., WATSON, G., LONSDALE, C. and SANDERSON, J. (2004a), 'Managing Appropriately in Power Regimes: Relationship and Performance Management in 12 Supply Chain Cases', *Supply Chain Management: An International Journal,* **9**(5), 357–71.

COX, A., LONSDALE, C., SANDERSON, J. and WATSON, G. (2004b), *Business Relationships for Competitive Advantage: Managing Alignment and Misalignment in Buyer and Supplier Transactions,* Palgrave Macmillan, New York.

COX, A., CHICKSAND, D. and PALMER, M. (2005), 'Facing change in the red meat industry: are agricultural co-operatives the way forward?', *21st Industrial Marketing and Procurement (IMP) Conference,* 1–3 September, 2005, Rotterdam.

COX, A., CHICKSAND, D. and YANG, T. (2006a), 'Collaboration in the red meat industry: understanding power, demand and supply characteristics in beef supply chains', *22nd Industrial Marketing and Procurement (IMP) Conference,* September, 2006, Milan.

COX, A., CHICKSAND, D. and YANG, T. (2006b), 'Developing a proactive marketing strategy for UK beef products: being ready for reduced export restrictions', *British Academy of Management (BAM) Conference,* September, Belfast.

COX, A., CHICKSAND, D. and YANG, T. (2007a), 'The proactive alignment of sourcing with marketing and branding strategies: a food service case', *SCM: International Journal*, **12**(5), 321–33.

COX, A., CHICKSAND, D. and IRELAND, P. (2007b), 'The power regimes perspective: choosing the appropriate competitive strategy in food and farming supply chains', *Management Online Review*, www.morexpertise.com, May 2007, 1–11.

CURRY, D. (2002), *Farming and Food: A Sustainable Future*, Report of Policy Commission on the Future of Farming and Food, UK Government, London.

CZINKOTA, M. R. and RONKAINEN, I. A. (1995), *International Marketing*, 4th edition, Dryden Press.

DEFRA (2002), *The Strategy for Sustainable Farming and Food: Facing the Future*, Department for the Environment, Food and Rural Affairs, London.

DEFRA (2006), *The Sustainable Farming and Food Strategy: Forward Look 2006*, Department for the Environment, Food and Rural Affairs, London.

DEFRA (2007a), *Review of Food Chain Initiatives*, Department for the Environment, Food and Rural Affairs, London.

DEFRA (2007b), *Farm Business Survey, England 2005/06*, Department for the Environment, Food and Rural Affairs, London.

DENVER POST (2003), *Shoppers See New Labels, Everyday Beef Becomes Rare, Area Stores Tout Specialist Brands*, Kelly Pate, June 16th.

DOYLE, P. (1994), *Marketing Management and Strategy*, Prentice-Hall, Hemel Hempstead.

EBLEX (2005), *Beef Action for Profit*, English Beef and Lamb Executive, London.

EFFP (2004), *Farming and Food: Collaborating for Profit*, English Farming and Food Partnership, London.

FARINA, E. (2001), 'Challenges for Brazil's food industry in the context of globalization and mercosur consolidation', *International Food and Agribusiness Management Review*, **2**(3–4), 315–30.

FCC and RMIF (2003), *Cutting Cost: Adding Value in Red Meat*, Food Chain Centre, Letchmore Heath, Watford.

FEARNE, A. (1998), 'The evolution of partnerships in the meat supply chain: insights from the british beef industry', *Supply Chain Management: An International Journal*, **3**(4), 214–29.

FEARNE, A. (2005), *Estimating the Impact of RMIF Business Improvement Techniques in the Competitiveness of the GB Meat Industry*, DEFRA, 1–2.

FEARNE, A., DUFFY, R. and HORNIBROOK, S. (2004), 'Measuring distributive and procedural justice in buyer/supplier relationships: an empirical study of UK supermarket supply chains', *88th Seminar, European Association of Agricultural Economics*, 5–6 May, Paris.

FRANCIS, M. (2004), 'Application of the food value chain analysis method in the UK red meat industry', *9th ISL Conference*, Bangalore.

GARRETT, B. (2006), 'Business Relationship Optimisation in UK Beef Supply Chain,' *A Best Practice Forum: Optimising Buyer and Supplier Relationships*, 26 January, London.

HARVEY, D. (2004), 'The UK livestock system', in *Food Supply Chain Management*, Bourlakis, M. and Weightman, P. (eds), Blackwell Publishing, Oxford, 62–82.

HENDRICKSON, M., HEFFERNAN, W., HOWARD, P. and HEFFERNAN, J. (2001), 'Consolidation in food retailing and dairy', *British Food Journal*, **103**(10), 715–28.

HIND, A. (1997), 'The changing values of the cooperative and its business focus', *American Journal of Agricultural Economics*, **79**(11), 1077–82.

HINES, P., LAMMING, R., JONES, D., COUSINS, P. and RICH, N. (2000), *Value Stream Management: Strategy and Excellence in the Supply Chain*, Prentice Hall, London.

HINGLEY, M. (2005), 'Power imbalance in UK agri-food supply channels: learning to live with the supermarkets?' *Journal of Marketing Management*, **21**(1), 63–88.

HINGLEY, M. and LINDGREEN, A. (2002), 'Marketing of agricultural products: case findings', *British Food Journal*, **104**(10), 806–27.

HIRST, P. and THOMPSON, B. (1996), *Globalisation in question*, Blackwell, Cambridge, MA.

HORNIBROOK, S. and FEARNE, A. (2003), 'Managing perceived risk as a marketing strategy for beef in the UK foodservice industry', *International Food and Agribusiness Management Review*, **6**(3), 70–93.

KATZ, J. and BOLAND, M. (2000), 'A new value-added strategy for the US beef industry: the case of US Premium Beef Ltd', *Supply Chain Management: An International Journal*, **5**(2), 99–110.

KOEHN, N. (1999), 'Henry Heinz and brand creation in the late nineteenth century: making markets for processed food', *Business History Review*, **73**(3), 349–93.

LEVITT, T. (1986), *The Marketing Imagination*, The Free Press, Collier Mackillar.

LOCKE, R. (2003), *The Promise and Perils of Globalisation: The Case of Nike*, MIT, Industrial Performance Centre Working Paper 02-008, MIT Press, Cambridge, MA.

LOUGHLIN, D. (1999), 'A study of the degree of branding standardisation by irish food and drink export companies', *Irish Marketing Review*, **12**(1), 46–54.

MAFF (2001), *England Rural Development Programme 2000–2006*, Ministry of Agriculture, Fisheries and Food, London.

MASON-JONES, R., NAYLOR, J. and TOWILL, D. (2000), 'Engineering the leagile supply chain', *International Journal of Agile Management Systems*, **2**(1), 54–61.

MURDOCH, J., MARSDEN, T. and BANKS, J. (2000), 'Quality, nature, and embeddedness: some theoretical considerations in the context of the food sector', *Economics Geography*, **76**(2), 107–25.

NIKE INC (2005), *Annual Report 2005, Evolve Immediately*, Nike, Inc.

OHNO, T. (1988), *The Toyota Production System: Beyond Large Scale Production*, Productivity Press, Portland, OR.

ORDONEZ, H., BASSO, L., PALAU, H. and SENESI, S. (2004), 'Beef and pork agribusiness in Argentina. Design and implementation of origin and quality assurance systems, June 12–15, Montreax, Switzerland. Comparative discrete structural analysis', *IAMA's 14th Annual Conference*.

PALMER, C. (1996), 'Building effective alliances in the meat supply chain: lessons from the UK', *Supply Chain Management: An International Journal*, **1**(3), 9–11.

PORTER, C. (2006), 'Brainstorming predicts a bright future for UK beef', *FarmBusiness*, 7 April, 18.

POWER, D. (2005), 'Supply chain management integration and implementation: a literature review'. *Supply Chain Management: An International Journal*, **10**(4), 252–63.

RAMSEY, B. (2003), 'Whither global branding? The case of food manufacturing', *Journal of Brand Management*, **11**(1), 9–21.

SANS, P., FONTGUYON, G. and BRIZ, J. (2005), 'Meat safety as a tool of differentiation for retailers: Spanish and French examples of meat supply chain brands', *International Journal of Retail and Distribution Management*, **33**(08), 618–35.

SHAW, S. and GIBBS, J. (1995), 'Retailer-supplier relationships and the evolution of marketing: two food industry case studies', *International Journal of Retail and Distribution Management*, **23**(7), 7–16.

SHOCKER, A., SRIVASTAVA, R. and RUEKERT, R. (1994), 'Challenges and opportunities facing brand management: an introduction to the special issue', *Journal of Marketing Research*, **31**(2), 149–58.

SIMOES, C. and DIBB, S. (2001), 'Rethinking the brand concept: new brand orientation', *Corporate Communications*, **6**(4), 217–24.

SIMONS, D. and ZOKAEI, K. (2005), 'Application of lean paradigm in red meat processing', *British Food Journal*, **107**(4), 192–211.

SIMONS, D., FRANCIS, M., BOURLAKIS, M. and FEARNE, A. (2003), 'Identifying the determinants of value in the U.K. red meat industry: A value chain analysis approach', *Chain and Network Science*, 109–21.

SINCLAIR, D. (2004), *UK Beef Sector Outlook*, Meat and Livestock Commission, London.

SMITH, W. (1956), 'Product differentiation and market segmentation as alternative marketing strategies', *Journal of Marketing*, **21**(1), 3–8.

STEIGER, C. (2006), 'Modern beef production in Brazil and Argentina', *Choices- The Magazine of Food, Farm and Resources*, **Issue 2**, 105–10.

TAYLOR, D. (2006), 'Strategic considerations in the development of lean agri-food supply chains: a case study of the UK pork sector', *Supply Chain Management: An International Journal*, **11**(3), 271–80.

TAYLOR, D. and SIMONS, D. (2004), *Food Value Chain Analysis in the Red Meat Sector*, Food Process Innovation Unit, Cardiff University, Cardiff.

USDA (2003), *Branded Livestock and Meat Programs*, USDA, Marketing and Regulatory Programs, Agricultural Marketing Service, Washington.

VAN DER VORST, J., BEULENS, A., DE WIT, W. and VAN BEEK, P. (1998), 'Supply chain management in food chains: improving performance by reducing uncertainty', *International Transactions in Operational Research*, **6**, 487–99.

VRONTIS, D. (1998), 'Strategic assessment: the importance of branding in the european beer market', *British Food Journal*, **100**(2), 76–84.

WOMACK, J. and JONES, D. (1996), *Lean Thinking: Banish Waste and Create Wealth in Your Organisation*, Simon Schuster, New York.

ZYLBERSZTAJN, D. and FILHO, C. (2003), 'Competitiveness of meat agri-food chain in Brazil', *Supply Chain Management: An International Journal*, **8**(2), 155–65.

ZOKAEI, K. and SIMONS, D. (2006), 'Performance improvements through implementation of lean practices: a study of the UK red meat industry', *International Food and Agribusiness Management Review*, **9**(2), 30–53.

Part III

Managing processes efficiently and effectively in food supply chains

9

Value chain analysis of the UK food sector

K. Zokaei, Cardiff University, UK

Abstract: The chapter reports on the findings from a major research project into the UK agri-food industry commissioned by the Department for Environment, Food and Rural Affairs over a period of four years. Several researchers were involved in this project which mapped 33 extended supply chains in detail and produced a portfolio of findings across four primary food sectors, that is cereals, red meat, horticulture and dairy. This chapter puts forward the data collection protocol deployed during this extensive project which should serve as a practical step-by-step guide to analysis of supply chains for readers. It is explained that there is a dearth of methodologies for supply chain analysis and improvement. Therefore one of the key contributions of this chapter is to put forward this successfully applied chain improvement guideline. Moreover, the chapter explains the subtle difference between chain analysis for improving efficiency and effectiveness. The proposed 'value chain analysis' method is distinctive in that it emphasises improvement in chain effectiveness while also delivers considerable efficiency gains. Finally, the method and its subtle differences are explained through an explanatory case study.

Key words: supply chain management, value chain, value stream mapping, lean thinking, UK food sector, supply chain improvement.

9.1   Introduction

Analysing and improving food supply chains have been topical issues over the past decade. Globalization of trade, sophistication of ever more demanding customers and various animal disease outbreaks have greatly contributed to this trend. Surprisingly there have been few methodologies documented in the literature for implementing successful food value chain improvement initiatives. This chapter reports the findings of a major research project and puts forward a tested data collection protocol for food value chain analysis (VCA) which should serve as a practical step-by-step guide for the reader. The proposed VCA method and data collection protocol was developed and adapted in several primary agri-food sectors in the UK.

In discussing the concept of value chain analysis, this article emphasises that 'doing things right' in the chain should become subordinated to 'doing the right things'. Many value chain improvement initiatives focus on efficiency improvements while falling short of addressing the overall objective of the system. Some initiatives forget that the value chain is no more than a channel for delivering what the end consumer demands. This chapter looks at the evolution of supply chain management theories from 'supply chain' to 'demand chain' to 'value chain' and explains their differences. Finally a case study is provided to explain how the proposed VCA method is deployed in practice to improve both efficiency and effectiveness of chains.

The notion that key processes across the supply chain form a value chain and the method of analysing the value chain for competitive advantage was introduced by Michael Porter of Harvard Business School (Porter, 1985). Subsequently, VCA has been developed in the management accounting literature (Shank, 1989; Shank and Govindarajan, 1993) and more recently in the operations management literature (Rainbird, 2004, Zokaei and Simons, 2006) following on from previous claims that supply chain management should go beyond a narrow focus on efficiency management to deliver superior value to the end consumer (Christopher, 2005). Value chain analysis refers to a structured method of analysing the effects of all core activities on cost and/or differentiation of the value chain. According to Dekker (2003, p 5) VCA analyses where in the supply chain the 'costs can be reduced or differentiation can be enhanced'.

Therefore, in an operational sense VCA is a subset of supply chain management (SCM). The essence of the VCA methodology developed by the author and colleagues at Cardiff Business School is to produce a systemic map of the value chain and a systematic method of analysing each strategic activity in relation to the consumer value. In this sense, the proposed VCA method draws extensively upon business process re-engineering (BPR) (Hammer and Champy, 1993), lean thinking (Womack and Jones 1996), value stream mapping (Hines and Rich, 1997; Rother and Shook, 1998; Jones and Womack, 2000) and Porter's value chain model. A key attribute of the proposed method is that analyses and metrics are based on determinant attributes such as quality and time, not on output financial attributes (Fitzgerald et al., 1991). The advantages proposed for deployment of operational measures are that they are the leading indicators of financial attributes and that from a change management perspective operational measures are more easily shared across company boundaries than sensitive financial data. The application of the proposed VCA method is explained in detail in the following. But first the evolution of the SCM body of knowledge is discussed to show the importance of the concept of VCA.

9.2   From supply chain management to value chain management

Value Chain Analysis (VCA) is a different approach from the conventional supply chain improvement approaches in that it emphasizes the concept of consumer orientation and consumer value. Whereas, conventional supply chain improvement initiatives predominantly focus on waste elimination and cost reduction (i.e. chain efficiency), VCA is concerned with differentiation and value enhancement in the supply chain as well as cost reduction.

Supply chain management (SCM) is a fairly new concept which only started to make a significant appearance in the management literature in the 1980s (Oliver and Webber, 1982; Houlihan, 1985; Stevens, 1989) and has since been popularized by several authors as an independent field of study (Cooper and Ellram, 1993; Davenport, 1993; Christopher, 2005; Mentzer *et al.*, 2001; Gibson *et al.*, 2005, Cousins *et al.*, 2006). Nonetheless, much of the underlying thinking dates back several decades. In fact, the roots of SCM can be traced to systems dynamics and analysis (Forrester, 1958), integrated logistics management (Bowersox *et al.*, 1959) and the idea of forming cooperative relationships with suppliers (Farmer and Macmillan, 1976).

Arguably, SCM is not developed enough to be regarded as an independent discipline. But the general consensus amongst academics is that SCM is a general problem domain represented by a significant – yet diverse – body of knowledge in the literature (Special Issue, 2006).

Table 9.1 provides a rounded understanding of the evolution of SCM concepts from logistics to supply chain to demand chain to value chain. It reviews some of the most frequently cited definitions of chain management presented in chronological order. The review illustrates that although there is little consensus on the scope and meaning of the value chain or value chain management, an evolutionary trend is evident in the literature. The latter chain management contributions are much more strategic and broader in scope in conceptualizing chain management.

The above literature review confirms that there is little consensus on the scope or meaning of SCM or VCM. It also explains that chain management has evolved from a narrow focus on physical aspects to a broad multifaceted theory. The early conceptions of SCM (Houlihan, 1985; Stevens, 1989) emphasize the importance of a holistic approach as opposed to single firm optimization. In fact, chain management theory begins by showing the potential which lies beyond the boundaries of a single firm. The original supply chain contributions largely focus on the physical aspects in the supply chain, for example the dynamics of information and material flows (Forrester, 1958) and inventory management and transportation (Jones and Riley, 1985). The narrow focus of the early SCM literature has inspired several authors to compare and contrast SCM with integrated materials and

Table 9.1 Overview of some key contributions concerning the value chain and value chain management

Contributor	Proposed delineation for value chain, VCM or related constructs (quotes in italic)	Key features of the definition
Forrester (1958)	Forrester promoted the dynamic study of whole systems as opposed to the study of separate functions or companies. *Company [and value chain] will come to be recognized not as a collection of separate functions but as a system in which the flows of information, materials, manpower, capital equipment, and money setup forces that determine the basic tendencies towards growth, fluctuation, and decline. I want to emphasise the idea of movement here because it is not just the simple three-dimensional relationships of functions that counts, but the constant ebb and flow of change in these functions – their relationships as dynamic activities.*	Forrester aims to show the importance of the interrelationships between company functions and between the company and its network of suppliers and customers. Forrester emphasizes that the dynamics of relationships between the flows of information, materials, human-power, finances and capital equipment should be studied and standard management methods should be extracted from such studies.
Houlihan (1985)	The whole supply chain is a single business process. SCM can be defined as having the following key characteristics: 1 *The supply chain is viewed [and managed] as a single process . . .* 2 *SCM calls for and in the end depends on strategic decision making. Supply is a shared objective of every function in the chain . . .* 3 *SCM calls for a different perspective on inventories which are used as balancing mechanism of last, not first, resort.* 4 *A new approach to systems is required – integration rather than interfacing.* (p. 26)	Houlihan characterizes the differences between SCM and traditional materials and manufacturing control science. Houlihan argues that, on account of the new economy, the traditional logistics and materials management approaches, which sought trade-offs between various conflicting key functional objectives of purchasing, production, distribution and sales, do not work very well any longer. A new approach needs to be adopted – supply chain management.

Porter (1985)	Michael Porter proposed the concept of value chain and the value chain model as a means of analysing intra-firm competitiveness. In addition he introduced the value system model, which effectively is an extension of the value chain model to the whole supply chain, to analyse inter-firm competitiveness. The value chain and value system models are activity based views of the firm and the chain. According to Porter, every firm/chain is a collection of value activities performed to make a product valuable to buyers.	The value system model is probably, today, recognized as the 'value stream map'. The value system model disaggregates the supply chain into strategically relevant activities (processes) in order to understand the sources of competitive advantage. More importantly, the value system and value chain models emphasize the importance of the linkages between the activities along the chain. Process re-engineering's approach to SCM is the extension of his work (Davenport, 1993)
Stevens (1989)	Stevens defines the supply chain and SCM as: *The supply chain is the connected series of activities which is concerned with planning, coordinating and controlling material, parts and finished goods from suppliers to the customer. It is concerned with two distinct flows through the organisation: material and information* (p. 3). *The objective of managing the supply chain is to synchronise the requirements of customer with the flow of materials from suppliers in order to effect a balance between what are often seen as conflicting goals of high customer service, low inventory management, and low unit costs.* (p. 3)	Stevens (1989) provides one of the earliest clear cut definitions of supply chain and supply chain management. He puts customer service at the heart of SCM and defines it as a bundle of delivery service, pre- and post-sales service, technical support and financial packages. Stevens (1989) proposes a structured framework for developing an integrated supply chain strategy which is even applicable to today's supply chains. This framework has three stages: 1 identifying the customer needs 2 diagnosing supply chain opportunities 3 developing an action plan for implementation.

Table 9.1 *Cont'd*

Contributor	Proposed delineation for value chain, VCM or related constructs (quotes in italic)	Key features of the definition
Hewitt (1994)	Hewitt defines the supply chain as a single business process which should be managed as a whole. Hewitt contends that this approach is sharply distinct from conventional logistics management since it simultaneously addresses all aspects of the operation in the whole chain. Hewitt regards this level of logistical evolution as *'integrated intra-company and inter-company supply chain management'*. (p. 4)	Hewitt defines SCM as the final stage in the evolution of logistics management. Successful SCM depends on the recognition and management of three critical dimensions in the chain: 1 physical flow (work activity) 2 information flow 3 decision/authority flow.
Womack and Jones (1996)	*The value stream is the set of all the specific actions required to bring a specific product (whether a good or service or increasingly a combination of the two) through the three critical management tasks of any business: the problem-solving task running from concept through detailed design and engineering to production launch, the information management task running from order taking through detailed scheduling to delivery, and the physical transformation task proceeding from raw materials to a finished product in the hands of the customer.* (p. 19)	The first principle of lean thinking (Womack and Jones, 1996) is consumer value. Womack and Jones bring consumer satisfaction, and subsequently the new product development process, to the heart of SCM argument. Womack and Jones introduce the notion of 'value stream' which is essentially a 'value system' looked at from a single product point of view (Porter, 1985). Whereas Porter contends that *'the relevant level for constructing a value chain [and value system model] is a firm's activities in a particular industry (the business unit)'* (Porter, 1985, p. 36), lean thinkers (Womack and Jones, 1996; Hines and Rich, 1997) propose that the appropriate level of analysis is disintegration of the chain into processes/activities at product level.

| Council of Logistics Management (CLM) (1998) | Supply chain management is the systemic, strategic coordination of the traditional business functions and the tactics across these business functions within a particular company and across businesses within the supply chain for the purposes of improving the long-term performance of the individual companies and the supply chain as a whole. | CLM distinguished SCM from logistics management and acknowledged that logistics is one of the aspects of SCM in 1998. This distinction led CLM to change its name to the Council of Supply Chain Management Professionals (CSCMP) in 2004. According to CSCMP, SCM extends the research on logistics to take into account issues of governance, multi-firm relationships and innovation in the chain to create consumer value. Thus, SCM is a philosophy for synchronization of all activities/capabilities (not just logistics) to create consumer value. |
| Croom et al. (2000) | ... the supply chain should be seen as the central unit of competitive analysis ... In short, the contention that it is supply chains, and not single firms, that compete is a central tenet in the field of supply chain management. (p. 68)

Supply chain management and other similar terms such as network sourcing, supply pipeline management, value chain management, and value stream management have become subject of increasing attention in recent years. (p. 67) | The paper sets out to 'establish the general problem domain of supply chain management'. (p. 67) It maps and evaluates SCM research and provides a topology of the domain which confirms a profound lack of theoretical research, i.e. SCM is not theoretically and conceptually well researched. The paper points out the central role of the supply chain for competitiveness and argues that VCM and SCM are the same.

The authors contend that 'whilst supply chain management as a concept is a recent development, much of the literature is predicated on the adoption and extension of older, established theoretical concepts' (Croom et al., 2000, p. 68) such as transaction cost economics and competitive strategy. |

Table 9.1 Cont'd

Contributor	Proposed delineation for value chain, VCM or related constructs (quotes in italic)	Key features of the definition
Mentzer et al. (2001)	*A supply chain is defined as a set of three or more entities (organizations or individuals) directly involved in the upstream and downstream flows of products, services, finances, and/or information from a source to a customer. (p. 4)* *SCM is defined as the systemic, strategic coordination of the traditional business functions and the tactics across these business functions within a particular company and across businesses within the supply chain, for the purpose of improving the long-term performance of the individual companies and the supply chain as a whole (p. 18).* *SCM is concerned with improving both efficiency (i.e. cost reduction) and effectiveness in a strategic context (i.e. creating customer value and satisfaction through integrated SCM) to obtain competitive advantage that ultimately brings profitability. (p. 15)*	Implicit in this definition of supply chain is that supply chains – as business phenomena – exist whether they are managed or not. Thus, the authors draw a definitive distinction between the 'supply chain' as a given phenomenon and 'supply chain management' as the science and art of managing supply chains. Mentzer et al. (2001) Distinguish between SCM philosophy and implementation. The philosophical view that companies across the supply chain constitute a potentially coordinated entity is regarded as a management philosophy and branded as 'supply chain orientation'. Subsequently, SCM is defined as the implementation of a 'supply chain orientation' vision and an upshot of 'supply chain orientation'. This contribution addresses the importance of both chain efficiency and effectiveness.
Gibson et al. (2005, p. 22)	*Supply chain management encompasses the planning and management of all activities involved in sourcing and procurement, conversion, and all logistics management activities. Importantly, it also includes coordination and collaboration with channel partners, which can be suppliers, intermediaries, third-party service providers, and customers. In essence supply chain management integrates supply and demand management within and across companies.*	This contribution reports on the results of a Council of Supply Chain Management Professionals (CSCMP, formerly Council of Logistics Management) survey of its members' views of SCM. The survey proposed two potential definitions for SCM. Based on the results of this study, the authors conclude that the verified definition (in the left box) is not definitive and further refinement and evolution of the definition are both possible and desirable.

logistics management (Cooper, Lambert and Pagh, 1997; Hewitt, 1994; Houlihan, 1985). These authors have generally come to the same conclusion that SCM is a much broader concept encompassing issues beyond the boundaries of the logistics sub-system.

The accounts of chain management in Table 9.1, however, transcend this narrow focus by taking account of broader issues such as long-term performance of the whole chain (CLM, 1998), supply chain competitiveness (Christopher, 2005), consumer enrichment (Ross, 1998) and new product development (Womack and Jones, 1996). Therefore, it is concluded that over the past two decades, SCM has evolved from a one-dimensional subject with a rather narrow focus on logistics and physical aspects of material flow into a multi-faceted theory encompassing a broad range of subjects. For example, Hewitt (1994, p. 7) contends that modern chain management simultaneously addresses 'all aspects of the operation of the supply chain, including work practices, information flows and authority/ decision making structures'. Modern SCM can be more appropriately described as value chain management since it encompasses value enhancement and strategic differentiation as well as cost reduction and operational efficiency improvements. Internationalization of trade, sophistication of technology and markets, increased global competition and the rise and dominance of the Japanese production philosophies (Womack and Jones, 1996; Hines, 1994) have contributed immensely to the evolution of SCM and its core concepts.

VCM is best described in Ross (1998) as the synchronization of competencies along the whole chain to create unique, innovative and individualized sources of consumer value. Another notable conjecture by Womack and Jones (1996) describes 'value stream management' as the integration of the problem solving and new product development task, management of the information task, and the physical transformation and transportation task. Christopher (2005) is more direct in describing SCM as 'demand chain management' to reflect the market orientation required in today businesses.

Chain management is a domain where efficiency improvements are the prime objective perhaps because of its origin in logistics and operations management, for example time-based competition (Stalk and Hout, 1990; Womack and Jones, 1996; Christopher, 2005), cash-to-cash time (Bowersox and Closs, 1996), quality-based competition (Womack et al., 1990) and cost-based competition (Shank and Govindarajan, 1993). Few recent publications, however, emphasize the importance of enhanced consumer satisfaction in the context of chain management (Zokaei and Hines, 2007; Zokaei and Simons, 2006; Hines et al., 1998). In this context, understanding consumers' attributes and jointly striving for augmentation of consumer satisfaction are imperative to successful VCM. Value chain analysis is a methodological approach both for identifying best solutions to add superior value to the end customer and eliminating operational waste. The following section provides a step-by-step practical guide to VCA.

9.3   Methodology for value chain analysis

9.3.1   Ten days data collection protocol

As mentioned earlier, the following VCA method and data collection protocol was developed during the Food Chain Centre's VCA programme which looked at four key sectors within the UK agri-business industry, dairy, cereals, red meat and fresh produce. This major piece of research consisted of in-depth analysis of 33 value chains from primary production to the point of consumption covering different routes to market and different raw materials (e.g. pork, beef and lamb in the case of red meat industry). Each one of the chains was studied independently as a separate project facilitated by a lead researcher and on average took three to five months to complete.

Following a recommendation by the Policy Commission on the Future of Farming and Food (also known as the Curry Commission), the Food Chain Centre (FCC) was established in 2002 to '. . . bring together people from each part of the food chain'. FCC then linked up with several industry and levy bodies in the UK fast moving consumer goods (FMCG) and agri-food sectors such as the Institute for Grocery Distribution (IGD), Efficient Consumer Response (ECR), the National Farmers Union (NFU), Home Grown Cereals Authority (HGCA), Dairy Industry Forum (DIF) and Red Meat Industry Forum (RMIF) to launch an extensive research initiative widely known as the VCA programme. FCC received several million pounds joint funding from Department for Environment, Food and Rural Affairs (DEFRA) and the Department for Trade and Industry (DTI) to improve supply chain performance and vertical collaboration in the UK agri-business sector and, subsequently, commissioned the Lean Enterprise Research Centre (LERC) at Cardiff Business School to analyse 33 value chains covering different routes to market as well as different raw materials (e.g. pork, beef and lamb in the case of red meat).

A standard data collection protocol was developed during the VCA programme referred to as the 'ten days VCA method' (Zokaei and Simons, 2006; Francis, 2004). The ten days VCA protocol is rooted in lean thinking (Womack and Jones, 1996) and, similar to lean value stream mapping techniques, it begins with the selection of a single product family for analysis. A product family is defined as products or stock keeping units (SKU) that flow through similar processes in their value chains. The method requires establishment of a value chain team to follow the flows of the selected product across firms and functions.

Generally, a cross-company team of various stake holders (e.g. primary producer, processors, retailer, food service, etc) were pulled together in each project and on average 8–12 contact days were spent with the team members in the field. All data was treated as confidential unless cleared by companies and a confidential report was published for each project containing a detailed analysis of the selected product. There were also various case

Table 9.2 Overview of value chain analysis programme activities

No.	Sector	Chains analysed	Reports (non-academic publications)	Timeframe
1	Red meat	8 chains through RMIF 1 in Wales	Sector specific final report Individual chain confidential reports and case studies	2003–06
2	Dairy	8 chains through DIF	Sector specific final report Individual chain confidential reports and case studies	2004–07
3	Cereals	8 chains through HGCA (3 sub-contracted to Cranfield University)	Sector specific final report Individual chain confidential reports and case studies	2004–07
4	Fresh produce	7 Defra funded chains (including 2 organic chains) with HDC 1 joint private and public funding	Sector specific final report Individual chain confidential reports and case studies	2004–07

studies and sector specific final reports published for the sponsors (see Table 9.2).

Broadly, the data collection protocol consisted of four stages:

1 Team building and introduction: At this stage, the team was familiarized with the mapping and analysis techniques deployed during the project and also the basic principles of SCM. At least one representative from each participating firm was committed to 'walk' the whole value chain from farm to the end point of sales or point of consumption depending on the project. A benefit-share agreement was put in place to ensure that the potential gains were fairly shared. Early team building activities had a great impact on the overall success of the projects and were recognized as significant events later on during the project.

2 Inter-firm and intra-firm data collection: During this stage a current state map of the physical and information flows along the whole supply chain was constructed with a specific focus on the time data (Rother and Shook, 1998). Rother and Shook were first to document the lean mapping technique for a single firm which was later extended to the whole supply chain by Jones and Womack (2000) both heavily borrowing from Toyota's approach to value chain analysis. The team walked the whole value chain and collected the necessary information over a period of on average 3–4 months. Some common units of analysis during

the mapping stage are time, delivery and quality recorded for each echelon along the chain. Also, the team looked at operations and logistics efficiency measures such as demand amplification, on-time/in-full delivery performance, lead times and defective parts. Financial data were not investigated to ensure maximum buy-in from all participant companies. Mapping of the end-to-end supply chain and collection of the current state data often required 4–5 days in the field depending on the size and complexity of the chain.

3   Evaluation of the current state and suggestions for the future state of the supply chain: Having gained a clear understanding of the current state, the team debated the data to identify potential improvement opportunities both at the whole chain and individual firm levels. Also, there was an opportunity to compare and contrast the current state against the team's understanding of the consumer value. The VCA was an opportunity for the team members to connect their role in the chain with the ultimate satisfaction of consumers. Clearly, consumer satisfaction can either come at a basic level where known requirements are met or it can come at a much more advanced level where consumer expectations are exceeded.

Normally various improvement opportunities were identified ranging from easily achievable fruits to very difficult.

4   Action planning: In the final stage of the project an action plan was developed to take the supply chain from the current state to the future state based on the immediacy of the actions, the size of the prize, availability of change resources and the relevance of the identified improvement opportunities to the consumer needs. The VCA project did not go further into a detailed implementation phase.

The data collection protocol (i.e. the ten days activity plan) is described in Table 9.3.

9.3.2   Food value chain analysis programme: generic findings

There were a number of generic findings across the four sectors. The following compiles the findings in different sectors into a single table providing comparisons and discussions. Interestingly, the VCA project in different sectors identified remarkably similar results. Also, it is evident that the emphasis of the VCA work has predominantly been on efficiency factors such as transport, quality, demand management and inventory levels. What's more, the VCA reports, in all four sectors, call for greater supply chain integration and collaboration. Clearly supply chain collaboration is required for improvement of both efficiency and effectiveness of supply chains. The following table demonstrates common themes across different sectors (where findings have been similar they have been bundled under one single theme).

Table 9.3 Value chain analysis: the ten days activity plan

Session	Event	Activities
1	Initial team building workshop	Explanation of key concepts such as flow, pull, demand amplification, etc. Explaining value stream mapping techniques and tools deployed during VCA, e.g. process activity mapping, product variety funnel, etc. Explaining principles of collaboration along the chain. Discussing a benefit sharing agreement. Identifying the core team members.
2	Workshop: current state	Selecting a suitable product group for mapping. For example, choosing the largest mutual flow or a product with the biggest potential for improvement. Creating a generic big picture (current state) map of the value chain.
3, 4, 5, 6	On-site mapping	Creating detailed current state maps for individual firms along the chain, e.g. farm, food processor, distribution centres and retail store. The current state maps cover both the physical and information flows. Also, the current state maps bear all the relevant operational (determinant) performance indicators. Identifying internal operational improvement opportunities at each facility.
7	Workshop: whole chain ideal state map	Discussing and creating an ideal state map so that the whole team can aspire towards a single shared vision, e.g. an ideal lean value chain. Identifying, discussing and categorizing consumer value. Identifying key performance indicators (KPIs) for the whole chain.
8, 9	Workshop: future state map	Discussing and creating a future state for the whole chain. The ideal state is a vision whereas the future state is an achievable target. At this workshop the ideal state map is rationalized to the future state map. Identifying key projects towards the future state. Linking key projects (opportunities for improvement) with the measures of consumer value to identify the vital few projects for improvement. Creating a clear action plan where all key stake holders and people responsible for implementation are identified.
10	Presentation of final results	Team presentation of recommendation for improvement and findings to top tier management of all companies involved. Confirm proposal with all stake holders and various project owners. Discussions concern benefit allocations and milestones. Final decisions taken about which improvement projects to progress.

Six key concerns can broadly be identified as common themes amongst all 33 chains in the four sectors analysed during the VCA programme (see Table 9.4). The other three issues each appear in at least two sectors. Five of the six common concerns directly relate to supply chain efficiency, whilst the sixth one is the *lack of* understanding of consumer value which is related to supply chain effectiveness.

Identification of consumer value is the first principle of lean and the VCA project created a rare opportunity for the team members to connect supply chain activities with the actual requirements of the end consumers. Although all the sector specific reports (even if only briefly) touch on a lack of understanding of consumer needs, none puts forward a practical solution for improving consumer alignment along the chain, let alone enhancing

Table 9.4   Cross-sector findings from the UK food VCA programme

No.	Concern	Red meat	Dairy	Fresh produce	Cereals
1	Quality issues and product loss	√	√ (mentioned in the report but not as an independent finding)	√	√
2	Demand management and waste in information flows	√	√	√	√
3	In-house operational inefficiency at different levels in the chain	√	√	√	√
4	Transportation inefficiencies	√	√	√	√
5	Lack of understanding of consumer requirements	√	√	√	√
6	Lack of trust and collaboration / lack of contracts and risk share agreements / opportunistic trading relationships	√	√	√ (in the report but little attention)	√ (in the report but little attention)
7	Lack of consistent measurement of operational performances at the whole chain level	–	√	√	–
8	Inventory management issues	√	–	–	√
9	Supply chain structural issues	√	–	–	–

consumer experience (FCC, 2007). More recently, the Food Chain Centre published its completion report reiterating the same problem: 'It is rare to find an understanding of consumer needs shared through a product chain but when achieved, it helps greatly to promote trust and innovation' (FCC, 2007 p. 19). Yet again, the report does not fully explain the issue or to show how realignment can be achieved and why it boosts chain performance and competitiveness.

9.3.3   Food value chain analysis: a case study

This case study looks at the entire supply chain from a medium size pig farm through a consolidated abattoir and processing plant to a public sector canteen. The fieldwork for this case study was conducted as part of the VCA programme in the UK. The companies involved were a farm, a meat packer, a food service company and the catering department of a public sector company. A team of senior company representatives and two academic facilitators followed pork legs and loins from farm to canteen. Table 9.5 lists the companies involved and representatives from each company.

The mapping exercise showed that the UK canteens have a relatively steady demand owing to almost a fixed number of people being catered for daily with occasional spikes in demand. Notably, the main reason for the team members participating in the VCA improvement was to improve logistical efficiency and that the chain tightly adheres to the target budget. So, at the outset, the VCA participants were predominantly focused on improving whole chain efficiencies and reducing cost. The food service company supplied a food range of around 1600 products across three temperature bands, ambient, frozen and chilled, delivered to around 1000 delivery points, making 150 000 deliveries and assembling 21 million food items per year. The total value of this catering contract in 2005 was just less than £100 million per annum, where approximately £15 million was spent on red meat procurement only (including pork).

Table 9.5   Supply chain improvement core team members

No.	Company	Representative
1	Farmer	No representation in the core team
2	Abattoir and meat packer	General sales manager
3	Food service company	Director of 'public catering' supply Senior buyer – fresh foods
4	Public organisation catering department	Operations manager
5	Cardiff University	The author

This case looks at the supply of frozen pork loins and legs from a meat packer plant in East Anglia. The farm has an integrated system of cereals, potatoes and pigs located in Lincolnshire with a long-term relationship with the meat packer. The following shows how the supply chain team was able to identify the disconnectedness of consumer value both with the product attributes and the supply chain activities. Also, there are discussions about how processes along the supply chain were potentially realigned with the consumer requirements and why supply chain effectiveness was partially improved. This is followed by a description of the subsequent efficiency gains.

Current state findings and analysis
As explained above, varieties of mapping techniques are deployed during the VCA such as a quality filter map, a delivery adherence map and demand amplification (Hines and Rich, 1997). The relevant tools were introduced to the team members at the outset of the project. The most basic tool deployed was process activity mapping (PAM). PAM is a means of recording every step along the chain and a platform for creating current state maps. It captures the details of all the tasks required for completion of each process including time taken to complete each task, distances moved and the number of times operators touch the product during each task. A separate PAM sheet was created for every part of the chain and activities were categorized as value adding (VA) and non value adding (NVA) along the process. Only a fraction of activities were considered to be value adding. The aim was to increase value adding operational time where possible. In the lean approach the ultimate arbiter of value was the end consumer and the yardstick for determining VA and NVA activities was the consumers' willingness to pay for the service.

It must be noted that some NVA are necessary given the technical and practical constraints. For example, if a product is waiting in stock it is recorded as NVA in the lean approach; nonetheless a certain amount of inventory is inevitable in any supply chain. For example, in this case study the total time at the meat cut operation was 15 178 s (or ~253 minutes), of which only 49 s were value adding, in other words 0.3% VA time.

Subsequently, all PAM data were pulled together to create a current state map of the physical flows for the whole chain and then the information flows were added to generate the current state map as illustrated in Fig. 9.1 (map of loin). The current state map shows the physical flows, the information flows, total lead time and percentage value adding time. In this case it shows that the total lead time for loin is 276 days and 11 hours, of which 233 days are spent at the farm (animal breeding and rearing). So lead time excluding the time at the farm is 43 days and 11 hours, whereas the value adding time was just less than 25 hours (i.e. 24 hours cooling at the meat packer, 15 minutes value adding operation in slaughter and cutting, 15 minutes value adding operation in the distribution centre and 20 minutes

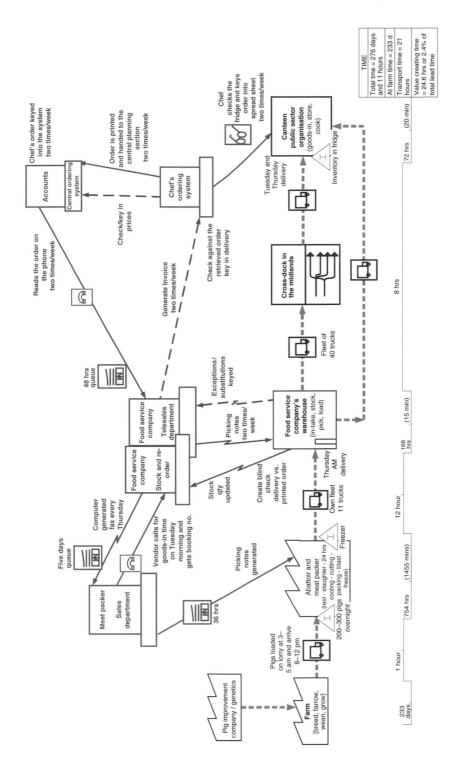

Fig. 9.1 Current state map for pork loins from farm to canteen.

value adding time during cooking in the canteen's kitchen), that is 2.4% of the total lead time excluding time spent in animal rearing.

The analysis of the chain identified several opportunities for improvement ranging from quick fixes to long-term changes. Value stream mapping often leads to the exposure of several 'low hanging fruits' which are in most cases dealt with during the course of the project. During the future state mapping sessions, a full list of all opportunities were generated and they were ranked through discussion and consensus. Discussions centered on identifying the cost/benefit of implementing each improvement opportunity. The team identified the following five key opportunities to be taken forward:

1. Review of the product specifications: the product specifications were not revised since established in 1963 and were by and large outdated.
2. Setting up electronic data interchange (EDI) between the food service company and the public sector organization. A telesales system was in operation with 20 staff dedicated to the telesales department at the food service company.
3. Backhaul opportunities between the supplier and the distribution company: both the processor and the food service company operated their own fleet. The team established that, in addition to this value stream, plenty more opportunities existed for backhaul to and from the central warehouse through better planning with various suppliers.
4. In-house improvement opportunities at the processing plant (such as an improved layout, work balance and packing equipment performance).
5. Work standardization at the farm: for example, reducing the variance in the performance of stockmen. Historic records showed that the skilled stockmen achieved a piglet mortality rate four times lower than the poor stockmen. It was endeavoured to standardize the skilled stockmens' operations when training new staff.

Generating a future state map

Following analysis of the current state map, the team worked towards generating a collective vision of a supply chain that operates as an integrated entity focusing on the enhancement of the supply chain value proposition and elimination of all non-value adding activities. The shared point amongst all team members was the satisfaction of the end consumer. Also, as already mentioned, consumer value is the first principle of lean thinking. The team members brainstormed various attributes of the consumer value, categorized them and related them to a set of supply chain key performance indicators (KPI) as illustrated in Table 9.6. Through consensus, five factors were identified as the key constituents of value, reflecting the requirements set by the public sector organization, the chefs and the actual consumers. The brainstorm session was a rudimentary way of capturing and discussing 'voice of customer'. The outcome of the discussion session was largely

Table 9.6 Translating consumer value into supply chain KPIs

Consumer value	Supply chain KPIs
Cost efficient distribution	Total cost to serve for the whole job
Quality and consistency	Produce to exact product specifications as required by the final customer
Value for money	Top 40 products' buying effectiveness (examined by independent third party company)
Delivery on-time/in-full	99.7% right quantity delivered (substitution allowed)
	97% perfect order (measured by comparing credit notes with invoices – no substitutions)
Strategic reserve	21 days to feed

influenced by and depended on the public organization sharing their knowledge of the consumer needs acquired through focus groups and direct contact with chefs.

At this stage it was obvious that, even though the foodservice company and the public sector organization were separately measuring and analysing consumer satisfaction, they had never coherently linked together the requirements of the consumers, the product features and the supply chain activities. The future state workshop provided the opportunity to link the three together for the first time. Altogether, a lack of consistent understanding about the consumer requirements was observed suggesting that, from a supply chain effectiveness perspective, the chain was in a state of 'unconscious incompetence'.

In contrast, the efficiency levels along the supply chain were good. For example, the distribution service achieved 99.7% lines delivered in full against lines ordered (substitutions permitted). In the abattoir, a state of the art slaughter line was observed, which had excellent ergonomics leading to better consistency in quality and the first Autofom application in the UK that ultrasonically scanned each carcass immediately after slaughter. A three-dimensional picture was built up of the muscle and fat present, which allowed accurate payment to the producer for the actual meat delivered, when fully implemented.

Table 9.6 shows different aspects of value in the ranking of importance as agreed upon by the team members. The team's perception was that the most important feature of value was cost efficient distribution. This was linked into an overall measure for the total cost of delivering the pork product. There was a cost-plus contract in operation between the food service company and the public sector organization whereupon any saving in the cost of distribution was to be shared equally between the two parties. The team's perception was that the aim of the VCA project is to deliver cost savings and an obvious area for cost saving – which could be equitably shared – was the distribution cost. The second most important facet of value was the quality and consistency of the product which already was being measured through rigorous methods such as customer direct feedback to the suppliers and random quality checks. The evidence acquired from the end customer suggested that the supply chain consistently met the specifications and the quality criteria. However, this was later proved to be wrong owing to the lack of understanding of consumer needs. The third attribute of the consumer value was the cost efficient purchase of the raw material. This was being measured through independent third-party monitoring of the procurement of the top 40 products (which included the pork lines). The fourth aspect of the consumer value was on-time/in-full (OTIF) delivery into the canteen. The food service company achieved 97% OTIF (calculated by checking the credit notes against the invoices). The measure was closer to 99.7% when substitutions were taken into account (the contract allowed for substitutions within reason). Last but not least, the end customer required a strategic reserve of at least 21 days stock to be kept in the distribution pipeline at anytime (i.e. inventory anywhere between the distribution company's warehouse and the canteen). It was not clear whether this was actually needed or just a legacy of past systems. However, keeping a strategic reserve was not a big issue since the products were delivered via a frozen chain. Then again, a chilled chain would have meant supply of cheaper, tastier and fresher produce.

The key improvement opportunities and the issues related to consumer value have so far been discussed in this case study. Moreover, value attributes were related to a set of supply chain KPIs. In order to understand the extent to which the key projects deliver against the supply chain objectives, a sanity check against the supply chain KPIs was carried out during the future state workshop.

In Fig. 9.2, each box is scored on a scale of 0–3 where 0 denotes no impact on the relevant KPI and 3 shows very high impact. As illustrated, implementation of the EDI link and product specification review equally had the highest impacts against the supply chain KPIs. The implementation of EDI could result in significant efficiency gains estimated at around £400 000 per year. Nevertheless, it required relatively large capital investment and hence the need for a long lasting cost/benefit sharing agreement between the food service and the public sector organizations. The two companies could not reach an agreement mainly because the remaining length of the contract did not cover the pay back period for the required investment. On the other hand, the review of product specifications required zero investment while potentially improving both effectiveness and efficiency of the chain. The following explains how chain effectiveness was improved and what efficiency gains were obtained as a result.

The future state discussions revealed that the food service company and the public sector procurement organization were both active in understanding the consumer needs through focus groups. Moreover, the processor and its suppliers took great care to produce to the correct specification. Even

Consumer value attributes	Supply chain KPI	Review of product specification	EDI	Backhaul between abattoir and food service	Farm work standards	Meat packer in-house efficiency gains	
Cost effective distribution	Total cost to serve for job	2	3	2	0	0	7
Quality and consistency	Product specification	3	1	0	0	0	4
Value for money	Check top 40 products' price	3	3	1	3	3	13
Delivery on-time/in-full	97% perfect order	0	1	0	0	0	1
Strategic reserve	21 days to feed	0	0	0	0	0	0
		8	8	3	3	3	

Fig. 9.2 Key improvement projects impact against supply chain key performance indicators (Scale: 0–3).

though the product was reasonably priced, had good quality and was delivered 99.7% in right quantity (allowing for substitutions), it did not match the customer attributes. The supply chain analysis connected all aspects of the supply chain together and revealed that the product specifications were outdated and did not reflect the true consumer needs. The team identified opportunities for changing the product attributes and supply chain activities as described below:

Boneless loin – The current state map showed that the loin was being supplied with bone. In the kitchen the product needed to be boned out and the disposal of the bone also incurred additional cost. The impact of aligning the chain to the consumer value in this case were twofold:

1. Effectiveness gains: a questionnaire was sent out by the public sector organization following the team's suggestion to see whether delivering boneless loin is aligned to the customers' preference. One limitation of the study was in consulting the chefs rather than the actual end consumers about their preferences. Moreover, the telephone interview did not ask exactly why chefs liked or disliked the boneless product; nor did it follow-up when the answer was not specified. This issue increases the possibility of type II error in analysing the results of the questionnaire. Answers were received from 23 chefs responsible for fairly similar size

canteens supplied from the same source and through similar channels. In all cases the public sector organization had followed up by telephone and in few cases had even obtained the results through telephone interviews.

The results showed that the final customer preferred the boneless product since there was no need for boning in the kitchen. Of the 23 respondents, 14 preferred boneless loin, three were indifferent and six said no, that is 61% in favour, 14% indifferent and 25% against boneless loin. The statistical question is whether the proportion of the 'yes' answers is significantly higher than would be expected by chance. To find an answer to this question, the researcher used the binomial distribution to identify the probability of finding six or less negative answers in a sample of 20 when the random probability of a 'no' in each trial is 50%. The probability of six or less respondents disliking the boneless product from the 20 respondents is 0.057, that is B (20, 6, 0.5) = 0.057. It must be noted that three respondents did not specify their preference and therefore their data were regarded as meaningless. Assuming an alpha level of 0.05, it can be (marginally) concluded that the customers preferred boneless loin.

2. Efficiency gains: realignment of the supply chain with the consumer value (i.e. boning at the cut and pack stage and delivering boneless loin) leads to a number of efficiency improvements. First, the boning operation was more time consuming and labour intensive when carried out at the canteen as opposed to being done at the processor on an industrial scale. The processor produced boneless loin for other customers and could batch products together. Second, there was a small residual value to the bone at the processor. Third, there were logistical savings to be made along the chain. Four bone-in loins were fitted in a box compared with six boneless products after the modification. Therefore 33% fewer boxes and delivery pallets were needed which amounted to 96 full pallet deliveries saved in a single year. Total potential savings were around 1.75% of the final price delivered to the canteen. When potential savings are repeated over time and in a range of products, at some point there may be the potential to redeploy resources to other activities. This is the continuous improvement principle of lean thinking (Womack and Jones, 1996). Total immediate savings were at least 0.51%, equal to about 1% profitability on sales against a backdrop of only 3–5% average chain profitability across the whole red meat sector. The savings relating to labour in the canteen were partly offset by the extra labour required at the processing end to bone-out loins; however this is not included in the above calculations owing to a lack of data.

Case study discussions
There is a tendency for food supply chain improvement efforts to focus solely on the efficiency factors. This chapter shed light on the great need

to address consumer value in the context of supply chain management by explaining *how* and *why* the above supply chain was disconnected from consumer needs although being reasonably efficient. For example, pork products were being delivered 99.7% correctly; yet a huge amount of waste existed since the product specifications had essentially not been revisited since it was established in 1963. In other words the supply chain was delivering the wrong product (bone-in loins) 99.7% on-time/in-full.

The above case study shows that efficiency measurement and improvements *per se* fall short of meeting the consumer requirements. The value chain improvement exercise threw-up many improvement opportunities; the team opted for the 'review of product specification' which delivered both supply chain effectiveness and efficiency improvements while requiring almost nil investment. On the other hand, opting for efficiency improvements such as the 'implementation of EDI' would have required hefty capital investment while not necessarily securing consumer satisfaction since the same out-of-specification product would have been delivered. One limitation of the study is that the real requirements of consumers were not captured, that is, only a post events questionnaire was sent to chefs and the actual consumers were not surveyed. As explained in Section 9.2, it is imperative for the industry to reconnect with consumer values, to realign processes to deliver the basic requirements and to find ways to enhance consumer value beyond the basic needs at different stages along the value chain.

9.4   Conclusions

This chapter has provided a practical step-by-step guide for implementing a successful value chain analysis project. It also reported on the generic findings of the VCA project in the UK food industry and explained that (consumer) value and system effectiveness should be the starting premise of value chain improvement endeavours. Subsequently, the case study provided insight into the practicalities of the proposed method and the challenges ahead.

In the case study, different participants in the chain had differing opinions of what was meant by value leading to conflicting behaviour and poor overall delivery of value for the end consumers. Also the role of SCM was perceived as limited to delivering operational/logistical services only (i.e. quality, cost and delivery). This perception was countered by obtaining consumer information (capturing VoC) and by steering the supply chain improvement initiative towards greater supply chain effectiveness. Moreover, the case study showed how inter-organizational potential can be leveraged to improve overall supply chain consumer satisfaction.

9.5   References

BOWERSOX, D. J., LALONDE, B. J. and SMYKAY, E. W. (1959). (eds.) *Readings in Physical Distribution Management*, MacMillan, New York.

BOWERSOX, D. J. and CLOSS, D. C. (1996). *Logistical Management: The Integrated Supply Chain Process*, McGraw-Hill Series in Marketing, The McGraw-Hill Companies, New York.

CHRISTOPHER, M. (2005). *Logistics and Supply Chain Management*, 3rd Edition, Pitman, London.

CLM (1998). *Logistics vs. Supply Chain Management*, Council of Logistics Management, (cited in) Bruzzone (2004) *Industrial Logistics* [online] http://st.itim.unige. it/cs/logistics/ (accessed 15.01.2006).

COOPER, M. and ELLRAM, L. (1993). 'Characteristics of supply chain management and the implications for purchasing and logistics strategy', *International Journal of Logistics Management*, **4**(2), 13–24.

COOPER, M., LAMBERT, D. and PAGH, J. (1997). Supply chain management: More than a new name for logistics, *International Journal of Logistics Management*, **8**(1), 1–14.

CROOM, S., ROMANO, P. and GIANNAKIS, M. (2000). 'Supply chain management: an analytical framework for critical literature review', *European Journal of Purchasing and Supply Management*, **6**, 67–83.

COUSINS, P. D., LAWSON, B. and SQUIRE, S. (2006). 'Supply chain management: theory and practice – the emergence of an academic discipline?', *International Journal of Operations and Production Management*, **26**(7), 697–702.

DAVENPORT, T. H. (1993). *Process Innovation: Reengineering Work through Information Technology*, Harvard Business School Press, Boston, MA.

DEKKER, H. C. (2003). Value chain analysis in interfirm relationships: a field study, *Management Accounting Research*, **13**(1), 1–23.

FARMER, D. H. and MACMILLAN, K. (1976). 'Voluntary collaboration vs. disloyalty to suppliers', *Journal of Purchasing and Material Management*, **12**(4), 3–8.

FCC (2007). *Food Chain Centre, Best Practice for Your Business: Completion Report*, IGD, London.

FITZGERALD, L., JOHNSON, R., BRIGNALL, T. J., SILVESTRO, R. and VOSS, C. (1991). *Performance Measurement in Service Businesses*, The Chartered Institute of Management Accountants, London.

FORRESTER, J. W. (1958). Industrial dynamics: a major breakthrough for decision makers, *Harvard Business Review*, July–August.

FRANCIS, M. (2004). 'Application of the food value chain analysis method in the UK red meat industry', *Proceeding of the 9th International Symposium on Logistics*, Bangalore, India, 11–14 July, 104–09.

GIBSON, B. J., MENTZER, J. T. and COOK, R. L. (2005). Supply chain management: the pursuit of a consensus definition, *Journal of Business Logistics*, **26**(2), 17–25.

HAMMER, M. and CHAMPY, J. (1993). *Reengineering the Corporation: A Manifesto for Business Revolution*, Harper Business, New York.

HEWITT, F. (1994). 'Supply chain redesign', *International Journal of Logistics Management*, **5**(2), 1–9.

HINES, P. and RICH, N. (1997). The seven value stream mapping tools, *International Journal of Operations and Production Management*, **17**(1), 46–64.

HINES, P., RICH, N. and HITTMEYER, M. (1998). 'Competing against ignorance: advantage through knowledge', *International Journal of Physical Distribution and Logistics*, **28**(1), 18–43.

HOULIHAN, J. (1985). International supply chain management, *International Journal of Physical Distribution and Logistics*, **15**(1), 22–38.

JONES, D. T. and WOMACK, J. (2000). *Seeing the Whole*, LEI, Massachusetts.

JONES, T. and RILEY, D. W. (1985). 'Using inventory for competitive advantage through supply chain management', *International Journal of Physical Distribution and Materials Management*, **15**(5), 16–26.

MENTZER, J. T., DEWITT, W., KEEBLER, J. S., SOONHONG, M., NANCY, W. N., SMITH, C. and ZACHARIA, Z. G. (2001). 'Defining supply chain management', *Journal of Business Logistics*, **22**(2), 31–50.

OLIVER, R. K. and WEBBER, M. D. (1982). 'Supply chain management: logistics catches up with strategy', *Outlook*, reprinted in Christopher, M. (1992), *Logistics: the Strategic Issues*, Chapman & Hall, London, UK.

PORTER, M. E. (1985). *Competitive Advantage*, The Free Press, New York.

RAINBIRD, M. (2004). A framework for operations management: the value chain, *International Journal of Physical Distribution & Logistics Management*, **34**(3/4), 337–45.

ROSS, D. F. (1998). *Competing Through Supply Chain Management, Creating Market-Winning Strategies Through Supply Chain Partnership*, Kluwar Academic Publishers, Boston.

ROTHER, M. and SHOOK, J. (1998). *Learning to See: Value Stream Mapping to Create Value and Eliminate Muda*, The Lean Enterprise Institute, Brookline.

SHANK, J. K. (1989). 'Strategic Cost Management: new wine or just new bottle?', *Management Accounting Research*, **1**, 47–65.

SHANK, J. K. and GOVINDARAJAN, V. (1993). *Strategic Cost Management*, The Free Press, New York.

SPECIAL ISSUE (2006). 'Supply chain management', *International Journal of Operations & Production Management*, **26**(7), 697–844.

STALK, G. and HOUT, T. (1990). *Competing Against Time: How Time-based Competition is Reshaping Global Markets*. The Free Press, New York.

STEVENS, G. C. (1989). 'Integrating the supply chain', *International Journal of Production Distribution and Materials Management*, **19**(8), 38.

WOMACK, J. and JONES, D. T. (1996). *Lean Thinking: Banish Waste and Create Wealth in Your Corporation*, Simon and Schuster, London.

WOMACK, J. P., JONES, D. T. and ROOS, D. (1990). *The Machine that Changed the World*, Rawson Associates, New York.

ZOKAEI, A. K. and HINES, P. A. (2007). 'Achieving consumer focus in supply chains', *International Journal of Physical Distribution and Logistics Management*, **37**(3), 223–47.

ZOKAEI, K. and SIMONS, D. W. (2006). 'Value chain analysis in consumer focus improvement: a case study of UK red meat industry', *International Journal Logistics Management*, **17**(2), 141–62.

10

Improving responsiveness in food supply chains

A. Harrison, Cranfield University, UK

Abstract: What happens when demand becomes very difficult to plan and the risk of wastage and stockouts increases? This chapter begins by considering the role of agility in supply chains today and compares it with lean thinking, addressing these challenges by reviewing the supply chains for two products – *bacalao* (salted fish) and bagged salads. The *bacalao* supply chain allows a comparison of agile and lean supply chains, while the Vitacress bagged salad case shows the principles of coping with rapid response and freshness integrated with 6-week growing cycles. The chapter ends with a summary of the nature of agile supply chains in terms of logistics focus, partnerships, key measures, process focus and logistics planning.

Key words: agile supply chains, food supply chains, short lead times, short throughput times.

10.1 Introduction

In Chapter 9, we saw how lean principles can be applied to food supply chains, where the emphasis is on the elimination of waste, leading to productivity and high quality. Agility is concerned with improving responsiveness to end-customer demand and mastering market turbulence (van Hoek and Harrison, 2001). This policy can be especially relevant when demand fluctuations take place under very short timeframes (hours, minutes), and/or where product lifecycles are relatively short. Both of these conditions – especially when accompanied by wide product ranges – make it difficult to forecast demand accurately.

Smoothed and levelled schedules – advocated by the lean thinking mindset – help to achieve high productivity and a stable inbound supply chain. But the downside is that such schedules often erode alignment between market needs and supply capabilities, leading to stocks of finished products that are not what the end customer wants. 'Lean production' in the trim and final assembly plant can lead to very 'un-lean' stocks of vehicles parked in disused airfields around Europe. Agility aims to forge a close alignment between needs and capabilities and is a particularly effective strategy under conditions of rapid demand change, high product variety and short product life cycles. The payoff may be that on-shelf availability (OSA) is improved without recourse to high levels of stock write-offs. The limitations may be that the supply chain is not as efficient in terms of resource utilisation and that manufacturing and distribution costs may be increased as a result. Overall, the aim is to improve value to the end customer while improving shareholder value at the same time (Harrison *et al.*, 2007). The concept of the agile supply chain is an important perspective from which to propose how supply chains of the future will evolve.

This chapter begins with a comparison between lean and agile thinking. The comparison is then developed by means of a case study on the Norwegian bacalao (salted fish) supply chains serving the Dominican Republic and Portugal, respectively. While the basic commodity is similar, the market priorities are strikingly different, resulting in quite distinctive logistics strategies. We then turn to an example from the fresh food supply chain (bagged salads) to illustrate the current challenges and the promising practices that are being incorporated into future supply chain design.

10.2   The concept of agility

The 'agile supply chain' is an essentially practical approach to organising logistics capabilities in order to be able to meet end-customer demand instantaneously. It is concerned with moving from supply chains that are structured around a focal company and its operating guidelines (e.g. 'Ford production system') to supply chains that are responsive to the needs of the end customer. The important change in mindset is to organise logistics from end-customer demand back upstream ('outside in'), as opposed to pushing product/service offerings into the market ('inside out'). Four key principles are involved, illustrated in Figure 10.1 (Harrison and van Hoek, 2008):

1. The agile supply chain is customer responsive. Most firms are forecast driven because they have inadequate real time data and would be unable to respond fast enough to real time demand even if they had access to it. So they are forced to make forecasts based on past sales history or shipments and to convert these forecasts into sales and operations plans. Breakthroughs of the last decade, in terms of such advances as efficient consumer response (ECR) and the use of information technology to capture demand data from point-of-sale or point-of-use, are transforming the capability to listen to the voice of the end customer. In turn, this creates the need to respond to real time demand data.

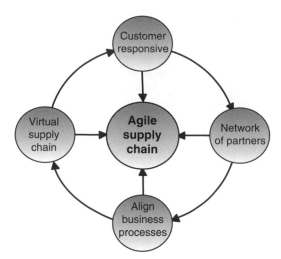

Fig. 10.1   Four principles of the agile supply chain (after Harrison and van Hoek, 2008).

2. The supply chain should be viewed as a network of partners who have a common goal to collaborate together in order to respond to end-customer needs. While individual partners may compete for market share in different networks, they collaborate with other members of each network in which they operate. Competitive advantage derives from focusing the efforts of the network of partners. In turn, this creates the need to coordinate the efforts of all partners through consistent objectives.

3. End-customer needs are met by aligning the business processes of network partners. Stand-alone process objectives that appear to make sense internally for a given firm, such as low manufacturing costs, may not make sense in the context of the network. Low manufacturing costs may mean large batch sizes that create high inventories and long lead times for other partners in the network. Significant challenges remain in order to coordinate the diversity of business processes that characterise food networks, for example, dairy farm, milk cooperative, cheese manufacturer, warehousing, distribution, back of store, front of store. In turn, this creates the need for partners to share common data, both internally (within a firm) as well as externally (between firms).

4. Network partners need to create a virtual supply chain in which key planning and control data are shared between partners. Virtual supply chains are information based rather than inventory based. While conventional logistics systems seek to identify optimal quantities of inventory and their spatial location, electronic point of sale (EPOS) and the internet have enabled network partners to act upon real-time demand

Table 10.1 Comparison of characteristics of lean and agile supply

Characteristic	Lean	Agile
Logistics focus	Eliminate waste	Customers and markets
Partnerships	Long-term, stable	Fluid clusters
Key measures	Output measures like productivity and cost	Measure capabilities and focus on customer satisfaction
Process focus	Work standardisation, conformance to standards	Focus on operator self-management to maximise autonomy
Logistics planning	Stable, fixed periods Smoothed, levelled schedules	Instantaneous response

data. The opportunity is to develop practices which avoid the distorted and noisy picture that evolves when orders are transmitted from one business process to another in an uncoordinated way.

Table 10.1 compares characteristics of lean and agile supply to show the differences in emphasis between the two.

We will now illustrate these characteristics in practice by means of our first food supply chain case, that of Norwegian bacalao (Jahre and Refsland Fougner, 2005).

10.3   Two supply chains for bacalao

Bacalao has been produced in Norway since about 1640. It is salted fish that has also been dried, traditionally in the open air on rocks, today in a drier. It can be kept refrigerated for several years and is said to improve over time. It has developed a strong position in the food cultures of many Latin countries, such as Brazil, the Dominican Republic and Portugal, where consumers often follow the Catholic tradition of eating more fish on Fridays and in the run-up to Easter. Marketing over many years has created the association with Norway as 'the land of bacalao', or 'bacalhau da Noruega' as it is called. It is a matter of great pride among consumers to master a variety of recipes for serving bacalao.

The overall supply chain is illustrated in Fig. 10.2. It takes at least 4 weeks to make the end product. The best fish is wild and taken by line, but trawled fish is also good, while nets give the lowest quality because the fish can be dead for a while before being hauled up. Today, the fish is increasingly farmed as well. The raw material is the major cost item – prices are set by the Råfiskelaget (the Norwegian raw fish association) and can vary a lot, for example from NOK 26/kg to NOK 15/kg within a year. Electricity

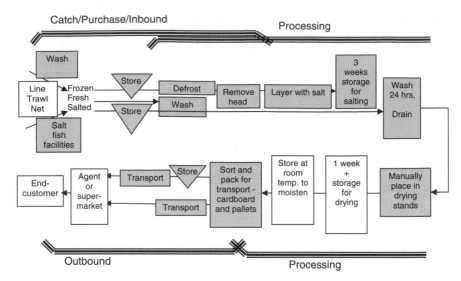

Fig. 10.2 Overall supply chain for bacalao.

and insurance are the other two major cost items. The fish is slaughtered (and bled on boat for the best quality), then matured in salt for 2–3 weeks. After salting it is dried, sorted, packed and distributed. There are no reliable ways of measuring salt and water content, so manual methods of touching and feeling the fish during each stage are used to ensure consistent quality and weight.

Bacalao is mainly produced from cod, which is preferred by Portuguese customers. But consumers in the Dominican Republic prefer pollock, which is a darker-fleshed fish that is more abundant in the North Atlantic. Cod is up to three times more expensive than pollock. The Norwegian fish industry is highly fragmented, with many small-scale fish farmers, fishermen and producers. Marketing activities are coordinated by the Norwegian Seafood Export Council.

Consumers are very quality conscious when buying bacalao. Quality is determined by colour, texture and firmness, water content and size. Portuguese consumers prefer smaller cod around 2.5 kg, while consumers in the Dominican Republic are less concerned with size, but prefer pollock. Note that 'quality' here refers to grade of fish rather than to conformance quality; both grades are fit for purpose in the markets they serve.

10.3.1   Bacalhau de Noruega

Company Noruega (CN) company has 150 employees and built its bacalao production facility in 1997 in the port of Ålesund, which has one of the largest harbours in Norway and one of the most modern fishing fleets in

Europe. The company focuses on volume in order to benefit from economies of scale. Production is stabilised throughout the year by ensuring a stable supply of fish through sourcing a combination of frozen and fresh fish, creating a buffer of some 3–4 months supply. The company only trades in full truckloads, which are distributed via Hamburg or Rotterdam. Product is sold under the generic brand name of 'Bacalao de Noruega' in standard transport packaging. While CN serves most Latin markets, 80% of its sales go to the Dominican Republic as pollock bacalao. This market is relatively stable throughout the year, which matches CN's stable production policy. CN is experimenting with pollock farming further to improve supply reliability.

10.3.2   Bacalao Superior

Company Superior (CS) is also based in the Ålesund area and accounts for 15–20% of Norwegian bacalao exports to Portugal. Only cod bacalao is exported to this market, which commands a 10–15% price premium over other Norwegian bacalao. The product is popular with consumers, which creates a strong relationship with the single supermarket chain that sells it. Fish are sold whole, with a CS tag showing guarantee of origin from fresh Norwegian cod, which was an idea that came from its supermarket customer. This ensures that CS bacalao stands out from other offerings. Joint marketing campaigns are funded by both CS and its supermarket customer and include TV promotions. Only fresh cod is used in bacalao superior, caught by the coastal fleet in small boats. Supply is heavily dependent during the winter on quotas that are permitted in the famous Lofoten fishing field in the far north. CS buys from three fresh cod suppliers and from 15–20 suppliers of salt fish. Processing follows traditional routes, but some technology has been introduced into cutting and drying. The finished product is transported to Portugal in 22 tonne truck loads three times per week. The finished product is stored in Lisbon at the customer's warehouse.

10.3.3   Comparing de Noruega and Superior

Table 10.2 summarises some of the major differences between these two products.

CN accepts more variation in its raw material source to enable continuous supply. This applies to type of fish as well as where and how it is caught. Farming and a healthy stock of frozen fish help to reduce further supply variations. On the other hand, CS seeks the best quality with minimum variation. The only inbound stock that is permitted is small quantities of salted cod.

While the raw materials and end product have many similarities, there are substantial differences in inbound and outbound logistics as well as

Table 10.2 Comparing de Noruega and Superior

Characteristic	da Noruega (Dominican Rep)	Superior (Portugal)
Raw material	Fresh/frozen pollock Different sizes Line/trawl/net/farm Continuous supply 3–4 months inbound stocks	Fresh cod, some salted Size specific Mostly line Seasonal supply Small inbound stock

Production process	High volume
	Single facility
	All types of fish processed in a
	   single factory: more efficient
	Undifferentiated packaging
		Customised
		Many dedicated facilities
		Cod only in single, focused
		   factory
		Fish individually tagged

Marketing	Continuous consumption
	Generic marketing with
	   Seafood Export Council
	Low price
	Generic packaging
	Little differentiation
		Special occasions
		Joint promotion with
		   supermarket customer
		Premium price
		Tagged to show origin
		Differentiated by market

processing and distribution strategies. These differences are fundamental to the need to support the brand (raising consumer expectations) by means of logistics strategy (meeting consumer expectations). We can conclude as follows:

- Two fundamentally different inbound strategies: CN focuses on secure, continuous supply and accepts greater variation in terms of type of fish, where and how caught, so farming is encouraged. They buffer and store extensively. CS goes for consistently high quality by not accepting much by way of variation: size, line catching and location are all important requirements. They do not store fresh fish or use frozen fish.
- Internally consistent marketing and logistics: CN matches the low price, continuous availability marketing mix by means of efficient sourcing and continuous availability and of 'lean' production and distribution methods. This enables high and consistent production volumes supported by a flexible product mix. There is less to go wrong in terms of supply, but the generic nature of the product mitigates against better margins or customer loyalty. CS matches the high price, seasonal availability marketing mix by means of highly selective sourcing and by focused factory production that is seasonal and relatively inefficient. Production is possible only when high quality, line-caught fresh fish are available. Limited and sporadic availability mean that the product has to reassert itself following supply interruptions, so the marketing pull must be consistent and strong. Traceability through tagging reinforces the superior quality image in consumers' minds, supported by joint marketing with the major retail customer.

The way that the two supply chains have evolved illustrates the tradeoffs at stake – more of one thing means less of another. The CS supply chain has become focused on top quality (grade) product, but at relatively high cost and sporadic availability. The CN supply chain has become focused on the opposite – low cost and continuous availability, but at average quality (grade). While we have used the terms 'lean' and 'agile' to typify these opposing mindsets, they are really only labels to describe the logistics responses to two very different market situations.

10.4   Vitacress: the challenge of very short shelf life

Vitacress Salads was formed in 1951 by UK entrepreneur Malcolm Isaac, and has become the world's leading producer of watercress. It specialises in top end, highly perishable, washed and ready to eat watercress and babyleaf salads. It was acquired by RAR Group, a private Portuguese company, in the summer of 2008. There are 600 employees in the UK and 1000 in total. Today, the company is one of the UK's largest growers, packers and distributors of high quality salads with 15% of the market. Vitacress has a 40% market share in Portugal and is developing its market position in Spain. The company 'has a commitment to total integration, growing a proportion of its seed requirements, drilling and cultivating the seeds, harvesting and transporting its crops and ... processing, packing and distributing its produce' (see http://www.vitacress.com/uk/about_vitacress.htm).

The reasoning behind the need for integration is that the shelf life is very short: the product has a shelf life of 5–6 days from packing and retailer requirements are such that no packed stock can be held by Vitacress. The order lead time is also very short – six hours prior to receipt of goods in the retail customer's DC.

10.4.1   Inbound supply chain

The inbound strategy is climate driven. In the summer, 95% of the product is sourced from the UK, while in the winter, 100% comes from abroad. Vitacress sources from its own farms in the UK, Portugal and Spain. Close relationships with partners in Kenya, Italy and the USA (Florida, Arizona and California) enable additional quantities to be purchased, along with sourcing from spot markets in Italy and France. Transport is by road from Spain and Portugal and by air from Kenya and the USA. Given the very short customer order lead time, winter harvesting must be carried out before receiving the customer's next sales forecast. This creates serious issues in planning the supply chain, as Fig. 10.3 illustrates.

The supplier has to make decisions 6–8 weeks (depending on crop and temperature) before the customer's order is received. So the whole of the

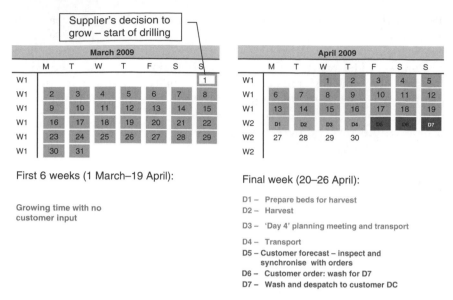

Fig. 10.3   Supply chain from Portugal.

period from 1st March to 24th April in this example is based on forecasts which have to be made far in advance of orders.

10.4.2   Forecast accuracy

Forecast detail only becomes available during the final week (days 1 to 7 in Fig. 10.3). By using the same source of weather forecasts as its major customers, Vitacress is able to take an educated guess at the day 5 customer forecast in time for its 'day 4' planning meeting on day 3. Sales forecasts are only 80% accurate at best, so Vitacress must be prepared for some significant changes when the orders are made on day 6. Orders that are made by 07.00 am on day 6 are received in store 36 hours later. The cycle can be tightened up during the UK summer period, when it is possible to harvest at 05.30 am and to have an order in the customer depot on the same day. The vertically integrated nature of the Vitacress inbound supply chain makes this far more practicable; most competitors do not source from their own farms, so could not be as precise about when their raw materials were harvested.

As if long growing periods and short order lead times were not sufficiently challenging, demand variability is also relatively high. Demand during peak weeks is about 35% higher than average and within each week demand for the peak day is about 40% higher than the average. Because of the drive for freshness, less than 24 hours of stock is held to buffer any changes in demand or supply and 98% of factory output is made against customer orders. Post-farm waste is another control figure, to be kept within 'low single figures'.

10.4.3   Coping with very short throughput times

What techniques does a company like Vitacress have available to cope with the need for ever shorter throughput times? We can speculate what these might be under a number of generic headings, some of which have been informed by the descriptions above:

- Work to improve forecast accuracy: the long growing season and the need for very fast response to customer orders means taking 'bets' on raw materials at a relatively long range. Devices that help improve the accuracy of these 'bets' all help to reduce risk. Seasonal planning and promotional planning are two, and working from the same weather forecasting models as customer is a third.
- Maintain flexible supply capabilities: by ensuring flexibility in regard to growing times, by retaining ownership of sources of supply and by selected strategic supply partnerships. If it is not available in Europe, then fly it in from Kenya or the USA.
- Keep the product formulation flexible: an element of flexibility can be designed into product specifications. Examples of this would include products where the description is not specific to a particular leaf and also where, at times of high sourcing risk, the customer agrees to a substitute leaf being included in the recipe.
- Speed up the planning process: there must be no delay between new information (for example an order from customer) and response. So there is no room for bureaucracy or for ponderous planning and control systems. Decision making needs to be delegated and local. The aim should be to get closer to the real time situation, and to put in place systems that facilitate this objective. This could include following EPOS sales at major customers for each Stock Keeping Unit (SKU) each hour.
- Speed up the response process: by allocating new orders in 'waves' each shift and by reducing the time between waves. It's not about fire fighting; it's about a smooth, continuous response to an ever-changing market place.
- Recognise the risks and plan counter measures: a high-speed operation is exposed to what is happening immediately and must respond accordingly. Most risks, such as the impact of the weather on product availability and quality, are known, even if their timing is not.
- Improve coordination between all steps in the supply chain: the vision is for the end customer to create demand in each store and for that demand to be seamlessly transferred to each process in the supply chain.

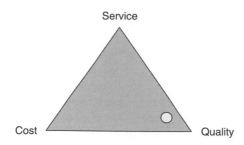

Fig. 10.4   Visualising the trade-offs.

There is no room for lack of understanding of the 'big picture' at any stage.

- Use uncompromising service KPIs: for example, if 100% performance is not obtained for the on-time, in full (OTIF) measure of performance against customer order, then a 'zero' is allocated.
- Behavioural capabilities must be developed in parallel: the above technical changes demand new time-constrained mindsets in all employees, both within the focal firm and in its supply chain partners. There needs to be a formal link between technical change and behavioural change.

The above techniques imply that there can often be a trade-off between service levels (OTIF), cost and quality. This trade-off can change from one product to another, from one shift to another. It can therefore help operating personnel to visualise what is wanted by being explicit. Figure 10.4 is designed to help in this visualisation process.

If the priority 'ball' is situated next to quality, for example, this dictates that quality is the top priority for this formulation for this customer. That could mean that cost suffers, for example, it may be necessary to scrap current (non-conforming) stock and to fly in replenishments from the USA. The delay could mean overtime in packing. It could also mean that there are delays in shipment and that service levels suffer. In order to make up for lost time, it may be necessary to courier the delivery to the customer at a premium price.

10.5   Conclusions

These case studies illustrate different stages in the route to agility. Let us compare the CS bacalao and Vitacress supply chains, using the characteristics listed in Table 10.3.

A transition can be perceived between the two cases. CS bacalao has developed a responsive supply chain that is geared to processing fresh line-caught cod quickly when it is available. Logistics planning is helped by use

Table 10.3 Characteristics of CS bacalao and Vitacress supply chains

Characteristic	CS bacalao supply chain	Vitacress supply chain
Logistics focus	Top quality product to premium market in Portugal Productivity and capacity utilisation are not top priorities	Top end product to markets in UK and Portugal Productivity and capacity utilisation are not top priorities
Partnerships	Raw material supply from selected reliable sources of fresh or salted fish Quality not compromised by freezing or farming Close partnership with retail customer	Total integration, but fluidity provided by well-established sources of backup supply to fill in the gaps. Occasional use of spot markets. Increasingly close partnerships with major customers
Key measures	Product traceability, conformance to uncompromising raw material and process quality standards	Focus on customer satisfaction (OTIF and quality), but help employees to recognise the trade-offs
Process focus	Focused factory for processing Subjective controls (touch and feel), but automated drying	Focus on operator self-management Help operators to cope with dynamic environment
Logistics planning	Process when conforming product available Stocks of finished product held by customer	Very fast reaction speeds to customer orders, supported by flexible supply sources Minimum inbound and outbound stocks

of salted cod from selected sources and by the ability to store finished product at the customer, but facility and labour utilisation are both subject to peaks and troughs as a result of uneven product availability. So there is still plenty of waste (in the form of finished product inventory and inconsistent productivity). Vitacress has moved further along the experience curve towards coping with market and supply uncertainty. It has been developing capabilities for very quickly responding to customer orders by developing relevant technical and human systems. This will also mean making compromises in terms of waste, but Vitacress is in the business of growing and processing a highly perishable product. These exigencies are driving Vitacress to become a truly agile company, able to respond to constantly changing customer demand by mastering uncertainty.

10.6   Acknowledgement

I am grateful to Ann-Karin Refsland Fougner, former PhD student at BI Norwegian School of Management and Marianne Jahre, Professor of Lund University and BI Norwegian School of Management for providing the basis for the Norwegian bacalao case.

10.7   References

HARRISON, A.S. and VAN HOEK R. I. (2008). *Logistics Management and Strategy: competing through the supply chain*, 3rd edition, FT/Prentice Hall, Harlow, Essex, UK.

HARRISON A. S., GODSELL, J., SKIPWORTH, H., WONG, C. Y., JULIEN, D. and ACHIMUGU, N. (2007). *Developing Supply Chain Strategy*, Cranfield University, UK.

JAHRE, M. and REFSLAND-FOUGNER, A-K. (2005). *Logistics – The Missing Link in Branding – Bacalhau da Noruega vs. Bacalhau Superior*, ISL – Logistics Conference Proceedings 2005, Lisbon.

VAN HOEK, R. I. and HARRISON, A. S. (2001). 'Measuring agile capabilities in the supply chain', *International Journal of Operations and Production Management*, **21** (1/2), 126–47.

11

Reducing product losses in the food supply chain

P. A. Chapman, University of Oxford, UK

Abstract: Shrinkage, the loss of products from the supply chain, warrants senior management, and in some cases government, attention by virtue of the scale of losses that can occur along a supply chain. Shrinkage is also a topic that should be of interest to supply chain and operations management professionals as it serves as an indicator of supply chain dysfunction, which means it is prudent to track performance and act to address the underlying causes even when the scale or value of direct losses does not appear at first glance to support such an investment. In order to introduce shrinkage and outline approaches to tackling it, this chapter sets shrinkage in context, describing the scale and nature of this issue and then moves on to explain a structure for ameliorative action.

Key words: performance management, shrinkage, stock loss.

11.1 Describing the shrinkage problem

The focus for this chapter is on shrinkage, an issue that relates to products being lost from the supply chain and also the loss of value, summarised as: 'Intended sales income that was not and cannot be realised.' (Chapman and Templar, 2006).

This definition includes lost products, that is products that cannot be sold as they are no longer in the system and lines that are unsaleable because of damage or being out-of-date. Also included is the difference between the intended reduction in selling price planned over a product's life cycle compared with products being sold at an unplanned discount. This is the case when a product's efficacy is compromised, such as when a fresh product begins to deteriorate and also when the market demand for a product diminishes, such as when a fashion line loses its appeal.

11.2   Scale of the problem

The level of shrinkage experienced by a product is influenced by a wide range of factors, which means it is not only difficult but also unwise to suggest an 'average' level of shrinkage. For example, shrinkage rates differ considerably both between and within product groups, between countries and even for the same type of product when sold through different but seemingly similar channels. The point to draw from this observation is the strong influence of context on shrinkage levels, as a result of the interaction of product attributes with local conditions.

11.2.1   Shrinkage in grocery retailing

The main context in which shrinkage will be discussed in this chapter is in the grocery sector, primarily in relation to fast moving consumer goods / consumer packaged goods. This sector benefits from research interest going back a number of decades and from a variety of perspectives, including criminological, accounting and operational.

In terms of its scale, research shows that in this sector shrinkage rates along the supply chain average 2.3%, with about 1.8% being identified in stores and 0.5% upstream in distribution (Beck, 2004). In a sector with a turnover in Europe of about one trillion Euros, this means European shrinkage is worth €23 billion each year, a figure equal to the GDP of Luxembourg. Losses in the USA are estimated as being US$27 billion (Beck and Peacock, 2008).

Within this sector, considerable differences in shrinkage can be observed across product categories. An illustration of this can be seen in the data from ten retailers who were known to the author. None of the retailers were able to supply data for every category, partly because their data collection method did not align exactly with the categories used and partly because not every retailer stocked every category of item. For each category, the responses were totalled and basic descriptive statistics calculated for the minimum, average and maximum rates of shrinkage. These results are shown in Fig. 11.1 where the top of the bar represents the maximum value, the bottom of the bar the minimum value and the diamond shape the average value.

Naturally care needs to be given when interpreting this data, including for the following reasons:

- The underlying data is, in a number of instances, unaudited.
- Respondents may have had to approximate values in order to align internally collated measures for the categories.
- The sample size is relatively low and therefore this sample is not necessarily representative of retailers in general. For example, the average of the five companies that provided data on their total shrinkage was 1.52%, which is lower than the industry average reported elsewhere, e.g.

1.84% by Efficient Consumer Response (ECR) Europe (see www. ecrnet.org).

- Respondents operate different business models. This means they take different approaches when balancing sales against losses and therefore it may be more profitable to incur a level of shrinkage under one model than under a different model.

With these cautions in mind, the variation in shrinkage rates can be observed between the categories, with the highest shrinkage found in deli (4.7%), in store bakery (4.0%) and floral (3.5%), all of which were significantly higher than the categories with the lowest shrinkage, that is frozen (0.5%), tobacco (0.6%) and dry grocery (0.7%). Another significant observation is the range of shrinkage rates within a category. For example, the categories with the highest range between lowest and highest levels of shrinkage are shown in Table 11.1.

This suggests that shrinkage rates are significantly affected by retailer specific factors. Another observation that points to this conclusion is that the highest level of shrinkage in each category occurred in six of the ten

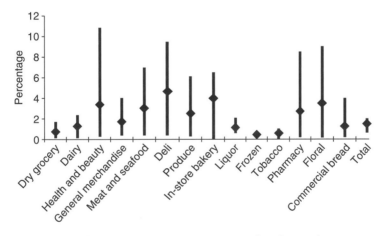

Fig. 11.1 Shrinkage rates amongst grocery retailers by product category.

Table 11.1 Grocery categories with a high range of shrinkage between retailers

	Low (%)	High (%)	Range (%)
Health and beauty	0.3	10.9	10.6
Floral	0.2	9.0	8.8
Pharmacy	0.2	8.5	8.3

retailers. The lowest level of shrinkage in each category occurred in two of the ten retailers. None of the retailers with at least one lowest shrink result in a category had any of the highest shrink results in other categories, and *vice versa*. This means that within the ten retailers that provided data, there were two separate sets of retailers, one with high shrinkage and the other with low shrinkage.

11.2.2   Shrinkage in other sectors

In the food sector, across the whole supply chain from the farm to the home, shrinkage is estimated to be 30–35%. Research from the USA attributes these losses to various stages of the food chain, although the amounts vary considerably depending on crop, type of retail outlet and household demographics. Results from two general studies are shown in Table 11.2, together with a specific example, losses for apples.

Research in the UK suggests that 30% of all food bought by consumers is thrown away, half of which was edible. The annual price paid from this wasted food was in the order of £10 billion and constituted 20% of domestic waste, which cost local government £1 billion to dispose of (Ventour, 2008).

These figures clearly highlight a major issue for the households and organisations involved but at this scale the implications are also societal. Losses at these levels present significant considerations in terms of the amount of additional inputs, including land, water, fertiliser and energy, required to grow the food, the waste involved in its distribution and the cost of disposal. Comparable levels of loss are also reported in the food chains of developed countries such as USA and UK and developing countries such as India, where the societal impacts of such loss is exacerbated by food scarcity. Where this is the case, this means that shrinkage is more than just a waste of resources but is also a cause of poor nutrition and the effects on health and life expectancy that result.

Table 11.2   Levels of shrinkage across food chains

Food chain stage	Overall food losses (1) (%)	Overall food losses (2) (%)	Apples (%)
Total[a]	35	27	39
Farm	18	–	4
Retail	5.6	2	12
Household	14	26[b]	28

Sources: Jones, 2006; Kantor *et al.*, 1997; Wells and Buzby, 2007.
[a] The values in the columns are not totalled by adding them because some losses are omitted and losses from farm, retail and household are not cumulative.
[b] Includes food service and consumer losses.

Losses are also known to exist in a range of other supply chains (although some of this loss may fall outside the scope of the definition offered above) including:

- cash, for example banking deposits from retail outlets, for example cinemas (Fithian, 2007)
- spare parts for capital equipment: civilian, such as farm equipment (McCall, 2003) and military (Kutz, 2008).
- energy: oil and gas (Ugochukwu and Ertel, 2008) and (more than a little worryingly) nuclear (Jameson, 2005).
- utilities, including water (Parks, 2004) where distribution loss is typically 20% (Hunaidi and Wang, 2006).

Clearly shrinkage is a major issue that affects a broad range of supply chains and their constituent organisations. However shrinkage has proven to be a resilient problem, despite concerted efforts to address it spanning more then 40 years (see for example Bernstein, 1963; Donnell, 1975). For these reasons, shrinkage deserves a greater effort to understand its nature, including its implications and causes and to identify approaches to its reduction.

11.2.3   Classifying the causes of shrinkage

Shrinkage occurs for a variety of reasons, including product being damaged, going out of date, lost in transit or stolen, where a common classification of the causes uses the following four categories:

- external theft
- internal theft
- intercompany fraud
- process failure.

There are concerns about this classification, not least because three of the four categories are 'malicious' and this has the effect of encouraging the view that shrinkage is predominately a crime issue. While there are clearly some categories of product where theft is a major issue, there are many others where this is not the case. This said, this approach to classifying shrinkage has wide acceptance with practitioners and it prevails particularly amongst retailers.

One way to achieve a more balanced approach to would be to take a classification system used to develop scenarios of losses in oil and gas production (Sklet, 2006) and expand 'process failure' into the following five categories:

1. Human and operational errors, e.g. failure to follow due process
2. Technical failure, e.g. mechanical failure of equipment
3. Process upsets, e.g. process parameters out of range

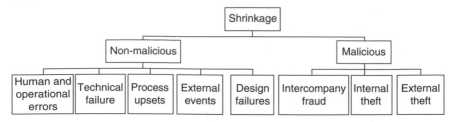

Fig. 11.2 Classification of causes of shrinkage.

4. External events or loads, e.g. collisions
5. Design failures, i.e. latent failures introduced during design.

This would lead to the classification of shrinkage shown in Fig. 11.2.

The challenge of using this and any other classification structure is the difficulty of correctly attributing shrinkage to a particular cause. In the case of non-malicious shrinkage, this requires root-cause analysis of each shrinkage event, which has a cost implication and it may be that several causes conspire together, making the primary cause difficult to isolate. In the case of malicious loss, much of this goes unnoticed at the time it occurs so there is difficulty in tracking it, a point that is discussed below.

11.2.4   Concerns over limited knowledge and misunderstandings

Efforts to address shrinkage are hampered by the ease with which it can be viewed incorrectly. Indeed, the degree to which shrinkage is misunderstood has led current understanding of shrinkage to be described as 'myopic' (Beck *et al.*, 2003). It has long been known that shrinkage is frequently attributed to the wrong cause (Bernstein, 1963) and retailers often emphasise external theft as a cause of shrinkage as they are in denial of the level of theft by their own employees. Therefore retailers underreport internal theft (Oliphant and Oliphant, 2001).

A key culprit in the misunderstanding of shrinkage is the widespread misreporting of findings from industry surveys. Surveys are a valuable research instrument, but they have their limitations and caution needs to be applied when conducting surveys and reporting their findings. When conducting surveys, accurate data can only be achieved if there are accurate processes generating the data (Biemer and Lyberg, 2003). Beck (2004) reports that most shrinkage (51%) is unknown. Therefore the values attributed by survey respondents to various causal categories of total (known and unknown) shrinkage lack rigor. The findings reported from such surveys need to be considered against a question fundamental to research design, 'Is it the things themselves or the people's views about them that are important?' (Easterby-Smith *et al.*, 1991).

With most shrinkage being unknown, research into this topic lacks a foundation of measured, quantified material which means that surveys have little value as a means of describing the nature of shrinkage, the 'thing' in this instance. Instead the value of industry surveys lies in measuring the respondent's views on the causes of shrinkage. Taken from this perspective, these surveys are a valuable source of data for studying the attribution process as a topic in itself and exploring shortcomings such as self-motivational biases, where failures are attributed to external factors in order to maintain self-esteem (Ross and Anderson, 1982) which may help explain why external theft is often emphasised, as reported above. However this is not the perspective generally taken when analysing and reporting the findings from such work. Instead, the lack of understanding of the extent and causes of shrinkage amongst practitioners and researchers is overlooked and various surveys are undertaken (for example Bamfield, 2004; Beck, 2004; Grasso, 2003; Hollinger and Langton, 2005) which use respondents' estimates to apportion shrinkage into causal categories. These findings are then widely reported with little attention given to their weaknesses.

One consequence of this situation is that policy makers in retail organisations use these findings to direct their shrinkage reduction strategies and budgets without realising their investment is based on received wisdom at best and collective falsehood at worst. The poor quality of the evidence underpinning decision making and action in the area of shrinkage serves only to highlight this area as a microcosm of poor practice in management research.

11.2.5   Local effects of systemic issues

One of the issues that adds to the difficulties of understanding shrinkage is that incidents that cause shrinkage may be considered as separate and unrelated local problems or the result of a series of disparate factors coming together. That is, viewed in isolation, shrinkage incidents appear unconnected from one another and the response is to deal with them locally as and when they occur. However, if incidents were viewed more broadly, then common causes can be identified and addressed through changes at a policy level. This type of systemic approach views shrinkage as the symptom of a range of causes and by taking this view highlights the need to judge the influence of each cause using clear-cut criteria to inform decision making and the selection of corrective action. This view parallels research into high reliability systems which identifies the role of the 'failure chain' in catastrophic failures, for example the 2000 crash of Concorde in Paris when a tyre burst on takeoff, rupturing a fuel tank, setting fire to an engine resulting in the plane crashing. In this event there were a range of issues that occurred and while none can be said to be the single cause of the disaster, they all contributed to it in some way. These included (Donnelly, 2001):

- A component was missing from the landing gear, causing take off speed to be reduced.
- The aircraft was taking off with a tailwind instead of into a headwind, which reduced its lift.
- The aircraft was overweight, reducing its ability to climb.
- The aircraft was unbalanced. This caused its path to deviate during takeoff, where it hit a landing light that broke and appeared to damage one of the four engines.
- A strip of metal had fallen from another plane onto the runway. Allegedly this strip had not been authorised by aviation authorities to be fitted to aircraft.
- The Concorde's tyres were not puncture resistant and one burst when it hit the metal strip.
- Pieces of the tyre ruptured a fuel tank, setting fire to an engine.

The plane tried to takeoff but with problems in two engines, a low takeoff speed, being overweight and a tailwind it was unable to maintain altitude long enough to return to the airport and crashed with the loss of 113 lives. This tragedy illustrates the way a set of events cumulatively cause a catastrophic event. The absence of any one of these events in the 'failure chain', that is had something not happened, could have prevented the disastrous outcome.

A less dramatic example of how thinking in terms of cumulative events applies to shrinkage can be illustrated by considering the response when fresh food, for example fish, is written off in a grocery store because it has 'gone off'. Taken in isolation, a localised view of shrinkage will see this as the fish having become unsaleable and having to be disposed of. First, it is vital that the event is recorded otherwise it will appear as 'unknown shrinkage' in the next inventory. Then a systemic perspective needs to be taken of the incident to understand what caused it and whether it is the consequence of several factors combining in a critical way.

This investigation uses the category headings for non-malicious shrinkage provided in Fig. 11.2, starting with whether due process been followed by store staff, such as ensuring the fish was kept at the correct temperature? Were there any technical issues, such as problems with the performance of the shipping boxes or other packaging? Were the demand planning and forecasting process and logistics and operations processes appropriate for these products? For example, demand for fish can be variable so there may simply have been too much ordered for a flat sales period. Had external events contributed to the waste, for example high or low temperatures, or products being damaged? Finally, a 'design failure' in this case would be to consider whether an appropriate species of fish is stocked. Some fish are particularly sensitive and therefore unsuitable to be sold in a 'grocery' store environment.

The ability to evaluate this incident and others like it in this way provides the means for making decisions that can remove the underlying causes of shrinkage, preventing future loss from occurring. The perspective of shrinkage as a systemic issue recognises that there can be a significant distance and time lag between where and when the causes of shrinkage were introduced, where the loss occurred and where and when the effect materialises. Therefore in order to understand and manage shrinkage, it is necessary to look across a supply chain and the life cycle of the elements to be found there. The life cycle of the various elements in a retail business can be broken into three horizons of long-term, medium-term and short-term. Long-term issues are those that are strategic in nature, typically designed into the infrastructure of the business and are very difficult to change retrospectively, such as the location and layout of a building. In the medium term there are some significant decisions that are made within the constraints set down by design or strategy. Short-term issues are tactical in nature and relate to the day-to-day running of operations. While some incidents can be attributed to a particular issue, there is merit in considering how the various aspects of the business combine to affect shrinkage. This means that when it comes to addressing issues, the right parts of the business can be targeted at the right time.

Looking across the business, there is a need to gather data on the performance of those activities that have an impact on shrinkage and this information needs to be brought together so it can be considered in its totality. Taking a store as an example, the measurement systems would assess long-term issues linked to store design, medium-term issues like the design of store procedures and short-term issues like how these procedures are being followed. The role of this measurement system is to report results, such as the use of good practice, track trends over time and direct resources to where they will be most effective. The likelihood is that these resources will be managed at a local level, for example in-store, so the information needs to be specific and advice on what actions to take needs to be specific and relevant to that operation, allowing macro issues to be deconstructed so they can guide timely intervention at the local level.

11.3   Strategies for reducing shrinkage

Taken from almost any angle, shrinkage is an undesirable attribute of any supply chain and all organisations would benefit from its reduction. However, as noted earlier, shrinkage is a stuborn problem so efforts to address it require a considered and broad platform to work from. This platform has as its foundation the recognition by senior management that shrinkage is both a significant problem and that its resolution is an opportunity to improve profitability directly and operational performance in general. Efforts to address shrinkage benefit from taking a holistic view, one that engages stakeholders within an organisation and also beyond through collaborating with supply chain partners. Bringing together a team to work across functions and organisations has a number of important advantages but with these come conflicts that arise from competing agendas and an increasingly complex team dynamic. In order to harness the advantages and mitigate the weaknesses of engaging a range of stakeholders, this work benefits from using a structured approach to organising and focussing effort. This effort benefits from the application of good practice associated with project management and change management, not least investing time in planning. This said, a structured approach enables an action-oriented approach that increases the efficiency of effort as well as the likelihood of delivering an effective outcome and is not an excuse for delay and procrastination. Prior research can provide a useful input to this effort by highlighting likely points of leverage. These points are expanded upon below.

11.3.1   Governance

Shrinkage transcends organisational and functional boundaries both in terms of where the problem manifests itself and where the solutions need to be applied. The scope of the work to compile information, diagnose the underlying issues and address shrinkage means that a strategic response is required, one where senior management commitment is established, organisational commitment is ensured and loss prevention is embedded amongst employees at all levels of the organisation (Beck and Peacock, 2008).

Senior management commitment is necessary as the only place in an organisation with a clear view is amongst those with executive responsibility. In a societal context this lies with the most senior government officials, while in organisations this is the executive board. To this end, several retailers operate a standing committee to address shrinkage chaired by their chief executive. While a significant act in itself, this is one part of the wider response required as there is a need for coordinated action between organisations, identifying shrinkage as an agenda item for 'top-to-top' meetings between executives along the supply chain who build shrinkage into their joint action plan. Appreciating that executives have many demands on their time, responsibility for ensuring organisational commitment and embedding loss prevention amongst all levels of the organisation should lie with a 'head of shrinkage' whose position benefits from a powerful mandate from the executive team.

To be effective, the head of shrinkage needs access to information on losses, including by product, by location and by time period. In practice this means having the support of a data analyst with access to a range of transaction and audit data and the ability to query this data through data mining. In FMCG/CPG retailing it is common to find losses concentrated on 'hot products' and in 'hot stores'. Identifying hot products allows losses on these specific lines to be investigated to determine the causes of these losses and to create an action plan to remove or mitigate the effect of these causes. Frequently the root causes of loss on these hot products are common to losses on other lines, which means that addressing these will deliver wider benefits. Losses in hot stores point to the underlying issue being one of ineffective management control. This highlights a need for a more people-focussed intervention, one that emphasises the need for the store manager to be effective in undertaking the tasks of leadership and to be rigorous about maintaining operational controls and proactive in addressing local issues faced in the store.

Armed with the knowledge of where shrinkage is manifesting itself, an analysis of root causes and an action plan for implementing solutions, the head of shrinkage plays several roles. The first of these is to engage stake-holders across the business to implement particular actions within their sphere of control. For example, if the issue was waste of fresh products in store, then one element of a programme could be to work with a logistics team to develop a rigorous regime for monitoring temperature control through the transportation network, to analyse this information and to address the root causes of failures to maintain products in their correct temperature regimes.

The second role is a facilitating one where the head of shrinkage engages functional heads from across and beyond their organisation to deliver actions within their respective domains. In the example of monitoring temperature control, this would be to ensure the head of logistics gives financial and personal support to the actions developed by the logistics teams. The head of logistics also needs to act should the logistics team fail to deliver the necessary improvements. In decomposing the plan in this way, specific individuals are made accountable for undertaking actions in their area of responsibility with functional heads retaining their authority and ensuring actions in their areas are implemented. The head of shrinkage is account-able for the effectiveness of the plan as a whole, which is reported regularly to the executive board.

11.3.2   Collaboration

There is a clear need for effort to address shrinkage in order to engage a wide range of stakeholders, to understand their perspectives and to ensure they are involved in the development and delivery of solutions. However, despite the widely reported importance of collaboration and its promise of significant interorganisational gains, it appears that reality falls somewhat short. Daugherty *et al.* (2006) cite Sabath and Fontanella (2002) as best capturing the underachievement of collaboration: '... collaboration is at the same time the most used, the most frequently misunderstood, the most popular and the most disappointing strategy that has come along to date'.

These words serve to highlight the general challenge that accompanies any effort by organisations along the supply chain to work together. In addressing shrinkage, the challenge is particularly acute as the causes of shrinkage and the success of specific solutions are often highly dependent on the context in which they were applied. Using Grint's categorisation of situations (Grint, 2005), this positions shrinkage as a 'complex' problem, as opposed to a 'simple' one. Complex problems are best addressed by analysing them in their context in order to arrive at the appropriate solutions, as they lack prescribable solutions that can be read across from apparently similar situations. As a result, this type of problem benefits from being addressed in a collaborative manner as understanding and the means of resolution is dispersed amongst stakeholders.

11.3.3   Approach

Recognising the difficulties organisations face when working together as described above, shrinkage can be addressed in a collaborative manner using a 'road map' (Beck *et al.*, 2003), a structured approach to problem solving depicted in Fig. 11.3. This approach advocates multifunctional, cross-organisational teams undertaking an 'end to end' review of the supply chain to understand the root causes of shrinkage and to develop interorganisational solutions that benefit all parties through reduced losses and increased sales.

This approach provides a guideline for project teams and charts the various steps that need to be taken in order for a more holistic and systematic approach to be taken. This roadmap supersedes conventional approaches to addressing shrinkage that are characteristically partial, piecemeal and for the most part poorly conceived practices. In doing so, the intention is to guide organisations to opportunities for dramatically improving their performance.

11.3.4   Points of leverage

Managing shrinkage is a challenge because incidences of loss are diffused across the supply chain and actions need to be undertaken by a range of

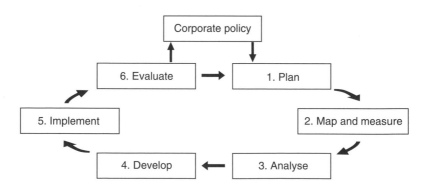

Fig. 11.3 The roadmap – a collaborative approach to reducing shrinkage (from Beck *et al.*, 2003).

stakeholders. Set against this is the encouraging news that there are a number of points of leverage where action can deliver significant benefits. These points of leverage are:

- Identify where high loss accumulates, e.g. hot products and hot stores.
- Understand and exploit the relationship between malicious and non-malicious loss.
- Design solutions into the system, to achieve sustainable, cost effective improvements.

As described earlier, shrinkage is not evenly distributed. By investing in data gathering and information reporting it is possible to identify where loss is concentrated. This provides the opportunity to prioritise efforts and address the biggest sources of shrinkage first.

In consumer goods retailing it is known that a significant (although not exactly quantified) amount of loss has a malicious cause. In addition, the complexity of this sector tends to result in a range of non-malicious losses. An important mechanism that connects these two types of loss is that the operational complexities and failures that lead to non-malicious loss provide the opportunity for malicious loss to occur and to go unnoticed for a period of time. For example, underpicking and mispicking in a distribution centre means that deliveries to stores have some level of inaccuracy. Where this is a common occurrence the effect is to condition the store team to expect delivery discrepancies. As a consequence the opportunity for malicious loss is introduced where items stolen from the delivery truck result in a 'normal' under delivery.

Another example of process failures presenting the opportunity for malicious loss is when too much product is picked in the distribution centre (DC) and delivered to store. This presents the opportunity for the excess to be stolen without anyone noticing an inventory discrepancy. Another consequence of issues such as this is to allow the manager in a hot store to deflect responsibility for their store's excessive loss. Where this is the case it is not possible, and would be unfair, to sanction the manager for high losses in their store. This means that efforts to ensure reliable, consistent supply chain operations deliver the multiple benefits of preventing non-malicious losses, removing the opportunity for malicious loss to occur and go unnoticed and isolating the cause of in-store shrinkage in the store itself.

Where companies address the sources of the majority of their operational failures, they benefit from a reduction in non-malicious losses and the opportunistic losses they had previously enabled. A residual amount of 'hard core' malicious loss will remain but this will now be much more visible, allowing a range of security solutions to be targeted against it.

When considering solutions, these will be most effective when applied to points of leverage. Several points of leverage have been termed 'cultural practices' and while originally conceived at the store level, these can be seen to apply across the supply chain. The nature of the points of leverage changes according to the strategic horizon, with Table 11.3 presenting examples of how these manifest themselves by temporal horizon.

In terms of considering the elements of a retail supply chain, solutions to shrinkage incidents in one stage of the supply chain may lie elsewhere, for example through improved product and packaging design, supplier selection, store design or supply chain network design. This means that the ability to work effectively across organisational boundaries will be essential in order to leverage these opportunities. Solutions also exist in the background activities of human resource management and information management, both of which provide the foundation that enables other improvements.

Table 11.3 A list of points of leverage and solutions for addressing shrinkage, categorised by elements of a retailing supply chain and by temporal horizon

Points of leverage	Temporal horizon		
	Long-term	Medium-term	Short-term
Prioritise procedural control in stores	Plan location Improve layout design	Introduce robust and secure processes and practices Respond to store specific risks	Adhere to process and procedures Audit performance and manage incidents
Prioritise procedural control in supply chain	Supply chain network design Improve distribution infrastructure	Introduce robust and secure processes and practices Develop supplier capabilities	Adhere to process and procedures Audit performance and manage incidents
Focus on hot products	Product design Packaging design	Pre-empt issues when making purchasing decisions	Maintain integrity of product from raw material to sale at checkout Audit performance and manage incidents
Engage people: provide strong leadership and develop a team	Create human resource policies.	Recruitment Training Innovate and experiment	Institute day-to-day attention to ensure: accountability; action; attitude
Use evidence-based management	ICT system design	New system implementation System upgrade	Monitor process and practice adherence Track shrinkage incidents Perform weekly performance updates

It is also known that the most cost effective time to implement solutions is at the design stage. This can be seen in the case of product design, where improvements or mistakes will have implications in all market channels, all retail outlets and all homes for years. Fixing issues through design has a relatively low cost and will deliver significant benefits, although there can be a long time lag until the benefits are realised.

11.4   Potential benefits

Through the aegis of ECR Europe, a pan-European trade association for grocery retailers and consumer goods manufacturers, since 2000 a range of companies have implemented the ideas presented in this chapter and their experience gives some indication of the actual and potential benefits associated with these ideas. The degree to which these ideas have been implemented spans a range where some companies have undertaken one or more projects with the aim of addressing a specific shrinkage problem while others have adopted these ideas and integrated them into their corporate policy. In addition to the differences in scope with which these ideas have been adopted, the nature of this work varies considerably between organisations, which is appropriate as their issues and the likely effectiveness of alternative approaches are context dependent. There are, however, the following common elements across this work:

- better measurement of shrinkage
- data sharing
- addressing all aspects of shrinkage
- external collaboration
- internal collaboration
- focus effort on hot products and hot stores
- focus effort on the removal of opportunity
- use the road map to guide project work.

A range of projects have been reported over the past eight years that have involved a range of organisations, addressing issues in different product categories. A sample of the results of projects that are in the public domain is shown in Table 11.4.

The sample of projects shown in Table 11.4 is representative of the several hundred projects that are known to have been undertaken around Europe and beyond in Latin America, North America, Asia and Africa. The results reported from these projects highlight the significant benefits that can be delivered. Of note is that many projects have focussed on a combination of shrinkage and sales. This combination helps engage a wider range of stakeholders and avoids the error of undertaking actions that improve one metric at the expense of the other.

Results from projects like the ones shown in Table 11.4 have encouraged several organisations to expand their efforts beyond a project-based

Table 11.4 Results of a selection of shrinkage projects

Organisations	Location	Results
Tesco and Gillette	Hungary	−74% losses + 288% sales
B&Q and Plasplugs	UK	−50% losses + 33% sales
Feira Nova and Danone	Portugal	−45% total shrink
Ahold and sausage suppliers	Poland	−42% losses
Sainsbury's spirits category	UK	−40% losses + 10% sales
Sonae and Colgate-Palmolive	Portugal	−29% losses
Sainsbury's and Menzies	UK	−25% losses + 10% sales
Wickes and GET	UK	−7% losses

Table 11.5 Organisations whose corporate shrinkage policy has been significantly influenced by ECR Europe

Organisation	Sector	Location
Adidas	Sports goods	Northern Europe
Ahold	Grocery	International
Boots the Chemist	Pharmacy	UK
Dollar General	Discount grocery	USA
DM	Pharmacy	Germany
Metro	Grocery	Belgium and Netherlands
P&G	Consumer goods	International
Tesco	Grocery	UK and International

approach to adopt these ideas into their corporate policy. A sample of such organisations is shown in Table 11.5.

For reasons of confidentiality and because it is difficult to isolate the direct effect of such efforts, few companies report the effect on total shrinkage of this work. One exception is Tesco. They reported (Sally and Peacock, 2005) that in 2003 their unknown losses were 1.01% of turnover. As a result of implementing a change in their approach to addressing shrinkage they were able to reduce this level to 0.69% in 2005. This improvement was valued at €150 million for 2005 and similar annual savings have been achieved subsequently. This saving was responsible for a significant proportion of the improvement in corporate performance that Tesco reported.

The work at Tesco provides a general lesson for managing stock loss in the supply chain. This is that a holistic, systematic approach uncovers opportunities to reduce shrinkage while the benefits are delivered by implementing the solutions and sustaining the improvements. Tesco proved to be particularly effective in realising and sustaining shrinkage savings, having implemented the governance structure described earlier, resourcing this with skilled and capable people and driving results by following the road map approach.

11.5 References

BAMFIELD J (2004), 'Shrinkage, shoplifting and the cost of retail crime in Europe: a cross-sectional analysis of major retailers in 16 European countries', *International Journal of Retail & Distribution Management*, **32**, 235–41.

BECK A (2004), *Shrinkage In Europe 2004: A Survey of Stock Loss in the Fast Moving Consumer Goods Sector*, ECR Europe, Brussels.

BECK A and PEACOCK C (2008), 'Lessons from the leaders of retail loss prevention', *Harvard Business Review*, **85**, 34.

BECK A, CHAPMAN P and PEACOCK C (2003), *Shrinkage: a Collaborative Approach to Reducing Stock Loss in the Supply Chain*, ECR Europe, Brussels.

BERNSTEIN J E (1963), 'Curbing losses and errors in retail store operations', *New York Certified Public Accountant*, October, 706–14.

BIEMER P P and LYBERG L E (2003), *Introduction to Survey Quality*, Wiley-Interscience, Hoboken, NJ.

CHAPMAN P and TEMPLAR S (2006), 'Scoping the contextual issues that influence shrinkage measurement', *International Journal of Retail & Distribution Management*, **34**, 860–72.

DAUGHERTY P J, RICHEY R G, ROATH A S, MIN S, CHEN H, ARNDT A D and GENCHEV S E (2006), 'Is collaboration paying off for firms?' *Business Horizons*, **49**, 61–70.

DONNELL J D (1975), 'Merchant vs. Shoplifter in the Courts', *Journal of Small Business Management*, **13**, 5–8.

DONNELLY S B (2001), 'Return of the Concorde', *Time*, V158 N21, 99.

EASTERBY-SMITH M, THORPE R and LOWE A (1991) *Management Research: An Introduction*, Sage Publications, London.

FITHIAN P (2007), 'Protecting the theatre's bottom line', *Boxoffice*, **143**, 2–5.

GRASSO S (ed.) (2003), *11th Annual Retail Crime Survey 2003*, British Retail Consortium, London.

GRINT K (2005), 'Problems, problems, problems: the social construction of "Leadership"', *Human Relations*, **58**, 1467–94.

HOLLINGER R C and LANGTON L (2005), *2004 National Retail Security Survey*, University of Florida, Gainesville, FL, USA.

HUNAIDI O and WANG A (2006), 'A new system for locating leaks in urban water distribution pipes', *Management of Environmental Quality*, **17**, 450–3.

JAMESON A (2005), 'Sellafield "lost" plutonium', *The Times*, 17/2/05.

JONES T W (2006), *Using Contemporary Archaeology and Applied Anthropology to Understand Food Loss in the American Food System*, Bureau of Applied Research in Anthropology University of Arizona. Available from: http://archaeologyand botanyresearch.com/downloads/Overview%20of%20Findings%205-04.doc [accessed 8 November 2008].

KANTOR L S, LIPTON K, MANCHESTER A and OLIVEIRA V (1997), 'Estimating and addressing America's food losses', *Food Review*, **20**, 2–12.

KUTZ G D (2008), *GAO-08–644T. Undercover Purchases on eBay and Craigslist Reveal a Market for Sensitive and Stolen US Military Items*, United States Government Accountability Office, Washington, DC.

MCCALL M (2003), *Results from the 2001–2002 National Farm Crime Survey. Paper No 266, ISBN 0 642 53820 3*, Australian Institute of Criminology, Canberra.

OLIPHANT B J and OLIPHANT G C (2001), 'Using a behaviour-based method to identify and reduce employee theft', *International Journal of Retail & Distribution Management*, **29**, 442–51.

PARKS J (2004), 'Water loss management and customer care', *American Water Works Association Journal*, **96**, 66–9.

ross l and anderson c a (1982), 'Shortcomings in the attribution process: on the origins and maintenance of erroneous social assessments', in *Judgement Under Uncertainty: Heuristics and Biases*, Kahneman D, Slovic P and Tversky A (eds), Cambridge University Press, Cambridge, 129–52.

sabath r and fontanella j (2002), 'The unfulfilled promise of supply chain collaboration', *Supply Chain Management Review*, **6**, 24–9.

sally d and peacock c (2005), 'Tesco case study', *ECR Europe Annual Conference*, Paris.

sklet s (2006), 'Hydrocarbon releases in oil and gas production platforms: release scenarios and safety barriers', *Journal of Loss Prevention in the Process Industries*, **19**, 481–93.

ugochukwu c n c and ertel j (2008), 'Negative impacts of oil exploration on biodiversity management in the Niger delta area of Nigeria', *Impact Assessment and Project Appraisal*, **26**, 139–47.

ventour l (2008), *The Food We Waste*, Waste & Resources Action Programme, Banbury, UK.

wells h f and buzby j c (2007), 'ERS food availability data look at consumption in three ways', *Amber Waves*, **5**, 40–1.

12

Methods for assessing time and cost in a food supply chain

S. Templar and C. Mena, Cranfield University, UK

Abstract: This chapter presents an overview of supply chain time – cost mapping (SCTCM), a framework which any organization can deploy which will enable them to gain greater visibility of both time and cost in their supply chain operation. The framework is underpinned by incorporating the tools and techniques of time-based process mapping and activity-based costing. It provides an organization with a holistic perspective of supply chain time and costs associated with a single product. The SCTCM can be used to identify and evaluate waste (non-value adding time) and provide additional insight into the costs associated with the supply chain activities consumed by a product. The framework provides a base case which can then be used to measure the impact of different improvement scenarios in terms of changes in total time and cost and highlights the impact of trade-offs between the individual activities that make up the supply chain process for a given product.

Key words: process improvement, cost, activity based costing (ABC), time, time-based process mapping (TBPM).

12.1 Introduction

Organizations are constantly involved in ventures to improve their operational effectiveness. These ventures can take many forms such as business process re-engineering, TQM, lean manufacturing and Six Sigma. Many of these efforts take place in a relatively localized manner without necessarily having the complete supply chain picture. This can lead to changes in the process that can benefit some functions in the organization but unfavourably affect others, leading to local optimization or situations where a project improves certain aspects of the operation (e.g. lead time) but adversely affects others (e.g. costs). To avoid this, it is necessary to have visibility of the supply chain and to understand the interactions between different functions and variables.

This chapter presents a framework designed to provide increased visibility of time and cost through the supply chain by reconciling the techniques of time-based process mapping (TBPM) and activity-based costing (ABC). The framework is intended to help companies evaluate the costs of waste in their processes and to identify key areas for process improvement for the supply chain as a whole. The framework was designed specifically for companies in the food and drinks industry and it is applicable to all users across the entire range of supply chain operations. The tools contained in the framework will enable different users to gain the appropriate level of supply chain information that they require to improve the efficiency and effectiveness of their supply chains.

The chapter is written as a 'how to' guide, providing the reader with the requirements to implement the framework. First we describe the framework, discussing each of the stages and providing advice on their use. A case study is then used to exemplify the use of the framework and to discuss some of the challenges and opportunities that arise from its application. Finally, we highlight the benefits from previous uses of the framework and present conclusions on its application.

12.2 Supply chain time – costing mapping (SCTCM) framework

The supply chain time – cost mapping (SCTCM) framework consists of six stages (see Fig. 12.1). Stage 1 is project definition and involves setting aims, objectives and scope for a project. Stage 2 involves a mapping of the processes in the supply chain. Stages 3 and 4, which can take place in parallel, involve the collection and analysis of time and cost data, respectively. The information arising from this analysis is used in stage 5 to create a cost–time profile. Finally stage 6 involves identifying opportunities and defining the next steps.

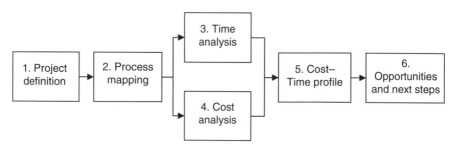

Fig. 12.1 Supply chain time – cost mapping framework.

12.2.1 Project definition

This first stage is crucial to the success of the process. The initial focus is to appoint a project champion, establish the principle aim of the project, develop a set of supporting objectives, agree the scope of the project and its boundaries and identify the resources needed to undertake the project.

The aim of the project needs to be unambiguous and concise, with success criteria to provide key reference points for the project team. The objectives (to reduce cost, to map time, to identify waste, etc.) must be aligned with the overriding aim of the project. The scope of the project should include the unit of analysis (e.g. a single product, a product family) and the timeframe of the study. The supply chain processes should also be included.

Once these fundamental factors have been established, the members of the project team can be appointed. The project team will generally comprise a permanent core team (around six in number including the project champion) and will be augmented with subject matter experts as the need arises. The team may be made up of both internal and external members depending on the supply chain processes included in the scope. If the scope of the study crosses organizational boundaries then appropriate confidentiality agreements will need to be put in place.

A key output of this first stage is the high-level project plan, which provides details of the key work packages, their owners, objectives with key deliverables and timescales.

12.2.2 Process mapping

Having agreed the scope, the next stage is to identify and record the relevant supply chain activities for the product in question. The key consideration at this point is the level of detail required; data should be captured at the lowest possible level (task) as the task data can then be aggregated at a later date into activities and process, depending on the level of visibility required. The output at this stage will be a supply chain map detailing the activities in the process. This map will provide the essential backbone of the study and act as a key enabler of the cost collection exercise.

The mapping process begins by physically walking through the process from beginning to end and it is recommended that you are accompanied in this stage by each process's owner. This is an extremely important activity, as by observing the processes you can see first hand and gain an understanding of the time and resources consumed by the product as it travels along the supply chain. It is important to capture and record the inputs (e.g. raw materials, equipment, people), controls (e.g. production schedule, budget) and outputs (e.g. work in progress, finished products) of every activity. This is crucial in terms of being able to cost each activity and identify the changes to product form as it passes along the supply chain. The latter point is extremely important in food and drink supply chains, as measurable units at each process stage may be different; for example, the unit of analysis may be a pack of cooked meat weighing 100 g, but at the beginning of the process it would have been a live animal. Once the map has been completed it should be validated with each process owner and changes made to ensure accuracy.

12.2.3   Time analysis

The key objective at this stage is to identify how time is spent in the supply chain and to assess where time is not adding value. It is important here to establish our operational definition of value, which is based on the Gregory and Rawling (1997, p. 31) definition which states that: 'Value is a property of a product or service that the customer cares about and would be willing to pay for'.

Blackburn (1991) argues that a value adding activity is one which achieves the three criteria noted in Table 12.1.

Those activities that do not match the above criteria will be recorded as non-value adding and the associated time and cost will be classified as waste.

The activities mapped from the process map together with the time split between value and non-value adding activities will be used to generate a time-based process map (TBPM). A TBPM is a stacked-bar chart that illustrates the sequence of activities; each bar represents an activity and for every bar the proportion of value and non-value adding time is identified. It is important to validate the TBPM with the activity owners as the split between value adding and non-value adding activities could be a contentious one based on the criteria used to define value. Figure 12.2 illustrates a TBPM for a distribution process.

Table 12.1 Value adding criteria

	Value-adding examples	Non value-adding examples
1. The customer cares about the change	Taking an order Assembling a product	Unnecessary transport Unnecessary packaging
2. Physically change the item	Most manufacturing processes (e.g. sterilization, blending, cooking)	Storage Waiting for decisions
3. Right first time	Maintaining quality standards Delivery on time in full	Rework Damages Over-ordering

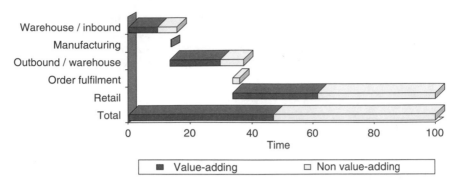

Fig. 12.2   Example of a time-based process map.

12.2.4   Cost analysis

This stage focuses on collecting and analysing both financial and non-financial data associated with the unit of analysis as it travels along the supply chain. It is essential to capture both direct and indirect costs and to identify the accounting systems used to produce the cost data as well as the accounting policy and procedures of the organization which are used to calculate the cost of the product.

It is important to identify the relevant sources and providers of financial information and a completed process map will help to enable and facilitate the collection of the data. Key data sources to consider are the bills of material for direct costs; departmental budgets and monthly operational reports will provide visibility of standard costs, indirect costs and production volumes. Assumptions underpinning the allocation, apportionment basis and indirect absorption rates will be contained in the accounting manuals and operating procedures of the organization.

The level of cost detail may also differ along the supply chain based on the cost unit for each process such as cost per tonne, per pallet, per case or stock keeping unit; it will be necessary here to collect conversion rates, for example the number of cases per pallet. These conversion rates are important, as equivalent units may be employed to cost the unit of analysis along the supply chain as in the case of meat price per kilogram. If the scope of the project extends beyond one organization, access to cost information may be difficult to obtain without an open book agreement between the parties involved. If this is not available then agreed contractual rates will have be used.

The purpose of the translation cost matrix is to translate cost information into a format to cost the supply chain processes identified by the project's scope and the associated activities which are consumed by a product. The major inputs to the translation cost matrix are the outputs from the cost collection exercise and the time-based process maps, together with additional financial information collected by interviewing operational and finance people, supported by on-site observations collected during the process mapping, to determine the appropriate resource drivers that will enable indirect costs to be pointed to supply chain activities. Cost drivers are then used to calculate the consumption of costs by the unit of analysis.

12.2.5   Cost–time profile

The cost–time profile (CTP) presents graphically the relationship between the cumulative time and cost consumed by the individual activities which make up a process. Time is measured on the X axis while cost is measured on the Y axis. The outputs from the TBPM and the cost translation matrix are the inputs needed to construct a CTP; an example is illustrated in Fig. 12.3.

The CTP can be used to illustrate the current situation, flagging up those activities which consume either a high proportion of total process time or cost, or both time and cost. The graph can also be used to compare a number of scenarios mapping them against the original curve to illustrate changes in cost and time for each scenario.

12.2.6   Opportunities and next steps

The aim of the SCTCM is to provide time and cost information to enable practitioners to make decisions with regard to the management of resources deployed in their supply chain operation. This information is presented in a very visual way using charts and tables which illustrate the supply activi-

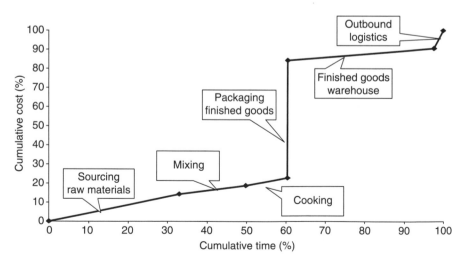

Fig. 12.3   Example of a cost–time profile.

ties that generate the waste and their proportion of the total cost of a product. This allows practitioners to identify and quantify those supply chain activities which are generating non-value adding time and cost, providing the opportunity for managers to focus and prioritise targeted initiatives that can then be deployed to reduce waste and cost. The framework can be used to highlight the potential trade-offs between activities by understanding the cause and effect relationships between activities and quantifying the potential implications for the whole supply chain if changes are made to individual activities.

12.3   Using the framework: a case study

The case study company is a large multinational with average sales exceeding €1.7 million. The company supplies spices, herbs and flavourings to retail, food service and food processing markets. Owing to confidentiality issues it was agreed to omit any direct reference to the company or its products.

The study covered two retail products sold in the French and UK markets. These products were blended ingredients packed in jars; one of the products is blended, packaged and sold in France (called product FR-1), while the other is blended in the UK, packaged in France and sold in the UK (called product GB-1). This study covered the processes involved in planning, sourcing, manufacturing and distributing each of these products, from the point of arrival of the materials through to the final delivery at the retailers' store. The main objectives of the study were the following:

- To map in terms of time and cost the entire supply chain operation for two selected products.
- To compare the processes for the two products in the UK and France.
- To identify areas of opportunity for improvement in the supply chain.

12.3.1   Project definition
The project, conducted by two researchers, involved mapping and analysis of the processes in the supply chain. In order to facilitate data collection and analysis, the researchers decided to divide the project into five stages: data collection, primary analysis, validation, secondary analysis and feedback. Table 12.2 presents each of these stages, showing the main methods used and the time required at each stage.

12.3.2   Process mapping
At a general level, the internal supply chain for the two products is essentially the same. The main difference is that packaging is only conducted in

Table 12.2 Project plan

Stage	1	2	3	4	5
Purpose	Data collection	Primary analysis	Validation	Secondary analysis	Feedback

Methods	• Inbound warehouse visits
	• Blending sites visits
	• Packaging plant visit
	• Outbound warehouse visits
	• Interviews
	  – Forecasting/ Planning
	  – Accounting
	  – Manufacturing
	  – Purchasing
	  – Distribution
	  – Sales
	  – Merchandizing
		• Flow charting processes in France and UK
		• Time based process mapping
		• Initial costing analysis
			• 1 Workshop in France
			• 1 Validation session in UK
			• Collect value and cause–effect information for relationships matrix
			• Further data collection of value adding time
				• Value adding/non value adding analysis
				• Cost–time profiling
				• Produce final report
					• Final report
					• Feedback workshop
					• Action plan definition

Duration	7 days	10 days	2 days	10 days	1 day

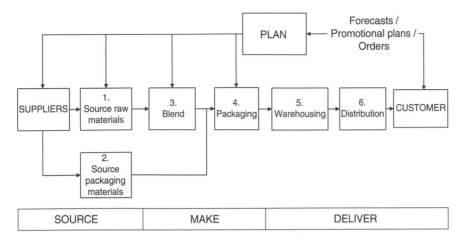

Fig. 12.4   Process overview.

France. Hence the UK product has to be transported to France for packaging. Figure 12.4 presents a general overview of the main stages required for both products.

12.3.3   Time analysis

At a more detailed level, the processes have some differences, particularly at the sourcing and distribution stages. For this reason the time consumed at each stage varies substantially between the two products as can be seen in Fig. 12.5 and Fig. 12.6, which show the top-level time-based process maps for GB-1 and FR-1, respectively, indicating value added time (VA), non-value adding time (NVA) and average storage time (stock). These figures show that the process for GB-1 has a total duration, from sourcing of materials to delivery to customer, of 138 days, compared to 95 days for FR-1. The main areas that account for this difference are warehousing and distribution.

Plan

Planning is essential for determining the flow of products along the supply chain. The planning processes for both products are very similar, as would be expected. These processes were not analysed from a time perspective because they take place at different timelines which are not necessarily aligned with the physical processes. In both processes there are a number of inputs that feed the forecast, such as seasonality, promotions and market inputs. Once the forecast is generated, specific plans are produced for the different stages in the process. In France the process includes the planning of the packaging lines which are not present in the UK.

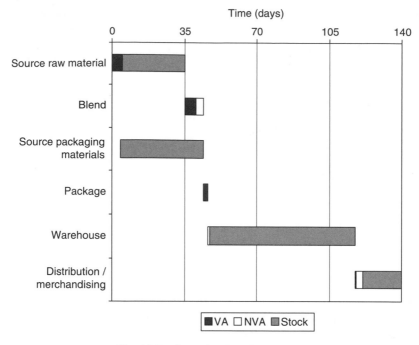

Fig. 12.5   Overview TBPM – GB-1.

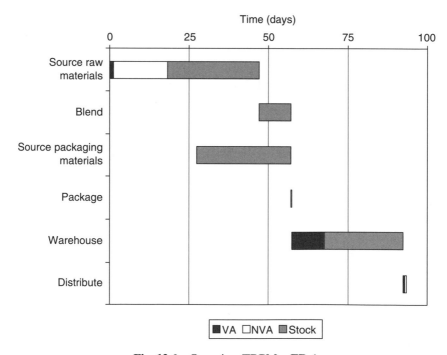

Fig. 12.6   Overview TBPM – FR-1.

Source

Raw materials are sourced from around the world. In the case of GB-1, the pre-sterilized raw materials are delivered directly to the factory; for FR-1 herbs and spices are delivered at the port and then transported to the factory. Both sourcing processes consume relatively large amounts of time as a proportion of the total supply chain, 22% for GB-1 and 50% for FR-1; the main reason for this is the amount of raw materials stock held for both products. The process for FR-1 is considerably longer than that for GB-1; the main reason for this is that it covers a larger scope with the delivery at the port, rather than delivery at the factory.

Packaging only takes place in France and all GB-1 blended in the UK is sent to France for packaging. Similar to the sourcing of raw materials, the vast majority of time in this process is consumed in storage. Different quantities are held for each of the four packaging materials, with labels being the highest, with about 40 days' stock on average.

Make

The blending processes for both products are very similar and consume only a few hours. Most of the time consumed at this stage is in waiting for quality control analysis or simply in storage. In the case of the product blended in the UK, transportation to France for packaging is required, consuming an estimated three days. Packaging for both products takes place in France. The actual packaging process is very swift, taking only a few minutes; the only difference between the two products is that GB-1 has to be returned to the UK for distribution.

Deliver

The delivery processes in France and UK are substantially different. In France, products are held in two distribution centres (DCs), from where they are transported by forwarders directly to the store. The sales force is responsible for laying out the product directly on the shelves. In the UK, the product is initially held in the DC and then moved to a network of merchandisers responsible for delivering the product in the stores and laying it out on the shelves.

12.3.4   Costing methods and analysis

To compare product costs, it is necessary to start by exploring the costing methods and accounting procedures employed for each product. This is presented below following the same framework used for the process analysis (source, make and deliver). It must be noted that the UK division does not have a single integrated supply chain costing system that tracks the products from receipt of raw materials through to the delivery of the finished product to customer. Table 12.3 summarizes the costing methods

Table 12.3 Comparison of costing methods and accounting procedures

	GB-1 (UK)	FR-1 (France)
Source	**Raw materials** This product is a blend of four ingredients which are bought from a range of suppliers presterilised and delivered into a warehouse in the UK. A standard costing approach is used for raw material inputs based on the quantities indicated in the bill of materials (BOM). The standards are reviewed quarterly, to adjust for factors such as currency fluctuations, seasonality and weather conditions that can affect the price of raw materials. **Procurement and inbound logistic** Procurement costs associated with raw materials are not included in the standard cost. They are accounted for as an indirect cost in the ABC calculations and charged to the blend during the blending process.	**Raw materials** FR-1 is a blend of seven different ingredients. Raw materials are not presterilised and delivered into the inbound warehouse. Standard costing is used to calculate the cost of the raw materials as specified in the BOM. The standard costs are reviewed quarterly as in the UK. **Procurement and inbound logistic** The total costs of both procurement and inbound logistics activities are averaged out over the total volume of raw materials received to give an average cost per kg for each activity.
Make	**Blending** The cost of the blend is calculated using the standard cost of the raw materials plus the cost of packaging. A standard 3% allowance cost is built into the blend cost to account for normal loss during the production process. The blending process costs are formulated using the product template in the ABC system for a standard batch size to give a cost per unit. **Transfer to France** The blend is transferred to France in large bags (400 kg), the transport is provided by a third-party logistics operator (3PL). The company has negotiated an average price per load with the haulier which allows calculating an average cost per bag.	**Blending** In France the product is blended, then sterilized, the standard cost of the blend at each processing stage includes the cost of raw material, interim packaging, labour cost and overhead costs. As the individual raw materials are combined into the blend by the production process, the costs are then tracked against a work in process (WIP) code. **Packaging** The sterilized blend and associated packaging costs, e.g. jar, cap, label etc, are combined at this stage to give the manufacturing cost of the product, line packaging overheads costs are distributed to the product using an activity-based approach as

Packaging
This product is packaged in France, where a similar ABC approach is used in the packing activity, except the cost unit is a single jar compared to a case in the UK. The packaged product is priced in € per jar and a standard exchange rate is used to convert the cost into £ per case.

Deliver
The delivery activity begins with the transportation of the finished product from France and concludes with the merchandiser placing a jar on the retailer's shelf.

Transport to United Kingdom
The product is palletized and transported back to the UK by a 3PL at a standard price per load for an average of 45 pallets.

Outbound warehouse
The product is received into a finished goods warehouse operated by a 3PL. A standard charge of per pallet received is made in addition to a fixed charge per month regardless of usage made for storage of finished goods.

Distribution to merchandiser
The 3PL provides three activities which have an agreed standard tariff i.e. picking an order, administering the picking order and delivery of the order to the merchandiser stock point.

Merchandiser
Sales support costs that are associated with this stage are included and absorbed to product at an average cost per case.

described earlier in the production process for GB-1. A standard change over allowance and a 3% standard loss for materials and packaging is included into the product cost calculation. The product cost is expressed as a cost per jar. The product is then palletized and transferred internally to the finished goods warehouse.

Deliver
The outbound logistic activity is shared with a 3PL forwarder, the finished product is stored and order processing, picking and consolidation are carried out in-house and then the product is delivered to a number of regional distribution centres (RDC) in France. The final leg to the retailer is undertaken by a 3PL forwarder who consolidates the products with other products destined for the store.

Outbound warehouse
The finished goods warehouse is shared with a sister operation. The costs of the warehouse facility are recovered using a combination of allocation, apportionment and absorption to calculate an average cost for either € per kg delivered or € per pallet depending on the distribution channel to market.

Distribution to store
The finished product is received at the point of sale via the 3PL forwarder. Regional sales personnel are responsible for placing the product on the shelf.

and accounting procedures used for both products at each stage of the process.

Source and make

The costing methodology used for product GB-1 combines standard costing for raw material inputs (including packaging) and activity-based costing (ABC) for the distribution of, sourcing and manufacturing activity costs (direct and indirect) of the products manufactured at their UK factory. This general approach has now been adopted by the French division for costing their products.

Deliver

Both divisions have adopted different supply chain approaches to the distribution of their retail products, which means they require different costing approaches. Key characteristics and considerations of the costing approaches employed for both products include:

- The objective of the product costing system is to maximize the distribution of manufacturing overheads to the products that consume the activities, i.e. move costs from below the gross profit line to include them in the cost of goods sold calculation.
- Both units operate process costing in the form of batch costing.
- Standard costing is used to calculate raw material and packaging inputs as specified in the bill of materials and these standards are reviewed quarterly.
- Activity-based costing is used to distribute direct and indirect manufacturing costs to products and the cost drivers are reviewed annually.
- Costs are attributed to both products as they pass through each processing stage.
- The output from one processing stage forms the input costs for the next activity; profit is only taken when sold to the retailer.
- Change-over allowances are incorporated into the manufacturing cost via the activity-based cost driver for each product.
- Each product has a cost template which specifies the consumption of activities to manufacture a single cost unit that can then be aggregated depending on batch size.
- Based on prior knowledge of the production process, an allowance of 3% is made for waste. This is treated as a normal loss and these costs are charged to the cost of production.
- There are multiple handling units for different supply chain activities and this makes it difficult to visualize the incremental cost of the finished product, unless a form of equivalent cost unit is applied to the supply chain.

- For activities which are delivered by third parties, an agreed rate is applied in the form of an average cost which may not reflect the true cost of service in respect of an individual product.
- The cost of financing inventory is not included in the product costing system and is accounted as a finance cost below the gross profit figure.

12.3.5   Product costing assumptions

Where possible, the budgeted costs for raw materials and in-house activities and the average rates for third party activities have been used to calculate the incremental product cost as it flows through the supply chain from receipt of raw materials to delivery at the point of sale. Where it was not possible to obtain a budget cost for an activity, assumptions were made. As there was no equivalent unit across the supply activities, a cost per case was used, as this is the common sales order unit.

It was found that there were some inequalities and exceptions between the accounting approaches used for each of the products, most notably:

- Inbound procurement and logistic costs are treated differently; in France, an average of these costs over volume received was estimated, while in the UK, these costs were included as indirect costs.
- Sales support costs for merchandisers in the UK were included in the delivery cost per case, while in France these costs are not recovered by product but are accounted as a central support activity.

The aim of this analysis was to follow the accounting culture/spirit of both organizations in terms of trying to trace the cost of activities consumed by products, so maximizing the cost of goods sold calculation and minimizing the amount of overheads that cannot be traced to a specific product or activity.

12.3.6   Allocating costs to products

When all the costing information from the various sources throughout the supply chain processes is combined and the different handling units (pallets, kg, jars and cases) are converted into an equivalent unit, it becomes possible to visualize and calculate a supply chain cost for both products. Merchandising and packaging costs account for around 81.1% of the cost of GB-1, while the actual cost of the raw materials represents 8.1%. In the case of FR-1, packaging costs account for 61% of cost, the next largest component is the raw materials, which equates to 13%. Figures 12.7 and 12.8 illustrate the cumulative cost of a case as it flows through the supply chain process for each product.

For GB-1 there are two significant increases in cost when the product is packaged and when it passes through the merchandising network to the

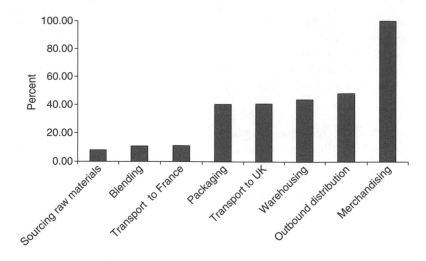

Fig. 12.7   Cumulative supply chain costs for GB-1.

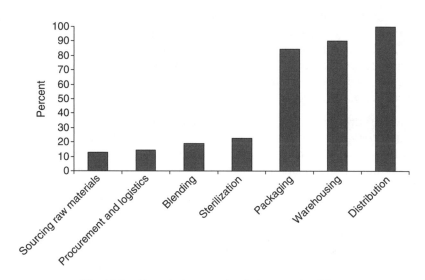

Fig. 12.8   Cumulative supply chain costs for FR-1.

point of sale. For FR-1 there is only one significant increase during the packing process.

12.3.7   Cost–time analysis
This section combines the time and cost analysis outlined in previous sections and explores the relationship between these variables for both GB-1 and FR-1 as they flow through their respective supply chains from

sourcing to delivery at the point of sale. This cost–time profile can be used to chart the complete supply chain and provide an insight into the relationship between the cost and time of individual supply chain activity clusters and can be used to highlight areas which require further investigation. Figures 12.9 and 12.10 illustrate the cost–time profile for each product

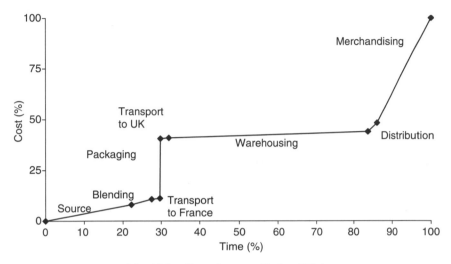

Fig. 12.9   Cost–time profile for GB-1.

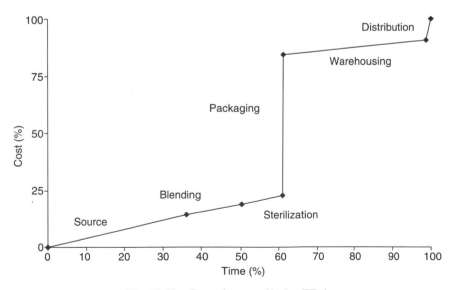

Fig. 12.10   Cost–time profile for FR-1.

at specific points along their supply chain combining both cumulative time and cost.

Two interesting time–cost couplings are highlighted: warehousing and packaging. Warehousing, where the percentage of total case cost for both products is relatively small compared to the percentage of time, but when the time is broken down into value added, non-value adding time and inventory time this confirms the need to investigate these activities further. On the other hand, the packaging process for both products accounts for a relatively high proportion of the total cost, but only accounts for a tiny proportion of total supply chain time. In the case of GB-1, 51% of the total case cost is attributed to the merchandising process compared to only 14% of supply chain time (see Table 12.4); a further investigation into the costing method and types of costs that are allocated to this activity may be worthwhile.

12.4   Areas of opportunity

The SCTCM framework does not take out time and cost in a supply chain, but it assists the process by presenting the current situation, identifying areas of opportunity and providing an important input to developing focused process improvement initiatives which are then targeted to reduce waste and cost in the supply chain. Table 12.5 highlights the outputs of the SCTCM framework which illustrates the area opportunities for both the French and UK supply chain operations.

Table 12.4 Cost–time comparison

GB-1 Cost/time per activity			FR-1 Cost/time per activity		
Activity	Cost (%)	Time (%)	Activity	Cost (%)	Time (%)
Sourcing raw materials	8	22	Sourcing raw materials	14	36
Blending	3	5	Blending	5	14
			Sterilization	4	11
Transport to France	1	2			
Packaging	29	1	Packaging	61	1
Transport to UK	1	2			
Warehousing	3	52	Warehousing	6	37
Outbound Distribution	4	2	Outbound Distribution	10	1
Merchandising	51	14			
Total	100	100	Total	100	100

Table 12.5 Summary of areas of opportunity

Process efficiency	The supply chain processes for GB-1 and FR-1 take on average 138 and 95 days, respectively. The proportion of value adding time (VAT) in each process was 12.4% for FR-1 and 9.8% for GB-1, indicating a substantial opportunity for lead-time reduction. The largest proportions of time are spent in warehouses either at a raw material or finished product state. In the case of raw materials, the reason for holding stock is the instability in supply and harvest periods, while in the case of finished products the main reasons are the use of large production runs and uncertainty in demand. However there was consensus in the company that the levels of stock were excessive. At the time this study was conducted, GB-1 was blended in the UK, transported to France for packaging and returned for distribution in the UK. This not only added six days to the product cycle time, but also incurred additional cost and could create disruptions in supply. At the time there were regulatory reasons for maintaining the blending process in the UK, however, from a process point of view it would be easier to have both blending and packaging in France. The cycle time for the GB-1 was 44 days longer than that for FR-1. Transportation for packing in France accounted for six of these days, but the majority could be attributed to the final warehousing and distribution stage. The average stock for the product in the UK warehouse was 70 days compared to only 35 days in France. Furthermore, the merchandising process carried another 19 days of stock that were not required in France.
Process costs	A direct comparison of the costs of the two products was difficult because there were some inequalities and exceptions between the accounting approaches used by the two divisions. These differences provide an opportunity because establishing common costing approaches would make comparison easier. The exploitation of this opportunity is already in progress, can be seen in the implementation of ABC in France. For the two products analysed, the packaging process represented a large proportion of the total cost. For GB-1 this process consumed 29.4% of the total cost and 61.5% for FR-1. The sourcing of raw material excluding packaging represented 8.1% of total supply chain cost for GB-1 and 12.9% for FR-1. A main reason for the difference was the accounting treatment of procurement and logistics costs between the UK and France rather than a major difference in the cost of materials. Merchandising represented a high proportion of the total cost of GB-1 (51.7%). This was substantially higher than the cost of distribution in France. However, outbound logistics costs in France did not recover any sales costs although those in the UK do. ABC was predominantly used in manufacturing in both the France and UK operations. There is an opportunity for expanding this method into sourcing and distribution activities where the costing process is currently based on averages. For example, in the case of GB-1, 67% of the total supply chain cost is calculated using traditional methods. This would include activities such as sourcing raw materials, transportation, warehousing and merchandising.

Table 12.5 *Cont'd*

Visibility of information	There is a lack of cost visibility when average costs are used, for example in third party logistics activities.
	The ABC system provides visibility of cost across processes. However, there are differences in the way things are accounted for, making comparisons difficult. As the systems in the two countries become more standardized, comparisons will become more transparent.
	Products do not take into account the cost of financing, since not all overhead costs are absorbed by the products. However, considering that the products spend a long time stored in warehouses, it would be relevant to estimate the costs of financing these stocks.
	Another issue of visibility of costs appears in the French finished goods warehouse. This warehouse is shared with another division and costs are allocated based on averages, which increases the chances of product cross-subsidisation.

12.5   Conclusions

The supply chain time and cost mapping framework provides organizations in a supply chain with enhanced visibility of both time and cost along the processes. A major finding that became apparent during the development of the framework was that the relationship between cost and time in processes is non-linear. This challenges the traditional view that reducing non-value adding time implies a proportional reduction in cost. The case studies that were used to develop the framework indicated that over 82% of supply chain time may be classified as non-value adding.

Feedback from industrial partners involved in the development of the framework was extremely positive. The approach enabled these organizations to implement 'in-house' projects to evaluate waste and its associated costs, which resulted in the introduction of various quantifiable process improvements. The outputs from these company-based projects increased the awareness of their people with regard to current supply chain issues and importantly the impact of cause and effect relationships in the process, especially when making decisions that have an impact on the wider supply chain.

The project teams consisted of people from different functions and a benefit which was not anticipated was the improvement of relationships between financial and operational personnel. This was important as during the development of the framework it became apparent that financial information was not usually shared between supply chain partners. The framework gave the organizations a structure and a set of tools for supply chain analysis and design which improved their ability to analyse the total cost of a product by activity and increased visibility of cost and time in their processes.

12.6   Sources of further information and advice

Process analysis

Bicheno, J. (2000), *The Lean Toolbox*, 2nd edition, PICSIE Books, Buckingham, England.

Time analysis

Gregory, I. and Rawling, S. (1997), *Profit from Time: Speed Up Business Improvements by Implementing Time Compression,* McMillan Business, Basingstoke, UK.

Stalk, G. Jr. and Hout, T. M. (2003), *Competing Against Time: How Time-Based Competition Is Reshaping Global Markets*, Free Press, New York.

Cost analysis

Kaplan, R. S. and Cooper, R. (1997), *Cost and Effect: Using Integrated Cost Systems to Drive Profitability and Performance*, Harvard Business School Press, USA.

Develin, N. and Bellis-Jones, R. (1995), *No Customer-No Business: The True Value of Activity Based Cost Management*, Accountancy Books, Milton Keynes, UK.

Supply chain perspective

Harrison, A. and van Hoek, R. (2008), 'Value and logistics costs' (chapter 3, pp 65–94) and 'Managing the lead-time frontier' (chapter 5, pp 139–70) in *Logistics Management and Strategy: Competing Through the Supply Chain*, 3rd edition, Financial Times/Prentice Hall, Harlow, Essex, UK.

12.7   Acknowledgements

The authors would like to acknowledge the support and guidance given to them by all the members of the Supply Chain Costs and Effective Swift Service (SUCCESS) Steering Group especially Linda Whicker and Mike Bernon. The steering group comprises representatives from our industrial partners, The Chartered Institute of Logistics and Transport (UK), Warwick Manufacturing Group, at Warwick University and Cranfield Centre for Logistics and Supply Chain Management at Cranfield University. The SUCCESS research project was jointly funded by the Engineering Physical Science Research Council (EPSRC) and industrial partners.

12.8   References

BLACKBURN, J. D. (1991), *Time Based Competition: the Next Battleground in American Manufacturing*, Business One Irving, Homewood, Ill.

GREGORY, I. and RAWLING, S. (1997), *Profit from Time: Speed Up Business Improvements by Implementing Time Compression*, McMillan Business.

13

Improving food distribution performance through integration and collaboration in the supply chain

C. Stephens, Consultant, UK

Abstract: Grocery distribution in the UK is based on a commonly adopted centralised and stockless model and is highly efficient. Further efficiency gains are achieved through vertical and internal process integration, with additional gains possible through cross-channel collaboration.

Key words: UK, grocery, distribution, integration, collaboration.

13.1 Introduction

The UK food retail market is often held up as a paragon of logistics efficiency, with unit costs and inventory levels significantly lower than most of the rest of the world (Fernie, 1995). Logistics developments as described above have taken place against the background of market concentration and intense competition.

Since 1960, the major multiples' share of the UK grocery market has grown from around a quarter to over three-quarters according to annual research published by the Institute of Grocery Distribution (IGD), at the expense of the independent and cooperative sectors. Indeed the 'big four', Tesco, Asda, Sainsbury's and Morrisons account for over 70% of all sales (TNS research, July, 2008). However, these four contain some very different companies (Seth and Randall, 1999), with Sainsbury's and Asda having achieved their share largely through organic growth, whilst Tesco and Morrison have acquired competitors along the way. All four, however, have very similar distribution and replenishment models. The general adoption of a centralised distribution model, with varying levels of contractor

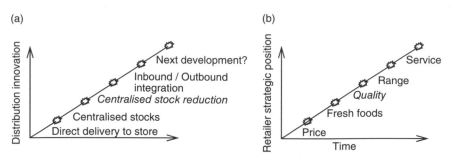

Fig. 13.1 Trends in distribution innovation (McKinnon, 1989) and strategic retail offering (Beaumont, 1987), after Greiner, 1972. — = Evolution; ✿ = Revolution.

involvement, occurred in parallel with the centralisation of retail power from the 1970s to the 1990s.

In the last decade, this model has been further developed to reflect vertical integration of control within retail supply chains. Although there is no single one-size-fits-all solution, food retail distribution operations in the UK all draw to some extent from each of the following:

- vertical integration (primary, nominated carrier, factory gate purchasing, vendor- or co-managed inventory and collaborative planning) to reduce costs and balance flows
- operational integration (backhaul, fronthaul and regional fleets) to reduce resources
- new technologies (including vehicle tracking, telematics, routing and scheduling systems and the use of double-deck trailers)
- cross-channel collaboration, facilitated by the use of specialist independent contractors.

13.2   Grocery distribution in the UK

The key trends and drivers in the retail distribution environment have been:

- centralisation of distribution: this has grown from 60% of total volume in the late 1960s (Pettit, 1967) to around 95%, or more, currently (IGD, 2006)
- concentration of retail power in the hands of a few major multiples (Akehurst, 1983; Fernie, 1992; 1997; Bourlakis, 1998) to the extent that the top four UK food retailers now control over three-quarters of the market
- the rise and fall in the use of third party distribution services providers (McKinnon, 1986; Fernie, 1990), with third party transport penetration having grown from 40% in 1984 to a high of 75% in 1998 before falling

back to 45% in 2006, and warehousing penetration from 14% to a high of 68% in 2000 falling back to 46% by 2006 (IGD, 2006).

Against the background of retail concentration, the arguments for the centralisation of distribution are so compelling that, whilst there have been differences in the rate of uptake, all major UK retailers had implemented these techniques almost universally by the 1990s. Having achieved parity in this respect, the next key trend has been the optimisation of physical distribution resources through the tools of:

* integration of primary and secondary distribution
* increased asset utilisation through multi-cycling and new handling techniques and technologies
* reduction in inventory through rapid replenishment, improved forecasting and co-managed or vendor-managed stocks.

Distribution of fresh foods to major UK retailers is thus highly integrated and centralised. Typically, a major retailer will operate around 10–15 regional distribution centres (RDCs), each collating the individual store orders for 50–70 stores and handling between one and two million cases per week. Some RDCs operate across a range of temperature regimes ('composite' RDCs), others are dedicated to a single product group (e.g. frozen or produce). Most retailers operate at least some of their RDCs themselves, with the balance contracted out to third party operators. Many RDCs have their own depot-base transport fleets for store deliveries, either operated in-house or by third parties.

Retailer-led initiatives, such as ECR (efficient consumer response), have had an impact on physical distribution systems. Examples include stock and processing activities being forced back up the supply chain, redefinition of exactly 'who does what' in the supply chain and further reduction in order lead times and extension of 'chill' disciplines to other temperature regimes.

13.3   Improved performance through vertical integration

Having achieved broad parity in the efficiency and cost effectiveness of their distribution systems through centralisation and the achievement of critical mass, the next step for all of the retailers was the internal optimisation of physical distribution resources, through fleet integration, multi-cycling and inventory reduction, amongst other things. This creates a climate of further change, in which many of the traditional assumptions about the way in which firms compete are being challenged. Among these is the notion that competitive advantage is created by supply chain excellence and thus by implication by physical distribution excellence. However, competitive advantage in physical distribution is gained in the short term only, with any emergent best practices easily copied and adopted by competitors.

Having taken more or less complete control of deliveries into stores, in the last ten years retailers have become increasingly involved in the supply chain from factory to RDC. Initiatives such as ex-factory buying (or factory gate pricing, also known as FGP), intermediate stock holding and retailer-controlled primary transport (i.e. transport from manufacturing sites to the retailers' regional or national distribution centres, as opposed to the 'secondary' leg from distribution centre to store) have been some of the manifestations of this trend and appear to demonstrate a willingness to examine every opportunity to drive costs out of the supply chain.

Finegan (2002) suggests that retailers have been slower to look at some of these areas of opportunity owing to a lack of understanding of costs, a preoccupation with service to stores, perceived complexity and the fragmented nature of the transport market. However, he concludes that pressure on margins and highly evolved central operations and supporting technologies have facilitated initiatives across the entire supply chain.

Although collaboration based on ECR has become ubiquitous in the last ten years, academic studies of such collaborations, and thus literature about them, are scarce (Corsten and Kumar, 2005). Based on a survey of 266 suppliers to Sainsbury (the identity of the retailer is not actually revealed in this paper, but can be inferred from Corsten and Kumar, 2003), the authors identify enablers for suppliers to collaborate with retailers in a vertical partnership, including relative scale of operations, perceived fairness and perceptions of the relationship between scale of effort invested and reward gained. They conclude that, while vertical collaborations in the context of ECR are generally a good thing, rewards are almost certainly not fairly apportioned. However, they still recommend that suppliers should seek to enter into such partnerships, although they should select potential partners carefully, on the basis of perceived levels of trust and the extent to which the supplier is 'smart' and can therefore contribute experience and learning. The power and aggression of retailers may be beneficial to them in the short term in that it allows them to hang onto a greater share of the benefits of collaborations with suppliers. However, this will be counter productive in the longer term, as cynical suppliers will simply invest less effort in the partnerships to the detriment of the total benefit, in an attempt to balance reward with perceived effort or investment (Corsten and Kumar, 2003). The absence of mutual trust, possibly caused by asymmetry of power, appears to be a significant inhibitor to the development of successful collaborations (Frankel *et al.*, 2002).

An empirical survey conducted amongst UK suppliers of fresh produce to supermarkets suggests that whilst levels of collaborative effort and economic factors are important in determining the value and quality of a vertical supplier–retailer relationship, relative interdependence is the most important enabler/inhibitor (Duffy and Fearne, 2004). Crudely, the more power the retailer has relative to the supplier, the greater the asymmetry independence between the two parties and thus a disproportionate sharing of the benefits of the partnership. Symmetry of dependence leads to higher levels of perceived 'fairness' and thus trust, which in turn leads to deeper cooperation and collaboration, resulting in greater accrued benefits to be shared between the two partners.

The use of transaction cost economics as a rationale for partnerships is flawed in that, whilst it recognises the need for control measures and costs to deal with opportunism by one of the parties to the arrangement, it does not adequately recognise the impact of power on the potential outputs (Duffy and Fearne, 2004). On the one hand, vertical partnerships within a chain, as promoted by ECR, will deliver some benefit compared with traditional adversarial models, as they push costs back up or down the chain between competing chain members, but ultimately leave the total sum of costs unchanged. On the other hand, partnerships may potentially reduce overall costs, but the extent to which benefits are allotted fairly can be compromised by power, which is, more often than not, a reflection of relative size. Having sought to extricate more value through such extended influence and control, it is reasonable to assume that the retailers would regard further innovation in new areas and directions as being fair game.

One of the most significant areas in which food retailers have sought to form relationships with logistics service providers in physical distribution is in primary distribution – the movement of goods from manufacturers or their warehouses into the retail distribution centres (DCs). Historically, retailers bought their goods from manufacturers on a 'delivered in' basis, that is, the manufacturer was responsible for organising transport to the retailer's distribution centre and the costs of this operation were included within the product price (Smith, 1999). During the 1990s, retailers identified three drawbacks to this method of supply:

- there was no visibility of the transport element within the total product price. Thus two similar bought-in prices for a particular product might hide variations in manufacturing efficiency, offset by distribution efficiencies. Retailers would ideally seek to source the most efficiently produced product without costs being distorted by distribution costs, either through relative inefficiency or distance from the centres of demand.
- smaller manufacturers, who might nonetheless be able to manufacture efficiently, were unable to buy distribution services economically, owing to the absence of economies of scale.
- increased volume pressure on the DC networks was leading to congestion by delivery vehicles, many of which were only delivering small

quantities of products from a single supplier or small groups of suppliers.

The perceived advantages arising from increased retailer intervention in primary distribution were thus:

- visibility of relative manufacturing costs, allowing buyers to concentrate on sourcing products from the most efficient manufacturers
- lower transport costs, achieved by the retailer either acting on behalf of groups of manufacturers in the third party market place or by directly placing distribution operations with a contractor on behalf of the retailer themselves
- managed intake profiles, allowing for better planning and allocation of resources at the DC and consequently improving capacity.

The conventional model of retailer centralised distribution (McKinnon, 1989) has the retailer in control of operations from the point of receipt of goods at the distribution centre through to delivery to the store and return of empty transit equipment. Although retailer controlled, some of these operations have been contracted out to third-party logistics providers, albeit on a dedicated basis. The new model, which emerged through the 1990s, sees retailers controlling the flow of goods all the way from factory to store, although with some differences in application and methodology. The key differences are:

- the nature of the commercial relationship between the retailer, the manufacturer and the provider of primary transport. Some retailers (for example Sainsbury) have experimented with 'ex-factory' or 'factory-gate' pricing. Under this methodology, the retailer pays for the goods excluding any transport elements and then engages a transport provider to work on its behalf. The retailer pays directly for the transport services. Other retailers have sought to encourage their suppliers to use certain nominated carriers on a 'pool' basis, with transport costs still paid for directly by the manufacturer, but on a basis overseen by the retailer.
- degree of integration with secondary (i.e. DC to store) operations. Some retailers have offered collection services to manufacturers, to individual DCs by vehicles on the way back from making store deliveries. Others have put in place a network of inter-DC movements, allowing store delivery vehicles to collect products for a number of different DCs. Some have even put product destined for competing retailers' networks through this system on a commercial basis.
- the extent to which the operations are actively managed by the retailer from day-to-day, with direct intervention in the planning of booking times, management of contingencies (lateness) and the pursuit of further cost saving measures. Sainsbury operate a primary operations team, based at their head office in London. Asda, on the other

hand, leave much of the management process to their nominated contractors.

Each of the main UK food retailers has become involved in primary operations during the last ten years, although to differing extents and at differing paces. In many cases, intervention in primary has been gradual and incremental and it has been hard to discern exactly when each of the retailers has actually achieved a significant scale of operations. Furthermore, some of the differences in methodology make it difficult to define whether a specific retailer has achieved control of its primary operations or not. Typical techniques have included:

- transhipment and consolidation hubs typically put out to tender to 3PLs on the basis of aggregated volumes creating distribution buying power
- consolidation of manufacturer stocks in intermediate or primary consolidation centres, again to leverage warehousing buying power as well as increasing visibility of availability
- regional pooling of volume to achieve procurement savings
- backhaul supplier collections, using store vehicles into a single RDC, possibly augmented by inter-depot trunking to allow for the collection of all national volume from suppliers
- nominated carriers, often separating temperature regimes or other product groupings.

Third-party service providers (contractors) were obviously active in the primary marketplace prior to the implementation of these primary initiatives. All but the very largest manufacturers had tended to dispose of their in-house transport assets at the time of the shift from direct store delivery to centralised distribution in the 1970s and 1980s. A number of primary contractors thus achieved critical mass by integrating the volumes of a large number of manufacturers and put in place the infrastructure to do this in a timely manner.

Retailer intervention in primary operations represented both a threat and an opportunity for these contractors. They could either work with the retailers and thus hope to grow their volume, or resist and see volume transfer to other contractors. However, whilst the contractors had achieved cross-channel efficiency savings by integrating volumes for a number of retailers, the retailers themselves generally approached primary initiatives on an intra-channel basis. Thus Sainsbury sought to bring all of its volumes together in order to reduce distribution costs, choosing to ignore the fact that the typical Sainsbury supplier also supplies many of the other retailers. Ultimately, this creates the possibility of goods from a single manufacturer being distributed by a number of parallel primary networks. Whilst this may provide a reduction in costs of the 'free-market' operation of primary, it may also institutionalise further inefficiencies, such as, for example, a factory having to despatch its products on vehicles belonging to three different contractors with each vehicle rarely being full.

Primary distribution initiatives, therefore, have so far been concerned with further vertical integration of discrete supply chains. The cross-channel efficiencies which had been established by contractors in open market arrangements prior to retailer intervention have, by and large, been superseded by other models. Although the retailers have undoubtedly made gains in terms of lower costs and operational controls, it is not clear whether these gains could have been made without subverting the cross-channel efficiencies which were already in existence, nor whether any clear gains have been made from concentrating on single-channel integration. Since all of the retailers who have involved themselves in primary distribution initiatives seem to have set out with broadly the same agenda and objectives, it is hard to see how these initiatives have made any significant contribution to competitive advantage.

13.4   Improved performance through internal process integration

Having considered vertical integration as the extension of retail control along the length of the supply chain, to include supplier and inbound activities, further finessing is possible through the integration of previously separate internal operations. This can perhaps be thought of as horizontal integration between separate activities within the operations of a single retailer. In other words, if supply chains are generally considered as vertical hierarchies from raw materials suppliers to end consumers, parallel supply chains can be considered horizontally at each hierarchical level, by viewing supplier versus supplier, distributor versus distributor or consumer versus consumer. For example, whilst retailers have implemented managed primary and FGP activities, as discussed above, over the past 30 years, they have tended to continue to manage different types of merchandise separately, for example foods and general merchandise or clothing.

General merchandise (GM) has tended to be handled through completely separate depot distribution networks to foods because of the nature of the goods, throughput levels and other operational constraints. Even within foods, some retailers continue to separate temperature regimes, even though they may operate a number of multi-temperature or composite sites. Frozen foods, for example, still sit outside the mainstream food networks of Tesco, Sainsbury's and Morrisons. The main reason for this is that whilst all the other temperature controlled groups (perishable, chill and produce) are almost entirely stockless flow-through operations, frozen foods are still generally held as stock items, with much slower replenishment cycles. Although this might create synergy with slower moving ambient lines from the point of view of stock levels and cycle times, the technology for vehicles and depots is clearly different. In another development, in the late 1980s, Asda opened a series of composite depots across the UK to cover all food groups but subsequently separated 'temperature controlled' from 'ambient' depots as a way of creating additional capacity in the network.

The opportunity for unlocking synergies between separate operations within a single retailer has been realised recently through developments in asset tracking and planning, driven by new technology. Whereas a retailer might operate two depots in the same area, one for general merchandise and one for foods, gaps in capacity and potential shared vehicles are only identified through the implementation of dynamic scheduling tied to GPS tracking.

The key discipline in grocery distribution is adherence to schedule. Resource optimisation at store level, together with an innate desire to keep shelves full on the part of all retailers, lead to ground rules about the order in which products tend to be received in store throughout the 24-hour day. At its simplest, the shortest shelf life fresh foods, such as bakery products, fresh meats and produce and prepared meals tend to be received just before opening each day, whilst slower moving products or products with more shelf life and even stock held in the system tend to be received later in the day. This defines the operational peaks and troughs of the distribution resources of the various parallel networks and there is then opportunity to use the same vehicles to deliver chilled foods in the morning and clothing or tinned foods in the afternoon. Within a single depot, such multi-cycling has always been possible. Across parallel neighbouring depots, the concept of regional fleets or internal sub-contraction is a relatively recent phenomenon.

This 'regional' approach can also be extended to create virtual fleets. For example, whether or not inbound primary distribution has been integrated vertically into a retailer's operations or not, the fact remains that third-party vehicles will continue to visit its sites, whether operated by suppliers or their contractors. These incoming vehicles can potentially, therefore, be used to make a store delivery on their way out of the depot (so called front-haul) or can be eliminated from the system altogether by using a depot-based vehicle to make a supplier collection and deliver into the depot after making a store delivery (back-haul). Once all opportunities for fitting together depot and supplier collections, depot and store deliveries have been taken, a series of circular or even figure-of-eight journeys can be created and it is immaterial whose vehicles operate them or even at which part of the circle or figure-of-eight they start and finish.

So, the boundaries between operations and indeed operators are becoming increasingly blurred through the implementation of shared, regional or virtual fleets, to achieve the best combination of internal and external resources under the control of a single planning and management system.

13.5   New technologies

Routing and scheduling software packages have been in general use in the UK since the early 1980s, both at the tactical and strategic level. Strategically, demand patterns and cost models can be used to identify optimum network designs and depot locations, as well as planning resources. Once resources are set, tactical plans can be rerun daily or even more frequently to match actual volumes. Public access to global positioning by satellite (GPS) technology in the 1990s created the possibility of accurate vehicle location tracking. The two technologies of scheduling and tracking can now be integrated to give real-time feedback on the actual execution of the plan, based on vehicle progress against planned times, so that the plan can be recast as necessary 'on the fly'. The most widely adopted package has been that created by the integration of vehicle tracking by Isotrak and scheduling by Paragon into a so-called integrated transport management system. Variants of this have either been or are being implemented by many of the major UK retailers and their 3PL contractors.

Telematics, or in-vehicle computing, broadly falls into two types. In the first, mobile wireless technology, generally using the global system for mobile (GSM) or GPRS (general packet radio service) telephone networks, is used to capture performance data from engine management systems to optimise the mechanical performance of the vehicle. This can cover engine and road speeds, use of gearboxes, fuel consumption, temperatures and pressures and general wear and tear on mechanical parts. In the second, the same phone and data networks are used to pass information to and from the drivers of vehicles, such as routes, work schedules, consignment details and even proofs of delivery. Clearly, both types of data flows can be combined into an integrated whole. Telematics can now also be combined with tracking so that the vehicle generates its own response to the planned schedule. For example, telematics can create 'geofences' or virtual boundaries around locations to be visited on a schedule and trigger a message to the scheduling software to indicate when a geofence has been broken, so that the plan can be reviewed and updated. This may be particularly important in the integrated schedules of regional or virtual fleets, when any variance from the plan in one part of the cycle may potentially be passed on to a completely separate part of the operation if not corrected.

Increased fuel, wage and other variable costs have also led operators to try and achieve greater efficiency for every mile travelled. Vehicle technologies have improved to the point where double-deck trailers are now highly reliable and cost effective over shorter journeys than was the case as little as ten years ago. A typical double-deck trailer will have more than 60% additional capacity than a single deck, but with increased capital costs, longer loading and unloading times and higher fuel consumption.

Whilst double decks were only cost effective for long distance motorway trunks a few years ago, they are increasingly being used for shorter store delivery radial journeys, particularly for low value high cube products, including bread and general merchandise. Improvements in refrigeration technology have also made temperature controlled double decks a viable proposition.

13.6   Improved performance through cross-channel collaboration

Whilst there is widespread agreement that collaboration between competitors and competing supply chains is a logical step towards further efficiency gains and reduced environmental impact, organisational politics and structures can make this difficult in practice. Third party contractors can act as facilitators in this regard, not only providing the necessary 'impartial' shared resources, but also the mechanisms to ensure equitable division of the spoils. The prevalence of contractors in the UK food distribution market is therefore a key enabler for this initiative.

There has been a marked increase in interest in the environmental impact of supply chains in the last two years. In the first instance, public and media attention focused on an oversimplified understanding of the food miles issue, although there is now some recognition that importing foods by air may have some offsetting environmental benefits in terms of lower energy and chemical use nearer home for growing foods out of their normal environment, allied with the potential social benefits of encouraging third world economies. However, the UK grocery industry has accepted that it needs to be seen to be taking action nearer home. Government sponsored initiatives have joined up with trade bodies such as ECR Europe and the Institute of Grocery Distribution (IGD) to promote standard key performance indicators for measuring environmental impact, together with a range of initiatives broadly assembled under the heading 'fewer, friendlier miles'.

Whilst such initiatives encourage retailers and their contractors and suppliers to consider the actions which they can take to optimise distribution systems and physical resources, the notion of competitive collaboration is being actively encouraged on the basis that the potential prize, both in terms of the environment and cost savings, is worth the possible commercial or organisational risks. This has brought some retailers together, more often than not encouraged by 3PL contractors who have spotted a commercial opportunity for brokering deals. Thus, book and stationery chain WH Smith has been working with 3PL Wincanton to identify opportunities for teaming up with other retailers to service 'far flung places' more efficiently. Manufacturers Kellogs and Kimberley Clarke are sharing vehicles, DHL has been promoting a model for urban consolidation, based on trials and demonstration projects at Heathrow airport and shopping centres in Bristol and Sheffield.

One of the reasons cited for not integrating operations across retailers is the potential for access to commercially sensitive information about suppliers and their volumes. Historically, of course, primary transport contractors had access to all of this information as they integrated volumes in the open market. Having made the initial efficiency gains through managed intake and cost visibility, it remains for retailers to be persuaded that further gains are available through pooling with contractors, without compromise of sensitive information. A vision of further cost savings through pooled transport is based on the assumption that the contractors could once again, as they did prior to the primary initiatives, broker these operations and thus safeguard sensitive information.

Whiteoak (1999) describes a scenario where a focus on efficient use of vehicles will create demands for customer collections (by store vehicles), use of consolidators for small volumes and pooling with third parties in intermediate warehouses. These factors will, in turn, lead to ex-factory buying, 'cherry-picking' of routes and the potential for the creative use of shared fleets to overcome cost problems. Furthermore, different arrangements will be brought into play to deal with seasonal fluctuations and promotions. The complexity inherent in this ought to be more closely allied with the core competencies of specialist distribution contractors than with those of the retailers themselves. He proposes six focus areas as the basis for the necessary collaboration between retailers and contractors in cross-channel initiatives, all of which have, to a large extent, been addressed as part of the initial phase of retailer primary initiatives.

These areas, together with examples of their application in retailer primary initiatives, are:

- commercial principles (for example ex-factory arrangements, or nominated carriers)
- network strategy (for example, integration between primary and secondary movements)
- warehouse facilities (for example primary hubs and intermediate warehouses)
- full vehicle trunking (to support, for example, store back-haul to local RDC)
- store deliveries (either integrated with primary collections or carried out by supplier vehicles after delivery into DC)
- IT support (for example, systems already in place for pre-advice of deliveries and tracking of product).

The establishment of a common mind set and an effective framework for communications between competitors is currently being frustrated by a preference for cross-channel competition as opposed to cross-channel cooperation. As Whiteoak (1999) concludes, this frustration can be overcome by third party involvement in two areas:

- Logistics service providers should be directly involved in new ECR initiatives, on the basis of the prevailing practice of using contract logistics, to drive opportunities for synergy and consolidation
- The providers of value-added (communications) network services, such as electronic data interchange (EDI), have huge opportunities in offering the communications infrastructure and management software to facilitate shared-resource operations.

The pre-existence of contractors in the market place now affords an opportunity for a further step change in the way in which the supply chain is configured within UK retailing.

Fernie (1999) describes third party logistics providers as the 'missing piece in the ECR jigsaw'. He observes that whilst a great deal has been written about the development of relationships within the supply chain, particularly in the contexts of ECR and supply chain management, there has been little consideration of the physical processes of getting goods from manufacturers to stores. Fernie concludes that, as companies move to become 'virtual organisations', defined by a series of relationships (such as those described within the ECR framework), those companies will tend to concentrate on their core competencies and outsource those functions which lie beyond those competencies. This, he predicts, will have the further outcome of enabling further cooperation between competing firms in fully implementing the principles of ECR, as contractors can be used to facilitate transport pooling (as described by Whiteoak, 1999) in a 'hands off' manner.

Fernie noted that UK retailers have been 'at the forefront of fostering partnerships with professional distribution companies' to the extent that, of £1.9 billion spent on distribution services by retailers, some £1.3 billion was contracted out (Fernie, 1990). However, he also notes later that those retailers which are still carrying out their own distribution operations believe that they are providing a better service than contractors and that organisational history and inertia may play a role in defining which operations are contracted out and which are retained in-house (Fernie, 1995).

The idea of potential collaboration in logistics operations between competitors is not a particularly new one. Heskett (1973) identified the opportunity, explaining it as being driven by vast technological changes in physical distribution systems and capabilities since the Second World War. For example, traditional technologies have increased their capacities: ships and trucks are generally larger and can carry larger payloads than they could 20 years ago. At the same time, new technologies have either become available for the first time, such as containerisation, or have become accessible to a wider market as costs have fallen, for example with air freight.

Because of the maturity and fairly homogenous nature of retail distribution systems (Fernie, 1995), it can be argued that innovation in logistics might offer at least a short-term competitive benefit. However, the complex interrelationship between logistics and manufacturing which has developed over the last few decades requires collaboration and cooperation between supply chain members in order to facilitate innovation. Conversely, therefore, a willingness to innovate and change can be seen as an enabler, or explaining factor, for partnerships (Frankel *et al.*, 2002). The same authors note that, in spite of the widespread belief of the supply chain benefits driven by ECR implementation, the actual physical impact on distribution systems may not be that great, with, for example, inventory levels in the USA generally as high or even higher than before ECR was first promoted.

Almost 30 years after Heskett (1973) suggested two sets of circumstances under which logistics collaboration would be possible, Cuthbertson and Collet (2001) explore the relationships between potentially competing partners within a single supply chain in more detail. Specifically, the relationship between supplier and retailer is described as being both collaborative and competitive, in that both parties are also highly likely to be members of other competing supply chains. The need for information exchange and the alignment of key resources (money, people and technologies) is discussed, but an opportunity for much wider collaboration networks is identified. Retailers deal with many suppliers and suppliers deal with many retailers, and therefore there may be a role for sector-wide organisations to act as intermediaries and develop standards and platforms across industries. Whilst the examples discussed focus on information and technology standards, it does not require much of a leap of imagination to apply this thinking to physical distribution standards and processes.

More recently (Hoffman, 2005), it has been recognised that this potential for collaboration across supply chains faces an uphill struggle, because firms are likely to fear the loss of some competitive advantage grounded in their logistics systems. Having recognised the role of third parties in facilitating information exchange, not only by setting common standards, but also by developing the technologies which allow only limited access to data by the appropriately authorised parties, it is suggested that there may be a role for third parties in acting as agents to facilitate collaboration and integration elsewhere. Whilst recognising that vertical coordination within a single chain is likely to be much easier than horizontal collaboration, because of prejudices and fears among potential participants, it is suggested that the template for integration has been developed in the field of information exchange. This would, however, be more than just an extension of traditional third party service provision, in that it requires the participants to join more actively in setting common standards, in the full knowledge that they are actively joining with their competitors in these operations, rather than operating at arm's length. 'When senior management starts to view the supply chain as a strategic initiative, then collaboration will really start taking off.'

ECR is occasionally believed to have fallen far short of its promised efficiency and value. Many believe that unrealistic expectations among grocery industry participants are the root cause of this shortcoming. The level of internal and external change required to make desired outcomes a reality has been underestimated and poorly understood by prospective participants (Frankel *et al.*, 2002; Stank *et al.*, 1999).

Some of the original components of ECR (such as electronic data inter-change: EDI) or later enhancements and applications (collaborative planning and forecasting: CPFR) can be seen as prerequisite technologies which have to be in place, but which do not, of themselves, drive collaborative arrangements, which can only exist when these technologies are complemented by trust and interdependence (Stank *et al.*, 1999).

Three key points, therefore, emerge:

- There is evidence of collaboration between competitors in supply chain activities other than logistics and distribution. Of these, the key activity is information exchange, both to drive transactional efficiency (EDI, CPFR) and to design more agile and responsive chains (goal sharing, conflict resolution mechanisms, performance measurement).
- Although there has been resistance to cooperation in logistics because of perceptions of compromise of competitive advantage, there may be circumstances which override these prejudices, particularly where a third party can facilitate the initiative.
- More importantly, a number of prerequisites or enablers are identified in an attempt to explain, if not the circumstances in which collaborations will be established, at least the conditions under which they will perform best and flourish, or in which they will underperform. These include an imbalance of power between the parties leading to an imbalance of interdependence, relative size and maturity of aligned operations and processes, as well as the perceived distance of a supply chain activity from the end consumer. Because of the relative maturity and thus homo-geneity of physical distribution systems, innovation could be perceived as a source of competitive advantage, albeit possibly only in the short term, and thus willingness to innovate and change might be a significant enabler of the formation of collaborative partnerships.

13.7   Conclusion

Grocery distribution in the UK is highly efficient, both in terms of replenishment speed and cost. This has been achieved by the application of common distribution models in a highly concentrated marketplace, with the associated critical mass of volume. A standard distribution model, based on centralised RDCs, some of which are operated by third parties, has been universally adopted. The efficiencies thus gained have been built on in the first instance by vertical integration of the distribution chain, through tools such as inbound and outbound integration and factory-gate pricing. Further efficiency gains have then been achieved by integrating all internal processes, through the creation of virtual and regional fleets and other synergies between previously separate networks. To some extent, this is facilitated by technological developments, particularly in hardware and software for tracking, routing and scheduling, which allows for the constant re-optimisation of complex networks and resources in real time. Finally, traditional barriers between competing supply chains are being challenged, particularly in respect of the need to respond to the growing environmental agenda. This creates the climate and tools for the integration of previously competing distribution processes and ultimately to shared resources.

If anything, it seems likely that progress towards more general supply chain collaboration between competitors will accelerate in response to economic, social and political pressures. The threat of economic recession and increases in material costs, growing public awareness of and support for green initiatives and the threat of fiscal and other government interventions will all encourage traditional combatants to see past their points of difference and explore their areas of potential synergy.

13.8   Sources of further information and advice

For more information on current trends in retailing, retail distribution and collaborative initiatives in the UK, the web sites of both the IGD and ECR Europe are useful starting points: www.igd.com and www.ecrnet. org. Although the content will change, at the time of writing the IGD site contained a fact sheets section, with articles on opportunities for reducing empty running through collaboration and other sustainable supply chain initiatives. A number of commercial web sites claim to collate information on the latest developments and applications of GPS tracking technologies, including geo-fencing: www.uktelematicsonline.co.uk is typical. Vehicle and equipment manufacturer web sites (for example www. volvo.com/trucks/global/en-gb/home.htm, http://www.scania.co.uk/trucks/ environment/technology/ and http://www.isotrak.com/ among others) although not impartial will detail the latest technology and applications. Government and trade body sites increasingly feature content on the role of collaboration in developing sustainable distribution strategies, for example www.dft.gov.uk/pgr/freight/sustainable/sustainabledistributionas-trategy and http://www.ciltuk.org.uk/pages/envforum.

Finally, academic research in this area continues to develop: see www. cuimrc.cf.ac.uk/s_logistics (from Cardiff), http://www.greenlogistics.org/ PageView.aspx?id=97 (consortium of six UK universities) and http://www. som.cranfield.ac.uk/som/research/centres/lscm/about.asp (Cranfield).

13.9   References

AKEHURST, G (1983). 'Concentration in retail distribution: measurement and signifi-cance', *Service Industries Journal*, **3**(2), 161–79.

BOURLAKIS, M (1998). 'Transaction costs, internationalisation and logistics: the case of European food retailing', *International Journal of Logistics: Research and Applications*, **1**(3), 252–64.

CORSTEN, D and KUMAR, N (2003). 'Profits in the pie of the beholder', *Harvard Business Review*, **81**(5), 22–23.

CORSTEN, D and KUMAR, N (2005). 'Do suppliers benefit from collaborative relation-ships with large retailers? An empirical investigation of efficient consumer response adoption', *Journal of Marketing*, **69**(3), 80–94.

CUTHBERTSON, R and COLLET, F (2001). 'The collaboration network', *European Retail Digest*, **Issue 32**, 7.

DUFFY, R and FEARNE, A (2004). 'The impact of supply chain partnerships on supplier performance', *International Journal of Logistics Management*, **15**(1), 57–71.

FERNIE, J (1990). 'Third party or own account: trends in retail distribution', in *Retail Distribution Management*, Fernie, J. (ed.), Kogan Page, London.

FERNIE, J (1992). 'Distribution strategies of european retailers', *European Journal of Marketing*, **26**(8/9), 35–47.

FERNIE, J (1995). 'International comparisons of supply chain management in grocery retailing', *Service Industries Journal*, **15**(4), 134–47.

FERNIE, J (1997). 'Retail change and retail logistics in the United Kingdom: past trends and future prospects', *Service Industries Journal*, **17**(3), 383–96.

FERNIE, J (1999). 'Relationships in the supply chain' in *Logistics and Retail Management*, Fernie, J and Sparks, L (eds), Kogan Page, London, 23–46.

FINEGAN, N (2002). *Backhauling and Factory Gate Pricing*, IGD, Letchmore Heath.

FRANKEL, R, GOLDSBY, T J and WHIPPLE, J M (2002). 'Grocery industry collaboration in the wake of ECR', *International Journal of Logistics Management*, **13**(1), 57–71.

HESKETT, J L (1973). 'Sweeping changes in distribution', *Harvard Business Review*, **51**(2), 123–32.

HOFFMANN, W (2005). 'Hanging together', *Journal of Commerce*, **6**(40), L10–L12.

IGD (2006). *Retail Logistics*, Annual survey report published by the Institute of Grocery Distribution, Letchmore Heath, UK.

MCKINNON, A C (1986). 'The physical distribution strategies of multiple retailers', *International Journal of Retailing*, **1**(2), 49–63.

MCKINNON, A C (1989). *Physical Distribution Systems*, Routledge, London.

PETTIT, D E A (1967). *New Trends in Delivery Operations*, Wentworth, F (ed.), Physi-cal Distribution Management.

SETH, A and RANDALL, G (1999). *The Grocers: The rise and rise of the supermarket chains*, Kogan Page, London.

SMITH, D (1999). 'Logistics in Tesco: past, present and future', in *Logistics and Retail Management*, Fernie, J and Sparks, L (eds), Kogan Page, London, 154–83.

STANK, T, CRUM, M and ARANGO, M (1999). 'Benefits of interfirm coordination in food industry supply chains', *Journal of Business Logistics*, **20**(2), 21–42.

WHITEOAK, P (1999). 'Rethinking efficient replenishment in the grocery sector', in *Logistics and Retail Management*, Fernie, J and Sparks, L (eds), Kogan Page, London, 110–30.

Part IV

Maintaining quality and safety in food supply chains

14

Enhancing consumer confidence in food supply chains

M. Garcia Martinez, University of Kent, UK

Abstract: Understanding consumer confidence in the safety of food products and the regulation thereof is critical for effective implementation of risk management and communication strategies aimed at restoring consumer trust in food chains. The perceived risk and mistrust of food safety systems influences public concerns. Consumer responses to risk messages, contribute to the unacceptability of proposals for activities perceived as risky, stimulate social and political action to reduce or avoid risk, lead to questioning of the work and decisions of risk regulators and authorities, and promote the selective use of information sources. This chapter looks at public attitudes towards food, its safety and the risks and uncertainty associated with it. It considers the impact of food traceability and corporate social responsibility in increasing transparency throughout the food chain and enhancing consumer confidence in food and food producers. The chapter concludes with a discussion of the most effective and desirable mechanisms for achieving an appropriate level of food safety in the food supply chain.

Key words: food chains, food safety, perceived risk, supply chain management, trust.

14.1   Consumers' risk perception and trust in food chains

A number of food scares and controversies in recent years have left consumers confused about exactly what is and is not safe and uncertain about whether or not the government is acting in their best interests (Davies, 2001). This has resulted in increasing public distrust and reduced consumer confidence in the safety of food, the food industry and the government's ability adequately to regulate, manage and communicate food risks (Cantley, 2004; Caduff and Bernauer, 2006; Halkier and Holm, 2006).

   As a result, consumers want to know more about the motivations behind policy decisions regarding food safety (Davies, 2001). They are no longer content to accept a passive role when it comes to food and its safety and are increasingly looking for information from much further up the food supply chain and further back along the policy decision-making process in order to make more informed food choices. Thus, understanding consumer perceptions of food safety and its regulation is critical to the restoration of consumer confidence (Latouche *et al.*, 1998).

Indeed, the literature on perceived risk highlights that consumer confidence and trust in regulatory institutions and players in food chain is critical for effective implementation of policies regardless of their efficiency merits (De Jonge *et al.*, 2004; Walls *et al.*, 2004; De Jonge *et al.*, 2007). Research shows differences between players regarding the level of trust conferred to them by consumers (Frewer and Miles, 2003; Lang and Hallman, 2005), with consumer trust in some players having a greater impact on general consumer confidence in the safety of food than consumer trust in other players. Overall, consumer confidence in the safety of food is most strongly enhanced by trust in food manufacturers compared to trust in the government, farmers or retailers (De Jonge *et al.*, 2008) as they are perceived to have more responsibilities for the safety of food than farmers and retailers, though less responsibility than the government (De Jonge *et al.*, 2004).

The occurrence of a food safety incident and its reporting in the media significantly influences consumer confidence in regulatory institutions and players in the food chain (De Jonge *et al.*, 2004) even when there is no medical or scientific evidence (Verbeke and Van Kenhove, 2002). However, consumer confidence in the safety of different food product categories varies. Consumers were highly confident about the safety of agricultural products such as diary and vegetables, but expressed less confidence in processed foods like ready-to-eat meals and those products perceived to have chemical character, namely vitamin supplements and energy drinks. This indicates that food safety incidents that are related to products already perceived as risky may have a higher impact on consumer confidence compared to those perceived as safe.

Consumers' behaviour and associated attitudes towards a particular hazard are driven more by psychologically determined risk perceptions than the technical risk estimates provided by experts (see for example Slovic, 2000). Under uncertainty, people's reactions to risk frequently depart from the behaviour predicted by the expected utility theory, with serious implications for choice and more importantly demand for risk mitigating interventions (see Arrow, 1982; Starr and Whipple, 1984; Smith and Desvousges, 1987; Viscusi *et al.*, 1987; Lichtenberg and Zilberman, 1988; Viscusi and Hamilton, 1999). Particularly, incidents that are associated with increased consumer concerns can have significant negative consequences for the food industry and more critically for the regulatory authorities, limiting their ability to develop effective consumer protection policies (De Jonge *et al.*, 2007). In the recent past, food safety incidents have resulted in trade bans, price fluctuations, culling of animals, decreased consumption of affected products and reputational damage of both the particular industry perceived to be responsible for such incidents and the wider food industry in general (Buzby, 2001; Verbeke, 2001; De Jonge *et al.*, 2007). Such economic losses from perceived risk and the resulting irrational response to risk may not also be limited to the immediate time period following an incident, but potentially have long-term effects and reach beyond local and domestic markets (Shepherd and Saghaian, 2008).

In the past, expert groups have expressed frustration with public attitudes towards food, its safety and the risks and uncertainty associated with it (Macfarlane, 2002). The general public display behavioural patterns and make choices that seem irrational or illogical or at least inconsistent with expert opinion and scientific knowledge (Hansen *et al.*, 2003). Consumers place more importance on factors that may not contribute to technical risk estimates, while underestimating other factors, which potentially represent a substantial threat to human health (Miles and Frewer, 2001). For example, microbiological hazards in food are judged by scientists to be one of the main risks to health, but the public is far more concerned about potential hazards from pesticide residues and food additives (Miles *et al.*, 2004). Results suggest that people are more worried about risks caused by external factors ('technological hazards') over which they feel they have no control, while being much less concerned about personal factors or factors linked to their own behaviour or lifestyle ('lifestyle hazards') over which they feel they have more knowledge or more personal control (Frewer *et al.*, 1994), although there are individual differences in the extent to which this holds true (Miles *et al.*, 2004).

These differences have been further fuelled by biased media coverage of food safety issues, which tends to focus on relatively infrequent but highly sensational incidents (Baker, 1998). There is ample empirical evidence to suggest that the public is not ignorant and/or irrational (Slovic, 1992; Frewer *et al.*, 2002; Macfarlane, 2002) and the differences in perception are the result of consumers including other (non-scientific) factors in their judgements of 'acceptable' risks (Fischhoff, 1995). Consumers tend to personalise risk and take a multidimensional view that incorporates certain personal values (Macfarlane, 2002). Thus, scientific evaluation is not necessarily the principal determinant of consumer choice. If consumers are not convinced of the utility of a product, they are unlikely to accept the risks associated with its consumption, however small the scientific risk might be, particularly when they perceive that the benefits of such products accrue mostly to industry. This argument has been demonstrated particularly in the case of gene technology where attitudes towards technological issues are strongly related to other more general sociopolitical attitudes, including attitudes towards the environment and nature (Frewer *et al.*, 1997), attitudes towards science and modern technology and social (dis) trust (Siegrist, 1998).

14.2   Impact of food traceability in restoring consumer trust in food chains

Producers and regulatory institutions in Europe have attempted to restore consumer confidence in policy making and industry practices by introducing food and ingredient traceability systems (i.e. General Food Law, Regulation (EC) 178/ 2002). Traceability systems and subsequent quality and origin labelling schemes are expected to increase transparency throughout the food chain and to result in the development and maintenance of consumer trust in food and food producers (Van Rijswijk *et al.*, 2008). Quality signalling can transform credence attributes into search attributes and strengthen consumer trust, allowing a reduction in consumer perceived risk towards food quality and safety and information asymmetry between consumers and producers (Mojduszka and Caswell, 2000). The credence characteristic of a product (i.e. food safety, animal welfare, organic production) cannot be observed or inferred by direct inspection, on consumption or even after consumption, whereas search characteristics (i.e. appearance or size) are known by consumers prior to purchase (Darby and Karni, 1973).

However, consumer interest in traceability information cannot be taken for granted (Hobbs *et al.*, 2005; Verbeke and Ward, 2006). Research shows that people have little understanding of what traceability is (Giraud and Amblard, 2003) and that they are not particularly interested in cues directly related to traceability and product identification, despite uncertainty following a number of meat safety crises (Giraud and Halawany, 2006; Verbeke and Ward, 2006). Direct indications of traceability (e.g. bar codes and license numbers) and the provision of technical information associated with it are unlikely to increase consumer confidence given the low quality or safety inference potential of such information cues (Verbeke *et al.*, 2007). Conversely, quality labels and safety guarantees are much more valued by consumers. Hence, the presentation of simpler information on traceability systems accompanied by labels (quality and safety related) would increase the probability of being valued by consumers (Verbeke *et al.*, 2007; Van Rijswijk *et al.*, 2008).

Research shows that consumers across Europe have divergent associations, perceptions and expectations regarding traceability (Giraud and Halawany, 2006; Van Rijswijk *et al.*, 2008). Cultural values influence consumer food decision making (Briley *et al.*, 2000; Hoogland *et al.*, 2005), so expectations and attitudes towards food safety issues and risk management may vary depending on consumers' cultural backgrounds (Van Kleef *et al.*, 2006; Houghton *et al.*, 2008). The Van Rijswijk *et al.* (2008) study on consumer perceptions of traceability in Europe shows similarities between countries regarding the benefits consumers associate with traceability (i.e. health, quality, safety and control). Across-country differences were also identified. German consumers are more sensitive about the processing conditions, while French consumers care more about origin and give a greater importance to past product experience such as taste. Spanish consumers perceive a high quality product as a sign of trust and Italian consumers are more sensitive to safety conditions. In other words, the consumers' background greatly affects their attitude towards traceability-related attributes.

Consumers tend to associate traceability with higher product prices (Giraud and Halawany, 2006). Some consumers will be willing to pay higher beef prices to reassure safety while others will trade off price against safety improvements. In addition to socioeconomic characteristics, consumers' risk perception associated with beef is one of the main driving forces for price premiums (Loureiro and Umberger, 2004).

Hobbs *et al.* (2005) in Canada and Dickinson and Bailey (2002) in the USA found that consumers' willingness to pay (WTP) is higher for traceable meat compared to non-traceable meat. The WTP rises further for traceability provided characteristics (e.g. additional meat safety and humane animal treatment guarantees). Conversely, despite the importance of food safety to Spanish consumers, the majority (73%) are not willing to pay a price premium for traceable beef (Angulo and Gil, 2007). Traceability alone plays a very small role in Spanish consumer choices compared to products labelled with a protected designation of origin (PDO) label linked to a particular region with a reputation for food safety or food quality.

14.3   Consumers and corporate social responsibility in the food chain

Corporate social responsibility (CSR) and sustainable consumption have become a widespread topic in business and public discussion. Firms are increasingly engaging in socially responsible behaviour, not only to fulfil external obligations such as regulatory compliance and stakeholder demands, but also owing to enlightened self-interest considerations such as increased competitiveness and improved stock market performance (Bansal and Roth, 2000; Waddock and Smith, 2000). Governments, non-governmental organisations (NGOs) and the media have put corporations in the spotlight to account for the social consequences of their activities, often publicly shaming and stigmatizing businesses for their undesirable behaviour (Winston, 2002). As a result, CSR has emerged as an important area of action for companies globally.

In the food sector, which is dependent on natural, human and physical resources, CSR is gaining importance owing to the complex, labour intensive nature of food supply chains (Forsman-Hugg *et al.*, 2008). Increasingly, companies in both the food industry and retailing are aware of consumers' and stakeholders' interest in CSR issues and are taking initiatives and making efforts to consider their values and actions from the CSR perspective. Global consumers are increasingly willing to pay premiums for safe, organic sustainable products that address their health concerns, as well as their interests in preserving the environment and fighting poverty. Consumption has become a means by which people's non-material views about the nature of society and the future of the environment can be manifested in a tangible and measurable way (The Cooperative Bank, 2005).

While there is no universal agreement on what CSR is, or how it should be measured, the general idea is that corporate behaviour should reflect the 'triple bottom line' of economic, social and environmental performance (Elkington, 1997). CSR implies a wider perspective than the view that companies act in compliance with legal norms and produce safe food that meets basic quality criteria.

However, it is difficult for consumers to use CSR dimensions as selection criteria in the food purchase situation since they lack adequate and easily available information on CSR-related issues. Hence, the increasing demand by social activists for food companies and retailers to implement standards and certification programmes as tools to promote sustainable development through their supply chains by influencing suppliers to adopt more environmentally and socially responsible practices (Hatanaka *et al.*, 2005). Standards consist of a series of criteria, or rules, with which third-party suppliers are asked to comply (although the number, content and stringency of these criteria can vary substantially between schemes) (Genier *et al.*, 2008). In many cases, they represent an attempt to bridge the gap between legal and social norms in producer and consumer countries and, particularly on social issues, may be developed in response to perceived weaknesses in laws and law enforcement.

To encourage code adoption, companies may offer a 'carrot', such as a price premium, to suppliers in return for compliance. More commonly they wield a market entry 'stick', whereby non-compliant producers are excluded from the supply chain. In a highly competitive retail environment, private standards are very likely to increase in severity as firms attempt to 'outcompete' each other on social/credence attributes associated with their food offer (Garcia Martinez and Poole, 2009). One outcome of this is that it becomes increasingly challenging for producers, and particularly low-resourced small-scale producers from emerging/developing countries, to be able to meet the increasingly exacting standards (García Martinez and Poole, 2004).

In recent years, standards and codes have proliferated; references to over 100 schemes have been found (Genier *et al.*, 2008). In addition to those developed by food retailers, a number of NGOs have developed their own independent schemes as a way to promote alternative production and consumption systems that are more socially and environmentally sustainable (e.g. IFOAM, the Rainforest Alliance, the Marine Stewardship Council and the Fairtrade Labelling Organisations International). The market for these products remains small; however, sales are steadily increasing as food retailers and manufacturers recognise that there is a lucrative market niche willing to pay a premium for ethical attributes. The UK ethical food market was valued at £4.8 billion in 2006 (+17% over 2005) (The Cooperative Bank, 2007). This represents just 5.1% of the total grocery market but is becoming increasingly important, growing at 7.5% per annum (or 50% above the rate for the conventional grocery market).

However, research shows that consumers may overstate their propensity for purchasing ethically. Reconciling claimed behaviour with actual behaviour is a pertinent subject, especially when it comes to moral issues, and ethical consumption is one of these. There is still an imbalance between positive attitudes and purchasing behaviour, to a large extent due to consumers being confused about the end benefit (i.e. which is more ethical, 'organic' or 'Fairtrade'?). There is little leadership taking the message about organic forward, providing clarity in order that more consumers can make informed decisions, rather than taking blind decisions out of a sense of guilt or duty (Fearne, 2008). Hence the challenge for all involved is to induce a positive predisposition prior to the point of purchase. Appropriate information about ethical attributes is part of the augmented product that ethical consumers are seeking. Suppliers must assume their responsibility to deliver quality products with the added information attributes (Garcia Martinez and Poole, 2009).

Organic and fairly traded produce has considerable potential for improving the welfare of communities (Browne *et al.*, 2000). Nevertheless it is an important empirical question whether and to what extent the price premiums paid by consumers are transmitted to primary producers and their communities. While improved prices are an important opportunity for smallholders, the costs of meeting accreditation standards are also considerable, such that the net benefits must be analysed. It may be that the primary benefit for smallholders is access to valuable export markets rather than better prices and that these benefits, according to a growing body of evidence, accrue mainly to better-off producers (Garcia Martinez and Poole, 2009).

14.4   Improving communication with consumers

In response to increasing public distrust and reduced consumer confidence both in policy making and industry practices, government oversight of food safety has undergone profound regulatory reforms in recent years with the establishment of dedicated and 'independent' food safety agencies (Vos, 2000; Flynn *et al.*, 2004; Ansell and Vogel, 2006). These reforms towards a more transparent and inclusive process of food safety governance aim to remove the inherent historical conflict of interest in dual responsibilities of a single government department, traditionally agricultural ministries, both for regulating food safety and the interest of the agri-food sector (Ansell and Vogel, 2006; Borraz *et al.*, 2006; Steiner, 2006), which largely tended to favour industry over consumer interests in food safety decision making (GAO, 2005).

The outbreak of the BSE crisis in 1996 clearly revealed significant dysfunctions, both in industry practices and their supervision by governments, undermining general public trust in food safety governance (Vos, 2000; Borraz, 2007). The institutional failures put pressure on the European Commission to expand its scope of influence in regulating food safety across the European Union (as the breakdown in food safety controls were in the first instance a result of failure of national government control agencies) (Caduff and Bernauer, 2006) and forced the Commission to present a more coherent approach to food safety based on true principles of separation of the responsibility for legislation and scientific advice, responsibility for legislation and inspection, and greater transparency and information throughout the decision-making process and inspection (Vos, 2000).

Current risk management effort tries to restore public confidence by involvement of relevant stakeholders at an early stage in the regulatory decision making process (Rowe and Frewer, 2005). However, the effects of proactive participatory processes on public trust are presently unclear (Rowe and Frewer, 2000). If stakeholders are to engage positively in the process of setting standards, such processes require mutual trust and understanding on the part of government, industry and other stakeholders in order that 'quality' information is collected and assimilated into the regulatory process and confidence is built in the value of consultation (Garcia Martinez et al., forthcoming).

Research has stressed the importance of effective communication on food safety risks to facilitate informed decision making by consumers and to change consumers' health-related behaviour (Frewer, 2004; Fisher et al., 2005; Verbeke, 2005). Information about food risk controlling measures by responsible authorities is likely to increase perceptions of control among consumers, which in turn may decrease risk perceptions (Fischhoff et al., 1978; Redmond and Griffith, 2004). Consumers particularly appreciate communication that focuses on 'what is being done'. Easily accessible information about hygiene inspections of food premises is greatly appreciated by consumers (Worsfold, 2006). The posting of inspection results outside restaurants (the so-called 'scores on doors') has been shown to have a significant impact on both consumer patronage and business performance (Jin and Leslie, 2003; Greenstreet Berman, 2008). This suggests that information by enforcement authorities on food control measures and enforcement procedures offers reassurance to consumers that food safety is being monitored and business practices scrutinised. Consumers show a preference for preventive risk management as opposed to the adoption of a reactive approach and regard this approach as more indicative of good management (Van Kleef et al., 2006).

Despite experts' belief that providing information about scientific uncertainty in risk assessment to the general public will undermine trust in scientific institutions and will cause unnecessary panic and confusion (Frewer *et al.*, 2003), consumers still prefer this information to be made available in a user-friendly format (Frewer *et al.*, 2002). Communicating uncertainty in risk assessment is increasingly seen as highly relevant to ensure consumer confidence in regulatory institutions (Millstone and Van Zwanenberg, 2002; Shepherd *et al.*, 2006).

Research highlights the importance of cultural context regarding the impact of potential risk communication strategies (Van Dijk *et al.*, 2008). Hence, specific cultural characteristics may determine the need to adapt risk communication strategies to cross-country variations. This finding is particularly relevant within the EU context where the establishment of the European Food Safety Authority (EFSA) should lead to a standardised pan-European approach to risk communication. The Van Dijk *et al.* (2008) cross-country study shows that while communication of uncertainty had a positive impact in Germany, the same information had a negative impact in the UK and Norway. UK consumers have been found to be more sceptical about the efficiency of risk assessment practices compared to those in Germany and Greece (Van Kleef *et al.*, forthcoming). Cultural differences were also found in the perceived quality of food risk management associated with different hazards (Van Dijk *et al.*, 2008). The evaluation of risk management of mycotoxins was rated the highest in all countries, while the risk management of pesticide residues and GM potatoes differed between countries.

14.5 Managing food safety in food supply chains

Food safety is a credence characteristic and hence the credibility of the food product needs to be established by some form of food safety policy, if the market fails to provide sufficient information about this attribute (Cho and Hooker, 2002). There is an ongoing debate about the most effective and desirable mechanisms for achieving an appropriate level of food safety in the food supply chain. While there are some mandated food safety practices for firms in the food supply chain, the issue of economic incentives for firms actively to address food safety throughout the supply chain is unclear. Food safety controls often require significant investments in capital and labour, but do not have tangible returns. It is difficult to estimate the value of preventing a safety incident. However, a risk that is realised can potentially bankrupt the firm.

Private response to the implementation of food safety controls reflects not only regulatory incentives but also a wide range of market-based incentives related to customer audit requirements (Sperber, 1998), customer product specifications (Mehta and Wilcock, 1996; Henson and Northen, 1998), regulatory requirements for exports (Hobbs *et al.*, 2002), risk of product recall (Skees *et al.*, 2001) and liability laws (Buzby and Frenzen,

1999; Buzby *et al.*, 2001). In turn, these market incentives reflect the characteristics of the firm, its objectives and strategies, the supply chains and markets in which it operates, its products and the broader economic and commercial environment.

As a result, a supply chain manager's 'best practice' model today is to strive to achieve not only a fully integrated and efficient supply chain, capable of creating and sustaining competitive advantage (Christopher and Towill, 2002), but also one with sufficient flexibility and redundancy to enable the firm to respond to extreme events (Sheffi, 2005).

Evidence from many countries suggests that so-called 'channel captains' are a major driving force for privately motivated adoption of enhanced food safety controls (Henson and Northen, 1998; Garcia Martinez and Poole, 2004; Golan *et al.*, 2004). Major buyers, such as multiple food retailers and caterers/food service operators, frequently require their suppliers to implement food safety controls based on hazard analysis and critical control points (HACCP) principles but in some instances these controls may be inappropriate for smaller, independent operators (Buchweitz *et al.*, 2003), for whom a statutory minimum level of control is adequate.

Farm assurance schemes can be an effective mechanism for managing the transaction costs of buyers, including the costs associated with an information search, negotiation and monitoring. For large supermarkets with a multitude of suppliers scattered globally, such schemes have become a key means through which procurement costs are managed. The transaction cost 'savings' from farm assurance schemes depend, however, on the trust that buyers have in their associated systems of oversight, as reflected in levels of compliance by individual producers. While all such schemes in the UK are now governed by third party independent systems of accreditation that conform to the applicable international standards, there are still perceived to be differences in rigour (Food Standards Agency, 2002). Thus, retailer-driven assurance schemes are often perceived to provide higher levels of food safety than the generic codes of practice developed by producer organisations or government agencies, simply because the commercial motivations for compliance are generally much greater.

In the United States, it is the large technically proficient buyers of major food service chains that have led the implementation of stringent food safety requirements, particularly for meat (Golan *et al.*, 2004), the sector estimated to cause more than 40% of human illnesses in the USA that are associated with common pathogens (Roberts, 2005). These buyers monitor food safety along their supply chains, successfully creating markets for enhanced food safety through their ability to enforce safety standards using testing and process audits, rewarding suppliers who meet these standards through a price premium or guaranteed sales whilst punishing those that do not by excluding them from lucrative markets. Through contracts with these large buyers, meat processors are able to appropriate the benefits of their investments in food safety, securing market access and preferential status (Ollinger and Moore, 2004). Until recently, more generic systems of assurance, such as those seen in the UK, were underdeveloped in the USA. However, in 2003, the Food Marketing Institute acquired the Safe Quality Food (SQF) series of standards from the Ministry of Agriculture of the State of Western Australia (Henson, 2006), which may signal a shift towards the UK 'model' of private food safety governance.

Arm's length, transactional supply chain relationships place high information and monitoring costs on buyers in identifying and monitoring suppliers in order to minimise the risk of food safety failures (Hobbs, 1996). To minimise the associated transaction costs, there is a tendency for firms to engage in closer partnership arrangements with their key suppliers. As a governance structure, sharing of information serves to correct the potential distortions generated by asymmetric information, lowering transaction costs in relationships affected by moral hazard and adverse selection problems. This cooperation also creates the potential for capture of additional rents from consumers if participants in a closely coordinated supply chain can offer greater security and more credible guarantees than their competitors (Loader and Hobbs, 1999).

Private regulation plays a bigger role where supply chain governance structures are present and functional. For example, the establishment of food safety control mechanisms, such as traceability in the meat sector in the UK, can be viewed as a reciprocal multi-stage agency relationship where farmers and meat processors act as both agents and principals in vertically coordinated contractual agreements. Governance structures, in combination with strong channel leaders, may act as the catalyst for a cultural change towards transparent and traceable food supply chains. Failure to comply with the food safety demands of these systems and buyers results in a price discount as alternative channels are sought with lower food safety expectations, but also lower prices (Fearne and Walters, 2004). This tends to result in a 'two tier' system with enhanced food safety controls in supply chains to exacting buyers, and supply chains operating only at minimum food safety standards selling to 'residual' markets (Fearne et al., 2006). However, public regulation (as seen with current EU regulations) may be implemented where there is an insufficient food safety 'culture' upstream and/or where the industry is more fragmented.

Another example of the dynamic effect of supply chain governance structures is the progressive development of farm assurance schemes in the UK. Here, the shift to stricter food safety standards has been uneven across sectors, with the establishment of stringent food safety systems in concentrated and/or integrated markets such as eggs, poultry and pork (Food Standards Agency, 2002). Again, retailer pressure has been a strong motivator for adoption of these systems (Food Standards Agency, 2002). Conversely, in the beef and lamb sectors, strong retailer influence has been diffused by the large number of producers and the length and complexity of the food chain (Food Standards Agency, 2002). In these sectors there is a tension between the desire to recruit the majority of producers into prevailing farm assurance schemes and the desire to see standards improving throughout the chain.

One of the major potential motivating factors for firms to adopt private standards is the potential cost of product recalls. On the one hand, enhanced food safety controls may reduce the risk of product failure and in turn the risk that a recall will occur. On the other hand, private standards may be an effective defence in the case of regulatory action or litigation. Ollinger and Ballenger (2003) indicate that the number of Class I recalls in the USA has increased significantly, from 24 per year over the period 1993–96 to 42 per year for the period 1997–2000. Despite quite significant costs of compliance, rates of market exit are greater for meat processors with lower food safety controls. This suggests that the management of recall-related costs is probably becoming a greater issue and that the adoption of private standards is likely to become more widespread.

Ollinger and Mueller (2003) suggest that contractual arrangements between buyers and sellers is a potent mechanism for the enhancement of food safety controls, particularly in the context of food safety metasystems that imply close and coordinated interrelationships between all levels of the supply chain. This can be further enhanced by the use of private standards and effective mechanisms of conformity assessment, most commonly based on third party certification, and branding as a means of communicating product safety/quality at each level of the supply chain. However, although branding provides a commercial advantage to firms in market competition, it also results in greater exposure to the risks of product failure. Thus, for example, Bredahl and Holleran (1997) suggest that many of the transaction costs imposed on food retailers in the procurement of own-branded (private label) products are associated with the control of product safety.

14.6   References

ANGULO, A.M. and GIL, J.M. (2007). 'Risk perception and consumer willingness to pay for certified beef in Spain', *Food Quality and Preference*, **18** (8), 1106–17.

ANSELL, C. and VOGEL, D. (2006). 'The contested governance of European food safety regulation', in *What's the Beef: The Contested Governance of European Food Safety Regulation*, C. Ansell and D. Vogel (eds), MIT Press, Cambridge, Mass, 3–32.

ARROW, K.J. (1982). 'Risk perception in pyschology and economics', *Economic Inquiry*, **20**, 1–9.

BAKER, G.A. (1998). 'Strategic implications of consumer food safety preferences', *International Food and Agribusiness Management Review*, **1** (4), 451–63.

BANSAL, P. and ROTH, K. (2000). 'Why companies go green: A model of ecological responsiveness', *Academy of Management Journal*, **43** (4), 717–36.

BORRAZ, O. (2007). 'Governing standards: the rise of standardization processes in France and in the EU', *Governance: An International Journal of Policy, Administration, and Institutions*, **20** (1), 57–84.

BORRAZ, O., BESANCON, J. and CLERGEAU, C. (2006). 'Is it just about trust? The partial reform of French food safety regulation', in *What's the Beef: The Contested Governance of the European Food Safety Regulation*, C. Ansell and D. Vogel (eds), The MIT Press, Cambridge, Massachussetts, 125–52.

BREDAHL, M.E. and HOLLERAN, E. (1997). 'Food safety, transaction costs and institutional innovation', in *Quality Management and Process Improvement for Competitive Advantage in Agriculture and Food*, G. Schiefer and R. Helbig (eds), University of Bonn, Germany.

BRILEY, D.A., MORRIS, M.W. and SIMONSON, I. (2000). 'Reasons as carriers of culture: Dynamic versus dispositional models of cultural influence on decision making', *Journal of Consumer Research*, **27** (2), 157–78.

BROWNE, A.W., HARRISA, P.J.C., HOFNY-COLLINSB, A.H., PASIECZNIKB, N. and WALLACEA, R.R. (2000). 'Organic production and ethical trade: definition, practice and links', *Food Policy*, **25** (1), 69–89.

BUCHWEITZ, M.R.D., SALAY, E, CASWELL, J.A. and BACIC M.J. (2003). 'Implementation and costs of good manufacturing practices norms and hazard analysis and critical control points systems in foodservices in the campinas region, SP, Brazil', *Foodservice Research International*, **14** (2), 97–144.

BUZBY, J.C. (2001). 'Effects of food-safety perceptions on food demand and global trade', in *Changing Structure of Global Food Consumption and Trade*, Anita Regmi (ed.), US Department of Agriculture, Economics Research Service, Agriculture and Trade Report, Washington, DC, WRS-01-1, 55–66.

BUZBY, C. and FRENZEN, P.D. (1999). 'Food safety and tort liability', *Food Policy*, **24** (6), 637–51.

BUZBY, C., FRENZEN, P.D. and RASCO, B. (2001). *Product Liability and Microbial Foodborne Illness*, U.S. Department of Agriculture, Economic Research Service, Washington, DC.

CADUFF, L. and BERNAUER, T. (2006). 'Managing risk and regulation in European food safety governance', *Review of Policy Research*, **23** (1), 153–68.

CANTLEY, M. (2004). 'How should public policy respond to the challenges of modern biotechnology?' *Current Opinion in Biotechnology*, **15**, 258–63.

CHO, B.-H. and HOOKER, N.H. (2002). *A Note on Three Qualities: Search, Experience and Credence Attributes*: Department of Agricultural, Environmental, and Development Economics, The Ohio State University.

CHRISTOPHER, M.G. and TOWILL, D.R. (2002). 'Developing market specific supply chain strategies', *International Journal of Logistics Management*, **13** (1), 1–14.

DARBY, M. and KARNI, E. (1973). 'Free competition and the optimal amount of fraud', *Journal of Law and Economics*, **16**, 67–88.

DAVIES, S. (2001). 'Food choice in Europe – the consumer perspective', in *Food, People & Society: A European Perspective of Consumers Food Choice*, L.J. Frewer, E. Risvik and H. Schifferstein (eds), Springer, Berlin, 365–80.

DE JONGE, J., FREWER, L.J., VAN TRIJP, H.C.M., RENES, R.J., DE WIT, W. and TIMMER, J. (2004). 'Monitoring consumer confidence in food safety: an exploratory study', *British Food Journal*, **106**, 837–49.

DE JONGE, J., VAN TRIJP, H.C.M., RENES, R.J. and FREWER, L.J. (2007). 'Understanding consumer confidence in the safety of food: its two-dimensional structure and determinants', *Risk Analysis*, **27** (3), 729–40.

DE JONGE, J., VAN TRIJP, H.C.M., VAN DER LANS, I.A, RENES, R.J. and FREWER, L.J. (2008). 'How trust in institutions and organizations builds general consumer confidence in the safety of food: A decomposition of effects', *Appetite*, **51**, 311–17.

DICKINSON, D.L. and BAILEY, D. (2002). 'Meat traceability: are US consumers willing to pay for it?' *Journal of Agricultural and Resource Economics*, **27** (2), 348–64.

ELKINGTON, J. (1997). *Cannibals with Forks. The Triple Bottom Line of 21st Century Business*, Capstone Publishing, Oxford.

FEARNE, A. (2008). 'Organic fruit and vegetables – who buys what and why . . . and do we have a clue?', Dunnhumby Academy of Consumer Research, Kent Business School, University of Kent.

FEARNE, A. and WALTERS, R. (2004). *The Costs and Benefits of Farm Assurance to Livestock Producers in England*, Imperial College, London.

FEARNE, A., GARCIA MARTINEZ, M., CASWELL, J.A., HENSON, S. and KHATRI, Y. (2006). *Exploring Alternative Approaches to Traditional Modes of Regulation of Food Safety*. Document prepared for the FSA under the contract D03004. Imperial College, London.

FISCHHOFF, B. (1995). 'Risk perception and communication unplugged – 20 years of process', *Risk Analysis*, **15** (2), 137–45.

FISCHHOFF, B., SLOVIC, P., LICHTENSTEIN, S., READ S. and COMBS, B. (1978). 'How safe is safe enough? A psychometric study of attitudes towards technological risks and benefits', *Policy Sciences*, **9**, 127–52.

FISHER, A.R.H., DE JONGE, A.E.I., DE JONGE, R., FREWER, L.J. and NAUTA, M.J. (2005). 'Improving food safety in the domestic environment: the need for a transdisciplinary approach', *Risk Analysis*, **25**, 503–17.

FLYNN, A., CARSON, L., LEE, R., MARSDEN, T. and THANKAPPAN, S.S. (2004). *The Food Standards Agency: Making a Difference?*, The Centre for Business Relationships, Accountability, Sustainability and Society (BRASS), Cardiff University.

FOOD STANDARDS AGENCY (2002). *Review of Food Assurance Schemes*, Food Standards Agency, UK.

FORSMAN-HUGG, S., KATAJAJUURI, J.-M., PESONEN, I., MÄKELÄ, J. and TIMONEN, P. (2008). 'Building the content of CSR in the food chain with a stakeholder dialogue', paper presented at the *12th Congress of the European Association of Agricultural Economics*, Ghent, Belgium.

FREWER, L.J. (2004). 'The public and effective risk communication', *Abstracts of the 44th Congress of the European Societies of Toxicology*, **149**, 391–7.

FREWER, L.J. and MILES, S. (2003). 'Temporal stability of the psychological determinants of trust: Implications for communication about food risks', *Health, Risk and Society*, **5** (3), 259–71.

FREWER, L.J., SHEPHERD, R. and SPARKS, P. (1994). 'The interrelationship between perceived knowledge, control and risk associated with a range of food related hazards targeted at the self, other people and society', *Journal of Food Safety*, **14**, 19–40.

FREWER, L.J., HEDDERLEY, D., HOWARD, C. and SHEPHERD, R. (1997). 'Objection' mapping in determining group and individual concerns regarding genetic engineering', *Agriculture and Human Values*, **14**, 67–79.

FREWER, L.J., MILES, S., BRENNAN, M., KUZNESOF, S., NESS, M. and RITSON, C. (2002). 'Public preferences for informed choice under conditions of risk uncertainty', *Public Understanding of Science*, **11** (4), 363–72.

FREWER, L.J., HUNT, S., BRENNAN, M., KUZNESOF, S., NESS, M. and RITSON, C. (2003). 'The views of scientific experts on how the public conceptualise uncertainty', *Journal of Risk Research*, **6**, 75–85.

GAO. (2005). *Food Safety. Experiences of Seven Countries in Consolidating their Food Safety Systems*, United States Government Accountability Office, Washington, DC.

GARCIA MARTINEZ, M. and POOLE, N. (2004). 'The development of private fresh produce safety standards: Implications for developing Mediterranean exporting countries', *Food Policy*, **29** (3), 229–55.

GARCIA MARTINEZ, M. and POOLE, N. (2009). 'Ethical consumerism: development of a global trend and its impact on development', in *Standard Bearers: Horticultural*

Exports and Private Standards in Africa, A. Borot de Battisti, J. MacGregor and A. Graffham (eds), IIED and NRI, London, 18–21.

GARCIA MARTINEZ, M., FEARNE, A., HENSON, S. and CASWELL, J.A. 'Co-regulation of food safety: towards a diagnostic framework for more optimal allocation of regulatory resources', *Food Policy*.

GENIER, C., STAMP, M. and PFITZER, M. (2008). *Corporate Social Responsibility in the Agrifood Sector: Harnessing Innovation for Sustainable Development*, FSG Social Impact Advisors, Geneva, Switzerland.

GIRAUD, G. and AMBLARD, C. (2003). 'What does traceability mean for beef meat consumer?' *Food Science*, **23**, 40–64.

GIRAUD, G. and HALAWANY, R. (2006). 'Consumers' perception of food traceability in Europe', Paper presented at the *98th EAAE seminar: Marketing Dynamics within the Global Trading System: New Perspectives*, Chania, Crete, Greece.

GOLAN, E., ROBERTS, T., SALAY, E., CASWELL, J., OLLINGER, M. and MOORE, D. (2004). *Food Safety Innovation in the United States. Evidence from the Meat Industry*, United States Department of Agriculture, Economic Research Service http://www.ers.usda.gov/publications/aer831/aer831.pdf.

GREENSTREET BERMAN, (2008). *Evaluation of scores on the doors: Main Report*, Greenstreet Berman Ltd http://www.food.gov.uk/multimedia/pdfs/sotdmain report.pdf.

HANSEN, J., HOLM, L., FREWER, L.J., ROBINSON, A. and SANDOE, P. (2003). 'Beyond the knowledge deficit: recent research into lay and expert attitudes to food risks', *Appetite*, **41**, 111–21.

HALKIER, B. and HOLM, L. (2006). 'Shifting responsibilities for food safety in Europe: An introduction', *Appetite*, **47** (2), 127–33.

HATANAKA, M., BAIN, C. and BUSCH, L. (2005). 'Third-party certification in the global agrifood system', *Food Policy*, **30** (3), 354–69.

HENSON, S. (2006). 'The role of public and private standards in regulating international food markets', paper presented at the *Summer Symposium of the International Agricultural Trade Research Consortium (IATRC) on Food Regulation and Trade: Institutional Framework, Concepts of Analysis and Empirical Evidence*, Bonn.

HENSON, S. and NORTHEN, J.R. (1998). 'Economic determinants of food safety controls in the supply of retailer own-branded products in the UK', *Agribusiness*, **14**, 113–26.

HOBBS, J.E. (1996). 'A transaction cost analysis of quality, traceability and animal welfare issues in UK beef retailing', *British Food Journal*, **98** (6), 16–26.

HOBBS, J.E., FEARNE, A. and SPRIGGS, J. (2002). 'Incentive structures for food safety and quality assurance: an international comparison', *Food Control*, **13** (2), 77–81.

HOBBS, J.E., BAILEY, D.V., DICKINSON, D.L. and HAGHIRI, M. (2005). 'Traceability in the Canadian red meat sector: do consumers care?' *Canadian Journal of Agricultural Economics*, **53** (1), 47–65.

HOOGLAND, C.T., DE BOER, J., and BOERSEMA, J.J. (2005). 'Transparency of the meat chain in light of food culture and history', *Appetite*, **45**, 15–23.

HOUGHTON, J.R., ROWE, G., FREWER, L.J., VAN KLEEF, E., CHRYSSOCHOIDIS, G., KEHAGIA, O., KORZEN-BOHR, S., LASSEN, J., PFENNING, U. and STRADA, A. (2008). 'The quality of food risk management in Europe: Perspectives and priorities', *Food Policy*, **33** (1), 13–26.

JIN, G.Z. and LESLIE, P. (2003). 'The effect of information on product quality: evidence from restaurant hygiene grade cards', *The Quarterly Journal of Economics*, **118** (2), 409–51.

LANG, J.T. and HALLMAN, W.K. (2005). 'Who does the public trust? The case of genetically modified food in the United States', *Risk Analysis*, **25** (5), 1241–52.

LATOUCHE, K., RAINELLI, P. and VERMERSCH, D. (1998). 'Food safety issues and the BSE scare: some lessons from the French case', *Food Policy*, **23** (5), 347–56.

LICHTENBERG, E. and ZILBERMAN, D. (1988). 'Efficient regulation of environmental health risks', *The Quarterly Journal of Economics*, **103** (1), 167–78.

LOADER, R.J. and HOBBS, J.E. (1999). 'Strategic responses to food safety legislation', *Food Policy*, **24** (6), 685–706.

LOUREIRO, M.L. and UMBERGER, W.J. (2004). 'A choice experiment model for beef attributes: What consumer preferences tell us', Selected paper presented at the *American Agricultural Economics Association Annual Meetings*, Denver, Colorado.

MACFARLANE, R. (2002). 'Integrating the consumer interest in food safety: The role of science and other factors', *Food Policy*, **27** (1), 65–80.

MEHTA, S. and WILCOCK, A. (1996). 'Quality system standards in the Canadian food and beverage industry', *Quality Management Journal*, **4** (1), 72–96.

MILES, S. and FREWER, L.J. (2001). 'Investigating specific concerns about different food hazards', *Food Quality and Preference*, **12** (1), 47–61.

MILES, S., BRENNAN, M., KUZNESOF, S., NESS, M., RITSON, C., and FREWER, L.J. (2004). 'Public worry about specific food safety issues', *British Food Journal*, **106** (1), 9–22.

MILLSTONE, E. and VAN ZWANENBERG, P. (2002). 'The evolution of food safety policy making institutions in the UK, EU and Codex Alimentarious', *Social Policy and Administration*, **36**, 593–609.

MOJDUSZKA, E.M. and CASWELL, J.A. (2000). 'A test of nutritional quality signaling in food markets prior to implementation of mandatory labeling', *American Journal of Agricultural Economics*, **82** (2), 298–309.

OLLINGER, M. and BALLENGER, N. (2003). 'Weighing incentives for food safety in meat and poultry', *Amber Waves*, 34–41.

OLLINGER, M. and MOORE, D. (2004). 'Food safety investments in the meat and poultry industry survey results', in *Food Safety Innovation in the United States. Evidence from the Meat Industry*, E. Golan, T. Roberts, E. Salay, J. Caswell, M. Ollinger and D. Moore (eds), AER-831, United States Department of Agriculture, Economic Research Service, Washington, DC, 13–20.

OLLINGER, M. and MUELLER, V. (2003). *Managing for Safer Food: The Economics of Sanitation and Process Controls in Meat and Poultry Plants*, US Department of Agriculture, Economic Research Service, Washington, DC.

REDMOND, E.C. and GRIFFITH, C.J. (2004). 'Consumer perception of food safety risk, control and responsibility', *Appetite*, **43**, 309.

ROBERTS, T. (2005). 'Economics of private strategies to control foodborne pathogens', *Choices*, **20** (2), 117–22.

ROWE, G. and FREWER, L.J. (2000). 'Public participation methods: a framework for evaluation', *Science Technology and Human Values*, **25** (1), 3–29.

ROWE, G. and FREWER, L.J. (2005). 'A typology of public engagement mechanism', *Science, Technology and Human Values*, **30** (2), 251–90.

SHEFFI, Y. (2005). *The Resilient Enterprise. Overcoming Vulnerability for Competitive Advantage*, The MIT Press, Cambridge, MA.

SHEPHERD, J.D. and SAGHAIAN, S. (2008). 'Consumer response to food safety events: an interaction between risk perception and trust of information in the chicken and beef markets', selected paper presented at the *Southern Agricultural Economics Association Annual Meeting*, Dallas, TX, February 2–6.

SHEPHERD, R., BARKER, G., FRENCH, S., HART, A., MAULE, J. and CASSIDY, A. (2006). 'Managing food chain risks: integrating technical and stakeholder perspectives on uncertainty', *Journal of Agricultural Economics*, **57**, 313–27.

SIEGRIST, M. (1998). 'Belief in gene technology: the influence of environmental attitudes and gender', *Personality and Individual Differences*, **24** (6), 861–6.

SKEES, J.R., BOTTS, A. and ZEULI, K.A. (2001). 'The potential for recall insurance to improve food safety', *International Food and Agribusiness Management Review*, **4** (1), 99–111.

SLOVIC, P. (1992). 'Perceptions of risk: reflections on the psychometric paradigm', in *Theories of Risk*, D. Golding and S. Krimsky (eds), Praegor, London.

SLOVIC, P. (2000). *The Perception of Risk*, Earthscan Publications Ltd, London.

SMITH, V.K. and DESVOUSGES, W.H. (1987). 'An empirical analysis of the economic value of risk changes', *The Journal of Political Economy*, **95** (1), 89–114.

SPERBER, W.H. (1998). 'Auditing and verification of food safety and HACCP', *Food Control*, **9** (2), 157–62.

STARR, C. and WHIPPLE, C. (1984). 'A perspective on health and safety risk analysis', *Management Science*, **30** (4), 452–63.

STEINER, B. (2006). 'Governance reform of German food safety regulation: cosmic or real?', in *What's the Beef: the Contested Governance of European Food Safety*, C. Ansell and D. Vogel (eds), The MIT Press, Cambridge, MA, 181–210.

THE COOPERATIVE BANK. (2005). *The Ethical Consumerism Report*, The Cooperative Bank, UK.

THE COOPERATIVE BANK. (2007). *The Ethical Consumerism Report 2007*, The Cooperative Bank, UK.

VAN DIJK, H., HOUGHTON, J.R., VAN KLEEF, E., VAN DER LANS, I., ROWE, G. and FREWER, L.J. (2008). 'Consumer response to communication about food risk management', *Appetite*, **50**, 340–52.

VAN KLEEF, E., FREWER, L.J., CHRYSSOCHOIDIS, G.M., HOUGHTON, J.R., KORZEN-BOHR, S., KRYSTALLIS, T., LASSEN, J., PFENNING, U. and ROWE, G. (2006). 'Perceptions of food risk management among key stakeholders: results from a cross-European study', *Appetite*, **47**, 46–63.

VAN KLEEF, E., UELAND, Ø., THEODORIDIS, G., ROWE, G., PFENNING, U., HOUGHTON, J., VAN DIJK, H., CHRYSSOCHOIDIS, G. and FREWER, L.J. 'Food risk management quality: consumer evaluations of past and emerging food safety incidents', *Health, Risk and Society*.

VAN RIJSWIJK, W., FREWER, L.J., MENOZZI, D. and FAIOLI, G. (2008). 'Consumer perceptions of traceability: a cross-national comparison of the associated benefits', *Food Quality and Preference*, **19** (5), 452–64.

VERBEKE, W. (2001). 'Beliefs, attitude and behaviour towards fresh meat revisited after the Belgian dioxin crisis', *Food Quality and Preference*, **12**, 489–98.

VERBEKE, W. (2005). 'Agriculture and the food industry in the information age', *European Review of Agricultural Economics*, **32** (3), 347–68.

VERBEKE, W. and VAN KENHOVE, P. (2002). 'Impact of emotional stability and attitude on consumption decisions under risk', *Journal of Health Communication*, **7** (4), 455–72.

VERBEKE, W. and WARD, R.W. (2006). 'Consumer interest in information cues denoting quality, traceability and origin: An application of ordered probit models to beef labels', *Food Quality and Preference*, **17** (6), 453–67.

VERBEKE, W., FREWER, L.J., SCHOLDERER, J. and DE BRABANDER, H.F. (2007). 'Why consumers behave as they do with respect to food safety and risk information', *Analytica Chimica Acta*, **586** (1–2), 2–7.

VISCUSI, W.K. and HAMILTON, J.T (1999). 'Are risk regulators rational? Evidence from hazardous waste cleanup decisions', *The American Economic Review*, **89** (4), 1010–27.

VISCUSI, W.K., MAGAT, W.A. and HUBER, J. (1987). 'An investigation of the rationality of consumer valuations of multiple health risks', *The RAND Journal of Economics*, **18** (4), 465–79.

vos, E. (2000). 'EU food safety regulation in the aftermath of the BSE crisis', *Journal of Consumer Policy*, **23**, 227–55.

waddock, s. and smith, n. (2000). 'Corporate social responsibility audits: Doing well by doing good', *Sloan Management Review*, **41** (2), 75–83.

walls, j., pidgeon, n., weyman, a. and horlick-jones, t. (2004). 'Critical trust: understanding lay perceptions of health and safety risk regulation', *Health, Risk and Society*, **6** (2), 133–50.

winston, m. (2002). 'NGO strategies for promoting corporate social responsibility', *Ethics and International Affairs*, **16** (1), 71–87.

worsfold, d. (2006). 'Consumer information on hygiene inspections of food premises', *Journal of Food Service*, **17**, 23–31.

15

Quality and safety standards in food supply chains

R. Baines, Royal Agricultural College, UK

Abstract: Private standards have been developed in many countries and some standards have progressed beyond national status to become global standards, the most important of which are evaluated here. The standards compared here include pre-farm gate standards (SQF1000, GlobalGAP and ISO22000) and food industry standards (SQF2000, BRC Global, IFS and ISO22000). Like most national standards, all claim to address food safety but only some also consider product quality and extrinsic quality factors like environmental care or worker conditions of employment. What each standard actually delivers is discussed here and summarised in two decision trees, one for food safety and quality and the other for environment, worker care and food defence.

Key words: private standards, benchmarking, food safety and quality, environment, worker care, food defence.

15.1 Introduction

Modern food supply chains have become increasingly globalised and complex (Manning and Baines, 2004) and have also become the domain of the private sector with governments mainly providing a regulatory and hygienic oversight. This chapter will take the reader through the evolution of private standards around the world in the context of the changing governance of food supply and the rise in global sourcing of raw materials to finished products. With the decline in direct government inspections of food crossing national borders, allied to the increasing pressure of the food industry to be responsible for food safety, and liable in the event of any breakdowns, many national private standards have been established but only a few operate internationally. The chapter will provide the reader with a systematic analysis of the main international standards and includes a decision tree to assist in selecting the right standard for those involved in, or planning to engage in, international food trade.

The term private standards is used here to describe food and farm assurances that have been developed by interested parties (e.g. producers, food industry sectors, manufacturers, retailers, non-governmental organisations, and similar) with the specific purpose of providing sets of rules for production, processing, manufacturing and distribution that can be independently checked so that the next stages in supply chains can be 'assured' that the products coming forward meet the conditions laid down in that standard. As such, the term 'private standard' is synonymous with terms like food and farm assurance schemes, systems or codes.

15.2   Private standards and their evolution

Before embarking on an evaluation of current standards, it is important to explore what is meant by private standards. Many small and medium sized processing and distribution organisations have a long history of quality assurance (QA) systems. Generally these systems were either promoted by trade organisations, industry bodies, or were privately owned and branded like the systems employed by multinational food companies. Such schemes were either regulated by the operator (1st party), audited by the purchaser (2nd party), or they were independently certified using agreed protocols (3rd party audit). Recognised management systems such as ISO 9001 were common (Baines and Davies, 1998, 1999; Varzakas and Jukes, 1997; Caswell and Henson, 1997) as these early private standards had evolved out of the business philosophy of total quality management. These approaches, however, were found to be deficient in managing food safety challenges as supply chains concentrated into fewer, bigger suppliers in the 1990s.

Early (1995) described quality assurance in the food industry as a strategic management function that embraces policies, standards and systems for the maintenance of quality. Further, Sparling et al. (2001) considered that the growth of QA standards would move suppliers away from commodity (and wholesale) markets into more speciality (or value added) markets, especially in the agriculture sector. However, Manning et al. (2006) argued that QA needs to move from qualitative measures of quality to more quantitative measures in order to demonstrate to external stakeholders and consumers that QA can actually deliver tangible benefits.

As a result of several benchmarking studies, Baines and Ryan (2002) concluded that a well constructed private standard had the potential to be a business management tool for:

- demonstrating regulatory compliance to governments
- driving internal business efficiencies through operators, internal auditors and managers knowing what they are doing and why and
- communicating to suppliers, customers and consumers the integrity that the business has placed in the products sold.

If the capacity exists independently to inspect and certify to the standard under ISO Guide 62 or 65, then the assurances given can be global and can thus facilitate international trade, as long as the standard is known by buyers and consumers in the countries of consumption.

15.2.1   Private standards, regulation and food safety

The rise in prominence of private standards is one element of a number of trends that define food production and supply over recent decades. Perhaps the most important of these trends is the continuing concerns over the safety of food allied to a recent history of major food safety breakdowns including BSE in cattle and the reported links to new variant CJD, mineral oil contamination of animal feeds, the discovery of a banned colouring, Sudan red, in many processed foods and the many examples of microbiological contamination of foods ranging from raw meats to fresh mixed salads.

Governments have responded to food safety breakdowns mainly through the introduction of consumer protection legislation that is grounded on risk assessments and the adoption of hazard analysis critical control point (HACCP) approaches, at least for higher risk food groups in the food processing and manufacturing industries. Australia and New Zealand went further and proposed to include HACCP-based risk assessment down to the farm level in 1995 which resulted in the Australia, New Zealand Food Standards Code (FSANZ online). Similarly, the European Union under their General Food Law in 2005 (Europa online) brought primary production under hygienic regulations; however, neither region has effectively embedded this aspect of these laws in practice to date. Over the same period, governments have generally reduced their resources for food inspections in favour of a lighter 'regulatory oversight'. Allied to this, greater responsibility is placed on industry to manage food hygiene and ensure that the food they sell is safe.

What does this mean to food businesses today? Should a food business cause harm or injury to consumers, then governments will apply sanctions, including prosecution, under consumer protection law. Therefore the onus is on businesses to set up systems to reduce the risks of food being unsafe and it is this that provided the stimulus to embed food safety management into private standards alongside quality management. Indeed, having a management system in place for food safety can be used as a defence by demonstrating due diligence, while in strict liability regions such systems can be used to lower the cost of insurance premiums taken out to cover the costs of litigation and liability.

Although worldwide regulators are still increasing their efforts to safeguard the foods we eat, there has been a continued weakening of consumer confidence in the capacity of regulators to assure safe food adequately. In response, the food industry has sought to regain the confidence of consumers, especially along discrete supply chains where chain captains (such as global retailers and the main brand name food companies), who are the main link between the chain and consumers, are increasingly insisting on third party certification of the private standards they have developed to assure the integrity of the food in terms of both safety management (the legal defence) and quality specifications (for market positioning, competition and price). The term 'chain captain' is used here in the same context as Hughes (1994) and others use it to denote the power and influence of such organisations to specify what the whole chain should do, as they have the economic power and influence to do so. As a result, third party or independent certification is seen by many in the industry to provide a more cost-effective method of surveillance than regulatory inspection alone. This is based on the following arguments:

* Internal audit systems within food businesses can be developed to provide the level of surveillance necessary for the level of food risk associated with different raw materials and food preparations.
* Independent or third party audits provide an independent check that the internal system is operating and is capable of detecting, removing or recalling defective products.
* The system is capable of providing traceability down the chain and provenance up the chain.

Regulatory controls are important though as they provide the necessary public check of private certification systems and are a necessary link between food chains and international agreements linked to trade.

15.2.2   Globalisation and private standards

Several other trends have contributed to the globalisation of food supply and thus the evolution of private standards, these include:

* Increasing trade in food and raw materials across national borders and between different parts of the world. Access to some trading blocks requires compliance with international grades and standards and import regulations. Where private standards prevail in the market place, then although country access may be governed by regulatory rules, access to markets may additionally require compliance with appropriate private standards.
* A consequence of this trade is that more food is purchased unseen and therefore requires appropriate description and documentation to ensure what is delivered is what was ordered. In other words, a system of identity preservation and traceability is required for both regulatory inspection and under various private standards. As more materials are sourced from around the world, businesses require the standard of inspection and certification to be consistent. This has led to most private standards

being inspected by accredited certification bodies under ISO Guide 62 or 65.

- Irrespective of the regulatory system, chain captains are increasingly dominating food trade. As a result, they are driving the trend for more vertical integration and discrete supply chains. In addition, they will often exceed regulatory food safety requirements through 'their' private standards and will define specific criteria related to product quality and safety while some will also define the systems of production (such as worker and animal welfare, environmental protection and conservation) all as conditions of supply.
- These chain captains also recognise that governments do not have the capacity (personnel and financial) to inspect food as frequently as would be desirable or necessary to address potential liability, nor do regulators inspect against specific quality and credence requirements. Therefore chain captains are increasingly demanding private inspection and independent certification against their own QA protocols and product specifications.
- Consumers have also changed over this period in that they (we) want local and exotic food, convenience and pre-prepared foods but they do not seem to want seasonality which further drives global sourcing of such products. Irrespective of the origin or preparation, they expect the food to be safe and of the right quality for the price they are prepared to pay. This can be communicated through private standards. Furthermore, some consumers are interested in how food is produced and are prepared to search out and even pay a premium for such products like organic or fair trade products (Baines and Davies, 2007). Again, private standards have been developed to meet these requirements either as the main focus of the standard (e.g. organic certification, Fair Trade, or Rainforest Alliance) or as an additional element of safety and quality focussed standards (e.g. safe quality food (SQF) optional modules for environmental and worker care (see www.sqfi.com) and the recently introduced Global Gap Risk Assessment on Social Practices).

Over the same time horizon, food and agriculture have become key areas in relation to trade liberalisation and have been linked with government to government trade and disputes that have required World Trade Organisation (WTO) arbitration. Should such arbitration include food safety concerns, then the case may be referred to the Codex Alimentarius Commission (a UN agency funded by the Food and Agriculture Organisation and the World Health Organisation). Furthermore, although Codex Standards are not compulsory, they are increasingly being embraced in national rules and guidelines, especially the adoption of HACCP (Codex, 1969).

It is against such food trade and regulatory developments that industry driven private standards evolved in the 1990s in some countries (e.g. Australia, New Zealand and Europe) but not in others (e.g. the USA, Canada and south America). As the pace of globalisation has increased, with multinational food companies and multiple retailers sourcing from further afield, some standards have been exported to these new regions, for example the expansion of GlobalGAP into Asia and Africa and the expansion of SQF1000 & 2000, British Retail Consortium (BRC) and International Food Standard (IFS) into Asia and the Americas. It should be noted here that private standards are not subject to WTO trade rules as they are classed as voluntary standards even though they may dominate supply chains and effectively be used as conditions for market access.

15.2.3   Private standard developments from retailer to producer

So why have QA systems evolved as they have? As stated earlier, the principal force behind the development of private standards has been, and remains, the legal requirement to demonstrate that food is safe. The stage in the supply chain that is exposed to the greatest liability is retail and food service, as these are the interface with consumers and hence consumer protection legislation. In response to this legislation, EU food retailers in particular developed their own quality and safety systems to demonstrate 'due diligence' in their distribution, handling and the sale of food. This meant that the greatest risk remaining for retailers was their suppliers, so they then demanded the same of them as a condition of continued supply. A logical development of these end-of-chain schemes was for individual or groups of retailers to examine supplier activity and set common hygienic requirements. Current examples of this approach are the British Retail Consortium Global Standard (BRC) and the International Food Standard (IFS) developed by German and French retailers.

This process has been repeated down the supply chain with food processors and manufacturers requiring their suppliers to implement QA systems that meet their requirements, or more accurately meet the conditions of their chain captains. In response to these demands, primary producers in a number of industry sectors and in particular regions developed their own farm assurance schemes, for example the range of farm assurance standards under the British Farm Standard, the Australian Care programmes for cattle, sheep flocks and fresh produce and Horticulture New Zealand.

Our reviews of various supply chains have shown that the adoption of HACCP to address food safety risks generally does not include the primary production level where codes of good agricultural practice (GAP) often prevail, for example, along beef chains (Baines, 2000; Buncic, 2000), fresh produce chains (Baines and Ryan, 2002; Baines, 2002) and with grain supply (Baines and Davies, 2003) even though food hazards can occur at this stage. The main arguments for relying on GAPs is that it is considered to be too difficult to implement HACCP on farms and that HACCP can be strategically placed further along vertically integrated chains. Indeed, adopting HACCP at strategic stages has been particularly important for the safer supply of livestock products (e.g. meat at the slaughter stage and milk from 'first buyers' onwards) and in the supply of fresh produce (fruit and vegetables at the pack house stage onwards).

When GAPs are considered within the HACCP system, they should be more accurately described as HACCP prerequisite programmes. Some farm level standards have progressed further in relation to addressing food safety risks, for example a number of poultry private standards have developed sector wide HACCP plans to guide producers (Manning et al., 2008). Furthermore, Baines and Ryan (2002) and Baines et al. (2004) have argued that correct HACCP adoption at the farm level is feasible and affordable and this is backed up by the fact that primary producers under the SQF1000 code (level 2 or 3) are doing this already, as are pig producers under the Australian Pig Industry Quality programme (APIQ).

15.3   Comparison of international standards

It would be difficult to describe adequately in this chapter the multitude of national and regional standards that have been established over the last 20 or so years. However, there has been a significant amount of harmonisation between standards within regions and as a result those that have become internationally recognised are generally indicative of the approach of national standards in that region. Furthermore, owing to the importance of global sourcing allied to the control of international supply chains, primarily through enforcing private standards as a condition of market access, it is important to understand what these standards actually deliver to customers, consumers and regulators.

15.3.1   Benchmarking international food and farm standards

The plethora of national private standards and the preference of some global retailers only to accept 'their' standards has led to a call either to rationalise standards or to provide some form of equivalence analysis through benchmarking. In response to this, the CIES (International Committee of Food Retail Chains) (who represent global retailers and whose members command some 65% of global food trade) developed a benchmarking tool, the global food safety initiative (GFSI) which is based on evaluating whether standards are underpinned by HACCP, have management systems to operate them and are supported by good guidance. Their argument was that any standard benchmarked would be accepted by all CIES member retailers (Byrnes, 2003).

To date the GFSI has recognised four food industry schemes (British Retail Consortium Global Standard, Dutch HACCP, International Food Standard and SQF2000) and two farm level schemes (SQF1000 and Horticulture New Zealand). Global GAP was submitted for benchmarking when it was known as EurepGAP but was found to be non-compliant. Since this initial submission, the CIES has decided to discontinue farm level bench-marking. The recently introduced ISO22000 (food and farm standards) are in the process of being submitted to GFSI for benchmarking.

As has already been stated, private standards address more than just food safety. For this reason a more detailed benchmarking approach has been developed. The main areas considered under this approach include:

- ownership of the standard and degree of stakeholder engagement in its development
- presence of a management system (as with GFSI)
- food safety management (as with GFSI)
- management of food quality and product credence
- supporting guidance (as with GFSI)
- audit approaches and decisions over compliance and non-compliance and
- standards body control over service providers (e.g. suppliers, consultants, trainers and training organisations, auditors and certification bodies).

The criteria used to select standards that are truly global were twofold; the standards are being used in more than one country and they have been, or will be, considered by the benchmarking activities of the CIES. Given these criteria, the private standards considered here are:

- at the farm level: SQF1000, GlobalGAP and ISO22000
- from the food industry: SQF2000, BRC Global, IFS and ISO22000.

The method of analysis used was to evaluate each standard along with any guidance documentation and the audit checklists. Once initial analysis was complete, the findings were reviewed by a certification body familiar with use of the standard, this was the verification stage of the analysis. Finally, the conclusions drawn from this analysis were compared to other bench-marking or evaluation studies of the standards in question. The main conclusions from these studies are given below.

15.3.2   Standard ownership and engagement with stakeholders

Details of the organisations that own the standards under consideration are given in Table 15.1 and from this it can be seen that there are different types of ownership and governance. The ISO standard 22000 is owned and governed through the International Standards Organisation and its various regional technical committees. As a result, stakeholder engagement is through high level representation resulting in limited access to development of the standard by those who use it on the ground. In contrast the remaining standards are effectively owned by food retail associations (the chain captains). The approach to governance varies, however, with the Food Marketing Institute (FMI) through the SQF Institute ensuring that a wide range of stakeholders, including those not directly involved with the

Table 15.1 Ownership and governance of global food and farm standards and the degree of engagement with supply chain stakeholders

Global standard	Standard owners	Stakeholder engagement in standard development
SQF 2000 and 1000	SQFi – a Division of the Food Marketing Institute, USA (Origin of standard – West Australian government and industry)	International Technical Committee from all stages of the chain including non-members of the standard Stakeholder forums at annual conferences and case study suppliers.
Global GAP	Foodplus, German Eurohandel Institute made up mainly of European retailers	Sector technical Committees drawn only from members of standard Managed stakeholder sessions at conferences with invited suppliers presenting
ISO 22000	International Standards Organisation	National experts from 23 countries plus liaison with international organisations (Codex, WFSO, CIAA and CIES) formed Working Group 8 of ISO TC 34 Little contact or interaction with supply chain businesses
BRC Global	British Retail Consortium representing most of the UK's major retailers	Standards governance drawn from members' technical directors and BRC Management Technical advisory – as above plus trade association and certification body representatives
International Food Standard	Initial standard – the German retail federation (HDE) and French counterpart (FCD) IFS Food, version 5, is now a collaboration of three retail federations from Germany, France and Italy	Predominantly German, French and Italian retailers Governance drawn from retail associations through technical committees and groups?

standard, have the opportunity to influence the standard's management and development. In contrast, the remaining retail owned standards allow little supplier involvement (BRC and IFS) or the suppliers invited to participate have to be members of the standard (GlobalGAP).

15.3.3   Food safety management approaches

Two contrasting approaches to addressing food risks and managing food safety were identified, namely the adoption of HACCP or the reliance on codes of practice. This study of the global standards has identified that all of the post-farm gate standards are grounded in HACCP adoption (Fig. 15.1), as is SQF1000 (level 2); in contrast, GlobalGAP relies on codes of good agricultural practice as prescribed in the guidance and audit checklist. This is considered less rigorous, as good agricultural practices are listed as HACCP prerequisite programmes. It is claimed that the pre-farm gate ISO22000 standard also requires HACCP adoption; however, documentation available to date does not make this clear although it is likely that the approach will be to develop sector HACCP plans for producers to follow. It should be noted, however, that the SQF codes and ISO22000 do potentially provide a whole chain-linked solution to food safety risk assessment and management (Fig. 15.3).

15.3.4   Food quality and credence

Chain captains have identified the lack of food quality as a risk associated with their suppliers, at least post-farm gate. As such the BRC, IFS and SQF2000 standards all include quality as key criteria (Fig. 15.2). The SQF1000 code (level 3) also requires quality to be addressed pre-farm gate, thus allowing quality management to be harmonised along the whole supply chain at the same time as risk-based food safety management is harmonised (Fig. 15.3). The ISO22000 standard does not include product quality criteria which may reflect the government-led development of this standard through ISO where quality has not been an issue for regulation.

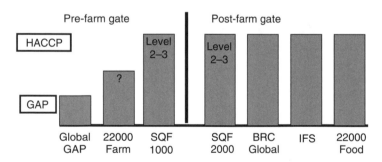

Fig. 15.1   Food safety approaches of the major global food and farm private standards where HACCP signifies full adoption of HACCP system at the business level and GAP refers to good agricultural practices. (Note: Level 2–3 refers to higher levels of compliance within the standard and ? refers to uncertainty about the approach of standard at this stage in the chain.)

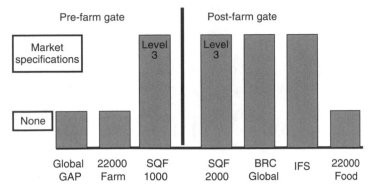

Fig. 15.2   Product quality specifications of the major global food and farm private standards where market specifications refer to customer requirements. (Note: Level 3 refers to the highest level of compliance under this standard.)

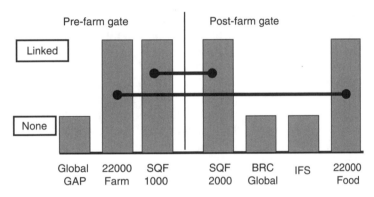

Fig. 15.3   Supply chain linkage in terms of standard's requirements pre- and post-farm gate for the major global food and farm private standards.

Furthermore, ISO argue that quality can be adequately managed by also adopting ISO9001 Total Quality Management, but this would add unreasonable costs to many small and medium sized enterprises (SME) food businesses. Finally, GlobalGAP does not include any audit criteria for product quality.

If we now consider extrinsic quality issues or credence, we see differences in approach between the standards (Fig. 15.4).

In terms of environmental care, both SQF and GlobalGAP address this; however, the approach is different. With GlobalGAP, environment and conservation are encouraged but not required to a point where major or minor non-compliances are raised though audit. Moreover, this investment in environment can be lost further along the chain if post-farm standards

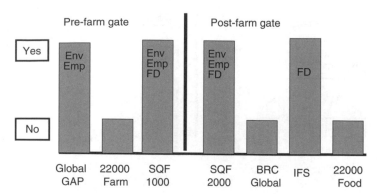

Fig. 15.4 Environmental (Env), worker care (Emp) and food defence (FD) requirements of the major global food and farm private standards. (Note: requirements may be recommendations or be optional requirements of the main standards – see text.)

do not include environmental considerations. In contrast, both SQF1000 and SQF2000 can include environmental care as an optional module if the market requires it (or if the supplier wants to do it), thus a whole chain solution to environmental care is realisable. None of the other standards address environment.

A similar approach to worker care is seen, with GlobalGAP encouraging worker care, while the SQF codes have an optional module in responsible social practice mainly aimed at worker care and employment relations. Another area of development is the recognition that food can be deliberately contaminated, either by dissatisfied employees or consumers, or as an act of terrorism. In response to this, both the SQF codes and IFS have introduced optional food defence modules to address the risks associated with deliberate contamination.

15.3.5 Training in standards

Understanding the standard and using it correctly are key elements in underpinning the quality and robustness of any standard, as any abuse of the standard by a stakeholder can undermine it for all. Therefore, it is important that those who use the standard, from supplier to auditor and consultant, are trained accordingly.

All of the standards reviewed required auditors to be trained in their standard and in addition to have been trained and competent in HACCP (Fig. 15.5). Thereafter, the retail owned standards SQF, BRC and IFS also require those who train in the standard to be trained themselves (Fig. 15.6). Finally, only the SQF codes require consultants and at least one person in the food business to be trained unless they retain the services of a trained and registered consultant.

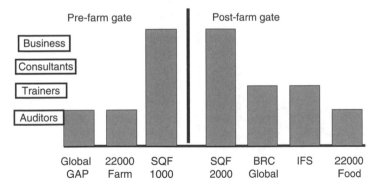

Fig. 15.5 Training requirements for suppliers and service providers of the major global food and farm private standards.

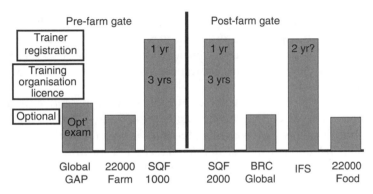

Fig. 15.6 Standard owner's control over training organisations and trainers for the major global food and farm private standards. (Note: yrs refer to registration period and ? denotes uncertainty of time.)

From this we can deduce that only the SQF codes require all users and service providers to be trained as part of the integrity of their standard. This means that every food and farm business certified under this standard has someone working with them who has been trained in its use. The SQF institute goes further by requiring training bodies and trainers to register with the institute so that any users of training services can check that providers are legitimate.

15.3.6 Standards body control over users and service providers
The next stage in exerting control over users of a standard is to know who is using it and how they are operating. The general approach for most standards is to manage the activities of auditors and certification bodies.

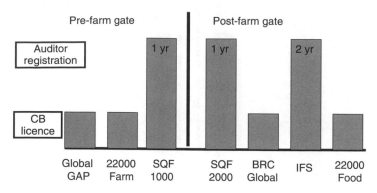

Fig. 15.7 Standard owner's control over certification bodies (CB) and auditors for the major global food and farm private standards. (Note: yrs refer to registration period.)

Standards owners can do this in two ways: by issuing licences first to certification bodies to operate in order to provide an audit and certification service (Fig. 15.7) and second, through the independent accreditation service provided by national government agencies under ISO Guide 62 or 65.

This means that the control of auditors, their training and activities are the responsibility of the employing certification bodies. Both IFS and SQF go further by requiring individual auditors to be registered either biennially or annually, respectively, after checks are carried out, including HACCP training, food sector category experience and competence in using the standard.

15.4   Conclusions

Before considering the current situation and likely future developments, it is worthwhile reflecting on the past. Prior to the 1990s and the on-going sequence of significant food safety incidents, many food processors relied on ISO 9000 to deliver total quality management. This approach was found to be deficient in managing food safety. In response, most private standards operating beyond the farm gate have embedded HACCP principles to address food hazard identification, control and management. In contrast, most pre-farm gate standards are based on sector legal requirements and good agricultural practice guidance.

Currently, food and farming businesses operating along modern supply chains are increasingly required to adopt one or more private food standards, in order to secure access to particular markets and customers. Most of these standards are grounded in national regulations and are additionally defined through codes of good practice; however a few also seek to be internationally recognised. This review has identified the differences and similarities between those private standards that are promoted as 'global' standards. However, it is important to note that selection of the right standard is not just an objective choice, the politics and power behind these standards needs to be considered.

15.4.1 Politics of standard selection

Although this chapter provides an objective review of standards, it is important to recognise the significant political influences associated with certain standards. Indeed, the best-known standards may not be the best! In considering the apparent popularity of standards, it is important also to consider who owns standards, as this will generally condition their use. For example, many suppliers to British, French and German retailers currently have to maintain both BRC and IFS standards to satisfy market access in spite of the CIES agreement of these same retailers to recognise all standards benchmarked under the Global Food Safety Initiative. It could be argued that this is another example of the use of economic or other powers by these chain captains in European and global markets.

European retailers are also seeking to popularise their standards in the countries that supply them. This is understandable, as they have a vested interest in the success of such standards and on imposing additional market access distinguishing requirements on suppliers. In contrast, the SQF Codes have grown out of Western Australia and have become popular with some primary producers and food processors in other counties, irrespective of the fact that the standard's intellectual property is now owned by the Food Marketing Institute, the US retailer and wholesaler association. Moreover, as private standards take off in the Americas, the SQF institute should have home advantage even though Global GAP, IFS, BRC and supporters of ISO22000 are seeking to take their market share.

In order to bring us up to date, we must also consider the introduction of another ISO standard, the ISO 22000 standard for food safety management, which was released in the latter part of 2005. Is this a new innovation, or merely an attempt by ISO and the major certification bodies around the world to win back the ground lost to other private standards when ISO 9000 lost its appeal and relevance to the food industry, mainly owing to the inability of the standard to provide a framework to address food safety? It is interesting to note that ISO22000 is to be benchmarked against GFSI and is being promoted as 'the one universal food safety standard' by at least one international certification body (SGS 2009 online); only time will tell whether chain captains will accept it as an equivalent standard or stay with those they effectively own and control.

The more recent emergence of European retailer requests for their suppliers to demonstrate responsible social and environmental practices in response to (or to create) an interest in credence purchasing in the market can further deflect suppliers away from the twin goals of food safety and product quality. This is yet another development for standards owners and members to ponder and this is reflected in those standards that have encouraged such activities (e.g. GlobalGAP for environment) or have provided optional modules for environmental and worker care (e.g. SQF codes and GlobalGAP GRASP). Finally, we have seen the risk of food terrorism being raised, particularly under the US Homeland Security Bill (Govtracks.US 2002). This has led both the SQF institute and IFS to introduce food defence modules in an effort to address 'unknown hazards and risks'.

15.4.2   Selecting the right standard

So which standard should you choose? Accepting the unpredictability of the politics of food chain captains, and indeed some governments, and acknowledging the speed at which supply chains evolve globally, the following may help you decide:

- If you are looking for a whole chain integrated solution to food safety management in all food categories, then both the SQF codes and ISO 22000 may meet your needs; however, ISO 22000 does not address food quality unless implemented in association with ISO 9000. We should not overlook the cost and resource requirements necessary to implement ISO standards, a major factor in the rejection of such standards by medium and small sized businesses in the past.
- Beyond the farm gate, SQF2000, BRC and IFS all address food safety through the adoption of HACCP and also address food quality issues; however, only SQF 2000 currently offers a seamless link between pre- and post farm gate through the linkages between SQF1000 and SQF2000.
- For countries, or regions, considering private assurance for the first time and who are targeting Europe as their export market, then GlobalGAP may (understandably perhaps) be their initial choice for the farm level. It is relatively simple to achieve, although farms will need to adopt procedures and good agricultural practices (GAPs) based on European conditions and legislation to meet what is essentially a very prescriptive audit checklist.
- If your target markets for agricultural products include the USA then you need to think again. GlobalGAP standards mainly focus on controls associated with pesticide residues and GAPs to protect the environment, whereas US retailers and food processors place greater emphasis on preventing pathogenic contamination of food (and more recently, the threat of deliberate contamination). FMI has recognised the need for private standards to complement government oversight and to address industry and consumer concerns and they have adopted the

SQF1000 and SQF2000 codes and these are effectively becoming the standards to be met by suppliers to the USA.

- If you are looking for quality and rigour in the management and delivery of private food standards, then all of the standards require auditors to have experience and be trained in HACCP and auditing. All offer training to other users of their standards; however, only the SQF codes require within business experts, consultants and trainers to be trained in the standard. This should ensure that all key players know what they should be doing to support food producers and suppliers.
- A key area of continuing concern for many food businesses is recognising and overcoming many of the problems associated with the actual (and often perceived) inappropriate actions of some auditors, certification bodies and consultants. Most standards rely on national accreditation activities to maintain standards within certification bodies. In addition, most require certification bodies to take out a licence with the standard owner. However, only the SQF Institute seeks to maintain control over these functions for the benefit of food chain businesses through a programme of certification body, auditor and consultant licensing and registration backed up by supplier feedback on the activities of these service providers.

15.4.3   Meeting end of chain requirements

It is recognised that multiple retailers around the world are increasingly setting the food supply agenda, even though they claim that this is consumer driven. Irrespective of the drivers and politics, the major retailers are all demanding the same level of food safety management based on the adoption of HACCP. They also require specified product quality attributes to be met. The convergence of retailer safety requirements can be seen in the supplier standards recognised by the CIES Global Food Safety Initiative. Currently the only post farm-gate standards recognised are:

- The BRC Global standard
- The Dutch HACCP standard
- The International Food Standard
- SQF2000

Although not benchmarked yet, we are of the opinion that the ISO22000 standard would also comply, especially as the CIES liaised with ISO TC34 WG8 on the standard's development. At the pre-farm gate level three, standards have been submitted to the GFSI but SQF1000 is the global standard is currently fully compliant (Horticulture New Zealand is classed as a national scheme).

Meeting retailer product quality specifications through private standards can be realised through the adoption of all post farm-gate standards except

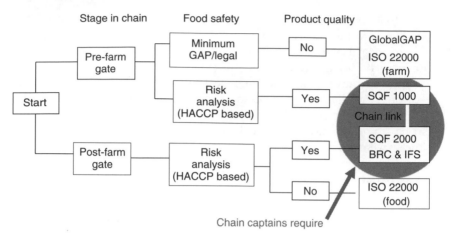

Fig. 15.8 Decision tree for selecting appropriate global food or farm private standard based on approaches to food safety risk management and product quality specifications. (Note: the grey zone denotes what most global retailers require as a condition of market access.)

ISO22000 which only focuses on food safety. At the farm level, only the SQF1000 code has a focus on product quality.

We should also consider the interests of some retailers in environmental and animal welfare issues as well as worker conditions (especially in agriculture). Some European retailers require pre-farm gate standards to address aspects of environmental responsibility and animal welfare. These areas, where they exceed legal requirements, are covered in the Global-GAP standards, but only as recommendations (which do not affect supplier compliance with the standard at audit). GlobalGAP introduced GRASP in 2008 for farmers who wish to demonstrate that they are implementing a social management system and this module will be audited separately. In contrast, the SQF Institute has developed optional modules in environmental care, worker care and, for the US market, biosecurity. These may be taken up by suppliers both pre- and post-farm gate where customers require such assurances. These optional modules form part of the auditor assessment.

Finally, the following decision trees may help in deciding which standard best suits specific current and future needs (Figs. 15.8 and 15.9).

15.5   Future trends

There is a dynamic tension operating in the field of international food and farm standards that is linked both to the contents and approach of various standards and to the politics of standards use in different parts of the world. What is clear though is that the cost to suppliers of meeting market stan-

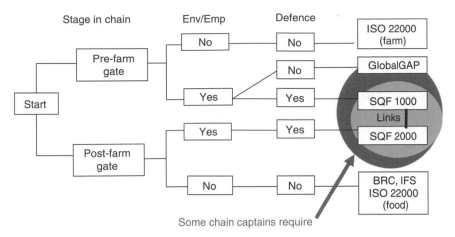

Fig. 15.9 Decision tree for selecting appropriate global food or farm private standard based on approaches to environmental and worker care and food defence specifications. (Note: the dark grey zone denotes what some global retailers require in relation to environment and worker care, light grey refers to requirements of some retailers for food defence or actions against deliberate contamination.)

dards is significant. Moreover, should suppliers be servicing more than one market, they may have to have more than one standard in place.

The costs of compliance have been recognised by standards owners and certification bodies and currently there are moves to harmonise audit protocols between SQF2000 and IFS and between SQF1000 and GlobalGAP. The former is a reasonable approach in terms of harmonising food safety and quality assessments and in terms of certification body licences and auditor registration but does not currently extend to the option of IFS compliant suppliers gaining access to SQF optional modules (other than a similar approach to food defence). Care should be taken however, in harmonising SQF1000 and GlobalGAP; as long as harmonisation of GlobalGAP is to level 1 of SQF1000 (i.e. food safety fundamentals and prerequisites) then they are equivalent. Equivalency does not extend, however, to the different approaches taken to environment by the two standards, nor to GlobalGAP suppliers gaining access to the SQF food defence module. Although not benchmarked as yet, the approach to worker care by the two standards is similar at the farm level but only extends further up the food chain under the SQF2000 code.

Although there are clear advantages in reducing certification costs for the industry by offering more than one certificate per audit where standards have been harmonised, there is a risk that too much harmonisation will result in these standards loosing their individuality and uniqueness. This could lead to one or more of the current international standards being taken over or closed in due course.

There is also a move within some international certification bodies to promote ISO22000 for food and farming, perhaps because this will take power away from the global retailers in terms of control over standards; however, it is difficult to see such retailers giving up on their own standards and the control they currently exert as chain captains.

Another area of future development being considered by some is the potential for the private sector to support and help deliver public regulation. It seems reasonable for the frequent contacts between auditors and food and farm businesses also to be used to provide oversight of regulatory requirements as long as confidentiality is maintained and public versus private market information is kept separated (Baines, 2002).

Finally, amidst this power struggle for private standard dominance, it is also feasible to see alternative approaches to risk management evolving that are not based on the prescriptive protocols and guidance of standards. It could be argued that self assessment of risk backed up by independent audit and risk ranking may be the way forward in a similar way to how insurance risks are calculated for businesses. As such, certification of standards could disappear in favour of a more general risk and insurance model for food and farming.

15.6   Sources of further information and advice

The development in private standards is a rapidly changing area with many stakeholders engaged at sector, national, regional and global levels. In order to keep up to date, it is advisable regularly to visit the home pages of major standards owners, certification bodies, the International Standards Organisation and the CIES. Some of these links are detailed below.

- CIES – http://www.ciesnet.com/1-wweare/index.asp
- SQFi – http://www.sqfi.com
- BRC – http://www.brc.org.uk/
- IFS – http://www.ifs-online.eu/index.php?
- ISO – http://www.iso.org/iso/catalogue_detail?csnumber=35466
- GlobalGAP – http://www.globalgap.org/cms/

15.7   References

BAINES, R. N. (2000). 'Food safety in meat – meeting international regulatory and market requirements', *Congreso de Produccion y Comercializacion de Carne 'Del Campo al Plato'*, Montevideo, Uruguay.

BAINES, R. N. (2002). 'Environmental benefits of farm assurance', *4th Langford Food Industry Conference*, Bristol UK.

BAINES, R. N. and DAVIES, W. P. (1998). 'Quality assurance in international food supply', in *3rd International Conference on Chain Management in the Agribusiness and the Food Industry*, G.W. Ziggers, J.H. Trienekens and P.J.P. Zuurbier (Eds), Wageningen, Netherlands, 213–23.

BAINES, R. N. and DAVIES, W. P. (1999). 'Building trust through farm assurance', *IAMA World Food Congress*, Florence, Italy.

BAINES, R. N. and DAVIES, W. P. (2003). 'HACCP at the farm level – reducing risks in grain and oilseed supply chains through quality assurance', *Grain Europe – Grains & Oilseeds in the Global Food Chain*, International Feed Industry Federation Conference, Rotterdam, NL.

BAINES, R. N. and DAVIES, W. P. (2007). 'Developing consumer trust in ethical food supply to meet increasing market interests in credence purchase', *Proceedings of the International Symposium on Fresh Supply Chain Management*, UN Food and Agriculture Organisation, RAPRA Publication 2007/21. Chiang Mai, Thailand, 314–25.

BAINES, R. N. and RYAN, P. (2002). 'Global trends in quality assurance', paper presented at the *Trade Partners UK and Ministry of Agriculture Modern Food Chain Seminar*, Kuala Lumpur, Malaysia.

BAINES, R. N., RYAN, P. J. and DAVIES, W. P. (2004). 'HACCP at the farm level – the missing link in food safety and security', *Proceedings of the International Food and Agribusiness Association (IAMA) Symposium 'Sustainable Value Creation in the Food Chain'*, Montreaux, Switzerland.

BUNCIC, S. (2000). 'Developments in food safety in the meat chain', *Langford Food Industry Conference*, University of Bristol, UK.

BYRNES, H. (2003). 'Introduction to the global food safety initiative', 36th AIFST Convention, Melbourne, Australia.

CASWELL, J. A. and HENSON, S. J. (1997). 'Interaction of private and public food quality control systems in global markets', in *Globalisation of the Food Industry: Policy Implications*, R.J. Loader, S.J. Henson and W.B. Trail (Eds), University of Reading, 217–34.

CODEX (1969). *Recommended International Code of Practice. General Principles of Food Hygiene*. CAC/RCP 1-1969, Rev. 4-2003.

EARLY, R. (1995). *A Guide to Quality Management Systems for the Food Industry*, Blackie Academic and Professional, London.

EUROPA (no date). General Food Law introduction. http://ec.europa.eu/food/food/foodlaw/index_en.htm (date accessed 06-06-09).

FSANZ (no date). Food Standards Code. http://www.foodstandards.gov.au/thecode/ (date accessed 06-06-09).

GOVTRACKS.US (2002). *National Homeland Security and Combating of Terrorism Act of 2002*, http://www.govtrack.us/congress/bill.xpd?bill=s107-2452 (date accessed 06-06-09).

HUGHES, D. (1994). *Breaking with Tradition: Building Partnerships and Alliances in the European Food Industry*, Wye College, Ashford, UK.

MANNING, L. and BAINES, R. N. (2004). 'Globalisation. A study of the poultry meat supply chain', *British Food Journal*, **1**(10/11), 819–36.

MANNING, L., BAINES, R. N. and CHADD, S. A. (2006). 'Quality assurance models in the food supply chain', *British Food Journal*, **108**(2), 91–104.

MANNING, L., BAINES, R. N. and CHADD, S. A. (2008). 'Benchmarking the poultry meat supply chain', *Benchmarking – an International Journal*, **15**(2), 1463–5771.

SGS (2009). ISO 22000. http://www.uk.sgs.com/foodsafety_uk_v2.htm (date accessed 07-06-09).

SPARLING, D., LEE, J. and HOWARD, W. (2001). 'Mungo Farms Inc: HACCP, ISO9000 and ISO14000', *International Food and Agribusiness Management Review*, **4**(1), 67–97.

VARZAKAS, T. and JUKES, D. J. (1997). 'The globilization of food regulation and quality: a study of the Greek food market', in *Globilisation of the Food Industry: Policy Implications*, R.J. Loader, S.J. Henson and W.B. Trail (Eds), University of Reading, 253–71.

16

Developments in quality management systems for food production chains

L. Theuvsen, Georg-August University of Goettingen, Germany

Abstract: This chapter discusses the state-of-the-art and recent developments in quality management in food supply chains. Special emphasis is put on the regulatory framework of quality management, the relationship between chain organization and quality management and recent attempts to improve chain-wide quality-related communication. These attempts include food law, certification schemes and the use of IT and quality management techniques. The implementation of tracking and tracing systems and total quality management approaches represent new directions that have gained growing relevance in the food sector. Finally, how firms perceive new developments in quality management systems and what this means for the implementation and acceptance of these systems are discussed.

Key words: certification schemes, quality-related communication, regulatory framework, Total Quality Management, traceability.

16.1 Regulatory framework

Food quality is a multi-faceted phenomenon including intrinsic – product safety and health, sensory properties and shelf life, reliability and convenience – as well as extrinsic – production system characteristics, environmental aspects – characteristics (Luning *et al.*, 2002). Markets for agricultural and food products are characterized by high information asymmetries since producers, processors and retailers are in most cases much better informed about the quality of their products than consumers (Henson and Traill, 1993). Often important quality attributes such as food safety, nutritional value, organic production or region of origin cannot be verified by consumers (or can be verified only at a prohibitively high cost). Such credence attributes can result in market failure owing to a lack of credible information in the market (Akerlof, 1970).

Against this background, attempts to protect consumers against food hazards, product adulteration and deception have a long history. Since trade in agricultural and food products was one of their main income sources and crucial for the supply of the local population, antique as well as medieval towns laid down regulations on food quality, food inspections, and metrics and weights. The oldest food laws were in force in the Babylonian, Egyptian and Roman Empires. In medieval times, food laws were in place from the 11th century (Mettke, 1979). In the 13th century, the King of England prohibited adulteration of bread (Assize of Bread and Ale). Other early quality regulations include the ban on butter dyeing laid down in Paris in 1396, the introduction of wine inspections in Germany in 1498 and the Bavarian Purity Law for beer (1516). In many countries the late 19th century marked a starting point for a much more systematic and comprehensive regulation of food quality based on more advanced natural science knowledge and improved measurement methods. Germany, for instance, introduced official food inspections in 1876 and the Food Law in 1879. The US government passed the Tea Importation Act in 1897 and the Food and Drugs Act and the Meat Inspection Act in 1906. In the UK, the Assize of Bread and Ale was repealed by the Statute Law Revision Act in 1863 (Scheuplein, 1999; Anonymous, 2008; Doltsinis, 2008). The following decades saw a systematic expansion of food quality regulations and inspections on a national basis.

Despite quite similar developments of food laws in many countries, the remaining differences resulted in considerable non-tariff trade barriers. Therefore, more recent decades have been characterized by efforts to harmonize food law internationally. An important milestone was the creation of the Codex Alimentarius Commission in 1963 by Food and Agriculture Organization (FAO) and World Health Organization (WHO). Its main purposes are health protection of consumers, ensuring fair trade practices and coordination of food standards (www.codexalimentarius.net). The World Trade Organization (WTO) Agreement on the Application of Sanitary and Phytosanitary Measures (SPS Agreement) has become the most important basis for the international harmonization of food law (Kastner and Pawsey, 2002). Article 3.1 of the SPS Agreement highlights the great importance attached to international harmonization: 'To harmonize sanitary and phytosanitary measures on as wide a basis as possible, Members shall base their sanitary or phytosanitary measures on international standards, guidelines or recommendations, where they exist, except as otherwise provided for in this Agreement ...'. Whereas least developed country members were allowed to delay implementation (Article 14), the SPS Agreement entered into force for most WTO members on January 1, 1995 (www.wto.org).

As a reaction to the BSE crisis and several other food incidents, food law has been undergoing major changes in the European Union (EU). General Food Law Regulation (EC) 178/2002 has strongly contributed to the ongoing international harmonization trend, for instance by providing definitions of technical terms such as food, feed or placing on the market. Before 2002 these terms were not used consistently throughout the EU. Furthermore, the General Food Law Regulation has introduced several new principles that can, at least to a certain degree, be considered typical of 21st century legislation on food quality (Streinz, 2007):

- From farm to fork: Typical of recent food laws is a whole supply chain approach that covers input industries such as feed manufacturers, agriculture, processors and retailers.
- No distinction between food and feed: The General Food Law Regulation does not distinguish between food and feed any more since feed becomes food in the end and several food crises started in the feed sector.
- Inclusion of agriculture: The broad definition of food in Article 2 means that farmers are considered food business operators and are no longer excluded from regulations on food quality.
- Risk analysis: EU food law includes distinct processes of risk assessment (hazard identification, hazard characterization, exposure assessment and risk characterization), risk management (process of weighing policy alternatives, considering risk assessment and selecting appropriate prevention and control options) and risk communication, that is, the interactive exchange of information and opinions throughout the risk analysis process.
- Precautionary principle: If available information does not allow exclusion of the possibility of harmful effects on health, provisional risk management measures may be adopted (Article 7).
- Private responsibility: Food business operators shall ensure that foods satisfy the requirements of food law. Member states are responsible for the enforcement of food law (Article 17).
- Traceability: Article 18 establishes the obligation to secure 'traceability of food . . . at all stages of production, processing and distribution'.

The General Food Law Regulation has been complemented by a considerable number of EU directives and regulations, for instance the so-called EU hygiene package (Regulations (EC) 852/2004, 853/2004 and 854/2004). Owing to the intensification of EU legislation on food quality, elbow room for national legislation has been reduced significantly (Streinz, 2007).

Although legislation has been very much intensified over the years, food laws have been more and more complemented by private regulation. In the UK, this development started in the 1980s when big food retailers filled in the gap that was left after deregulation and the retreat of the state (Marsden and Wrigley, 1996). Although remarkable differences between European and USA responses to food scares can be observed (Goldsmith et al., 2003), regulation of food production has evolved into a complex multi-level network of public and private interventions in many countries

(Marsden *et al.*, 1997; Harrison *et al.*, 1997). This development accelerated in the late 1990s when retailers tried to increase their control over food supply chains as a risk management strategy in the face of severe food crises such as BSE (Hornibrook and Fearne, 2002). Today certification schemes, especially those strongly influenced by big retailers (for instance, GlobalGAP, International Food Standard, BRC Global Standard), are major elements of the private regulation of food production (Theuvsen *et al.*, 2007a).

16.2   Quality management and the organization of food production chains

Quality management in food production is strongly influenced by the vertical coordination of food supply chains, that is, the way relationships between different actors in the supply chain (farmers, processors, retailers and so on) are organized. There is a broad spectrum of alternatives that supply chain actors can choose from when designing their business relationships (Peterson *et al.*, 2001). Most categorizations refer to the coordination intensity resulting from ownership and centralisation of decision making. With regard to the relationship between farmers and processors, spot markets, informal long-term relationships, marketing and production contracts, contract farming and vertical integration are often discussed as alternative approaches to the organization of food supply chains (Schulze *et al.*, 2007).

In spot market transactions, producers and processors negotiate only short-term contracts and are able to change their market partners very quickly. Therefore, 'control is fully located at the separate stages (...) and coordinated solely by market prices' (den Ouden *et al.*, 1996, p 281). The transactions are governed by classical contract law: 'sharp in by clear agreement, sharp out by clear performance' (Williamson, 1991, p 271). Although not bound by contracts, many supply chain actors actually prefer long-term relationships with a limited number of business partners. Stable relationships provide more reliability and lower search costs than spot markets without sacrificing opportunities to change market partners easily if this is considered necessary owing, for instance, to price movements.

Marketing contracts establish selling and buying obligations between contracting market partners. Besides secured supply for customers and guaranteed market access for suppliers, marketing contracts often shift price risks from suppliers towards processors, for instance through feed-price formula contracts in meat production (Roe *et al.*, 2004). Spot market-like transactions, informal long-term relationships and simple marketing contracts can still be found in many food supply chains, for instance pork production in Western Europe (Boston *et al.*, 2004).

Production contracts not only establish selling and buying obligations but also prescribe some details of the production process, for instance feed, breed or animal welfare aspects. Contract farming schemes represent an even stricter coordinated relationship between farmers and processors. Farmers remain legally autonomous but are heavily dependent on a contractor, for instance an abattoir, who provides all the critical resources (Mighell and Jones, 1963). In vertically integrated chains, there is joint ownership of resources in different parts of the food supply chain, for instance the farm and the processor level. Stricter vertically coordinated chains have a long history in poultry production (Furesi *et al.*, 2006). A World Bank study revealed that stricter vertically coordinated food supply chains are increasingly replacing open markets in many developing and transformational economies (Swinnen, 2005). This parallels a trend in US agriculture according to which nowadays well over a third of agricultural products are produced under contracts and in vertically integrated chains (Martinez, 2002a, 2002b).

The degree of vertical coordination strongly influences quality management in food supply chains. In a strictly vertically coordinated supply chain characterized by close contractual relationships or vertical integration, quality management is organized by a lead company. Although the details of these systems vary remarkably with regard to degree of centralization of decision making, degree of vertical integration or extent of contractual obligations, several typical characteristics can be observed (Spiller, 2004).

The system leader defines and enforces quality goals, often attuned to a centrally managed marketing concept. To do so, the lead company has inspection rights, for instance in the form of second-party audits of its supply chain partners and can exclude unreliable or opportunistically behaving partners from the chain. Very often contractually bound farms or firms have buying obligations for input factors supplied by the system leader (for instance, feed or chicks). Such concepts can often be found in the poultry industry but have also made their way into other agribusiness subsectors, for instance pork production in Denmark or the United States (Schulze *et al.*, 2007).

In food supply chains characterized by open markets, informal long-term relationships or simple marketing contracts, an alternative quality management approach has evolved that relies on certification standards and third-party audits (Spiller, 2004). 'Certification is the (voluntary) assessment and approval by an (accredited) party on an (accredited) standard' (Meuwissen *et al.*, 2003). A key feature of certification systems is that inspections are carried out by independent bodies beholden to standards laid down by external organizations (third-party audits; Luning *et al.*, 2002). By means of regular control and – whenever necessary – additional sampling, neutral inspection institutions, in many cases auditing companies, monitor the entire supply chain. After successfully passing the auditing procedure, companies are awarded a certificate that can be used as a quality signal in the market. System participants remain independent and do not enter contractual relationships with upstream or downstream industries. Far from it, owing to the reduction of quality uncertainties by certification systems, spot market transactions are favoured over alternative ways of organizing food supply chains (Schramm and Spiller, 2003).

Certification schemes have become widely prevalent in agriculture and the food industry. In Germany alone about 40 different such schemes are used for certifying farms and firms. For the EU, the (presumably somewhat exaggerated) number of more than 380 certification schemes is sometimes cited (Wesseler, 2006). A closer look at the systems in place reveals a broad spectrum that can be organized along different dimensions (Theuvsen and Spiller, 2007): standard setter, addressees, objectives, foci, geographical and supply chain coverage and number of participants. Table 16.1 summarizes the characteristics of six certification schemes: GlobalGAP, BRC Global Standard, the German Q&S system, the International Food Standard (IFS), the EU system of Protected Designation of Origin (PDO) and Protected Geographical Indication (PGI) (Regulation (EC) 510/2006), and the private Demeter standard for organic food products.

The effectiveness of these standards in relation to quality management is sometimes discussed critically. Whereas several authors report on, for instance, improved traceability (Hollmann-Hespos, 2008) and reduced risks as perceived by consumers (Franz et al., 2007), others consider these systems to be façades implemented only as a reaction of farm and firm managers to institutionalized expectations in their external environments (Walgenbach, 2007). Furthermore, the bureaucratic nature of certification systems is suspected to result in acceptance and implementation problems (Theuvsen, 2005; Jahn and Spiller, 2007).

16.3   Quality-related communication in food production chains

Food supply chains are characterized by a more or less intensive division of labour which results in efficiency gains through specialization and economies of scale, but also in a need for coordination through communication between supply chain partners (Theuvsen, 2004). Furthermore, in food supply chains, many firms do not have direct relationships with consumers. For these firms, dissemination of information, for instance about product quality preferred by consumers, through communication between customers and suppliers is a precondition for market orientation (Mohr and Nevin, 1990) and an important driver of product and process innovations (Dyer and Singh, 1998). A strong market orientation that directs all of a firm's efforts towards meeting customer demands is often considered a prerequisite for successful business operations (Kohli and Jaworski, 1990; Martin and Grbac, 2003). Last but not least, in the agribusiness sector improved documentation and information sharing are cornerstones of such important issues as improved traceability (Opara, 2003) and transparency (Hofstede,

Table 16.1 Characteristics of selected certification schemes (Gawron and Theuvsen, 2009a)

	Global GAP	BRC	Q&S	IFS	PDO/PGI	Demeter
Standard setter	Private	Private	Private	Private	State	Private
Addressees	Business	Business	Business, consumers	Business	Consumers	Consumers
Objective	Safety	Safety	Safety and product differentiation	Safety	Product differentiation	Product differentiation
Focus	Quality management system	Quality management system	Quality management system	Quality management system	Product	Production process
Geographical coverage	Global	International	National, increasingly international	International	International	International
Supply chain coverage	Agriculture	Processing	Whole supply chain	Processing	Agriculture, processing	Agriculture, processing
Number of participants	>71000	>13000	>100000	>8500	about 800	>3200

2003; Deimel *et al.*, 2008). Therefore, chain-wide quality-related communication is, for various reasons, very important for food supply chains.

Food supply chains are also characterized by reciprocal multi-stage agency relationships in which companies delegate tasks to each other (Theuvsen, 2004). Farmers, for instance, delegate processing to food manufacturers, who in turn delegate production of agricultural raw materials to farmers. It is often hypothesized that principals and agents in food chains behave opportunistically, that is, they act in self-interest with guile (Williamson, 1985). For this reason, the correctness and completeness of information transferred throughout food supply chains cannot be taken for granted. Besides opportunistic behaviour, agency relationships are characterized by information asymmetries. Opportunistic behaviour and information asymmetries result in agency problems known as hidden characteristics, hidden action and hidden intention (Arrow, 1985); these problems strongly influence the amount and quality of information shared in food supply chains (Theuvsen, 2004).

Comprehensive and reliable communication in food supply chains is considered highly relevant for quality assurance and quality management as well as to gain and sustain competitive advantages (Mack, 2007; den Ouden *et al.*, 1996) but, owing to agency problems, cannot always be taken for granted. Therefore, recent government (food laws) and private (certification schemes) regulation has more intensively focused on the quality-related information stored and transmitted in food supply chains. Although parts of the regulatory framework apply to all agricultural and food products, the main focus is on meat production which is highly susceptible to food safety incidents.

With regard to chain-wide communication, Article 18 of General Food Law Regulation (EC) 178/2002 is the basic rule since, according to Article 18, food business operators shall be able to identify any person from whom they have been supplied with inputs and any other business to which their products have been supplied. Nevertheless, since competent authorities are the main addressees of this information in case of food crises, the effect of Regulation (EC) 178/2002 on quality-related communication in food supply chains is very limited. More important effects emanate from the EU hygiene package (Regulations (EC) 852/2004, 853/2004 and 854/2004) through which the concept of food chain information was introduced. With regard to animal production, the food chain information approach includes the obligation of farmers to document certain information and the obligation of slaughterhouses to request this information from their supplying farmers. According to Regulation (EC) 853/2004, the quality-related information provided by farmers at least 24 hours before slaughtering has to cover:

* the farm's or the regional animal health status
* the animals' health status

- veterinary medicinal products administered to the animals within a relevant period and with a withdrawal period greater than zero, together with the dates of administration and withdrawal periods
- the occurrence of diseases that may affect the safety of meat
- the results of former analyses
- relevant reports about former inspections of animals
- production data, when this might indicate a disease and
- name and address of the private veterinarian.

Quality-related communication is also required by other EU legislation, for instance Regulation (EC) 2160/2003 on the control of salmonella and other foodborne zoonotic agents, and national legislation, for instance the German Pig Salmonella Regulation which stipulates obligatory information from farmers about their pig herds' salmonella status (Deimel *et al.*, 2009).

Many private certification schemes also require communication between supply chain partners. Table 16.2 gives an overview of quality-related information exchanges mandated in the German Q&S, the Dutch IKB and the Danish QSG systems. Interestingly, all three schemes restrict obligatory information exchange mainly to those areas already mandated by legislation. Furthermore, information transfer and data media are strongly influenced by supplier–customer relationships and the flow of goods (slaughter animals, fresh meat). Analyses of certification schemes in the organic farming sector revealed similar results (Theuvsen and Plumeyer, 2007).

To fulfil requirements stemming from legislation and certification schemes, but also to generally improve communication between supply chain partners, advanced IT solutions have been developed. This is strongly supported by research that focuses on information management and the development of reference models for quality information requirements in food supply chains (Schulze Althoff *et al.*, 2005).

Two different development directions can be distinguished with regard to the interface between farms and processors (Deimel *et al.*, 2009). In the first case, processors introduce IT systems for communication with and integration of suppliers and customers. These are typically proprietary systems that restrict access to current business partners of the system owner. Well-known examples are the Farming®Net software of the Dutch Vion Group and the Extranet software of the German cooperative Westfleisch. A second approach is proposed by software developers. Systems such as Desk-x or Farmer's Friend represent standard software solutions to communication problems in food supply chains. These systems allow economic analyses and benchmarking against competitors but also support the quality-related data exchange required by government and private regulation.

Despite these development trends, electronic communication in food supply chains is currently still in its infancy and mainly restricted to providing online access to or email transmission of slaughter documents. The current situation is characterized by media disruptions between the various

Table 16.2 Mandatory information exchange in three European certification schemes (Theuvsen *et al.*, 2007a)

	Q&S	IKB	QSG
		vertically integrated system – mentoring	
Feed industry			
→ ←	• Q&S-certified feed Information on feed ingredients	• GMP+-certified feed Information on feed ingredients	• QSG-certified feed Information on feed ingredients
Pig fattening farm			
→ ←	• Pig number/marking the pigs Documentation of transport Salmonella status Slaughter account	• Pig number/marking the pigs Documentation of transport Salmonella status Slaughter account	• Pig number/marking the pigs Documentation of transport Salmonella status Slaughter account
Slaughterhouse			
→ ←	• Q&S-certified meat Batch number	• IKB-certified meat Batch number	• QSG-certified meat Batch number
Processor			
→ ←	• Q&S-certified meat Batch number	• IKB-certified meat Batch number	• QSG-certified meat Batch number
Retailer			

stages of food supply chains and chain-wide communication is often considered insufficient and in need of major improvement (Deimel *et al.*, 2009). Against this background it has been argued that quality-related communication in food supply chains is not only a technical but also a management problem. Therefore, Gerlach *et al.* (2007) recommend supplier relationship management by processors as a way to improve inter-firm cooperation and communication.

Peupert and Theuvsen (2007) suggest quality function deployment (QFD) as a means of improving chain-wide communication. Traditionally, QFD is a product development tool that aims to translate the quality attributes required by customers into the quality functions of a product. The central device is the so-called 'house of quality' which allows customer demands to be related to product characteristics in a matrix (Akao, 1990). The basic ideas of the proposed, although not yet implemented, enhanced QFD approach are to translate the quality requirements of one firm, for instance a dairy that wants to introduce extended shelf-life products into the market, into the quality requirements of all chain members and, in this context, to use the house of quality as the central planning and communication instrument (Peupert and Theuvsen, 2007).

16.4   Tracking and tracing food products

In recent years, the traceability of food products has become increasingly relevant. Besides legislation (for instance, Article 18 of Regulation (EC) 178/2002 and Regulation (EC) 1830/2003 concerning the traceability and labelling of GMOs), there are other important drivers of investments in tracking and tracing systems: risk management, certification systems, business processes management, differentiation strategies and stakeholder demands (Theuvsen and Hollmann-Hespos, 2007).

Public product recalls are a major threat to food manufacturers and require elaborated risk management strategies. In the short run, product recalls mainly result in fewer sales owing to out-of-stocks and higher costs caused by backhaul and disposal of defective products, additional laboratory analyses, *ad hoc* process improvements, compensation payments and crisis communication with supply chain partners and consumers. In the long run, attenuation of brand value, lower customer loyalty and a weaker competitive position may result from product recalls. Additional long-term costs can result from the need to reposition brands, develop and implement new competitive strategies, intensify consumer communication, redesign business processes and introduce additional quality controls.

Improved traceability as part of an advanced crisis management system can contribute to cost savings and avoidance of sales and profit losses by allowing faster and more precise product recalls in the case of food safety incidents (Doeg, 2005).

Nearly all certification schemes nowadays widely disseminated throughout agribusiness supply chains include more or less detailed specifications with regard to improved documentation and traceability (Newslow, 2001). Insofar as certification has become a prerequisite for supplying large retailers, certification systems have become a major driver of improved traceability.

Improving internal and external business processes through advanced tracking and tracing systems may be another motivation for firms to invest in the improved traceability of food products. A study by Hardgrave *et al.* (2006) shows that radio frequency identification (RFID) systems – one of the most promising and rapidly developing tracking and tracing technologies (Bhuptani and Moradpour, 2005) – were able to boost sales in retail stores by 3.4% owing to their ability widely to eliminate out of stocks.

Differentiation strategies that allow food manufacturers to escape price competition to a certain degree (Porter, 1980) can be associated with improved tracking and tracing. This is most likely in businesses where firms deal with products, such as eggs, fresh meat and fish, that are subject to high food safety risks (Luten *et al.*, 2003). In these industries, customers and consumers may be willing to pay more for improved product safety resulting from more advanced tracking and tracing systems. Furthermore, traceability is of high relevance for those differentiation strategies that rely on product or process characteristics that cannot (or can hardly) be controlled by consumers (credence attributes; Akerlof, 1970). These include quality aspects such as organic, region-of-origin, GMO-free, free-range husbandry and other animal welfare aspects for which at least some consumers tend to show a higher willingness to pay (Bennett, 1996; Skuras and Vakrou, 2002; Krystallis and Chryssohoidis, 2005; Batte *et al.*, 2007). In these cases traceability provides the history of a product and is a precaution against fraud (for instance, labelling conventional products as organic) and a quality signal to consumers (Smyth and Phillips, 2002).

Last but not least, external stakeholders may force food manufacturers to improve traceability. Banks, for instance, may consider state-of-the-art tracking and tracing systems as a way to manage operational risks. This can influence a firm's capital costs owing to the high emphasis the so-called Basel II directive places on operational risks. Non-governmental organizations that question supply sources, the absence of GMOs or other quality traits may also motivate firms to improve their tracking and tracing systems.

A survey of 224 German food manufacturers between November 2005 and May 2006 revealed that predominantly three factors determine firms' investments in tracking and tracing systems: improvement of firms' processes, stakeholder requirements (including customer expectations) and legal requirements. Using these factors together with a single statement ('traceability is a precondition for successful certification') as cluster-building variables, five clusters of firms with different intentions to invest in tracking and tracing systems could be identified (Theuvsen and Hollmann-Hespos, 2007):

- Cluster 1 'Certified companies': Cluster 1 comprises 36 companies that have implemented tracking and tracing systems mainly in order to pass a third-party audit successfully and get a certificate. Risk management, process improvements and differentiation strategies are of minor relevance for these firms. Most of the companies in this cluster are small and specialize in producing retailer-owned brands. Producers of frozen foods, fish and beverages are frequently in this cluster. Only 15% of the respondents have ever suffered a public product recall. The implementation of tracking and tracing systems has not advanced very far; the technical capabilities of the systems implemented are considered rather low.
- Cluster 2 'Ignorants': The 28 companies in cluster 2 rank the relevance of traceability lowest in the sample and do not attribute high relevance to any of the potential reasons for improving traceability. In particular, stakeholder requirements and legislation are perceived as not very important. The companies in this cluster are very different in size. It is noteworthy that as many as 40% of these respondents have already undergone one or more product recalls. Nevertheless, their tracking and tracing systems are not very advanced.
- Cluster 3 'Lawful investors': 27 respondents give legal and stakeholder requirements as their main motives for implementing tracking and tracing systems. Most of the firms in this cluster are comparatively small. Only 13.4% produce retailer-owned brands. The tracking and tracing systems used by these firms are characterized by an advanced development status.
- Cluster 4 'Image-oriented firms': In this cluster stakeholder requirements are the main reason why tracking and tracing systems have been implemented. Improving traceability in order to meet the requirements of certification systems is also important. The 60 firms in this group are of above-average size and often produce retailer-owned brands. The companies in this cluster attribute high benefits to improved traceability.
- Cluster 5 'Versatile companies': The 73 firms in this cluster reveal several important reasons for investing in tracking and tracing systems and consider improved traceability to be very important. The companies are very different in size and have only rarely suffered public product recalls. The tracking and tracing systems are advanced and the capabilities of these systems are considered to be high.

All in all, the empirical results show that food manufacturers differ remarkably with regard to their motives for investing in tracking and tracing systems and their willingness to spend money on improved traceability.

16.5   Total Quality Management in food production chains

Total Quality Management (TQM) has evolved as the leading concept in quality management (Pfeifer, 2002). It is most widely spread in the automobile and automobile suppliers, electrical and electronics and machine industries, but it has also been discussed as a blueprint for the food industry (Beardsell and Dale, 1999; Luning *et al.*, 2002; Vasconcellos, 2003), which has been implemented in parts of it and used as a benchmark for assessing the effectiveness of quality assurance concepts in food supply chains (Theuvsen and Peupert, 2007).

TQM is said to be a 'management approach of an organization, centred on quality, based on the participation of all its members and aiming at long-run success through customer satisfaction, and benefits to all members of the organization and to society' (Hellsten and Klefsjö, 2000). Figure 16.1 reveals the most important conceptual cornerstones of TQM.

'Quality starts at the top' (Feigenbaum, 2004) is one of TQM's basic principles. It highlights that a quality improvement process always depends on top management's motivation and self-commitment (Hackman and Wageman, 1995), the integration of an organization's basic values concerning quality into its mission statement and the definition of a balanced and transparent goal system embracing operational quality goals. Furthermore, TQM attributes an important role to leadership and asks top management to initiate quality programmes, strive for continuous improvement and

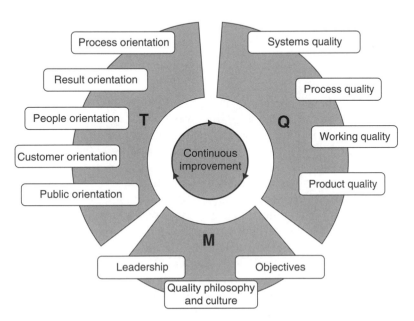

Fig. 16.1   Total quality management (after Kamiske and Malorny, 1992).

provide necessary resources for quality management (Pfeifer, 2002). The interplay between these aspects is considered paramount in influencing an organization's quality culture (Abraham *et al.*, 1999).

TQM considers people to be a decisive success factor and requires human resource management activities to be strictly focused on the organization's quality goals. The capabilities and motivation of employees are paramount in this respect (Hackman and Wageman, 1995). Furthermore, careful personnel selection, performance-based reward systems and organizational measures such as quality circles are important TQM elements.

Process orientation is another cornerstone of TQM. It includes the definition of process and quality responsibilities, improved process controls and continuous improvement processes (Pfeifer, 2002) and provides the precondition for improved result orientation that can have strong behavioural effects (Montes *et al.*, 2003). Therefore, TQM is often associated with the organization-wide use of quality-related key performance indicators (KPIs) (see, for instance, www.frieslandfoods.com).

Finally, TQM advises management to strengthen customer orientation through improving customer relationships and measuring customer satisfaction (Pfeifer, 2002). Other stakeholder groups' expectations should also be taken into account since in modern societies a widespread 'discontent with the industrialization of agricultural production and food provision systems has put agribusiness and the food industry at the core of societal debates' (Jansen and Vellema, 2004, p 4). Organizations that lose qualified acceptance of business operations by the public have difficulties entering processes of social exchange, as their partners have lost trust in their compliance with social rules (Palazzo and Scherer, 2006; Heyder and Theuvsen, 2008). Therefore, company policies and communication have to address customers' as well as other constituencies' requirements under a TQM framework.

TQM has quite a long history in the food industry (Gould, 1992), yet, it is still interpreted in diverse ways. In many cases it is used more or less as a synonym for an integrated quality management system that combines a HACCP concept with various certifications such as ISO 9001 (quality management), ISO 14001 (environmental management) and International Food Standard or Regulation (EC) 2091/92 (organic food products). Statistical quality control and the comprehensive application of management and planning tools are also deemed cornerstones of TQM in the food industry (Vasconcellos, 2003). Sometimes this approach is combined with participation in quality award contests such as the EFQM Excellence Award or the Malcolm Baldridge National Quality Award. Kahler Gewürze GmbH in Berlin (Germany) provides a vivid example how a food manufacturer uses a broad spectrum of certification systems on its way to a TQM approach (www.kahler-berlin.de). In this sense it is often advised to implement TQM in the food industry in a stepwise way (Hubbard, 2003).

Critical analyses show that recent developments in quality management in food supply chains do not fully meet TQM principles. Weaknesses can be observed with regard to most TQM principles. Therefore, it is recommended that goal, result and people orientation are strengthened. Furthermore, it should be stressed that TQM approaches in the food sector have to take into account a vast number of small and medium-sized enterprises. Therefore, quality circles, for instance, have to be organized as regional working groups meeting from time to time to enhance information exchange between members. Industry associations should also play a more active role in helping small and medium-sized enterprises to overcome their limited quality management capabilities which may hamper the implementation of TQM (Theuvsen and Peupert, 2007).

16.6 Chain uptake of quality management systems

In recent years there have been numerous analyses of chain uptake of recent developments in quality management (Theuvsen *et al.*, 2007b; Schulze *et al.*, 2008). In many cases these studies have focussed on farmers' acceptance and assessment of certification schemes (Jahn and Spiller, 2007; Enneking *et al.*, 2007), costs of certification procedures (Belletti *et al.*, 2007), traceability (Gellynck *et al.*, 2007) and reliability of auditing procedures (Jahn *et al.*, 2005).

A comprehensive analysis of managers' perceptions of certification schemes was published by Schulze *et al.* (2008) who surveyed all 1799 firms which were certified according to the International Food Standard (IFS) in February 2006. There were 389 food manufacturers mainly from Germany (55.0%), France (9.3%), Italy (6.9%) and Austria (6.4%) who participated in the study. On average, the firms surveyed have 346 employees. They represent 18 different subsectors of the food industry including beverages (20.7%), agricultural and horticultural produce (16.1%), meat products (13.2%), dried goods (12.9%) and dairy products (12.1%).

The results show that 74.6% of the companies are generally satisfied with the IFS. However, only 32.7% would have implemented the standard in the absence of any retailer requirements. There were 70.6% respondents who emphasized that from a firm's perspective the advantages of the IFS outweigh its disadvantages and 82.2% said that the standard provides some useful input for day-to-day business management. The survey showed that 51.3% report improved relationships with customers after IFS implementation and 56.2% agree that the IFS has contributed to improved food safety. The time spent for the certification process was the most important cost factor; 44.8% of the companies surveyed had to employ additional staff for the certification process.

A factor analysis revealed six factors underlying the firms' assessment of the IFS: cost/benefit ratio, quality of the IFS compared to other standards, communication of the standard owner, expertise of the auditor, costs of certification and effectiveness of the auditor. A regression analysis showed that it was perceived cost/benefit ratio not costs *per se* that provided the most influential factor for explaining firms' overall assessment of the standard. The perceived quality of the IFS compared to other certification systems and perceived communication of the standard owner came next.

Using the factors identified as cluster-building variables, three groups of firms were identified with regard to their evaluation of the IFS. The first group ('the unconcerned'; 29.1% of the companies) is generally satisfied with the IFS but does not perceive many advantages. Companies in the second cluster ('the satisfied'; 40.7%) have, all in all, a very positive attitude towards the IFS. They consider it a useful management instrument and observe improved relations with customers. The members of the third cluster ('the dissatisfied'; 30.2%) have been certified, but nevertheless do not see any positive effects on food safety, firm management or customer relations.

These results very much parallel previous studies with regard to the important role of perceived cost/benefit ratio (Zaibet and Bredahl 1997; Mumma *et al.*, 2002) and the very diverse assessments of certification schemes (Gawron and Theuvsen, 2006; Jahn and Spiller, 2007). An obvious managerial implication of the empirical findings is that those companies that do not yet perceive any advantages with regard to the implementation of the IFS should consider certification schemes more as quality management instruments. So far, companies already reporting an improvement of internal business processes or other advantages can serve as benchmarks for the more sceptical food manufacturers. Furthermore, the catalogue of requirements offers the standard setter a chance to enhance the satisfaction of certified companies by, for instance, integrating more subsector-specific requirements and benchmarking the standard against other certification schemes to avoid duplication of effort. Finally the standard setters should improve communication with certified companies by means of, for instance, a regular newsletter, an earlier announcement of upcoming changes of requirements, provision of more industry-specific information or invitation of suggestions for improvements from the companies' side.

16.7   Quality management in food production chains: some final remarks

Owing to the enormous importance of food hygiene and the prevalence of credence attributes that facilitate fraud and deception, the food sector was one of the early adopters of quality management. Nevertheless, despite the tremendous improvements triggered by a growing body of natural science knowledge since the 19th century, food production was not recognised as a worldclass benchmark in the field of quality management for quite a long time. Instead, the automobile and other high-tech industries took the leading role with regard to advancements in quality management. Several food scares in the late 20th century, in particular the BSE crisis in Europe, showed that quality management was in fact in need of major improvement. This insight led to big changes in public and private regulation of food production. Governments, in particular the European Commission, introduced new principles into the food law. The private sector also started remarkable initiatives, for instance in the field of certification systems. Often it was large processors or retailers that were at risk of losing their repuation when hit by a food crisis who strengthened their control over food supply chains and strongly proposed stricter vertically coordinated food supply chains and the introduction of third-party audits (Hornibrook and Fearne, 2002). As a consequence, the regulation of food production has evolved into a complex network of public and private actions on various governance levels (Harrison *et al.*, 1997).

Future trends include the further internationalization of quality management in the food industry. Recent studies show that certification systems once set up at the national level (for instance, the BRC Global Standard) have gained more and more relevance in other countries. The export orientation of countries delivering to markets where private standards have become gatekeepers for retailers seems to be an important determinant of how many farms and firms have been certified according to foreign standards (Gawron and Theuvsen, 2009b).

Another trend is the creation of more certification schemes that meet the highly diversified demands of consumers in modern society. Recently, the EU has strongly intensified research into animal welfare (www. welfarequality.net) as a basis for its future action in this field.

All in all, quality management in food production chains is strongly influenced by technical developments and societal demands, the latter often reflected through legislation or actions of non-governmental organizations. Owing to the rapid development of new technologies as well as changes in consumer preferences, quality management is likely to remain a field that will undergo continuous changes.

16.8   References

ABRAHAM, M, CRAWFORD, J and FISHER, T (1999), 'Key factors predicting effectiveness of cultural change and improved productivity in implementing Total Quality Management', *International Journal of Quality & Reliability Management*, **16** (2), 112–32.

AKAO, Y (1990), *Quality Function Deployment: Integrating customer requirements into product design*, Productivity Press, Cambridge, MA.

AKERLOF, G A (1970), 'The market for "lemons": Quality uncertainty and the market mechanisms', *Quarterly Journal of Economics*, **84** (3), 488–500.

ANONYMOUS (2008), *Milestones in US Food and Drug Law History*, www.fda.gov/opacom/backgrounders/miles.html (downloaded July 21, 2008).

ARROW, K J (1985), 'The economics of agency', in Pratt, J W, Zeckhauser, R J (eds), *Principals and Agents: The Structure of Business*, Harvard University Press, Cambridge, MA, 37–51.

BATTE, M T, HOOKER, N H, HAAB, T C and BEAVERSON, J (2007), 'Putting their money where their mouths are: Consumer willingness to pay for multi-ingredient, processed organic food products', *Food Policy*, **32** (2), 145–59.

BEARDSELL, M L and DALE, B G (1999), 'The relevance of total quality management in the food supply and distribution industry: A study', *British Food Journal*, **101** (3), 190–200.

BELLETTI, G, BURGASSI, T, MARESCOTTI, A and SCARAMUZZI, S (2007), 'The effects of certification costs on the success of a PDO/PGI', in Theuvsen, L, Spiller, A, Peupert, M and Jahn, G (eds), *Quality Management in Food Chains*, Wageningen Academic Publishers, Wageningen, 107–21.

BENNETT, R M (1996), 'People's willingness to pay for farm animal welfare', *Animal Welfare*, **5** (1), 3–11.

BHUPTANI, M and MORADPOUR, S (2005), *RFID Field Guide: Deploying Radio Frequency Identification Systems*, Prentice Hall, Upper Saddle River, NJ.

BOSTON, C, ONDERSTEIJN, C and GIESEN, G (2004), 'Using stakeholder views to develop strategies for the Dutch pork supply chain', paper presented at *14th Annual IAMA Conference*, Montreux, Switzerland.

DEIMEL, M, FRENTRUP, M and THEUVSEN, L (2008), 'Transparency in food supply chains: Empirical results from German pork and dairy production', *Journal of Chain and Network Science*, **8** (1), 21–32.

DEIMEL, M, PLUMEYER, C-H and THEUVSEN, L (2009), 'Stufenübergreifender Informationsaustausch in der Fleischwirtschaft: Recht und Zertifizierung als Einflussgrößen', *Berichte über Landwirtschaft*, **87** (1), 118–52.

DEN OUDEN, M, DIJKHUIZEN, A A, HUIRNE, R B M and ZUURBIER, P J P (1996), 'Vertical cooperation in agricultural production-marketing chains, with special reference to product differentiation in pork', *Agribusiness*, **12** (3), 277–90.

DOEG, C (2005), *Crisis Management in the Food and Drinks Industry: A Practical Approach*, 2nd edition, Springer, New York.

DOLTSINIS, S (2008), *Regulative Grundlagen der Lebensmittelsicherheit: Ein Abriss der Entwicklung bis zur Gegenwart*, www.vis.bayern.de/ernaehrung/lebensmittelsicherheit/ueberwachung/grundlagen.htm (downloaded: July 21, 2008).

DYER, J H and SINGH, H (1998), 'The relational view: Cooperative strategy and sources of interorganizational competitive advantage', *Academy of Management Review*, **23** (4), 660–79.

ENNEKING, U, OBERSOJER, T, BALLING, R, KRATZMAIR, M and KREITMEIR, A (2007), 'Enhancing the acceptance of quality systems by German farmers: The case of quality management and quality assurance', in Theuvsen, L, Spiller, A, Peupert, M and Jahn, G (eds), *Quality Management in Food Chains*, Wageningen Academic Publishers, Wageningen, 343–54.

FEIGENBAUM, A V (2004), *Total Quality Control*, 4th edition, McGraw Hill, New York.

FRANZ, R, ENNEKING, U and BALLING, R (2007), 'Perceived safety of organic and regional food from a perspective of uncertain consumers', in Theuvsen, L, Spiller, A, Peupert, M and Jahn, G (eds), *Quality Management in Food Chains*, Wageningen Academic Publishers, Wageningen, 255–63.

FURESI, R, MARTINO, G and PULINA, P (2006), 'Contractual choice and food safety strategy: Some empirical findings in Italian poultry sector', in: Fritz, M, Rickert, U, Schiefer, G (eds), *Trust and Risk in Business Networks*, ILB-Press, Bonn, 487–95.

GAWRON, J-C and THEUVSEN, L (2006), 'The International Food Standard: Bureaucratic burden or a helpful management instrument in global markets?' paper presented at 98th *EAAE Seminar*, Chania, Crete.

GAWRON, J-C and THEUVSEN, L (2009a), 'Certification systems in Central and Eastern Europe: A status quo analysis in the agrifood sector,' *British Journal of Food and Nutrition Sciences*, **59** (1), 5–10.

GAWRON, J-C and THEUVSEN, L (2009b), 'Certification schemes in the European agri-food sector: Overview and opportunities for Central and Eastern European countries', *Outlook on Agriculture*, **38** (1), 9–14.

GERLACH, S, SPILLER, A and WOCKEN, C (2007), 'Supplier relationship management in the German dairy industry', in Theuvsen, L, Spiller, A, Peupert, M and Jahn, G (eds), *Quality Management in Food Chains*, Wageningen Academic Publishers, Wageningen, 449–62.

GELLYNCK, X, JANUSZEWSKA, R, VERBEKE, W and VIAENE, J (2007), 'Firms' cost of traceability confronted with consumer requirements' in Theuvsen, L, Spiller, A, Peupert, M and Jahn, G (eds), *Quality Management in Food Chains*, Wageningen Academic Publishers, Wageningen, 45–56.

GOLDSMITH, P D, TURAN, N and GOW, H R (2003), 'Food safety in the meat industry: A regulatory quagmire', *International Food and Agribusiness Management Review*, **6** (1).

GOULD, W A (1992), *Total Quality Management for the Food Industries*, Woodhead Publishing, Abington.

HACKMAN, J R and WAGEMAN, R (1995), 'Total Quality Management: empirical, conceptual and practical issues', *Administrative Science Quarterly*, **40** (2), 309–42.

HARDGRAVE, B C, WALLER, M and MILLER, R (2006), *RFID's Impact on Out Of Stocks: A Sales Velocity Analysis*, University of Arkansas, working paper.

HARRISON, M, FLYNN, A and MARSDEN, T (1997), 'Contested regulatory practice and the implementation of food policy: Exploring the local and national interface', *Transactions of the Institute of British Geographers*, **22** (4), 473–87.

HELLSTEN, U and KLEFSJÖ, B (2000), 'TQM as a management system consisting of values, techniques and tools', *The TQM Magazine*, **12** (4), 238–44.

HENSON, S and TRAILL, B (1993), 'The demand for food safety: Market imperfections and the role of government', *Food Policy*, **18** (2), 152–62.

HEYDER, M and THEUVSEN, L (2008), 'Corporate social responsibility in the agribusiness: A research framework', in Berg, E, Hartmann, M, Heckelei, T, Holm-Müller, K and Schiefer, G (eds), *Risiken in der Agrar- und Ernährungswirtschaft und ihre Bewältigung*, Landwirtschaftsverlag, Münster-Hiltrup, 265–77.

HOFSTEDE, G J (2003), 'Transparency in netchains', in Harnos, Z, Herdon, M, Wiwczaroski, T B (eds), *Information Technology for a Better Agri-food Sector, Environment and Rural Living*, Debrecen University, Debrecen, 17–29.

HOLLMANN-HESPOS, T (2008), *Rückverfolgbarkeitssysteme in der Ernährungswirtschaft: Eine empirische Untersuchung des Investitionsverhaltens deutscher Unternehmen*, Verlag Dr. Kovac, Hamburg.

HORNIBROOK, S and FEARNE, A (2002), 'Vertical co-ordination as a risk management strategy: A case study of a retail supply chain in the UK beef industry', *Journal of Farm Management*, **11** (6), 353–64.

HUBBARD, M R (2003), *Statistical Quality Control for the Food Industry*, 3rd edition, Kluwer Academic/Plenum, New York.

JAHN, G, SCHRAMM, M and SPILLER, A (2005), 'The reliability of certification: Labels as a consumer policy tool', *Journal of Consumer Policy*, **28** (1), 53–73.

JAHN, G and SPILLER, A (2007), 'Controversial positions about the QS System in agriculture: An empirical study', in Theuvsen, L, Spiller, A, Peupert, M and Jahn, G (eds), *Quality Management in Food Chains*, Wageningen Academic Publishers, Wageningen, 355–67.

JANSEN, K and VELLEMA, S (eds) (2004), *Agribusiness and Society: Corporate Responses to Environmentalism, Market Opportunities and Public Regulation*, Zed Books, London.

KAMISKE, G F and MALORNY, C (1992), 'Total Quality Management: Ein bestechendes Führungsmodell mit hohen Anforderungen und großen Chancen', *Zeitschrift Führung und Organisation*, **61** (5), 274–8.

KASTNER, J J and PAWSEY, R K (2002), 'Harmonising sanitary measures and resolving trade disputes through the WTO-SPS framework. Part I: A case study of the US-EU hormone-treated beef dispute', *Food Control*, **13** (1), 49–55.

KOHLI, A K and JAWORSKI, B J (1990), 'Market orientation: The construct, research propositions and managerial implications', *Journal of Marketing*, **54** (2), 1–18.

KRYSTALLIS, A and CHRYSSOHOIDIS, G (2005), 'Consumers' willingness to pay for organic food: Factors that affect it and variation per organic product type', *British Food Journal*, **107** (5), 320–43.

LUNING, P A, MARCELIS, W J and JONGEN, W M F (2002), *Food Quality Management: A Techno-managerial Approach*, Wageningen Pers, Wageningen.

LUTEN, J B, OEHLENSCHLÄGER, J and ÓLAFSDÓTTIR, G (eds) (2003), *Quality of Fish from Catch to Consumer: Labelling, Monitoring and Traceability*, Wageningen Academic Publishers, Wageningen.

MACK, A (2007), *Nutzungskonzept für ein integriertes Audit- und Dokumentenmanagementsystem im überbetrieblichen Gesundheitsmanagement Schweine haltender Betriebe*, PhD thesis, University of Bonn.

MARSDEN, T and WRIGLEY, N (1996), 'Retailing, the food system and the regulatory state', in *Retailing, Consumption and Capital: Towards the New Retail Geography*, Wrigley, N, Lowe, M (eds), Longman, Harlow, 33–47.

MARSDEN, T, FLYNN, A and HARRISON, M (1997), 'Retailing, regulation, and food consumption: The public interest in a privatized world?' *Agribusiness*, **13** (2), 211–26.

MARTIN, J H and GRBAC, B (2003), 'Using supply chain management to leverage a firm's market orientation', *Industrial Marketing Management*, **32** (1), 25–38.

MARTINEZ, S W (2002a), *A Comparison of Vertical Coordination in the US Poultry, Egg and Pork Industry*, USDA, Economic Research Service, Agriculture Information Bulletin No. 747-05, Washington, DC.

MARTINEZ, S W (2002b), *Vertical Coordination of Marketing Systems: Lessons from the Poultry, Egg and Pork Industries*, USDA, Economic Research Service, Agricultural Economic Report No. 807, Washington, DC.

METTKE, T (1979), 'Die Entwicklung des Lebensmittelrechts', *Gewerblicher Rechtsschutz und Urheberrecht*, **81** (12), 817–24.

MEUWISSEN, M P M, VELTHUIS, A G J, HOGEVEEN, H and HUIRNE, R B M (2003), 'Traceability and certification in meat supply chains', *Journal of Agribusiness*, **21** (2), 167–81.

MIGHELL, R L and JONES, L A (1963), *Vertical Coordination in Agriculture*. USDA, Economic Research Service, Agricultural Economic Report No. 19, Washington, DC.

MOHR, J and NEVIN, J R (1990), 'Communication strategies in marketing channels: A theoretical perspective', *Journal of Marketing*, **54** (4), 36–51.

MONTES, J, JOVER A V and FERNÁNDEZ, L M M (2003), 'Factors affecting the relationship between Total Quality Management and organizational performance', *International Journal of Quality and Reliability Management*, **20** (2), 189–209.

MUMMA, G A, ALLEN, A J, WARREN, C C, ABDULLAHI, A and MUGALLA, C (2002), *Analyzing the Perceived Impact of ISO 9000 Standards on US Agribusiness*, paper presented at AEEA Annual Meeting, Long Beach, CA.

NEWSLOW, D L (2001), *The ISO 9000 Quality System: Applications in Food and Technology*, John Wiley, New York.

OPARA, L U (2003), 'Traceability in agriculture and food supply chain: A review of basic concepts, technological implications, and future prospects', *Food, Agriculture and Environment*, **1** (1), 101–6.

PALAZZO, G and SCHERER, A G (2006), 'Corporate legitimacy as deliberation: A communicative framework', *Journal of Business Ethics*, **66** (1), 71–88.

PETERSON, H C, WYSOCKI, A and HARSH, S B (2001), 'Strategic choice along the vertical coordination continuum', *International Food and Agribusiness Management Review*, **4** (2), 149–66.

PEUPERT, M and THEUVSEN, L (2007), 'Improving quality-related communication in food chains with Quality Function Deployment: The dairy industry', in Theuvsen, L, Spiller, A, Peupert, M and Jahn, G (eds), *Quality Management in Food Chains*, Wageningen Academic Publishers, Wageningen, 125–37.

PFEIFER, T (2002), *Quality Management: Strategies, Methods, Techniques*, Hanser, Munich and Vienna.

PORTER, M E (1980), *Competitive Strategy: Techniques for Analyzing Industries and Competitors*, Free Press, New York.

ROE, B, SPORLEDER T L and BELLEVILLE, B (2004), 'Hog producer preferences for marketing contract attributes', *American Journal of Agricultural Economics*, **86** (1), 115–23.

SCHEUPLEIN, R J (1999), 'History of food regulation', in Van der Heijden, K, Younes, M, Fishbein, L and Miller, S (eds), *International Food Safety Handbook: Science, International Regulation and Control*, Marcel Dekker, New York, 647–58.

SCHRAMM, M and SPILLER, A (2003), 'Farm-Audit und Farm-Advisory-System: Ein Beitrag zur Ökonomie von Qualitätssicherungssystemen', *Berichte über Landwirtschaft*, **81** (2), 165–91.

SCHULZE, B, SPILLER, A and THEUVSEN, L (2007), 'A broader view on vertical coordination: Lessons from German pork production', *Journal on Chain and Network Science*, **7** (1), 35–53.

SCHULZE, H, ALBERSMEIER, F, GAWRON, J-C, SPILLER, A and THEUVSEN, L (2008), 'Heterogeneity in the evaluation of quality assurance systems: The International Food Standard (IFS) in European agribusiness', *International Food and Agribusiness Management Review*, **11** (3), 99–139.

SCHULZE ALTHOFF, G, ELLEBRECHT, A and PETERSEN, B (2005), 'Chain quality information management: Development of a reference model for quality information requirements in pork chains', *Journal of Chain and Network Science*, **5** (1), 27–38.

SKURAS D and VAKROU A (2002), 'Consumers' willingness to pay for origin labelled wine: A Greek case study', *British Food Journal*, **104** (11), 898–912.

SMYTH, S and PHILLIPS, P W B (2002), 'Product differentiation alternatives: Identity preservation, segregation, and traceability', *AgBioForum*, **5** (2), 30–42.

SPILLER, A (2004), 'Qualitätssicherung in der Wertschöpfungskette – Vor- und Nachteile unterschiedlicher Organisationskonzepte', in Dachverband Agrarforschung (ed.), *Lebensmittelqualität und Qualitätssicherungssysteme*, DLG-Verlag, Frankfurt/M, 83–96.

STREINZ, R (2007), 'Das neue Lebensmittel und Futtermittelgesetzbuch vor dem Hintergrund des Verbraucherschutzes', in Calliess, C, Härtel, I, Veit, B (eds), *Neue Haftungsrisiken in der Landwirtschaft: Gentechnik, Lebensmittel- und Futtermittelrecht, Umweltschadensrecht*, Nomos, Baden-Baden, 47–77.

SWINNEN, J F M (ed.) (2005), *The Dynamics of Vertical Coordination in Agrifood Chains in Eastern Europe and Centra Asia*, The World Bank, Washington, DC.

THEUVSEN, L (2004), 'Transparency in netchains as an organizational phenomenon: Exploring the role of interdependencies', *Journal on Chain and Network Science*, **4** (2), 125–38.

THEUVSEN, L (2005), 'Quality assurance in the agrofood sector: An organizational-sociological perspective, in Hagedorn, K, Nagel, U J, Odening, M (eds), *Umwelt-*

und Produktqualität im Agrarbereich, Landwirtschaftsverlag, Münster-Hiltrup, 173–81.

THEUVSEN, L and HOLLMANN-HESPOS, T (2007), 'Investments in tracking and tracing systems: An empirical analysis of German food manufacturers', in Parker, C G, Skerratt, S, Park, C and Shields, J (eds), *Environmental and Rural Sustainability through ICT. Proceedings of EFITA/WCCA Conference 2007*, Glasgow.

THEUVSEN, L and PEUPERT, M (2007), 'Quality assurance in agribusiness: Evaluating 'Qualität und Sicherheit' and 'QM Milch' from a Total Quality Management perspective', in Theuvsen, L, Spiller, A, Peupert, M and Jahn, G (eds), *Quality Management in Food Chains*, Wageningen Academic Publishers, Wageningen, 149–61.

THEUVSEN, L and PLUMEYER, C-H (2007), 'Certification schemes, quality-related communication in food supply chains and consequences for IT infrastructures', in Parker, C G, Skerratt, S, Park, C and Shields, J (eds), *Environmental and Rural Sustainability through ICT. Proceedings of EFITA/WCCA Conference 2007*, Glasgow.

THEUVSEN, L and SPILLER, A (2007), 'Perspectives of quality management in modern agribusiness', in Theuvsen, L, Spiller, A, Peupert, M and Jahn, G (eds), *Quality Management in Food Chains*, Wageningen Academic Publishers, Wageningen, 13–19.

THEUVSEN, L, PLUMEYER, C-H and GAWRON, J-C (2007a), 'Certification systems in the meat industry: Overview and consequences for chain-wide communication', *Polish Journal of Food and Nutrition Sciences*, **57** (4 C), 563–9.

THEUVSEN, L, SPILLER, A, PEUPERT, M and JAHN, G (eds) (2007b), *Quality Management in Food Chains*, Wageningen Academic Publishers, Wageningen.

VASCONCELLOS, J A (2003), *Quality Assurance for the Food Industry: A Practical Approach*, CRC Press, Boca Raton, FL.

WALGENBACH, P (2007), 'Façade and means of control: The use of ISO 9000 standards', in Theuvsen, L, Spiller, A, Peupert, M and Jahn, G (eds), *Quality Management in Food Chains*, Wageningen Academic Publishers, Wageningen, 29–42.

WESSELER, G (2006), 'Qualitätssicherung in der Stufe Landwirtschaft – Was kommt auf die Bauern zu?' Paper presented at a workshop on *Qualitätssicherung in der Stufe Landwirtschaft* organized by the German Farmer Association (Deutscher Bauernverband), November 3, 2006, Berlin.

WILLIAMSON, O E (1985), *The Economic Institutions of Capitalism*, Free Press, New York, London.

WILLIAMSON, O E (1991), 'Comparative economic organization: The analysis of discrete structural alternatives', *Administrative Science Quarterly*, **36** (2), 269–96.

ZAIBET, L, and BREDAHL, M (1997), 'Gains from ISO certification in the UK meat sector', *Agribusiness*, **13** (4), 375–84.

Part V

Using technology effectively in food supply chains

17

Role of diagnostic packaging in food supply chain management

D. Hobday, S.P.J. Higson and C. Mena, Cranfield University, UK

Abstract: Interest in the use of intelligent packaging within supply chains has been increasing in recent years. The growth in awareness and continuing research in this area will provide better solutions and recommendations for food suppliers and retailers. This chapter focuses on two key diagnostic packaging technologies: time temperature indicators (TTIs) and freshness quality indicators (FQIs). Future developments in these technologies are discussed together with their potential impact on issues such as stock traceability, inventory management and waste production at all stages of the supply chain. This is followed by a discussion surrounding the collaboration of intelligent packaging with auto identification systems such as radio frequency identification (RFID) and the potential impact that these could have. A brief analysis is included of the benefits and limitations of these technologies and their return on investment.

Key words: intelligent packaging, food labels, supply chain, FQI, TTI, RFID.

17.1 Introduction

Modern-day food packaging is expected to provide containment and protection for food produce whilst being informative, attractive, easy to use and environmentally friendly. Consumer demands are primarily focussed towards food products that are as-fresh and unprocessed. Retailer demands are, by contrast, focused towards cost-effective measures that meet consumer demands whilst extending product shelf lives. These requirements have lead to a rapid development of innovations within the packaging industry. There are two key emergent branches that aim to meet these demands, namely, active packaging and intelligent packaging. Although there are products and technologies already available and in use, the focus of this chapter will be on future developments and the potential convergence of these two branches.

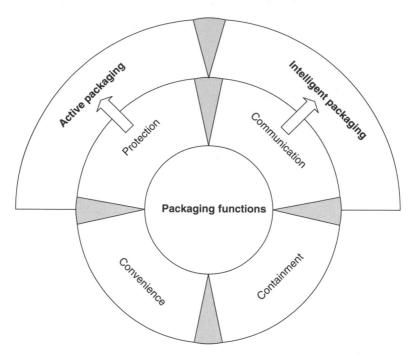

Fig. 17.1 Model of packaging functions (adapted from Yam *et al.*, 2005).

Figure 17.1 presents an overview of active and intelligent packaging and the areas that they can potentially improve. The diagram shows the previously mentioned four key functions of packaging. The terms active and intelligent packaging are situated above the roles that they aim to enhance. For example, active packaging provides a means of improving the protection of the contents, whereas intelligent packaging enhances the communication of data to the user. It is important to note that in some cases these categories overlap since some active packaging systems also include improvements based on user convenience, such as microwave susceptors (Yam *et al.*, 2005) which are, for example, used to allow better cooking performance in microwave foodstuffs. In the diagram, overlaps between packaging features are represented by the areas in grey.

The focus of this chapter is towards food packaging that facilitates improved communication with the user, whether the user is a warehouse computer system or a food consumer. The information that is to be communicated will be much more than just the sell-by date, the product brand or the ingredients. With future developments in packaging technology, it is hoped that the user will be able to determine quantitatively to what extent degradation of the food has occurred and how much time is left before the food becomes spoiled.

Table 17.1 Examples of active packaging applications for use within the food industry (from Kerry *et al.*, 2006)

Active property	Constituents
Absorbing/scavenging properties	Oxygen, carbon dioxide, moisture, ethylene, flavours, taints, UV light
Releasing/emitting properties	Ethanol, carbon dioxide, antioxidants, preservatives, sulphur dioxide, flavours, pesticides
Removing properties	Catalysing food components removal: lactose, cholesterol
Temperature control	Insulating materials, self-heating and self-cooling packaging, microwave susceptors and modifiers, temperature-sensitive packaging
Microbial and quality control	UV and surface treated packaging materials

17.1.1 Active packaging

Active packaging is defined as 'packaging in which subsidiary constituents have been deliberately included in or on either the packaging material or the packaging headspace to enhance the performance of the packaging system' (Robertson, 2006). The interaction of the active features can be through a chemical (modified atmosphere) or biological (antimicrobial agents) interface to provide an extended shelf life or an addition to the packaging that enhances its performance.

Examples of active technologies used in packaging are shown in Table 17.1. As previously mentioned, the underlying function of these technologies is to protect and increase the longevity of the product within the packaging. The examples given in the table act to preserve and protect the food product so that it is able to maintain the desired flavour and a customary appearance through delaying or hindering bacterial spoilage. This is achieved by modifying the atmospheric conditions of the packaging or by changing the surface of the packaging.

Although these technologies are not the focus of this chapter it is important to acknowledge their significance as they are often used in conjunction with intelligent packaging.

17.1.2 Intelligent packaging

Smart or intelligent packaging is a widely used term that often covers many different branches of technology and packaging design. Although there is no formal academic definition for the terminology 'smart/intelligent packaging', many agree that it can be defined as any packaging that goes beyond the use of simple materials in conjunction with printed barcodes or labels

(Kerry and Butler, 2008). The term 'intelligent packaging' is often used to describe improvements in existing materials or methods to extend shelf life by preventing microbial growth (Coma, 2008; Sivertsvik *et al.*, 2002). Intelligent packaging is also used to illustrate additional design features to packaging that are convenient and that may enhance the usability of a product.

The less stringent definition of intelligent packaging allows for a greater scope of technologies and products. Table 17.2 below summarises the main ideas considered in this topic of research and proposes potential or available technologies that could be used.

The focus for this chapter will be on the development of intelligent packaging systems that can be used alongside food products to facilitate better stock management and safety within the food industry. The key driver for this technology is an improvement in food safety and food quality assurance (Ahvenainen, 2003). The improvement for better food safety directly benefits the end-consumer, however, with this technology there may also be direct benefits for food retailers and suppliers.

Sustainability of food supply chains is a topic which is generating plenty of political interest and is another driving force for the development of this technology. High levels of waste can occur within the food industry, most of which is sent to landfill (C-Tech Innovation Ltd, 2004). Guidelines issued by government and environmental departments suggest that there needs to be long-term alternatives to landfill through the reduction of waste production (DEFRA, 2007). This could potentially drive optimal technologies that could be used in supply chains to determine the cause of high wastage levels. Consumers are adding to this pressure as they are becoming more environmentally aware as well as food 'savvy' when it comes to selecting products based on traceability and health benefits. The ultimate aim is to produce technologies that could communicate and assure consumers about issues of traceability and safety as well as reducing waste occurrence in supply chains.

For sealed packed produce, the only source of quality assurance at present is the packaging material and its integrity. This is communicated via

Table 17.2 Examples of intelligent packaging applications for use within the food industry (from Kerry *et al.*, 2006)

Area of research	Application
Tamper evidence and pack integrity	Breach of pack containment
Indicators of product safety/quality	Time temperature indicators (TTI), gas sensing devices, microbial growth, pathogen detectors
Traceability/anti-theft devices	Radio frequency identification (RFID) labels, tags, chips
Product authenticity	Holographic images, logos, hidden design print elements, RFID

the branding and the information available on the label, including a sell-by date, the source of contents and location of production. With the consumer trend for fresher, less preserved foods of high quality, a potential market is emerging for an intelligent packaging that can identify food spoilage or deterioration of quality without being invasive, destructive or expensive. For the purposes of this chapter, the focus will be on the interactivity of the packaging with the consumer and the retailer and how reliable the information provided is in terms of food quality and safety, in other words, producing packaging that conveys data to the user that is both easily comprehensive and accurate.

17.2   Importance of intelligent packaging

This section will highlight the key areas in which this technology could reduce costs in the future and discuss the problems that it can help to overcome. It is anticipated that intelligent packaging will play an important role in the following areas:

- food production and supply industry
- food retailing
- consumers' food awareness.

These areas share similar problems that intelligent packaging can help to solve or minimise (Hurme *et al.*, 2002). The first of these problems is the wastage of food. Intelligent packaging can aid the reduction of food waste via improved communication of food degradation to consumers and could potentially reduce the amount of food waste from poor food production and supply chain systems. In a similar vein is the issue of stock management, traceability and information flow within the food industry. Above all, consumer concerns about safety and quality must be considered, as well as overcoming confusion caused by over reliance on date codes (i.e. use-by and best-before dates).

In recent times, food wastage has become an important environmental, economical and political focal point for research and debate (BBC, 2008; Hogg *et al.*, 2007; Ventour, 2007). WRAP (Waste and Resources Action Programme) of the UK recently reported that as much as a third of food bought by consumers is thrown away and placed into landfill (Ventour, 2007). The food, drink and tobacco industry is responsible for 11% (7.5 megatonnes) of the total industrial and commercial waste in the UK with the retail and wholesale sector being responsible for a further 19% (12.9 megatonnes) (DEFRA, 2007). There are three key stages throughout the food supply chain which can be envisaged as a supply pipeline. Figure 17.2 shows the simplified stages of food production, supply and consumption together with the waste produced at each stage. It is important to note that some waste will be unavoidable. There are many examples in this context

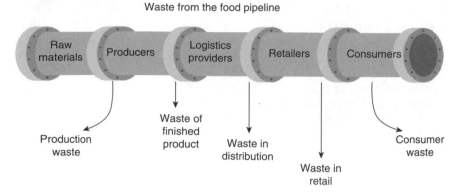

Fig. 17.2   Areas of waste production.

such as, for example, banana skins from a food production facility manu-facturing banana-containing goods.

A significant driver for the food industry to uptake this technology is the increase in production efficiency which in turn would aid in the reduction of production costs. As mentioned previously, governmental pressures on issues surrounding waste production and disposal are growing. Some types of wastage could be better understood if there was a more accurate means of determining their source or sources from a food supply chain. In terms of food traceability, packaging with sensors that follow a product from 'farm to fork' would provide key information about the producer and the route through the supply chain to the eventual sale to a consumer (Yam *et al.*, 2005). In a situation where a product needs to be recalled, traceability would mean that less drastic measures would need to be taken to purge the supply chain. For example, if there was an emergency food recall caused by a contaminant from one factory, the information obtained from an intelli-gent label would prevent the waste of a whole product line and would affect only the product which had passed through that factory. Another advantage mentioned previously would be enhanced consumer confidence in knowing where the item had come from if the label was able to display such information.

Another important driver that needs to be considered for the introduc-tion of intelligent packaging are the requirements of the consumer. A packaging technology that could prevent consumers buying and eating unsafe or poor quality produce would show obvious benefits. Research has shown, however, that there are several potential barriers to overcome when introducing new packaging technology (Hurme and Ahvenainen, 1996). Table 17.3 shows the potential barriers and challenges that need to be overcome when introducing packaging technology to be used by the general public.

Table 17.3 Problems and solutions encountered when introducing new products using active and/or intelligent packaging techniques (adapted from Hurme and Ahvenainen, 1996)

Problems	Solutions
Consumer attitude	Consumer research: education and information
Doubts over the performance	Storage tests before launching; consumer education and information
Increased packaging cost	Use in selected, high quality products; marketing tool for increased quality and QA
False sense of security, ignorance of date markings	Consumer education and information
Mishandling and abuse	Active compound/sensor incorporated into label or packaging film; consumer education and information
False complaints and returns of packs with indicators	Indicator automatically readable at the point of purchase
Difficulty of checking every indicator at point of purchase	Bar code labels: intended for QA for retailers only; RFID system within stores

These solutions would require substantial investment before any real returning benefit is seen by the consumer, retailer or food producer (Han *et al.*, 2005). The function of this technology for industrial use provides more opportunity and benefits in the short term. Using an indicator or sensor for monitoring food spoilage would require a design that integrates easily into existing supply chains and that can also interpret changes and potential hazards which the foodstuff is exposed to. The information that is obtained from the indicators has to be simple and accurate so that the data can be used to make key decisions. A sudden change in temperature in a cold chain would require a decision to be made about the safety of the food as well as an assessment of the scale of the problem.

The role of the intelligent label is to respond to changes in the external environment. The change could be a simple fluctuation in temperature along the cold chain or an increase in volume of a product owing to seasonality. Considering the example of temperature abuse, a time temperature indicator (TTI) (Taoukis and Labuza, 2003, Selman, 1995) label could be attached to a temperature-sensitive product, where a change in temperature affects the state of the TTI label. The change displayed by the TTI could then be fed into a data processing unit that could model and estimate the best business recommendations for how to manage the packaged food through the supply chain (Yam *et al.*, 2005). Figure 17.3 shows the potential feedback loop of a TTI where once a decision has been made from the data gathered from the labels and an action has been taken in real time, the data can be matched to an appropriate model.

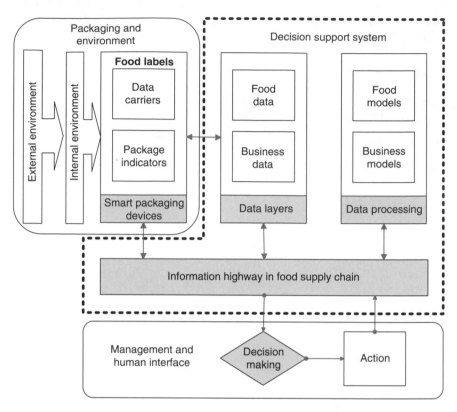

Fig. 17.3   Possible information flow diagram of a supply chain augmented with a FQI intelligent packaging system (adapted from Yam *et al.*, 2005).

Although the model is simplistic, it shows the data flow from the external environment, through the labelling technology and into a management decision. In this situation an indicator, such as a TTI, could highlight areas in supply chains which are not maintained or are inefficient at keeping food items chilled. As the data is recorded in real time, companies would get warning of failure or problems and be able to cope better with managing such a problem. Another use for TTIs has been to provide evidence in insurance claims caused by logistical negligence (Tsironi *et al.*, 2008). Transit vehicles that have failed to provide adequate temperature control for chilled products have been shown to be at fault from the data from these sensors.

This model is improved if measuring changes to the external environment is not relied upon solely. If we were able to measure directly the food and the impact of a change in external environment within the packaging, then a more robust food model could be generated and a more reliable business recommendation could be made on what decision to take. A food quality indicator (FQI) could be used in this case to measure direct food changes and spoilage in order to ascertain actual bacterial changes in the food. Figure 17.3 also shows the extra input of information into the food label. A change in temperature could be recorded (external factor) and then the spoilage of the food could also be monitored (internal factor).

The potential application of this technology, if used and harnessed correctly, is to provide a real time flow of information to food producers, retailers and consumers (Kerry et al., 2006). There are many examples of these technologies that are in current use (Hurme and Ahvenainen, 1996; Han et al., 2005; Pacquit et al., 2008). In the literature there are also countless examples of sensors and indicators that are being developed (Adhikari and Majumdar, 2004; Butler, 2001; Connolly, 2007). The next section will examine the impact of TTIs and FQIs as intelligent food labels and compare the two technologies.

17.3 Diagnostic packaging: time temperature indicators (TTIs) versus freshness quality indicators (FQIs) and new technologies

Indicators can be defined as devices or substances that inform the user about the presence, absence or concentration of another substance. They can also indicate the extent of a reaction between substances by means of a distinctive change, such as colour. They are said to differ from sensors as they do not require a receptor or transducer and can impart information through a direct change or reaction (Kerry and Butler, 2008). The subtle difference between TTIs and FQIs is what each one interacts with. A thorough overview of TTIs is presented elsewhere (Taoukis and Labuza, 2003) and several authors have discussed their reliability and function (Selman, 1995; Taoukis et al., 1991; Singh and Wells, 1985). There are also several examples of FQIs in the literature (Kerry and Butler, 2008; Hurme et al., 2002; Kerry et al., 2006; Pacquit et al., 2008; Smolander, 2003). Figure 17.4 shows the distinction between TTIs and FQIs when used as intelligent labels. TTIs rely on modelling and predictive behaviour of microbial growth when temperature abuse occurs. FQIs on the other hand, interact with the metabolites caused by microbial growth and do not require microbial growth models.

Here, the focus is on freshness quality indicators (FQI) and time temperature indicators (TTI) as methods for determining food spoilage and estimating shelf lives of differing products. As mentioned previously, FQIs provide direct food quality information by reacting to changes taking place within the foodstuff. These changes can be due either to the external or the internal environment. Internally, these changes are caused by chemical degradation or microbiological activity and metabolism as the food

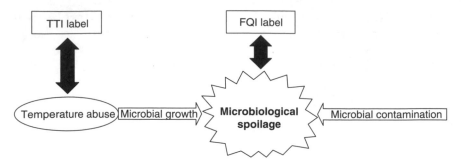

Fig. 17.4 Comparison of the information obtained by FQI and TTI labels (adapted from Smolander, 2003).

perishes. TTIs on the other hand monitor the temperature and time lapse over a predetermined amount of time so that they are able to estimate how degraded food items changed due to an external environment variation, which in this case would be temperature. For example, if a sensor is set to measure an item that is to be stored for 5 days at 7°C, then it would indicate when 5 days had elapsed. If this item had been subjected to a higher temperature then the indicator would behave differently and would react quicker to inform the user that the food has degraded sooner.

Temperature is deemed to be the most important external factor controlling food spoilage. Storage temperature has a direct influence on the kinetics of the chemical and biological changes that occur in food products. Currently there are three types of commercially available TTIs which are critical temperature indicators, partial history indicators and full history indicators (Singh, 2000). There has been extensive review of these indicators throughout the literature (Taoukis and Labuza, 2003; Selman, 1995; Taoukis *et al.*, 1991; Singh and Wells, 1985; Claeys *et al.*, 2002; Smolander *et al.*, 2004). Although the technology has been available for over 20 years, the rate of adoption or implementation has been very slow. This has been a result of the high cost of the technology and the lack of supportive information technology that will allow appropriate utilisation (Yam *et al.*, 2005). The information that is obtained by these sensors is also deemed to be limited, especially in the context of cheaper FQI technologies becoming available (Kerry and Butler, 2008).

Most FQIs respond to changes in the gaseous headspace of the packaging. Several techniques exist which correlate changes in certain gases to microbial growth or chemical spoilage. This technology is usually concerned with items such as meat and fish which give rise to distinctive aromas once they spoil or become unsafe to eat. There are also examples of this technology in use with other items such as fruit and vegetables, in addition to materials that are sensitive to changes in concentrations of oxygen and carbon dioxide (Ahvenainen, 2003). Mostly they are concerned

Table 17.4 Comparisons of the advantages and disadvantages of TTI and FQI technology (adapted from Pacquit *et al.*, 2008)

Time temperature indicators		Food quality indicators	
Advantages	Disadvantages	Advantages	Disadvantages
Easily read	No relation to actual food spoilage	Easily read	Not yet commercially available
Accepted QA and adaptable		Direct indication of quality	
	No concern for food safety		Affixed inside packs
Commercially available		Potential cheap unit cost	Potential limitation of materials, i.e. toxicity
Cheap unit cost	Storage and activation issues	Low developmental costs	
Science understood			No concern for food safety
Maturing technology	Not suitable for rapid spoilage products		Difficulty in applying to all food products

with food that has an extremely short shelf life. Table 17.4 summarises some of the points that have been raised by previous research (Hurme and Ahvenainen, 1996).

The main issue with both technologies is the limitated concern over food safety. In both cases the indicators are there to measure food spoilage which is followed as the food degrades and spoilage bacteria populations increase to unsafe levels. This does not take account of the situation where a small number of pathogenic bacteria are sourced onto the foodstuffs (e.g. *Salmonella* or E. coli; Hurme *et al.*, 2002). The presence of a very small count of these bacteria makes the food unsafe. A TTI is concerned only with the external environment and so it has no method of detecting the bacteria, whereas a FQI may not have the sensitivity or ability to detect such small populations especially if produced commercially on a small budget. Another potential pitfall of these technologies is their sensitivity to spoilage. If the lower limit of detection of spoilage is too high, then users may already be able to determine if the food is spoiled by visually checking it. On the other hand if the limit is too low, then food that is edible would be deemed by the sensor to be unfit for consumption. An advantage is that TTIs and FQIs allow certain levels of tailoring for the levels of detection as there are different standards of spoiled food throughout the world (Singh, 2000).

Another important aspect is the contrast between the kind of produce the user is labelling. The technology in a TTI label functions primarily with foodstuffs that have to be kept at a low and constant temperature. It is only when there are deviations from this temperature that the indicator behaves differently to one that has been attached to a correctly stored item. There is also the issue that the reliability of the indicator increases with the amount of time it is set to measure. This limits the use of TTIs to mainly chilled or frozen long-life goods (Singh and Wells, 1985; Riva *et al.*, 2001).

FQIs solely rely upon the change in headspace gas of the food. The data that they collect relies heavily on the bacterial growth and the metabolites that are produced. The gas has to be of the correct type and concentration to react effectively with the sensing element. This limits the deployment of this kind of food monitoring technology to aromatic and perishable foodstuffs. This is the main reason that research in this area focuses on meat, fish and other short-shelf life food (Kerry *et al.*, 2006; Pacquit *et al.*, 2008).

As previously stated, one of the main barriers preventing the integration of TTIs into supply chains over the last 20 years has been the cost and the requirement for a data and information handling system. Over recent years, production costs of TTIs and FQIs have been reduced and now it is almost unheard of to not have a data base or computing network setup for supply chain management purposes. These two factors have enabled a route to commercialisation for these products. There are still however intrinsic problems that exist when introducing and producing new products and services like these, together with attributing costs and reliance to the present system. A dramatic shift from one labelling type to another may be required in order to provide a pathway for either technology.

17.4   The use of radio frequency identification tags (RFID) in future supply chains

RFID (radio frequency identification tags) has long been heralded as 'the next big thing' in supply chain management and as the solution to a lot of inventory management problems. In this section, a consideration of a system that has benefited from RFID technology will be discussed. Comparisons between this case study and the food industry will then be drawn to demonstrate the potential impact of RFID technology in collaboration with smart packaging technology. This chapter will not discuss the working of or technical aspects of RFID in detail. An excellent overview has been written elsewhere which explains the workings and limitations of the technology (Clarke, 2008) and the reader is also referred to Chapter 21 of this book by Katerina Pramatari and co-authors.

RFID permits the transfer of electronic data and is therefore classified as a separate intelligent device and does not fall into either the sensor or indicator categories. The concept is that tags are attached to items (ranging from cattle, containers, pallets, individual packets etc) to give the user a real-time collection of data. This data is transmitted to an information system and allows analysis and tracking of the object to which the tag is attached. For some, RFID technology is seen as the natural evolution of the barcode in that it gives objects identification as well as a potential array of other information.

A case study of Marks and Spencer (Stafford, 2008) shows how RFID technology can be transferred to a clothing department. The original technology was used in the food supply chain as a measure of volume and inventory management. In the case study, intelligent labels were attached to individual items of clothing that were then tracked and registered during their journey through the supply chain. When transferring the technology from food to clothing, some of the key benefits were deemed to be:

- improved store service
- product visibility
- inventory accuracy
- improved processes.

In this case the tags were only used as a stock control and were unable to emit signals without the correct interrogating signal. This meant that they were able to avoid the issue of breaching customer privacy. An intelligent label used as a food monitoring device would need to be able to communicate with the user at any point to highlight any problems. This is a potential hurdle in light of the recent bad press that RFID has received for being a technology that intrudes into consumer privacy. Another problem is the cost of an integrated label which would require adequate power to monitor food and signal any problems if necessary. This would require a reliable power source and could place the cost of the technology out of the range of potential users.

It is often said that retail is detail. RFID technology in the first instance acts as a descriptor of what it is attached to. Consider a plastic tray in a food depot containing packaged portions of chicken breast; a written tag on the side of the tray would be able to confirm what the contents were and a few more key pieces of information (weight, date and place of origin etc) which would then be manually entered onto a stock management system. A barcode could provide a method of relaying this information, and maybe more, to a stock management system. If a real time device, such as an RFID tag with an attached FQI, were to be attached to the tray, much more information could be ascertained. This information could include more data on the source of the meat, the route taken so far by the tray through the supply chain pipeline, the predicted remaining shelf life of the meat and so forth.

For many, the idea of an RFID tag attached to an individual item, such as a single piece of chicken, for the purposes of stock management in the food industry, would be complex and expensive. If the tag were to provide more information about the remaining shelf life and possible contamination, the cost and information trade off would be better balanced. Figure

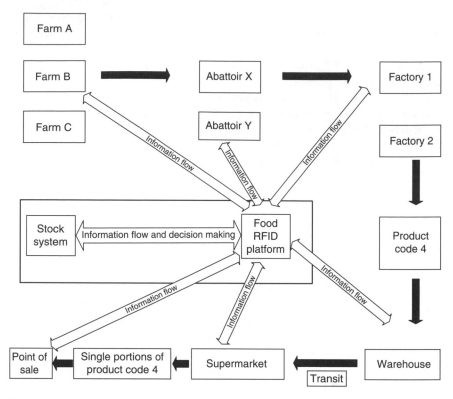

Fig. 17.5 Simplified map of a supply chain and how an intelligent tag could be used for stock control and product monitoring (adapted from Stafford, 2008).

17.5 shows a simple model of a supply chain using RFID technology in conjunction with a FQI tag. The route that the RFID chip has taken and its location is updated upon entry and exit at every stage of the diagram. A remote reading could be taken of the state of the foodstuff to which it is attached.

Following this simplified diagram, a generic meat product (here named 'product code 4') is followed from producer to point of sale. The emphasis here is on the different routes the product could take through a supply chain and the breakdown of original produce to the final finished product. The wealth of information that could be obtained from a system like this would help to find weakness in efficiency throughout the chain as well as helping manage inventory levels and supply. In the case of Marks and Spencer, there is already an RFID framework set up to work with food inventory management. Adding a sensor to estimate product safety and condition to this existing framework, would enhance the ability for managers to make stock management decisions.

17.5   Conclusions

Improvements in food packaging could potentially solve some key problems for retailers and food suppliers. These include understanding the process efficiency of food production, as well as monitoring high volumes of produce as it passes through from production to sale. Issues that have been raised about food waste in the current economic and environmental climate have highlighted potential uses for technologies that can have an impact on food safety and food wastage. Over the past 20 years, supply chain management approaches to inventory management have evolved to allow integration of information technologies. This has emphasised the importance of information sharing and keeping low inventories.

This chapter has only highlighted two potential products from the growing field of smart packaging. The breadth of research is providing substantial improvements and has already seen commercialisation of a number of products. The main hurdle is to provide a cost effective yet reliable sensor so that the benefits and the value of the information retrieved obtained far outweighs the initial cost of implementation.

One of the main reasons why sensors have not been more heavily promoted is due to the lack of cross collaboration on the technology functions that are presently available. RFID has been successfully rolled out in some form into various supply chain situations. A less costly route to sensor integration might include using an RFID network to display the data that the sensor acquires. Therefore the strategy of any food sensor company should be to work with an existing system to reduce the cost of initial implementation.

17.6   References

ADHIKARI, B. and MAJUMDAR, S. (2004), 'Polymers in sensor applications', *Progress in Polymer Science,* **29**(7), 699–766.

AHVENAINEN, R. (2003), 'Active and intelligent packaging: an introduction', in *Novel Food Packaging Technologies,* Ahvenainen, R. (ed.), 1st edition, Woodhead Publishing, Cambridge, UK, 5–21.

BBC (2008), *Stop wasting food, Brown urging,* available at: http://news.bbc.co.uk/1/hi/uk_politics/7492573.stm (accessed 07/07).

BUTLER, P. (2001), 'Smart packaging – intelligent packaging for food, beverages, pharmaceuticals and household products', *Materials World,* **9**(3), 11–13.

CLAEYS W. L., VAN LOEY, A. M. and HENDRICKX, M. E. (2002), 'Intrinsic time temperature integrators for heat treatment of milk', *Trends in Food Science and Technology,* **13**(9–10), 293–311.

CLARKE, R. (2008), 'The influence of product and packaging characteristics on passive RFID readability', in *Smart Packaging Technologies for Fast Moving Consumer Goods,* Kerry, J. P. and Butler, P. (eds), 1st edition, John Wiley and Sons, Chichester, UK, 167–95.

COMA, V. (2008), 'Bioactive packaging technologies for extended shelf life of meat-based products', *Meat Science,* **78**(1–2), 90–103.

CONNOLLY, C. (2007), 'Sensor trends in processing and packaging of foods and pharmaceuticals', *Sensor Review,* **27**(2), 103–8.

C-TECH INNOVATION LTD (2004), *United Kingdom Food and Drink Processing – Mass Balance,* Biffaward Programme on Sustainable Resource Use, UK.

DEFRA, DEPARTMENT FOR ENVIRONMENT, FOOD AND RURAL AFFAIRS (2007), *Waste Strategy for England 2007,* cm 7086, DEFRA, London.

HAN, J. H., HO, C. H. L. and RODRIGUES, E. T. (2005), 'Intelligent packaging', in *Innovations in Food Packaging,* Jung H. Han (ed.), Academic Press, London, pp. 138–55.

HOGG, D., BARTH, J., SCHELISS, K. and FAVOINO, E. (2007), *Dealing with Food Waste in the UK,* Eunomia Research and Consulting, London.

HURME, E. and AHVENAINEN, R. (1996), 'Active and smart packaging of ready made foods', in *Minimal Processing and Ready Made Foods,* Ohlsson, T., Ahvenainen, R. and Mattila-Sandholm, T. (eds), 1st edition, Goteburg, SIK, 169–82.

HURME, E., SIPILAINEN-MALM, T. and AHVENAINEN, R. (2002), 'Active and intelligent packaging', in *Minimal Processing Technologies in the Food Industry,* Ohlsson, T. and Bengtsson, N. (eds), 1st edition, Woodhead Publishing, Cambridge UK, 87–123.

KERRY, J. P. and BUTLER, P. (eds) (2008), *Smart Packaging Technologies for Fast Moving Consumer Goods,* 1st edition, John Wiley & Sons, Chichester UK.

KERRY, J. P., O'GRADY, M. N. and HOGAN, S. A. (2006), 'Past, current and potential utilisation of active and intelligent packaging systems for meat and muscle-based products: A review', *Meat Science,* **74**(1), 113–30.

PACQUIT, A., CROWLEY, K. and DIAMOND, D. (2008), 'Smart packaging technologies for fish and seafood products', in *Smart Packaging Technologies for Fast Moving Consumer Goods,* Kerry, J. and Butler, P. (eds), 1st edition, John Wiley & Sons ltd., Chichester UK, 75–98.

RIVA, M., PIERGIOVANNI, L. and SCHIRALDI, A. (2001), 'Performances of time-temperature indicators in the study of temperature exposure of packaged fresh foods', *Packaging Technology and Science,* **14**(1), 1–9.

ROBERTSON, G. L. (2006), *Food Packaging – Principles and Practice,* 2nd edition, CRC Press, Boca Raton, FL, USA.

SELMAN, J. D. (1995), 'Time temperature indicators', in *Active Food Packaging,* Blackie Academic and Professional, Rooney, M. L. (ed.), New York, 215–37.

SINGH, R. P. (2000), 'Scientific principles of shelf life evaluation', in *Shelf-Life Evaluation of Food,* Man, D. and Jones, A. (eds), 2nd edition, Aspen Publishers, Gaitherburg, MD, 3–22.

SINGH, R. P. and WELLS, J. H. (1985), 'The use of time-temperature indicators to monitor quality of frozen hamburger', *Journal of American Food Technology,* **39**(12), 42–50.

SIVERTSVIK, M., ROSNES, J. T. and BERGSLINEN, H. (2002), 'Modified atmosphere packaging', in *Minimal Processing Technologies in the Food Industry,* Ohlsson, T. and Bengtsson, N. (eds), 1st edition, Woodhead Publishing, Cambridge, UK, 61–86.

SMOLANDER, M. (2003), 'The use of freshness indicators in packaging', in *Novel food packaging Techniques,* Ahvenainen, R. (ed.), 1st edition, Woodhead Publishing, Cambridge UK, 127–43.

SMOLANDER, M., ALAKOMI, H., RITVANEN, T., VAINIONPÄÄ, J. and AHVENAINEN, R. (2004), 'Monitoring of the quality of modified atmosphere packaged broiler chicken cuts stored in different temperature conditions. A. Time–temperature indicators as quality-indicating tools', *Food Control,* **15**(3), 217–29.

STAFFORD, J. (2008), 'How Marks & Spencer is using RFID to improve customer service and business efficiency: A case study', in *Smart Packaging Technologies for Fast Moving Consumer Goods,* Kerry, J. P. and Butler, P. (eds.), 1st edition, John Wiley and Sons, Chichester, UK, 167–210.

TAOUKIS, P. S. and LABUZA, T. P. (2003), 'Time–temperature indicators', in *Novel Food Packaging Techniques*, Ahvenainen, R. (ed.), 1st edition, Woodhead Publishing, Cambridge UK, 103–26.

TAOUKIS, P. S., FU, B. and LABUZA, T. P. (1991), 'Time-temperature indicators', *Journal of American Food Technology*, **45**(10), 70–82.

TSIRONI, T., GOGOU, E., VELLIOU, E. and TAOUKIS, P. S. (2008), 'Application and validation of the TTI based chill chain management system SMAS (Safety Monitoring and Assurance System) on shelf life optimization of vacuum packed chilled tuna', *International Journal of Food Microbiology*, 128(1), 108–115.

VENTOUR, L. (2007), *Understanding Food Waste*, WRAP, Banbury, UK.

YAM, K. L., TAKHISTOV, P. T. and MILTZ, J. (2005), 'Intelligent packaging: Concepts and applications', *Journal of Food Science*, **70**(1), R1–R10.

18

Advances in the cold chain to improve food safety, food quality and the food supply chain

S. J. James and C. James, Food Refrigeration and Process Engineering Research Centre (FRPERC), UK

Abstract: This chapter begins by discussing the importance of maintaining the cold chain for the microbiological safety and quality of foods. The specific stages of the cold chain (chilling, freezing, storage, transport and retail display) are then discussed in detail. Environmental issues and advances in energy reduction are also discussed. Finally, a section is included on the practical and theoretical factors that need to be considered by food producers/suppliers when specifying refrigeration systems for the food supply chain.

Key words: chilling, cold chain, energy, freezing, refrigerated distribution, refrigerated retail display.

18.1 Introduction: the cold chain

Refrigeration stops or reduces the rate at which changes occur in food. These changes can be microbiological (i.e. growth of microorganisms), physiological (e.g. ripening, senescence and respiration), biochemical (e.g. browning reactions, lipid oxidation and pigment degradation) and/or physical (such as moisture loss). An efficient and effective cold chain is designed to provide the best conditions for slowing, or preventing, these changes for as long as is practical. Effective refrigeration produces safe food with a long high-quality shelf life.

To provide safe food products of high organoleptic quality, attention must be paid to every aspect of the cold chain from initial chilling or freezing of the raw ingredients, through storage and transport, to retail display. The cold chain consists of two distinct types of operation. In processes such as primary and secondary chilling or freezing, the aim is to change the average temperature of the food. In others, such as chilled or frozen storage, transport and retail display, the prime aim is to maintain the temperature of the food. Removing the required amount of heat from a food is a difficult, time and energy consuming operation, but critical to the operation of the cold chain. As a food moves along the cold chain it becomes increasingly difficult to control and maintain its temperature. This is because the temperatures of bulk packs of refrigerated product in large storerooms are far less sensitive to small heat inputs than single consumer packs in open display cases or in a domestic refrigerator/freezer. Failure to understand the needs of each process results in excessive weight loss, higher energy use, reduced shelf life or a deterioration in product quality.

18.2 Effect of refrigeration on microbiological safety and shelf life

Temperature is one of the major factors affecting microbiological growth. Microbiological growth is described in terms of the lag phase and the generation time. When a microorganism is introduced to a particular environment there is a time (the lag phase) in which no increase in numbers is apparent, followed by a period when growth occurs. The generation time is a measure of the rate of growth in the latter stage. Microorganisms have an optimum growth temperature at which a particular strain grows most rapidly, that is the lag phase and generation time are both at a minimum. They also have a maximum growth temperature above which growth no longer occurs. Above this temperature, one or more of the enzymes essential for growth are inactivated and the cell is considered to be heat injured. However, in general, unless the temperature is raised to a point substantially above the maximum growth temperature then the injury is not lethal and growth will recommence as the temperature is reduced. Attaining temperatures substantially above the maximum growth temperature are therefore critical during cooking and reheating operations.

Of most concern during storage, distribution and retail display of food is a third temperature, the minimum growth temperature for a microorganism. As the temperature of an organism is reduced below that for optimum growth, the lag phase and generation time both increase. The minimum growth temperature can be considered to be the highest temperature at which either of the growth criteria, the lag phase and the generation time, becomes infinitely long. The minimum growth temperature is not only a function of the particular organism but also the type of food or growth media that is used for the incubation. Although some pathogens can grow at 0°C, or even slightly lower (Table 18.1), from a practical point of view the risks to food safety are considerably reduced if food is maintained below 5°C.

Table 18.1 Generally accepted minimum growth temperatures for
pathogenic bacteria

	Minimum temperature (°C)
Campylobacter spp.	30
Clostridia perfringens	12
Staphylococcus aureus	7
Pathogenic *Escherichia coli* strains	7–6
Salmonella spp.	5–4
Bacillus cereus	5
Clostridium botulinum non-proteolytic	3
Listeria monocytogenes	0
Yersinia enterocolitica	−2

There are little data on the impact of the initial freezing process on the safety of foods. However, it is difficult to envisage any sensible freezing process that would result in most foods being held for substantial periods at temperatures that would support a dangerous growth of pathogens. Providing the food does not rise above −12 °C during frozen storage and display, there are no issues of food safety with frozen storage and display.

Food may also become microbiologically unacceptable as a result of the growth of spoilage microorganisms. Their growth can produce unacceptable changes in the sensory quality of many foods and their rate of growth is also very temperature dependent. The development of off odours is usually the first sign of putrefaction and in meat it occurs when bacterial levels reach approximately 10^7 cm^{-2} of surface area (Ingram, 1972). When bacterial levels have increased a further ten-fold slime begins to appear on the surface and meat received in this condition is usually condemned out of hand. At 0°C beef with average initial contamination levels can be kept for at least 15 days before any off odours can be detected. Every 5°C rise in the storage temperature above 0°C will approximately halve the storage time that can be achieved.

18.3   Effect of refrigeration on quality

Microbial safety and spoilage are not the only aspects of food quality that are temperature dependent. The rate of loss of vitamins from fruit and vegetables during storage also depends upon the storage temperature. It is of interest to note that it is not always a case of the lower the better, especially for citrus and tropical fruits. The optimum temperature for oranges is approximately 12 °C with the rate of vitamin loss increasing at temperatures warmer or colder than this value.

Some foods exhibit particular quality advantages as a result of rapid cooling. In meat the pH starts to fall immediately after slaughter and protein denaturation begins. The result of this denaturation is a pink pro-teinaceous fluid, commonly called 'drip', often seen in prepackaged meat joints. The rate of denaturation is directly related to temperature and it therefore follows that the faster the chilling rate the less the drip. Investiga-tions using pork and beef muscles have shown that rapid rates of chilling can halve the amount of drip loss (Taylor, 1972). Fish passing through rigor mortis above 17°C are to a great extent unusable because the fillets shrink and become tough (Morrison, 1993). A relatively short delay of an hour or two before chilling can demonstrably reduce shelf life.

However, chilling has serious effects on the texture of meat if it is carried out rapidly when the meat is still in the pre-rigor condition, that is, before the meat pH has fallen below about 6.2 (Bendall, 1972). A phenomenon known as cold shortening occurs, which results in the production of very tough meat after cooking. As 'rules of thumb' cooling to temperatures not below 10°C in 10 hours for beef and lamb and in 5 hours for pork can avoid cold shortening (James and James, 2002). Poultry meat is generally less prone to cold shortening. However, electrical stimulation can be utilised to enable more rapid cooling to be carried out without the occurrence of cold shortening.

The rate of sugar loss (sweetness) in freshly harvested sweet corn is very temperature dependent. After 20 hours at 30°C almost 60% of the sweet-ness is lost compared with 16% at 10°C and less than 4% at 0°C. Prompt cooling is clearly required if this vegetable is to retain its desirable sweet-ness. Similarly, the ripening of fruit can be controlled by rapid cooling, the rate of ripening declining as temperature is reduced and ceasing below about 4°C (Honikel, 1986).

For several fruits and vegetables, exposure to temperatures below a criti-cal limit, but above the initial freezing temperature, may result in chilling injury. Typical symptoms of chilling injury are internal or external browning, superficial spots, failure to ripen, development of off flavours, and so on. Fruits and vegetables from the tropical and subtropical zones are primarily susceptible to chilling injury, however several Mediterranean products are also susceptible (IIR, 2000). The extent of damage depends on the tempera-ture, duration of exposure and the sensitivity of the fruit or vegetable. Some commodities have high sensitivity, while others have moderate or low sensitivity. For each commodity, the critical temperature depends on the species and/or variety. In some cases, unripe fruits are more sensitive than ripe fruits.

The formation of ice crystals during freezing and frozen storage causes physical changes to the structure of foods. In most cases, these changes are perceived as reducing the quality of the thawed material. In extreme cases, such as cucumbers, freezing completely destroys the structure of the food. In meat and fish, the main result is increased drip on cutting. There is a general view that fast freezing offers some quality advantage, with 'quick frozen' appearing on many frozen foods in the expectation that consumers will pay more for a 'quick frozen' product. However, there are little data in the scientific literature to prove that, in general, the method of freezing, or the rate of freezing, has any substantial influence on the final eating quality of many frozen foods, with the possible exception of fruits, egg products, frozen deserts and products containing flour thickened sauces.

A final, but important, quality and economic advantage of temperature control is a reduction in weight loss, which results in a higher yield of saleable material. Meat, for example, has a high water content and the rate of evaporation depends on the vapour pressure at the surface. Vapour pressure increases with temperature and thus any reduction in the surface temperature will reduce the rate of evaporation. The use of very rapid chilling systems for pork carcasses has been shown to reduce weight loss by at least 1% when compared with conventional systems (James et al., 1983).

18.4   Chilling and freezing

During chilling and freezing, heat can only be removed by four basic mechanisms: radiation, conduction, convection or evaporation. To achieve substantial rates of heat loss by radiation, large temperature differences are required between the surface of the product and that of the enclosure. This occurs in the initial stages of chilling or freezing of cooked or warm products. Physical contact between the product and the source of refrigeration is required in order to extract heat by conduction. Plate conduction coolers are used for quick cooling of some packaged products and highly perishable products such as fish blocks. For the majority of foods, the heat lost through evaporation of water from the surface is a minor component of the total heat loss, although it is the major component in vacuum cooling. Evaporation from the surface of a food reduces yield and is not desirable in many refrigeration operations, but can be useful in the initial cooling of unpackaged cooked food products. Convection is by far the most important heat transfer mechanism employed in the majority of food refrigeration systems.

Most foods are cooled by the convection of heat into air or another refrigerated medium. The rate of heat removal depends on:

- the surface area available for heat flow
- the temperature difference between the surface and the medium
- the surface heat transfer coefficient.

For the majority of chilled and frozen foods, air systems are used, primarily because of their flexibility and ease of use. However, other systems can offer much faster and more controlled chilling or freezing.

From a hygiene/HACCP based approach, prepacking the food prior to chilling or freezing will lower the risk of contamination/cross-contamination during the chilling process. However, in most cases it will significantly reduce the rate of cooling and this may allow the growth of any microorganisms present. Provided that the cooling media (air, water, etc) and refrigeration equipment used are kept sufficiently clean, no one cooling method can be said to be intrinsically more hygienic than any other. For unwrapped food, a rapid cooling system allows less time for any contamination/cross-contamination to occur than slower cooling systems.

It is not unusual for food products (or ingredients found in food products) to be chilled or frozen a number of times before they reach the consumer. For example, during industrial processing, frozen raw material is often thawed or tempered before being turned into meat-based products, for example, pies, convenience meals, burgers, or consumer portions, fillets, steaks, and so on. These consumer-sized portions are often refrozen before storage, distribution and sale.

18.4.1   Chilling and freezing systems

There are a large number of different chilling and freezing systems for food based on moving air, wet air, direct contact, immersion, ice, cryogenic, vacuum and pressure shift.

Air systems range from the most basic, in which a fan draws air through a refrigerated coil and blows the cooled air around an insulated room, to purpose-built conveyerised blast chilling tunnels or spirals. Relatively low rates of heat transfer are attained from product surfaces in air-cooled systems. The big advantages of air systems are their cost and versatility, especially when there is a requirement to cool a variety of irregularly shaped products.

One of the principal disadvantages of air cooling systems is their tendency to dehydrate unwrapped products. A way around this problem is to saturate the air with water. Wet air cooling systems recirculate air over ice cold water so that air leaving the cooler is cold (0–1°C) and virtually saturated with water vapour (100% relative humidity, RH). An ice-bank chiller uses a refrigeration plant with an evaporator (plate or coil) immersed in a tank of water that chills the water to 0°C. Using off-peak electricity during times of low load and overnight, a store of ice is built up on the evaporator, which subsequently melts to maintain temperatures during times of high load.

Contact refrigeration methods are based on heat transfer by contact between products and metal surfaces, which in turn are cooled by either primary or secondary refrigerants. Contact cooling offers several advantages over air cooling, such as much better heat transfer and significant energy savings. Contact cooling systems include plate coolers, jacketed heat exchangers, belt coolers and falling film systems.

Immersion/spray systems involve dipping product into a cold liquid, or spraying a cold liquid onto the food. This produces high rates of heat transfer owing to the intimate contact between product and cooling medium. Both offer several inherent advantages over air cooling in terms of reduced dehydration and coil frosting problems (Robertson *et al.*, 1976). Clearly if the food is unwrapped, the liquid has to be 'food safe'. Cooling using ice or cryogenic substances are essentially immersion/spray processes. The freezing point of the cooling medium used dictates its use for chilling or freezing. Obviously any immersion/spray freezing process must employ a medium at a temperature substantially below 0 °C. This necessitates the use of non-toxic salt, sugar or alcohol solutions in water, or the use of cryogens or other refrigerants.

Chilling with crushed ice or an ice/water mixture is simple, effective and commonly used for fish cooling. Cooling is more attributable to the contact between the produce and the cold melt water percolating through it (i.e. hydrocooling) than with the ice itself. The individual fish are packed in boxes between layers of crushed ice, which extract heat from the fish and consequently melt. Ice has the advantage of being able to deliver a large amount of refrigeration in a short time as well as maintaining a very constant temperature, 0 °C to −0.5 °C where sea water is present.

Direct spraying of liquid nitrogen onto a food product whilst it is conveyed through an insulated tunnel is one of the most commonly used methods of applying cryogens. The method of cooling is essentially similar to water-based evaporative cooling, cooling being brought about by boiling off the refrigerant, the essential difference being the temperature required for boiling. As well as using the latent heat absorbed by the boiling liquid, sensible heat is absorbed by the resulting cold gas. Owing to very low operating temperatures and high surface heat transfer coefficients between product and medium, cooling rates of cryogenic systems are often substantially higher than other refrigeration systems.

Food products that have a large surface area to volume ratio and an ability to release internal water readily are amenable to vacuum cooling. The products are placed in a vacuum chamber (typically operating at between 530 to 670 N m^{-2}) and the resultant evaporative cooling removes heat from the food. Evaporative cooling is quite significant, the amount of heat released through the evaporation of 1 g of water is equivalent to that released in cooling 548 g of water by 1 °C. Suitable products, such as lettuce, can be vacuum cooled in less than 1 hour. In general terms, a 5 °C reduction in product temperature is achieved for every 1% of water that is evaporated. Since vacuum cooling requires the removal of water from the product, pre-wetting is commonly applied to prevent the removal of water from the tissue of the product.

High pressure freezing and in particular 'pressure shift' freezing is attracting considerable scientific interest (LeBail *et al.*, 2002). The food is cooled under high pressure to sub-zero temperatures but does not undergo a phase change and freeze until the pressure is released. Rapid nucleation results in small even ice crystals. However, studies on pork and beef have failed to show any real commercial quality advantages.

18.5   Storage

There are clear differences between the environmental conditions required for cooling, which is a heat removal/temperature reduction process and those required for storage, where the aim is to maintain a set product temperature. Three factors during storage, the storage temperature, the degree of fluctuation in the storage temperature and the type of wrapping/packaging in which the food is stored, are commonly believed to have the main influence on storage life. The storage life of most chilled foods is limited by the growth of spoilage microorganisms. The storage life of many frozen foods is limited by quality changes, primarily rancidity development in the fat of meat, for instance.

18.6   Transport

Over a million refrigerated road vehicles, 400,000 refrigerated containers and many thousands of other forms of refrigerated transport systems are used to distribute refrigerated foods throughout the world (Gac, 2002). All these transportation systems are expected to maintain the temperature of the food within close limits to ensure its optimum safety and high quality shelf life. Developments in temperature controlled transportation systems for chilled products have led to the rapid expansion of the chilled food market.

It is particularly important that the food is at the correct temperature before loading since the refrigeration systems used in most transport containers are not designed to extract heat from the load but to maintain the temperature of the load. Irrespective of the type of refrigeration equipment used, the product will not be maintained at its desired temperature during transportation unless it is surrounded by air or surfaces at or below the maximum transportation temperature. This is usually achieved by a system that circulates moving air, either forced or by gravity, around the load. Inadequate air distribution is probably the principal cause of product deterioration and loss of shelf life during transport. If products have been cooled to the correct temperature before loading and do not generate heat then they only have to be isolated from external heat ingress. Surrounding them with a blanket of cooled air achieves this purpose. Care has to be taken during loading to stop any product touching the inner surfaces of the vehicle because this would allow heat ingress by conduction during transport. In the large containers used for long distance transportation food temperatures can be kept within ± 0.5 °C of the set point. With this degree of temperature control, transportation times of 8 to 14 weeks (for vacuum packed meats stored at −1.5 °C) can be carried out and the food retains a sufficient chilled storage life for retail display.

Products such as fruits and vegetables that produce heat by respiration, or products that have to be cooled during transit, also require circulation of air through the product. This can be achieved by directing the supply air through ducts to channels at floor level or in the floor itself. In general it is not advisable to rely on product cooling during transportation.

18.6.1   Sea transport

Recent developments in temperature control, packaging and controlled atmospheres have substantially increased the range of foods that can be transported around the world in a chilled condition. Control of the oxygen and carbon dioxide levels in shipboard containers has allowed fruit and vegetables, such as apples, pears, avocados, melons, mangoes, nectarines, blueberries and asparagus, to be shipped (typically 40 days in the container) from Australia and New Zealand to markets in the USA, Europe, Middle East and Japan (Adams, 1988). If the correct varieties are selected and rapidly cooled immediately after harvest, the product arrives in good condition and has a long subsequent shelf life. With conventional vacuum packaging it is difficult to achieve a shelf life in excess of 12 weeks with beef and 8 weeks for lamb (Gill, 1984). However, a shelf life of up to 23 weeks at $-2\,°C$ can be achieved in cuts of lamb individually packed in evacuated bags of linear polyethylene and then placed in gas flushed foil laminate bags filled with a volume of CO_2 approximately equal to that of the meat (Gill and Penney, 1986). Similar storage lives are currently being achieved with beef primals transported from Australia and South Africa to the EU.

Most International Standard Organisation (ISO) containers are either 'refrigerated' or 'insulated'. The refrigerated containers have refrigeration units built into their structure. The units operate electrically, either from an external power supply on board the ship or dock or from a generator on a road vehicle. Insulated containers either utilise the plug type refrigeration units already described or may be connected directly to an air-handling system in a ship's hold or at the docks. Close temperature control is most easily achieved in containers that are placed in insulated holds and connected to the ship's refrigeration system. However, suitable refrigeration facilities must be available for any overland sections of the journey. When the containers are fully loaded and the cooled air is forced uniformly through the spaces between cartons, the maximum difference between delivery and return air can be less than 0.8°C (Heap, 1986). The entire product in a container can be maintained to within $\pm 1.0°C$ of the set point.

Refrigerated containers are easier to transport overland than the insulated types, but have to be carried on deck when shipped because of problems in operating the refrigeration units within closed holds. On board ship they are therefore subjected to much higher ambient temperatures, and consequently larger heat gains, which makes it far more difficult to control product temperatures.

18.6.2 Air transport

Air freighting is increasingly being used for high value perishable products, such as strawberries, asparagus and live lobsters (Sharp, 1988; Stera, 1999). However, foods do not necessarily have to fall into this category to make air transportation viable, since it has been shown that 'the intrinsic value of an item has little to do with whether or not it can benefit from air shipment, the deciding factor is not price but mark-up and profit' (ASHRAE, 2006).

There was a 10–12% per year increase in the volume of perishables transported by air during in the 1990s (Stera, 1999). Although air freighting of foods offers a rapid method of serving distant markets, there are many problems because the product is unprotected by refrigeration for much of its journey. Up to 80% of the total journey time is made up of waiting on the tarmac and transport to and from the airport. During flight, the hold is normally between 15 and 20 °C. Perishable cargo is usually carried in standard containers, sometimes with an insulating lining and/or ice or dry ice but is often unprotected on aircraft pallets (Sharp, 1988). Thus it is important that the product be: (1) transported in insulated containers to reduce heat gain, (2) be precooled and held at the required temperature until loading, (3) containers should be filled to capacity and (4) a thermograph should accompany each consignment.

18.6.3 Land transport

Overland transportation systems range from 12-m refrigerated containers for long distance road or rail movement of bulk chilled or frozen products, to small uninsulated vans supplying food to local retail outlets or even directly to the consumer. Some of the first refrigerated road and rail vehicles for chilled product were cooled by air that was circulated by free or forced systems, over large containers of ice (Ciobanu et al., 1976). Similar systems using solid carbon dioxide as the refrigerant have also been used to cool transport vehicles.

In a 1970/71 survey of vehicles in the UK used to transfer chilled meat from small abattoirs to shops, almost 70% were unrefrigerated and 20% had no insulation (Cutting and Malton, 1972). However, now the majority of current road transport vehicles for chilled foods are refrigerated using either mechanical, eutectic plates or liquid nitrogen cooling systems. Many advantages are claimed for liquid nitrogen transport systems, including minimal maintenance requirements, uniform cargo temperatures, silent operation, low capital costs, environmental acceptability, rapid temperature reduction and increased shelf life owing to the modified atmosphere (Smith, 1986). Overall costs are claimed to be comparable with mechanical systems (Smith, 1986). However, published trials on the distribution of milk have shown that the operating costs using liquid nitrogen, per 100 l of milk transported, may be 2.2 times that of mechanically refrigerated transport systems (Nieboer, 1988). The rise in supermarket home delivery services where there are requirements for mixed loads of products that may each require different storage temperatures is introducing a new complexity to local land delivery (Cairns, 1996).

18.7   Retail display

The temperature of individual consumer packs, small individual items and especially thin sliced products responds very quickly to small amounts of added heat. All these products are commonly found in retail display cabinets and marketing constraints require that they have maximum visibility. Maintaining the temperature of products below set limits while they are on open display in a heated store will always be a difficult task.

Average temperatures in chill displays can vary considerably from cabinet to cabinet, with inlet and outlet values ranging from −6.7 to +6.0 °C, and −0.3 to +7.8 °C, respectively, in one survey (Lyons and Drew, 1985). The temperature performance of an individual display cabinet does not only depend on its design. Its position within a store and the way the products are positioned within the display area significantly influences product temperatures. In non-integral (remote) cabinets (i.e. those without built-in refrigeration systems) the design and performance of the store's central refrigeration system is also critical for effective temperature control.

The desired chilled display life for wrapped meat, fish, vegetables and processed foods ranges from a few days to weeks and is primarily limited by microbiological considerations. Retailers of unwrapped fish, meat and delicatessen products normally require a display life of one working day, which is often restricted by appearance changes. Frozen food can potentially be displayed for many weeks.

Reducing energy consumption in a chilled multi-deck cabinet is substantially different from reducing it in a frozen well cabinet (James *et al.*, 2009). Improvements have been made in insulation, fans and energy efficient lighting but only 10% of the heat load on a chilled multi-deck comes from these sources, compared with 30% on the frozen well. Research efforts are concentrating on minimising infiltration through the open front of multi-deck chill cabinets, by the optimisation of air curtains and airflows, since this is the source of 80% of the heat load. In frozen well cabinets reducing heat radiation onto the surface of the food, accounting for over 40% of the heat load, is a major challenge.

18.7.1   Unwrapped products

Display cabinets for delicatessen products are available with gravity or forced convection coils and the glass fronts may be nearly vertical or angled up to 20°. Sections through three of the commonest types of delicatessen cabinet are shown in Fig. 18.1. In the gravity cabinet (Fig.18.1 a) cooled air

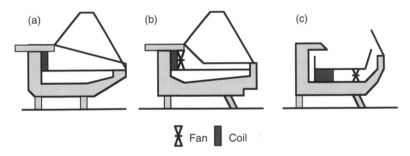

Fig. 18.1 Three types of retail display cabinet for unwrapped products. (a) Gravity cabinet, (b) and (c) forced circulation cabinets.

Table 18.2 Relationship between evaporative weight loss and the appearance of sliced beef topside after 6 hours on display

Evaporative loss (g cm^{-2})	Change in appearance
Up to 0.01	Red, attractive and still wet; may lose some brightness
0.015–0.025	Surface becoming drier; still attractive but darker
0.025–0.035	Distinct obvious darkening; becoming dry and leathery
0.05	Dry, blackening
0.05–0.10	Black

from the raised rear-mounted evaporator coil descends into the display well by natural convection and the warm air rises back to the evaporator. In the forced circulation cabinets (Fig. 18.1 b and c) air is drawn through an evaporator coil by a fan and then ducted into the rear of the display, returning to the coil after passing directly over the products (Fig. 18.1 b), or forming an air curtain (Fig. 18.1 c), via a slot in the front of the cabinet and a duct under the display shelf.

Changes in product appearance are normally the criteria that limit the display life of unwrapped foods with the consumer selecting newly loaded product in preference to that displayed for some time. Deterioration in appearance has been related to degree of dehydration in red meat (Table 18.2) and is likely to similarly occur in other foods.

Apart from any relationship to appearance, weight loss is of considerable importance in its own right. A small survey carried out in the 1980s found average relative humidity ranged from 41–73% and air velocity from 0.1–0.67 m s^{-1} in delicatessen cabinets. The lowest rate of weight loss was measured in a cabinet of the type shown in Fig 18.1c which achieved mean conditions over the products of 0.4°C, 0.14 m s^{-1} and 65% RH (James and Swain, 1986). The same study showed that relative humidity was more important than the air temperature or velocity. Reducing the relative humidity (RH) from 95% to 40% increased weight loss over a 6 hour display period by a factor of between 14 and 18. In further work, a model developed to predict the rate of weight loss from unwrapped meat under the range of environmental conditions found in chilled retail displays showed that it was governed by the mean value of the conditions (James *et al.*, 1988). Fluctuations in temperature or relative humidity had little effect on weight loss and any apparent effect was caused by changes in the mean conditions.

There is a conflict between the need to make the display attractive and convenient to increase sales appeal and the optimum display conditions for the product. High lighting levels increase the heat load and the consequent temperature rise dehumidifies the refrigerated air. The introduction of humidification systems can significantly improve display life (Brown *et al.*, 2005).

18.7.2 Wrapped products

To achieve the display life of days to weeks required for wrapped chilled foods, the product should be maintained at a temperature as close to its initial freezing point as possible to prevent microbial spoilage. In some cases, for example particular cheeses, dairy products and tropical fruits, quality problems may limit the minimum temperature that can be used, but for the majority of meat, fish and processed foods, the range −1 to 0 °C is desirable.

Air movement and relative humidity have little effect on the display life of a wrapped product, but the degree of temperature control can be important especially for transparent, controlled atmosphere packs. Large temperature cycles will cause water loss from the product and this water vapour will condense on the inner surface of the pack and consequently reduce consumer appeal.

Although cabinets of the type described for delicatessen products can be used for wrapped foods, most are sold from multi-deck cabinets with single or twin air curtain systems. Twin air curtains tend to provide more constant product temperatures but are more expensive. It is important that the front edges of the cabinet shelves do not project through the air curtain since the refrigerated air will then be diverted out of the cabinet. On the other hand if narrow shelves are used, the curtain may collapse and ambient air can be drawn into the display well.

To maintain product temperatures close to 0 °C, the air off the coil must typically be −4 °C and any ingress of humid air from within the store will quickly cause the coil to ice up. Frequent defrosts are often required and even in a well-maintained unit the cabinet temperature may rise to 10–12 °C and the product by at least 3 °C (Brolls, 1986). External factors such as the store ambient temperature, the position of the cabinet and poor pretreatment and placement of products substantially affect cabinet performance. Warm and humid ambient air, and loading with insufficiently cooled products, can also overload the refrigeration system. Even if the food is at its correct temperature, uneven loading or too much product can disturb the airflow patterns and destroy the insulating layer of cooled air surrounding the product.

An in-store survey of 299 prepackaged meat products in chilled retail displays found product temperatures in the range −8.0 to 14.0 °C, with a mean of 5.3 °C and 18% above 9 °C (Rose, 1986). Other surveys (Bøgh-Sørensen, 1980; Malton, 1971) have shown that temperatures of packs from the top of stacks were appreciably higher than those from below owing to radiant heat pick up from store and cabinet lighting. It has also been stated that products in transparent film overwrapped packs can achieve temperatures above that of the surrounding refrigerated air owing to radiant heat trapped in the package by the 'greenhouse' effect. However, specific investigations have failed to demonstrate this effect (Gill, 1988).

18.7.3 Frozen foods

Frozen foods are always packaged before being displayed and in the majority of cases the packaging obscures, and protects, the food on display. If packed in transparent film the surface of many frozen foods will discolour rapidly when illuminated.

Traditionally open-well cabinets were used to display frozen products, but increasingly multi-deck cabinets are used because of their increased display space and sales appeal. The rate of heat gain in a multi-deck cabinet, and consequently the energy consumption, is much higher than in a well cabinet. Owing to the increased costs of energy, multi-deck cabinets are now appearing on the market with double glazed doors that have to be opened to access the food on display.

18.8 Energy and the environment

The dominant types of refrigerant used in the food industry in the last 60 years have belonged to a group of chemicals known as halogenated hydrocarbons. Members of this group, which includes the chlorofluorocarbons (CFCs) and the hydrochlorofluorocarbons (HCFCs), have excellent properties, such as low toxicity, compatibility with lubricants, high stability, good thermodynamic performance and relatively low cost, making them excellent refrigerants for industrial, commercial and domestic use. However, their high chemical stability leads to environmental problems when they are released and rise into the stratosphere. Scientific evidence clearly shows that emissions of CFCs have been damaging the ozone layer and contributing significantly to global warming. With the removal of CFCs from aerosols, foam blowing and solvents, the largest single application sector in the world is refrigeration, which accounts for almost 30% of total consumption.

Until recently R12, R22 and R502 were the three most common refrigerants used in the food industry. R12 and R502 have significant ozone depletion potential (ODP) and global warning potential (GWP) and R22, although smaller, is still dangerous in the long term. Consequently, as a result of international agreements, pure CFCs (e.g. R12, R502) have been completely banned. Pure HCFCs (mainly R22) are banned in new industrial plant and are soon to be phased out completely. HCFC blends and HFC blends originally introduced as CFC replacements are covered by F-Gas regulations that limit leak rates.

Chemical companies are making large investments in terms of both time and money in developing new refrigerants that have reduced or negligible environmental effects. Other researchers are looking at the many non-CFC alternatives, including ammonia, propane, butane, carbon dioxide, water and air that have been used in the past.

Ammonia is increasingly a common refrigerant in large industrial food cooling and storage plants. It is a cheap, efficient refrigerant whose pungent odour aids leak detection well before toxic exposure or flammable concentrations are reached. The renewed interest in this refrigerant has led to the development of compact, low charge (i.e. small amounts of ammonia) systems that significantly reduce the possible hazards in the event of leakage. It is expected that ammonia will meet increasing use in large industrial food refrigeration systems. Carbon dioxide is being advocated for retail display cabinets and hydrocarbons, particularly propane and butane or mixtures of both, for domestic refrigerators.

As well as the direct affect of refrigerants on the environment, energy efficiency is increasingly of concern to the food industry. Worldwide it is estimated that 40% of all food requires refrigeration and 15% of the electricity consumed worldwide is used for refrigeration (Mattarolo, 1990). In the UK, 11% of electricity is consumed by the food industry (DBERR, 2005). However, detailed estimates of what proportion of this is used for refrigeration processes are less clear and often contradictory (James et al., 2009). Using the best available data, the energy saving potential in the top five refrigeration operations (retail, catering, transport, storage and primary chilling), in terms of the potential to reduce energy consumed, lies between 4300 and 8500 GWh/year in the UK (James et al., 2009).

It is clear that maintenance of food refrigeration systems will reduce energy consumption (James et al., 2009). Repairing door seals and door curtains, ensuring that doors can be closed and cleaning condensers produce significant reductions in energy consumption. In large cold storage sites, it has been shown that energy can be substantially reduced if door protection is improved, pedestrian doors fitted, liquid pressure amplification pumps fitted, defrosts optimised, suction liquid heat exchangers fitted and other minor issues corrected (James et al., 2009).

In the retail environment, the majority of the refrigeration energy is consumed in chilled and frozen retail display cabinets (James *et al.*, 2009). Laboratory trials have revealed large, up to six-fold, differences in the energy consumption of frozen food display cabinets in similar display areas. In chilled retail display, which accounts for a larger share of the market, similar large differences, up to five-fold, were measured. A substantial energy saving can therefore be achieved by simply informing and encouraging retailers to replace energy inefficient cabinets by the best currently available.

New/alternative refrigeration systems/cycles, such as trigeneration, air cycle, sorption–adsorption systems, thermoelectric, Stirling cycle, thermoacoustic and magnetic refrigeration, have the potential to save energy in the future if applied to food refrigeration (Tassou *et al.*, 2009). However, none appear to be likely to produce a step change reduction in refrigeration energy consumption within the food industry within the next decade.

18.9   Specifying refrigeration systems

In the author's experience, the poor performance of new refrigeration systems used to maintain the cold chain can often be chased back to a poor, non-existent, or ambiguous process specification. In older systems it is often due to a change in use that was not considered in the original specification. There are three stages in obtaining a refrigeration system that works:

1. determining the process specification, i.e. specifying exactly the condition of the product(s) when they enter and exit the system, and the amounts that have to be processed;
2. drawing up the engineering specification, i.e. turning processing conditions into terms that a refrigeration engineer can understand, independent of the food process;
3. the procurement and commissioning of the total system, including any services or utilities.

The first task in designing a system is the preparation of a clear specification by the user of how the facility will be used at present and in the foreseeable future. In preparing this specification, the user should consult all parties concerned. These may be officials enforcing legislation, customers, other departments within the company and engineering consultants or contractors, but the ultimate decisions taken in forming this specification are the users alone.

The process specification must include, as a minimum, data on the food(s) to be refrigerated, in terms of size, shape and throughput. The maximum capacity must be catered for and the refrigeration system should also be specified to operate adequately and economically at all other throughputs. The range of temperature requirements for each product must also be clearly stated. If it is intended to minimise loss, it is useful to quantify at an early stage how much extra money can be spent to save a given amount of weight. All the information collected so far, and the decisions taken, will be on existing production. Another question that needs to be asked is, 'will there be any changes in the use of the chiller in the future?'

The refrigeration system chiller, freezer, storeroom and so on is one operation in a sequence of operations. It influences the whole production process and interacts with it. An idea must be obtained of how the system will be loaded, unloaded and cleaned, and these operations must always be intimately involved with those of the rest of the production process.

There is often a conflict of interest in the usage of a chiller or freezer. In practice, a chiller/freezer can often be used as a marshalling yard for sorting orders and as a place for storing product not sold. If it is intended that either of these operations are to take place in the chiller/freezer, the design must be made much more flexible in order to cover the conditions needed in a marshalling area or a refrigerated store. In the case of a batch or semi-continuous operation, holding areas may be required at the beginning and end of the process in order to even out flows of material from adjacent processes. The time available for the process will be in part dictated by the space that is available; a slow process will take more space than a fast process, for a given throughput.

Other refrigeration loads, in addition to that caused by the input of heat from the product, also need to be specified. Many of these, such as infiltration through openings, the use of lights, machinery and people working in the refrigerated space, are all under the control of the user and must be specified so that the heat load emitted by them can be incorporated in the final design. Ideally, all the loads should then be summed together on a time basis to produce a load profile. If the refrigeration process is to be incorporated with all other processes within a plant, in order to achieve an economic solution, then the load profile is important. The ambient design conditions must be specified. This means that the temperatures adjacent to the refrigerated equipment and the temperatures of the ambient to which heat will ultimately be rejected. In stand-alone refrigerated processes this will often be the wet and dry bulb temperatures of the outside air. If the process is to be integrated with heat reclamation, the temperature of the heat sinks must be specified. Finally, the defrost regime should also be specified. There are times in any process where it is critical that coil defrosting and its accompanying temperature rise does not take place and that the coil is cleared of frost before commencing the specified part of the process.

Although it is common practice throughout the food industry to leave much of this specification to refrigeration contractors or engineering specialists, the end user should specify all the above requirements. The refrigeration contractors or engineering specialists are in a position to give good advice about this. However, since all the above are outside their control, the end user, with their knowledge of how well they can control their overall process, should always take the final decision.

The aim of drawing up an engineering specification is to turn the user requirements into a specification that any refrigeration engineer can then use to design a system. The first step in this process is iterative. First, a full range of time, temperature and air velocity options must be assembled for each cooling specification covering the complete range of each product. Each must then be evaluated against the user requirements. If they are not a fit, then another option is selected and the process repeated. If there are no more options available there are only two alternatives: either standards must be lowered (recognising in doing so that cooling specifications will not be met) or the factory operation must be altered.

A full engineering specification will typically include the environmental conditions within the refrigerated enclosure, air temperature, air velocity and humidity; the way the air will move within the refrigerated enclosure; the size of the equipment; the refrigeration load profile; the ambient design conditions and the defrost requirements. The final phase of the engineering specification should be drawing up a schedule for testing the engineering specification prior to handing over the equipment. This test will be in engineering and not product terms.

The specification produced should be the document that forms the basis for quotations and finally the contract between the user and his contractor and must be stated in terms that are objectively measurable once the chiller/freezer is completed. Arguments often ensue between contractors and their clients from an unclear, ambiguous or unenforceable specification. Such lack of clarity is often expensive to all parties and should be avoided.

18.10 Conclusions

In general, after initial chilling or freezing, as a chilled or frozen food product moves along the cold chain it becomes increasingly difficult to control and maintain its temperature. Temperatures of bulk packs of chilled or frozen product in large storerooms are far less sensitive to small heat inputs than single consumer packs in transport or open display cases.

If primary and secondary cooling operations are efficiently carried out then the food will be reduced below its required temperature before it is placed in storage. In this situation the cold store's refrigeration system is only required to extract extraneous heat that enters through the walls, door openings, and so on.

Even when temperature controlled dispatch bays are used, there is a slight heat pick up during loading. In bulk transportation the resulting temperature rise is small and the vehicle's refrigeration system rapidly returns the product to the required temperature. Larger problems exist in local multi-drop distribution to individual stores. There is a large heat input every time the doors are opened and product unloaded, small packs rapidly rise in temperature and the vehicle often lacks the refrigeration capacity or time to recool the food.

Temperature control during retail display is often poor owing to the retailers need to display as much product as possible in a way that is very assessable to the consumer. Increasing energy costs may be the key factor that persuades retailers to reduce consumer access and hence improve temperature control.

In recent years, energy conservation requirements have caused an increased interest in the possibility of using more efficient storage temperatures than have been used to date. Researchers, such as Jul (1982), have questioned the wisdom of storage below −20 °C and have asked whether there is any real economic advantage in very low temperature preservation. There is a growing realisation that storage lives of several foods can be less dependent on temperature than previously thought. Since research has shown that many food products, such as red meats, often produce non-linear time–temperature curves, there is probably an optimum storage temperature for a particular food product. Improved packing and preservation of products can also increase storage life and may allow higher storage temperatures to be used.

18.11   References

ADAMS G R (1988), 'Controlled atmosphere containers', *Refrigeration for Food and People,* Meeting of IIR Commissions C2, D1, D2/3, E1, Brisbane, Australia, 244–8.

ASHRAE (2006) *ASHRAE Handbook – Refrigeration,* American Society of Heating, Refrigerating and Air-Conditioning Engineers, Atlanta.

BENDALL J R (1972), 'The influence of rate of chilling on the development of rigor and "cold shortening"', *Meat Research Institute Symposium No. 2. Meat Chilling – Why and How?* MRI, Langford, Bristol, UK, 3.1–3.6.

BØGH-SØRENSEN L (1980), 'Product temperatures in chilled cabinets', *Proceedings 26th European Meeting of Meat Research Workers,* Colorado Springs, USA, n.22.

BROLLS E K (1986), 'Factors affecting retail display cases', *Recent Advances and Developments in the Refrigeration of Meat by Chilling,* Meeting of IIR Commission C2, Bristol, UK, Section 9, 405–13.

BROWN T, CORRY J and JAMES S J (2005), 'Humidification of chilled fruit and vegetables on retail display using an ultrasonic fogging system with water/air ozonation', *International Journal of Refrigeration,* **27**(8), 862–8.

CAIRNS S (1996), 'Delivering alternatives: Success and failures of home delivery services for food shopping', *Transport Policy,* **3**, 155–76.

CIOBANU A, LASCU G, BERCESCU V and NICULESCU L (1976), *Cooling Technology in the Food Industry,* Abacus Press, Tunbridge Wells.

CUTTING C L and MALTON R (1972), 'Recent observations on UK meat transport', *Meat Research Institute Symposium No. 2. Meat Chilling – Why and How?* MRI, Langford, Bristol, UK, 24.1–24.11.

DEPARTMENT FOR BUSINESS ENTERPRISE and REGULATORY REFORM (DBERR) (2005), Electricity Supply and Consumption (DUKES 5.2), http://stats.berr.gov.uk/energystats/dukes5_2.xls.

GAC A (2002), 'Refrigerated transport: what's new?' *International Journal of Refrigeration*, **25**, 501–3.

GILL C O (1984), 'Longer shelf life for chilled lamb', *23rd New Zealand Meat Industry Research Conference*, Hamilton.

GILL J (1988), 'The greenhouse effect', *Food*, (April), 47, 49, 51.

GILL C O and PENNEY N (1986), 'Packaging of chilled red meats for shipment to remote markets', *Recent Advances and Developments in the Refrigeration of Meat by Chilling*, Meeting of IIR Commission C2, Bristol, UK, Section 10, 521–5.

HEAP R D (1986), 'Container transport of chilled meat', *Recent Advances and Developments in the Refrigeration of Meat by Chilling*, Meeting of IIR Commission C2, Bristol, UK, 505–10.

HONIKEL K O (1986), 'Influence of chilling on biochemical changes and quality of pork', *Recent Advances and Developments in the Refrigeration of Meat by Chilling*, Meeting of IIR Commission C2, Bristol, UK, Section 1, 45–53.

IIR (2000), *Recommendations for Chilled Storage of Perishable Produce*, IIR, Paris.

INGRAM M (1972), 'Meat chilling – the first reason why', *Meat Research Institute Symposium No. 2. Meat Chilling – Why and How?* MRI, Langford, Bristol UK, 1.1–1.12.

JAMES S J and JAMES C (2002) *Meat Refrigeration*, Woodhead Publishing, Cambridge, UK.

JAMES S J and SWAIN M V L (1986), 'Retail display conditions for unwrapped chilled foods', *Proceeding of the Institute of Refrigeration*, 1985–1986, **82**, 1–7.

JAMES S J, GIGIEL A J and HUDSON W R (1983) 'The ultra rapid chilling of pork', *Meat Science*, **8**, 63–78.

JAMES S J, SWAIN M V L and DAUDIN J D (1988), 'Mass transfer under retail display conditions', *34th International Congress Meat Science Technology*, Brisbane, Australia, 652–4.

JAMES S J, SWAIN M J, BROWN T, EVANS J A, TASSOU S A, GE Y T, EAMES I, MISSENDEN J, MAIDMENT G and BAGLEE D (2009), 'Improving the energy efficiency of food refrigeration operations', *Proceedings of the Institute of Refrigeration*, in press.

JUL M (1982), 'The intricacies of the freezer chain', *International Journal of Refrigeration*, **5**, 226–30.

LEBAIL A, CHEVALIER D, MUSSA D M and GHOUL M (2002), 'High pressure freezing and thawing of foods: a review', *International Journal of Refrigeration*, **25**, 504–13.

LYONS H and DREW K (1985), 'A question of degree', *Food*, December, 15–17.

MALTON R (1971), 'Some factors affecting temperature of over-wrapped trays of meat in retailers display cabinets', *Proceedings 17th European Meeting of Meat Research Workers*, Bristol, UK, J2.

MATTAROLO, L (1990), 'Refrigeration and food processing to ensure the nutrition of the growing world population', *Progress in the Science and Technology of Refrigeration in Food Engineering*, Proceedings of meetings of commissions B2, C2, D1, D2-D3, September 24–28, 1990, Dresden (Germany), Institut International du Froid, Paris (France), 43–54.

MORRISON C R (1993), 'Fish and shellfish', in, *Frozen Food Technology*, Mallet C P (ed), Blackie Academic & Professional, London, 196–236.

NIEBOER H (1988), 'Distribution of dairy products', *Cold-Chains in Economic Perspective*, Meeting of IIR Commission C2, Wageningen (The Netherlands), 16.1–16.9.

ROBERTSON G H, CIPOLLETTI, J C, FARKAS, D F and SECOR G E (1976), 'Methodology for direct contact freezing of vegetables in aqueous freezing media', *Journal of Food Science*, **41**, 845–51.

ROSE S A (1986), 'Microbiological and temperature observations on pre-packaged ready-to-eat meats retailed from chilled display cabinets', *Recent Advances and Developments in the Refrigeration of Meat by Chilling*, Meeting of IIR Commission C2, Bristol, UK, Section 9, 463–9.

SHARP A K (1988), 'Air freight of perishable product', *Refrigeration for Food and People,* Meeting of IIR Commissions C2, D1, D2/3, E1, Brisbane, Australia, 219–24.

SMITH B K (1986), 'Liquid nitrogen in-transit refrigeration', *Recent Advances and Developments in the Refrigeration of Meat by Chilling*, Proceedings of the Conference of IIR Commission C2, Bristol, UK, 383–90.

STERA A C (1999), 'Long distance refrigerated transport into the third millennium', *20th International Congress of Refrigeration*, IIF/IIR Sydney, Australia, paper 736.

TASSOU S A, LEWIS J, GE Y T, HADAWEY A and CHAE I (2009), 'A review of emerging technologies for food refrigeration applications', *Applied Thermal Engineering*, in press.

TAYLOR A A (1972), 'Influence of carcass chilling rate on drip in meat', *Meat Research Institute Symposium No. 2. Meat Chilling – Why and How?* MRI, Langford, Bristol, UK, 5.1–5.8.

19

Simulation modelling for food supply chain redesign*

J.G.A.J. van der Vorst, Logistics, Decision and Information Sciences, Wageningen University, The Netherlands, D.-J. van der Zee, University of Groningen, The Netherlands, S.-O. Tromp, Wageningen University and Research Centre, The Netherlands

Abstract: Food supply chains are confronted with increased consumer demands for food quality and sustainability. When redesigning these chains, the analysis of food quality change and environmental load for new scenarios is as important as the analysis of efficiency and responsiveness requirements. Simulation tools are often used to support decision making in the supply chain (re)design when logistic uncertainties are in place, building on their inherent modelling flexibility. Mostly, the underlying assumption is that product quality is not influenced by or does not influence chain design. Clearly, this is not true for food supply chains, as quality change is intrinsic to the industry. In this chapter we discuss specific characteristics and modelling requirements of food supply chains. We propose a new integrated approach towards logistics, sustainability and food quality analysis, and implement the approach by introducing a new simulation environment, ALADIN™. This embeds food quality change models and sustainability indicators in discrete event simulation models. A case example illustrates the benefits of its use relating to speed and quality of integrated decision making, but also to creativity in terms of alternative solutions.

Key words: simulation, supply chain design, logistics, food quality, sustainability.

19.1 Introduction

Previous chapters show that the management of food supply chains has become a complex task and that support for decisions is needed. They make

*This chapter is largely based on an original article published in *IJPR*: van der Vorst, Tromp and van der Zee (2009), 'Simulation modelling for food supply chain redesign; integrated decision making on product quality, sustainability and logistics,' *International Journal of Production Research*, 47:23, 6611–6631.

clear how, for example, product assortments seem to grow without bound, product life cycles have decreased and consumer demands for product freshness and food safety have increased on the one hand, and have become more unpredictable at the product level on the other hand. This has resulted in a demand for short lead times and high delivery frequencies for small batches. Hence very flexible production organisation and supply chains are required to realize this performance, while keeping costs in line with company standards.

In promoting and building a flexible food supply chain, organizations need to understand the way competition is changing. Future competitiveness will depend on companies' abilities to join supply chains and effectively participate in them. Effective participation strongly relies on mutual coordination of activities and sharing of information. Ultimately, performance of the food supply chain is measurable via the acceptance of the end product by the consumer. Their acceptance builds on a combination of availability, price, quality and food safety, sustainability and other qualitative attributes. The highest added value for all participants can only be achieved when these aspects are optimized at the chain or network level. This is especially true for realizing a high quality, safe and sustainable product that possesses integrity. In these respects chain performance strongly depends on decisions made by chain actors and their interactions. Coordinated efforts should and can be facilitated by efficient logistics systems and exchange of information between chain/network participants. In particular, it is a great challenge effectively and efficiently to realize the coordination and information processes in chains, in the dynamic institutional context that they are in.

In order effectively and efficiently to develop and assess improvements in the food chain design in order to manage its increased complexity, one needs decision support tools. Typically, such decision support systems build on mathematical tools and simulation software. In this chapter we will focus on simulation modelling to support food supply chain redesign. The specific challenge being addressed here is to embed food quality models and sustainability issues together with logistics processes in discrete event simulation models, in order to facilitate an integrated approach towards logistic, sustainability and product quality analysis of food supply chains. We hypothesize that integrated decision making will result in overall better decisions compared to disciplinary decision making that takes only one of these aspects into account. The key contribution of discrete event simulation lies in its capability to model and trade off elementary uncertainties underlying product quality and chain logistics, as well as their interaction.

This chapter is organized as follows. First, Section 19.2 describes the essential characteristics of a food supply chain network (FSCN), in terms of the parties involved, process and product characteristics and alternative redesign strategies. Section 19.3 discusses the appropriateness of simulation and pitfalls in the modelling process. In Section 19.4, we highlight the requirements of model capabilities that are essential for successful FSCN simulation and we review existing modelling tools. Next, in Section 19.5, we will introduce the main features of a decision support tool specifically designed for modelling food supply chains, called ALADIN™. In Section 19.6, we will present a case study to illustrate the applicability and potential of integrated decision making for FSC redesign. We will summarize our main conclusions and highlight directions for future research in Section 19.7 and conclude with some main sources of further information.

19.2   Characteristics of the food supply chain network

In recent years, western European consumers have become more demanding about food attributes such as quality, integrity, safety, sustainability, diversity and the associated information services. At the same time, companies in the food industry are acting more and more on a global scale. This is reflected by company size, increasing cross-border flows of livestock and food products, and international cooperation and partnerships. Global competition, together with advances in information technology, have stimulated partners in the food industry to pursue a coordinated approach to establishing more effective and efficient supply chains, supply chain management (SCM). In line with the Global Supply Chain Forum (Lambert and Cooper, 2000), we define SCM as the integrated planning, coordination and control of all logistic business processes and activities in the supply chain in order to deliver superior consumer value at less cost to the supply chain as a whole, while satisfying the requirements of other stakeholders (e.g. the government or non-governmental organizations, NGOs) in the wider context of the total supply chain network (van der Vorst and Beulens, 2002). SCM should result in the choice of a supply chain scenario, that is, an internally consistent view on how a supply chain should be configured in terms of the choice of partners from the total supply chain network and the way their mutual activities of supply, production and distribution of goods are coordinated. Clearly, this is not an easy task, because of a great variety of policies, conflicting objectives, and the inherent uncertainty of the business environment (Alfieri and Brandimarte, 1997).

The design of food supply chains (FSCs) is further complicated by an intrinsic focus on product quality (van der Vorst and Beulens, 2002; Luning and Marcelis, 2006) and demand for environmental sustainability (Hagelaar et al., 2004; Srivastava, 2007). The way in which food quality is controlled and guaranteed in the network is of vital importance for chain performance.

Also, apart from being a performance measure of its own, product quality is directly related to other food attributes like integrity and safety. Recently, more attention has been given to sustainability by introducing the notion of 'green' SCM, that is, 'the set of SCM policies held, actions taken, and relationships formed in response to concerns related to the natural environment with regard to the design, acquisition, production, distribution, use, reuse, and disposal of the firm's goods and services' (Zsidisin and Siferd, 2001). Within the context of FSCs the sustainability discussion focuses on the reduction of product waste, that is, products that have to be thrown away because the quality is not suitable any more (e.g. van Donselaar *et al.*, 2006), number of miles a product has travelled before it reaches the consumers' plate (so-called 'food miles') and all greenhouse gas emissions related to the business processes in the supply chain network (so-called 'carbon footprint') (Edwards-Jones *et al.*, 2008). We conclude that investments in FSC design should not only be aimed at improving logistics performance, but also at the preservation of food quality and environmental sustainability.

19.2.1 Supply chain parties
The food industry is becoming an interconnected system with a large variety of relationships. This is reflected in the market place by the formation of (virtual) FSCs via alliances, horizontal and vertical cooperation, and forward and backward integration (van der Vorst *et al.*, 2005; see Fig. 19.1). Lazzarini *et al.* (2001) refer to a 'netchain' and define it as 'a directed network of actors who cooperate to bring a product to customers'. In a FSCN more than one supply chain and more than one business process can be identified, both parallel and sequential in time. As a result, organizations may play different roles in different chain settings and therefore collaborate with differing chain partners, who may be their competitors in other chain settings. We can conclude that supply chain networks are complex systems

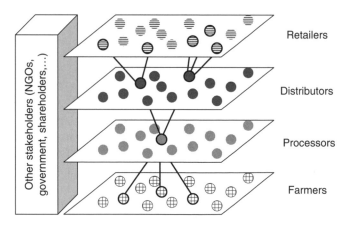

Fig. 19.1 Schematic diagram of a food supply chain network (van der Vorst *et al.*, 2005).

owing to the presence of multiple (semi)-autonomous organizations, functions and people within a dynamic environment.

A FSCN comprises organizations that are responsible for the production and distribution of vegetable or animal-based products. From a general perspective, we distinguish two main types. The first type is the FSCN for fresh agricultural products (such as fresh vegetables and fruit). In general, these chains may concern growers, auctions, wholesalers, importers and exporters, retailers and speciality shops and their logistics service suppliers. The main processes are the handling, (conditioned) storing, packing, transportation and trading of food products. Basically, all of these stages leave the intrinsic characteristics of the product grown or produced in the countryside unharmed, except for the product quality which depends on the environmental conditions in time. Over time, the product quality can either increase (e.g. ripening of fruits) or decrease, if harvested at a mature stage. The second type is the FSCN for processed food products (such as portioned meats, snacks, desserts, canned food products). In general, these chains comprise growers, importers, food industry (processors), retailers and out-of-home segments and their logistics service suppliers. In these chains, agricultural products are used as raw materials to produce consumer products with higher added value. Sometimes the consumer products are hardly perishable owing to conservation processes. This reduces the complexity of the FSC design significantly and largely eliminates the need for quality change models. This chapter focuses especially on those food products, either fresh or processed, that are subject to notable quality changes over time.

19.2.2   Process and product characteristics

Bourlakis and Weightman (2004), Jongen and Meulenberg (2005) and van der Vorst *et al.* (2005) discuss a list of specific process and product characteristics of FSCNs that have an impact on the redesign process, including the following:

- seasonality in production, requiring global sourcing
- variable process yields in quantity and quality caused by biological variations, seasonality, and random factors connected with weather, pests and other biological hazards
- keeping quality constraints for raw materials, intermediates and finished products, and quality decay while products pass through the supply chain. As a result there is a chance of product shrinkage and stock-outs in retail outlets when the product's best-before dates have passed and/or product quality level has declined too much
- requirement for conditioned transportation and storage means (e.g. cooling)
- necessity for lot traceability of work in process owing to quality and environmental requirements and product responsibility.

Owing to these specific characteristics of food products, the partnership thoughts of SCM in FSCs have already received much attention over the past two decades. It is vital for industrial producers to contract suppliers to guarantee the supply of raw materials in terms of the right volume, quality, place and time. Furthermore, they coordinate the timing of the supply of goods with suppliers, to match capacity availability. Actors in FSCNs understand that products are subject to quality decay as they traverse the supply chain, while the degree and speed of decay may be influenced by environmental conditions. For example, exposing a batch of fresh milk, fruit or meat to high temperatures for some time will significantly reduce product keeping quality (shelf life). Supply chain coordination is essential to make appropriate decisions about food conditioning.

19.2.3   Redesign strategies for food supply chains (FSCs)

The literature suggests several strategic, tactical and operational redesign strategies to improve the efficiency and effectiveness of supply chain processes. An extensive literature review by van der Vorst and Beulens (2002) identifies a generic list of SCM redesign strategies to facilitate the redesign process and attain joint supply chain objectives:

- Redesign the roles and processes performed in the supply chain (e.g. reduce the number of parties involved, reallocate roles such as inventory control and eliminate non-value-adding activities such as stock keeping).
- Reduce lead times (e.g. implement information and communication technology (ICT) systems for information exchange and decision support, increase manufacturing flexibility or reallocate facilities).
- Create information transparency (e.g. establish an information exchange infrastructure in the supply chain and exchange information on demand/supply/inventory or work-in-process, standardize product coding).
- Synchronize logistical processes with consumer demand (e.g. increase frequencies of production and delivery processes, decrease lot sizes).
- Coordinate and simplify logistical decisions in the supply chain (e.g. coordinate lot sizes, consolidate goods flows, eliminate human intervention, introduce product standardization and modularization).

The above strategies address the general case of supply chain design. Specifically, for FSC we can add the redesign strategy to alter the time-dependent environmental conditions, under which products are (re)packed (e.g. using modified atmosphere packaging), stored and transported (e.g. using reefer containers), in order to improve food quality. This will result in longer shelf lives and, therefore, provide room for the introduction of innovative logistics concepts. Furthermore, emphasis should be put on redesigning processes in order to reduce greenhouse gas emissions and energy consumption; see Linton *et al.* (2007) for an overview of this subject.

19.3   Rationale of modelling and simulation in food supply chain design

Before we discuss the specific modelling requirements for FSCNs in Section 19.4, we will first discuss the appropriateness of modelling and simulation for food supply chain design. Furthermore, we will present elementary principles of modelling and steps in the simulation study.

19.3.1   Why (simulation) modelling?

It is rarely feasible to experiment with the actual system, because such an experiment would often be too costly or too disruptive to the system, or because the required system might not even exist. For these reasons, it is usually necessary to build a model as a representation of the real system and to study it as a surrogate for the real system. Pidd (1999) defines a model as follows: 'A model is an external and explicit representation of part of reality as seen by the people who wish to use that model to understand, to change, to manage, and to control that part of reality in some way or another'. A model is a convenient world in which one can attempt to change things without incurring the possible direct consequences of such action in the real world. In this sense, 'models become tools for thinking' (Pidd, 1999).

Law and Kelton (1991) distinguish alternative ways in which a system might be studied (Fig. 19.2). Physical models refer, for example, to cockpit simulators or miniature super tankers in a pool. Mathematical models represent a system in terms of logical and quantitative relationships that are manipulated and changed to see how the model reacts and thus how the actual system *would* react, *if* the mathematical model is valid.

To study a system of interest, we often have to make a set of assumptions about how it works. These assumptions are used to constitute a model that

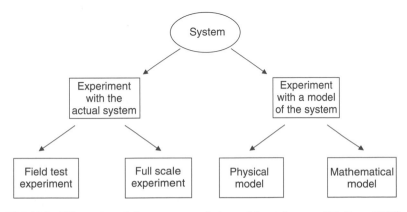

Fig. 19.2 Ways of studying a system (adapted from Law and Kelton, 1991).

in turn is used to try to gain some insight into the behaviour of the corresponding system. If the relationships that compose the model are simple enough, it may be possible to use analytical methods (such as algebra, probability theory or linear programming) to obtain exact information about questions of interest. In analytical models the relationships between the elements of the system are expressed through mathematical equations. Silver *et al.* (1998) state that if mathematical models are to be more useful as aids for managerial decision making, they must be more realistic representations of the problem; in particular, they must permit some of the usual 'givens' to be treated as decision variables. Moreover, such models must ultimately be in an operational form such that the user can understand the inherent assumptions, the associated required input data can be realistically obtained and the recommended course of action can be provided within a relatively short period of time. However, most real-world systems, including food supply chains, are too complex to allow for analytical modelling and these models are preferably studied by means of simulation (Law and Kelton, 1991).

19.3.2   When is simulation appropriate?

Simulation is a powerful tool that is applied frequently. The popularity of simulation has increased with the increase in computer power, development of sophisticated software and decrease in computer costs. This does not mean that simulation is the appropriate tool in each situation (Kettenis and Van der Vorst, 2007).

An advantage of simulation related to experimenting with the real-world system is that the speed of the simulation may be faster than real time. For example to perform a one-day simulation of a post office will take only a few seconds. Another advantage of simulation is in the effort, time and cost involved in studying alternative system designs. Finally, we mention the possibility of visualizing simulation models in terms of the (dynamic) logic adopted and estimated system performance. Typically, visualization may be helpful in verification and validation of models, next to fostering creativity in solution finding and credibility among problem owners (Bell *et al.*, 1999). The disadvantages of simulation may be in expensive and time consuming modelling efforts. Table 19.1 provides an overview of advantages and disadvantages of simulation.

19.3.3   Modelling and simulation of complex systems

Figure 19.3 shows the steps that will compose a typical, sound simulation study and the relationships between them according to Banks *et al.* (1996) and Law and Kelton (1991). A simulation study is not a simple sequential process. As the study proceeds and a better understanding of the system of interest is obtained, it is often desirable to go back to a previous step.

Table 19.1 Advantages and disadvantages of simulation (adapted from Law and Kelton, 2000)

Advantages	Disadvantages
• Most systems with stochastic elements are too complex for analytical evaluation. Thus simulation is the only possibility. • Simulation allows the performance of an existing system to be estimated under some projected set of operating conditions. • Alternative proposed system designs can be compared to see which best meets a specified requirement. • Better control is exercised over the experimental conditions than when experimenting with the system itself. • Allows study of a system with a long timeframe, in compressed time, or even in expanded time.	• Each run of a stochastic model produces only estimates of a model's true characteristics for a particular set of input parameters. Thus several independent runs of the model are required. An analytical model, if appropriate, can produce the exact true characteristics. • It is expensive and time consuming to develop. • The large volume of numbers produced or the persuasive impact of a realistic animation often creates a tendency to place too much confidence in a study's results. • If a model is not a valid representation of a system under study, the results are of little use.

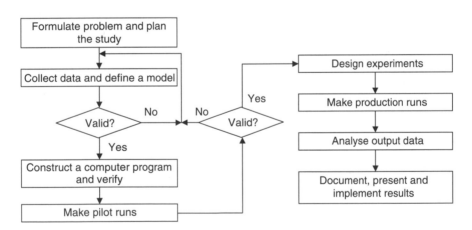

Fig. 19.3 Steps in a simulation study.

The first validation step concerns the involvement of people who are intimately familiar with the operations of the actual system. In the second validation step, pilot runs can be used to test the sensitivity of the model's output to small changes in an input parameter. If the output changes greatly, a better estimate of the input parameter must be obtained. Furthermore, if a system similar to the one of interest currently exists, output data from pilot runs of the simulation model could be compared with those obtained in reality.

Law and Kelton (1991) and Davis (1993) identify a number of generic pitfalls that can prevent successful completion of a simulation study and are – 15 years later – still appropriate. First of all, a set of well-defined objectives and performance measures should be defined at the beginning of the simulation study which suit all parties in the supply chain. Furthermore, key persons should be involved in the project on a regular basis. Next to this, some degree of abstraction is usually necessary. Some parts may be left out of the model completely; others may be aggregated. The summarized characteristics of the aggregated parts must be checked against expert opinion to see if they represent the situation fairly. Finally, data must often come from different disparate locations; to ensure the success of the modelling effort, it is necessary to obtain sufficient commitment of resources to ensure accurate, useful data about each of the links in the supply chain.

19.4   Modelling requirements for food supply chains

In the previous sections we characterized FSCNs in terms of parties involved, processes, products and alternative design strategies, and we discussed the rationale for modelling and simulation processes. Let us now relate these characteristics to requirements to be set for models specifically used for FSC simulation. We distinguish between requirements of simulation modelling that address the general case of SCM and requirements that are specific to the food industry. As far as the general case is concerned, we build on earlier work (van der Zee and van der Vorst, 2005). An overview of these requirements is meant to (1) support a review of current tools for supply chain simulation and (2) structure our discussion of the new tool.

19.4.1   General requirements of supply chain modelling

As stated, a typical supply chain involves multiple (semi)-autonomous parties, who may have several, possibly conflicting, objectives. Actions of one actor in the supply chain may influence product and/or process characteristics for the next actor. SCM requires, among others, the alignment of partner strategies and interests, high intensity of information sharing, collaborative planning decisions and shared IT tools. These requirements often represent major hurdles inhibiting the full integration of a logistics chain. Even when there is a strong partnership between logistics nodes, in practice there are potential conflict areas, such as local versus global interests, and a strong reluctance to share common information about production planning and scheduling, such as, for example, inventory and capacity levels (Terzi and Cavalieri, 2004). SCM requires trust and in-depth insight into each other's processes, which is difficult, since the widely followed competitive model suggests that companies will lose bargaining power and therefore the ability to control profits, as suppliers or customers gain knowledge (Barratt and Oliveira, 2001).

The aforementioned characteristics make clear that active participation and cooperation of all parties are essential ingredients for the effective design of new supply chain network scenarios. This is even more so since the complexity of the system and the solution space in terms of the number of alternative chain scenarios is significant. Involvement is therefore not only a prerequisite for solution acceptance, but also fosters creative minds in finding alternative and possibly better solutions, building on each other's expertise on specific chain operations. In order to facilitate an active involvement of decision makers in modelling and solution finding, high demands are set on model transparency and completeness. Transparency refers to the insight into model components and their workings, whereas completeness addresses a full overview of design parameters. This leads us to the following requirements for simulation model design (van der Zee and van der Vorst, 2005):

1. Model elements and relationships: Supply chains assume an integrated approach to physical transformation, data processing and decision making. Especially, the allocation of control policies to specific chain members and relationships, such as hierarchy and coordination, deserve explicit attention as decision variables. This requires the explicit notion of actors, roles, control policies, processes and flows in the model.

2. Model dynamics: The control of dynamic effects within the supply chain, as reflected in for example stock levels and lead times, is an important issue given the many parties involved. Therefore, the logistics of control, that is the timing and execution of decision activities, should be explicit. This requires the ability to determine the dynamic system state, calculate the values of multiple performance indicators at all times and, even more important, allocate performance indicators to the relevant supply chain stages.

3. User interface: The active and joint participation of the problem owners, that is, the supply chain partners, in the simulation study is required for two reasons (Hurrion, 1991; McHaney and Cronan, 1998; Bell et al., 1999; Robinson, 2002). First, as a means of creating trust in the solution and among the parties involved, so there is a better chance of acceptance of the outcomes of the study. Second, the quality of the solution may be improved. This refers to model correctness as well as the performance of the chain scenario. Clearly, it is almost impossible for the analyst to have all relevant information on chain dynamics. Therefore, the domain-related contribution of the problem owner in terms of alternative solutions is vital to the success of the project. Given the foreseen role of the problem owners, an explicit choice and

representation of decision variables that appeals to their imagination is important. This boils down to visibility and understanding of all supply chain processes in the model, see point 1 above – Model elements and relationships.

4. Ease of modelling scenarios: The execution of 'what if' analysis should be transparent, given the complexity of the supply chain, the large number of conceivable scenarios and the wishes and requirements of the problem owners. This concerns both the choice of building blocks and the time required for tailoring, and adapting them to the right format for model adoption. Another demand is model reuse, because of the combination of volatile business environments and the major modelling effort required. Reusable models may help to increase the speed of modelling and analysing alternative scenarios, while reducing costs of decision support.

19.4.2   Specific requirements of modelling food supply chains

Next to the general requirements of modelling, additional, more specific, requirements for modelling FSCs should be mentioned. Here we will address the issue of modelling food quality and environmental sustainability, being prime performance indicators for FSCs.

1. Model elements and relationships: Modelling food quality assumes the presence of attributes of model elements that, next to logistics cost and service aspects, express the actual product status on quality. Methods must also be defined for modelling quality decay owing to progress in time and environmental conditions. In turn, attribute values of food and their foreseen behaviour may be an input to dedicated (proactive) control policies for operating the supply chain, being responsive to, for example, the (estimated) best-before date. Clearly, quality preservation is a major issue in FSCN, which can be improved via the use of sophisticated environmental conditioning techniques (in transport and warehousing) and a reduction in lead times. Of course these new techniques and supply chain processes should be evaluated for energy use and environmental load to guarantee sustainability. Model elements should incorporate these specific characteristics of FSCN, especially the keeping quality constraints for products and the occurrence of quality decay while progressing through the supply chain under specific environmental conditions.

2. Model dynamics: Food quality tends to be a continuous variable. Here we consider the process of its decay at discrete moments in time. This assumes an event-related 'inspection' of relevant food attributes. Finally, the model should be able to deal with the aspect of uncertainty. FSCNs deal with biological products that are not homogeneous in product quality and yield (see Section 19.2). Control policies in the model should be able to distinguish between batches with different characteristics and

make (logistical) decisions based on this information. This will allow for the concept of 'quality controlled logistics' (van der Vorst *et al.*, 2007).

Recall that our choice of discrete event simulation is motivated by the type of problem studied, that is the design of FSCs, which are (1) characterized by uncertainties in product quality and logistics as well as their interaction, and (2) evaluated for logistic costs and service, product quality and sustainability.

19.4.3   Modeling food quality change

In the food science literature, much attention has been paid to food quality change modelling and the development of time temperature indicators (TTI) to monitor the temperature conditions of food products individually throughout distribution (Taoukis and Labuza, 1999; Schouten *et al.*, 2002a,b; Tijskens, 2004). Typically, next to biological variations, food quality is determined by time and environmental conditions (such as temperature, humidity and the presence of contaminants), see Fig. 19.4. Environmental conditions may be influenced by, for example, the type of packaging, way of loading and the availability of temperature conditioned transportation means and warehouses. Figure 19.4 shows an idealized pattern for product decay for a particular perishable product. Typically, realistic values, shown in the figure as individual measurements (+, x, o), deviate from this pattern

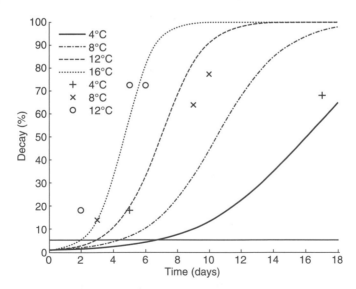

Fig. 19.4   Example of idealized food quality decay as a function of time for alternative temperature conditions for a specific product (+, × and o show significant outliers measured in a laboratory test) (Schouten *et al.*, 2002a).

to some extent. This uncertainty follows from, among others, biological variations (see above) and non-homogeneous conditioning. For example, temperature distributions within a batch of food products tend to be non-uniform as it tends to be warmer in the core.

The use of time-dependent quality information in the design of perishable inventory management systems is gaining increasing attention from researchers (e.g. van Donselaar *et al.*, 2006). However, using this information in the design of distribution systems is only sparingly addressed in the literature. We only found one reference in literature; Giannakourou and Taoukis (2003) consider the potential of a TTI-based system for optimization of frozen product distribution and stock management using Monte Carlo simulation techniques. TTI-responses are translated to the level of product deterioration, at any point in the distribution system, which enables the classification of products according to their remaining keeping quality (shelf life). Their results indicate that the number of rejected products in the market can be minimized using a TTI-based management system based on least-shelf-life-first-out (LSFO), in which products with the closest expiration date are advanced first.

19.4.4   Review of simulation tools for food supply chain design

Many types of models have been developed to support supply chain design (Min and Zhou, 2002; Gunasekaran, 2004; Meixell and Gargeya, 2005; Kleijnen, 2005). Kleijnen and Smits (2003) distinguish four simulation types for SCM: (1) spreadsheet simulation, (2) system dynamics (SD), (3) discrete event dynamic system simulation (DEDS) and (4) business games. They conclude that the question to be answered determines the simulation type needed; SD provides qualitative insights, whereas DEDS simulation quantifies results and incorporates uncertainties. Games can educate and train users. In many cases, discrete event simulation is a natural approach for supporting supply chain network design, as their complexity obstructs analytical evaluation, see for example Ridall *et al.* (2000) and Huang *et al.* (2003). Discrete event simulation tools, however, tend to stress logistics analysis rather than product quality or sustainability.

In the past, many simulation tools for supply chain analysis have been developed. Van der Zee and van der Vorst (2005) present a literature review in which they assess the modelling characteristics of these packages, given the previous requirements of FSC modelling. They conclude that current simulation approaches cannot fully cope with the demands on model and tool design for supply chain analysis. They mention an important shortcoming of available tools concerning the modelling of supply chain decision making. In line with earlier findings in the field of manufacturing (see for example, Mize *et al.*, 1992; and Karacal and Mize, 1996), they conclude that decision makers control rules and their interactions are mostly 'hidden'. A reason for this may be the analyst's choice of building blocks, which does not appeal to supply chain partners. Further, control elements may be dispersed throughout the model, being associated with various building blocks or with the time-indexed scheduling of events. Also, they may simply not be visualized. The 'hiding' of control is surprising as control structures are intrinsic to supply chains. This implicit modelling harms realism, as well as harming modelling flexibility and modularity. Essentially, the implicit modelling of decision making in simulation analysis can be traced back to the (implicit) reference models underlying simulation tool libraries and the analyst's activities in model building. As far as the embedding of food quality models in discrete event simulation models is concerned, we could find no examples in literature.

19.4.5   Contribution of this chapter

The challenge being addressed in this chapter is to embed food quality models and sustainability issues together with logistics processes in discrete event simulation models, in order to facilitate an integrated approach towards logistic, sustainability and product quality analysis of FSCs. We hypothesize that integrated decision making will result in overall better decisions compared to disciplinary decision making when taking only one of these aspects into account. A key contribution of discrete event simulation lies in its capability to model and trade off elementary uncertainties underlying product quality and chain logistics, as well as their interaction. Our focus in this paper is on exploiting this flexibility by the development of a simulation environment for FSC modelling, rather than specific models, like the aforementioned model by Giannakourou and Taoukis (2003). Typically, such an environment allows a variety of models to be built to evaluate a wide range of FSC issues, such as the incorporation of new chain actors, use of innovative and sustainable transport modes, consolidation practices and concepts like vendor managed inventory, whilst taking relevant uncertainties into account.

The foreseen advantages of the integrated approach would be in the speed and quality of decision making about FSC design. Decision speed may increase as many iterations may be avoided following from the separate consideration of food quality and chain logistics. But, probably more important, the quality of solutions may be improved as more and other innovative scenarios may be tested, following on from a total performance overview. One of those innovative scenarios is, for example, using quality information proactively to direct distribution processes to profitable markets, also called 'quality controlled logistics' (see van der Vorst et al., 2007).

Starting from the above observations of the needs, available means and opportunities for FSC design, we propose a new simulation environment, named ALADIN™ (Agro-Logistic Analysis and Design INstrument). ALADIN™ concerns a library of building blocks for simulation modelling and builds on the discrete event simulation tool Enterprise Dynamics™. Next to basic building blocks for modelling FSC infrastructures (producers, distributors etc), and flows of goods, information and so on, its library embeds food quality models. To show the potential of integrated decision making for FSC design, we discuss a case study concerning the import of pineapples from Ghana to the Netherlands.

19.5   Simulation environment

In this section we introduce the simulation environment ALADIN™. After a general characterization of the tool, we discuss it in some detail being guided by the classification of demands of simulation modelling for FSCs, see Section 19.4.

19.5.1   General description

ALADIN™ is a visual interactive simulation environment building on the logistics suite of the object oriented simulation package Enterprise Dynamics™ (ED). It concerns a library of generic building blocks for modelling FSCs and their behaviour. The choice of underlying concepts is largely based on the modelling framework proposed in our earlier work (van der Zee and van der Vorst, 2005; van der Zee, 2006). Food quality models for a range of fresh products are embedded in this library. They relate food quality to food logistics in terms of time and choice of resources for food production, transportation and storage.

19.5.2   Model elements and relationships

ALADIN™ is based on three key concepts: agents, jobs and flows. *Agents* represent supply chain network entities (such as planners, retail outlets, producers and distribution systems) as autonomous objects that are assigned decision making intelligence. All chain activities are defined as *jobs*, including activities related to decision making. Where physical jobs result in goods, control jobs result in job definitions for agents in the controllers' domain of control. *Flow items* (also called business entities) constitute the movable objects within a supply chain. We include four types of flow items in the modelling framework: product flows, information flows, resources that facilitate the transformation processes (assignment of capacity) and job definitions. Job definitions specify a job in terms of, for example its input, processing conditions and the agents to whom the resulting output should be sent. By introducing a demand controller in ALADIN™, physical and information and control layers can be separated (Fig. 19.5). In this way, model transparency is increased, as discussed in Section 19.3.

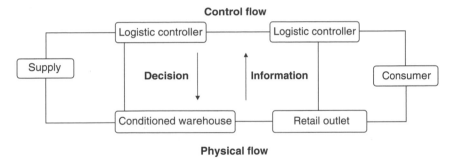

Fig. 19.5 ALADIN™ improves modelling transparency by making a distinction between the goods flow (physical flow) and its planning and control (control flow).

Table 19.2 Specific agents in ALADIN™

Agents	Representation
Production unit	Food factory or a grower, who produces products with biological variation in quality and quantity (seasonality)
Transportation unit	Climate controlled truck or vessel with specific temperature and modified atmosphere settings and related energy use and CO_2 emission per unit
Storing and distribution unit	Warehouse or retail outlet with specific climate control characteristics and related energy use and CO_2 emission per unit
Demand unit	Marketplace with demand for products with specific shelf lives, colours, etc
Food product	Specific food product (e.g. pepper, cut vegetable) with its specific quality decay model, related to the settings of environmental conditions in time
Demand controller	Explicit modelling of information flow and decision-making activity that activates the goods flow

In ALADIN™, specific agents have been developed, see Table 19.2. Supply chain network models are composed of a reusable set of software components (building blocks, called 'atoms' in ED) that represent agents (with multiple inputs and outputs), their control policies (e.g. inventory policies, routing policies) and their interaction protocols, that is message types that regulate the flow of information, goods and cash. Besides these supply chain building blocks, ALADIN™'s core consists of quality change models. These models describe quality behaviour, for example botrytis in strawberries or weight loss of bell peppers, under specified conditions (temperature, relative humidity, modified atmosphere, etc). They incorporate parameters that reflect stochastic biological variations in product quality change and are developed by experts in laboratory experiments under controlled conditions (see Schouten *et al.*, 2002a,b).

Alternative designs for perishable product supply chains (see the redesign strategies in Section 19.2.3) can be simulated, visualized and analysed. ALADIN™ adds the indicators of product quality or product freshness (remaining keeping quality and product waste) and energy use and CO_2 emissions to classical performance indicators such as transportation costs, stock levels and delivery reliability (e.g. Gunasekaran *et al.*, 2000). In this way, ALADIN™ helps the decision maker to trade off logistics costs and service (product quality, sustainability and availability), when assessing specific (re)designs of the FSC.

19.5.3   Model dynamics

Model dynamics is realized by job execution. We capture the dynamic behaviour of the chain processes by modelling the FSC as a network of agents, jobs and flows with precedence relationships; the jobs can be triggered by multiple causes and have outcomes and processing times that depend on the entities processed and available resources. This includes the calculation of (variations in) product quality aspects (such as weight, colour and firmness) related to the specific conditions, to which the products have been exposed, and sustainability indicators.

19.5.4   User interface and ease of modelling

In our choice of concepts we tried to adopt basic logistic terminology and developed a library of recognizable building blocks, starting from experiences of several industrial projects. This includes an explicit representation of supply chain coordination in terms of decision makers, their activities and their mutual tuning of activities, also see Section 19.4.2.

19.5.5   Applications

ALADIN™ has been successfully applied in several case studies in which new supply chain scenarios have been evaluated. For example, we compared alternative distribution systems (e.g. warehousing, cross docking and different transport modes under different environmental conditions) for the export of fresh products such as peppers and tomatoes. Furthermore, new ordering policies for fresh products have been evaluated, in which a balance is sought between stock-outs and product waste (shrinkage) in retail outlets. ALADIN™ visualizes and quantifies the consequences of design choices for the remaining shelf life of the product and the level of environmental load. In order to illustrate the advantages of integral decision making and the capabilities of ALADIN™ in somewhat more detail, we discuss one of the case studies in the next section.

19.6   Case study: pineapple supply chain

To illustrate the added value of an integrated analysis of alternative FSC designs, we consider a case study concerning the import of pineapples from Ghana to the Netherlands. In this case two import supply chain scenarios have been compared for logistics costs, product quality decay, energy use and CO_2 emissions. First we will consider the background of the case and the scenarios that were chosen for further analysis. Next we consider the data collection and modelling process. We conclude with a discussion of the simulation results and a brief evaluation of the contributions made by ALADIN™ in modelling and analysing alternative supply chain scenarios.

19.6.1   Background

The market for fresh pineapple in Europe is increasing; European consumers demand ready-to-eat products with a sweet taste and golden colour. The import of fresh pineapples to the Netherlands predominantly from Ghana, Costa Rica, Ivory Coast and South Africa amounts to about 35 tonnes on a yearly basis, although almost 75% of that is redistributed mainly to Germany and Russia. Pineapples intended for shipping are harvested when green, while those intended for immediate eating are harvested in the semi-ripe state and those intended for canning in the ripe state. Only sound fruit may be approved for transport; pineapples require particular temperature, humidity/moisture and ventilation conditions. Intact pineapples can be kept for several weeks, whereas cut pineapple has a much more restricted shelf life. Based on discussions with two product experts (who have performed studies on the keepability of cut pineapples under specific laboratory conditions, see Tijskens 2004) and participating chain partners, a generic quality decay model was developed in order to estimate the quality decay of cut pineapple (see Fig. 19.6). This model uses the yeast concentration as limiting quality attribute, starting after cutting the whole pineapple.

It can be seen that the keepability of cut pineapple varies from 6 to 9 days at a fixed temperature of 4 °C. This is a result of biological variation in the initial quality of the product. Biological variation within the same batch causes differences in the initial quality of cut pineapple, such that different packages of cut pineapple may have a different pattern of quality decay at the same temperature. Each package of cut pineapple is provided with a guaranteed best-before date (BBD) at a maximum storage temperature of 4 °C, which is equal to 'the current date + 6 days'.

Many fresh pineapples reach the Netherlands by costly air transport. This is motivated by the fact that, so far, alternative ways of transportation, like over sea, have resulted in significant quality decay and product shrinkage, owing to lengthy transportation times. Major developments in

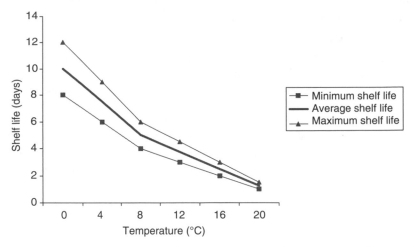

Fig. 19.6 Average and variability in shelf life of cut pineapple depending on the temperature.

quality preservation via the use of modified atmosphere packaging and sophisticated chilling techniques, however, challenge Dutch importers and retailers to reconsider their means of transportation. Could transport by sea now be an option, to reduce overall chain costs? A project group including all supply chain members identified several alternative FSC designs of which we will discuss two for illustrative purposes. These scenarios are (see Fig. 19.7):

1. Producing pineapples in Ghana, cutting in Ghana, air transporting cut pineapple to the Netherlands and distributing the cut pineapples to retail outlets ('the air chain').
2. Producing pineapples in Ghana, sea transporting intact pineapples, cutting in the Netherlands and distributing the cut pineapples to retail outlets ('the sea chain').

To measure the effectiveness and efficiency of alternative designs, the project team formulated three key performance indicators for this FSC: the distribution costs along the supply chain (we only focus on transport and warehousing and leave out the costs of the cutting process), the energy and emissions during distribution (regarding emissions only CO_2 emissions are considered, where 73 g CO_2 is calculated per MJ direct energy use) and the product quality when arriving at the retail store. This last factor is measured by three sub-indicators:

- the remaining number of days until the predetermined BBD. In other words, the remaining selling time at the retail outlet

(a)

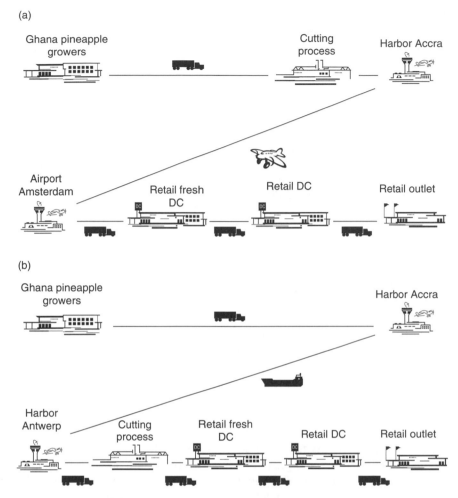

(b)

Fig. 19.7 Two supply chain scenarios for importing pineapples from Ghana:
(a) air transport of sliced pineapples, (b) sea transport of whole pineapples.

- the remaining keepability of the product at a storage temperature of
  4 °C according to the expert model in Fig. 19.7. In other words, for how
  long will the yeast concentration still be acceptable?
- the percentage of products for which the BBD is not reached yet, but
  has a yeast concentration which is not acceptable any more.

Note that the definition of performance indicators is case dependent and
relates to the business strategies of participating companies, product and
process characteristics.

19.6.2   Collecting data

In order to be able to model the scenarios, chain data was collected using document analyses and experts interviews (see Tables 19.3 and 19.4). Table 19.3 shows all distribution activities (transport and storage) from harvest to retail outlet for the air chain. For each activity data are collected about the duration, temperature, cost, direct energy use and emissions. Each supplied batch triggers the activities represented in the tables. Table 19.4 presents data for the sea chain as far as they are different from those of the air chain.

At the time of the research there were six flights per week from Ghana, each distributing 160 kg of cut pineapples. By sea transport, only two shipments per week were taking place, each distributing 1200 kg intact pineapples. Note that 2.5 kg intact pineapple gives about 1 kg cut pineapple. Therefore, both scenarios are comparable in volume, because in each scenario 960 kg cut pineapple is supplied to the retailer.

19.6.3   Modelling and analysis: ALADIN™

Evaluating the two scenarios on the defined performance indicators required modelling and analysis of several supply chain scenarios. We did so using ALADIN™. Note that in the project much more complex scenarios were evaluated; the two presented here are just for illustration purposes.

We modelled the supply chain, applying some of the reusable building blocks and designed scenarios by setting the model elements, for example, applying air transport versus transport by ship. Alternative designs of the product supply chains were simulated, visualized and analysed. By changing the environmental conditions to which the pineapples are exposed (e.g. by using new packaging materials or conditioned reefer containers) we could simulate the impact of changes in the distribution system on keeping quality and sustainability indicators. Applying new logistical concepts changes the control and product flows which have an impact on costs and, via a change in the duration or processes, changes the keeping quality of the pineapples and the environmental load.

19.6.4   Simulation results

Table 19.5 presents an overview of the main model outcomes based on the data and assumptions described before. It shows that from a cost and sustainability perspective, the sea chain provides the best results, when looking at product quality the air chain performs slightly better. Note that the interpretation and weighting of the outcomes of the study is left to the decision makers. Simulation will not provide this answer. Further decision support may come from alternative techniques, like multi criteria analysis (see, for example, Quariguasi Frota Neto et al., 2008). We do not discuss these techniques here.

Table 19.3 Data for the air chain from Ghana

Activity	Time (h)	Temperature (°C)	Logistic cost (Euro cents)	Energy (MJ)	Remarks
1. Transport grower–producer	4	25	2 per kg	16.5 per tonne km	Open truck
2. Storage producer (whole)	44	7	52 per pallet (780 kg) per day	0.001677 per second	Cold storage
3. Storage producer (cut)	5	4	52 per pallet (320 kg) per day	0.001677 per second	Cold storage
4. Transport producer–airport	0.5	4	4 per kg	2.27 per tonne km	Climate controlled
5. Storage and handling airport	4.5	5	100 per pallet (320 kg) per day + 2.5 per kg	0.0049 per second	Stored in reefer container
6. Air transport	7	8	80 per kg + 22 per kg	9.76 per tonne km	
7. Storage airport Schiphol	6	7	15 per kg	0.001677 per second	
8. Transport Schiphol–retail fresh DC	1	5	5 per kg	2.27 per tonne km	
9. Storage retail fresh DC	20	4	52 per pallet (320 kg) per day	0.001677 per second	
10. Transport–retail DC	1	4	5 per kg	2.27 per tonne km	
11. Cross docking retail DC	4	4	52 per pallet (320 kg) per day	0.001677 per second	
12. Transport retail DC–retail outlet	1	4	5 per kg	2.27 per tonne km	

Table 19.4   Data for the sea chain from Ghana

Activity	Time (h)	Product temp. (°C)	Logistic cost (Euro cents)	Energy (MJ)	Remarks
2. Storage seaport	6	25	0	0	Wooden barn
3. Sea transport	288	8	19 per kg	0.1 per tonne km	
4. Storage Antwerp	24	7	52 per pallet (780 kg) per day	0.001677 per second	Cold storage
6. Storage producer (whole)	9, 33 or 57*	7	52 per pallet (780 kg) per day	0.001677 per second	
7. Storage producer (cut)	4	4	52 per pallet (320 kg) per day	0.001677 per second	
8. Transport producer–retail fresh DC	1	4	5 per kg	2.27 per tonne km	
9. Storage retail fresh DC	19	4	52 per pallet (320 kg) per day	0.001677 per second	

* Retail outlets are ordering every day, while produce is supplied only once per three days. This implies that some pineapples are stored for a longer time than others.

Table 19.5   Comparing the overall results of the two scenarios

Scenario	Average quality	Logistics costs	Energy and emission
Air chain Ghana	+ BBD 3.9 days; bad before BBD 5.9%	– 1192 Euro/week	– – 77,227 MJ/week 5638 kg CO_2
Sea chain Ghana	+/– BBD 3.7 days; bad before BBD 10.6%	+ 793 Euro/week	+ 14,508 MJ/week 1059 kg CO_2

When we look closer at the simulation results of the air chain, the following issues come to the prominence. Air transport is responsible for over 70% of all logistic costs. Energy use happens mostly during air transport (85%), but the open truck from harvest to producer also uses a lot of energy (10%). Looking at remaining keepability (according to the expert model) at the moment the products arrive at the retail outlet, the average is below 5 days, with a variation from less than 3.5 days to more than 5.5 days. The average remaining selling time (according to the BBD) at the time of arrival at the retail outlet is equal to 3.9 days. The 6 days cutting BBD seems to be realistic for this chain. A rather small percentage of all products –5.9% on average, according to the expert model – has a keepability that is less than the remaining selling time according to the BBD. Typically, they reflect products with a bad initial quality.

Results for the sea chain indicate that sea transport is responsible for almost 60% of all logistic costs. Energy use occurs mostly during transport from grower to sea port (over 50%). At the moment the products arrive at the retail outlet, the average remaining keepability at 4 °C (according to the expert model) is about 4 days, with a variation from less than 3.5 days to more than 4.5 days. Setting the BBD at 6 days after cutting is not realistic in this case. Five days after cutting 10.6% of all products has a keepability which is less than the BBD-code indicates (based on 6 days).

19.6.5   Evaluation

Let us now address the role of ALADIN™ in FSC design, building on the experiences of the case study. First, the integration of facilities for modelling product quality, sustainability and product logistics provides an improved means of analysing FSCs. Instead of studying effects of alternative scenarios on product quality, sustainability and logistics using separate tools, a single tool suffices. More effective solutions may result from this approach, as interaction effects for logistics and product quality may be studied. The speed and quality of decision making clearly benefited from the integrated approach. Furthermore, the definition of model building blocks and relationships in line with the modelling framework proposed by van der Zee and van der Vorst (2005) and van der Zee (2006), resulted in transparent models. In this way they contributed to the communicative value of visual simulation models. A final remark concerns the need for screening candidate solutions. Typically, an FSC allows for configuring a multitude of alternative supply chain configurations. Using simulation for modelling all configurations may simply be too time consuming. We therefore advocate the use of a screening procedure for preselecting alternative configurations. This may include, for example, deterministic models, or expert consultation.

19.7   Conclusions and future trends

This chapter has dealt with modelling and simulation of food supply chain scenarios to facilitate redesign projects. The challenge addressed is to embed food quality models and sustainability indicators in discrete event simulation models, in order to facilitate an integrated approach towards logistic, sustainability and product quality analysis of FSCs. By introducing a new discrete event simulation tool, ALADIN™, which answers this challenge, we aim to provide a new and improved means of analysing and redesigning

FSCs. Its core consists of the combination of reusable process building blocks and quality decay models that facilitate the modelling of FSCs. As such, it contributes to improved decision making with respect to FSC design. Specific strengths of the tool relate to:

- The integration of logistics, quality decay and sustainability modelling: The presence of these models makes it possible to use simulation in workshop settings as a transparent tool for trading off FSC performance with respect to all respective elements.
- The explicit modelling of control structures, building on an explicit modelling framework: Rather than relying on the implicit mental reference models of the analyst and the availability of standard building blocks in the library of Enterprise Dynamics™, new building blocks were developed in ALADIN™ to offer the analyst guidance in modelling specific FSCs. This provides communication via an explicit and well-defined notion of concepts, and helps to reduce the modelling efforts of the analyst, because of the possibilities for reuse of model classes, i.e. agents, flow items and jobs.
- The capabilities for more effective and efficient decision support of FSC design: The case example of a pineapple supply chain showed that the tool provides an integrated means for participants to generate transparency in the supply chain network and jointly develop and evaluate innovative supply chain scenarios.

Future research will focus on the further development of ALADIN™, extending the complexity of quality decay models and quality interaction effects of multiple products distributed together. Furthermore, in line with van der Zee and Slomp (2009) and van der Zee (2007) we are researching a promising possibility for using ALADIN™ as a basis for a simulation gaming and training tool. Such a tool should enable managers jointly to evaluate alternative decision making scenarios in FSCs using multiple performance indicators, such as costs, product quality and sustainability.

19.8   Sources of further information and advice

Simulation literature may be classified according to three perspectives: model building and coding for simulation, statistics for simulation and doing projects using simulation. Typically, the main focus of (course) books is tailored to one of these angles. Most books on model building and coding start by assuming the use of a specific simulation language. As a simulation language mostly sets its own standards on modelling and features, it is necessary to consider the choice of language prior to choosing a book, examples include Rosetti (2010), Al-Aomar et al. (2009) and Hauge and Paige (2004). A main entry for statistics of simulation is the work by Law and Kelton (2000) and Law (2007). It addresses all the main issues and supplies adequate references to address more specific questions. Whereas previous work tends to be of a rather technical nature, addressing model building and statistics, the work by Robinson (2004) starts from a project view of simulation use. Being non-software specific, it guides the reader through all the main stages in the simulation study by giving details of questions to be addressed and indicating ways of answering them. Further, we mention the book by Chung (2003). The case studies that are part of this work may be particularly helpful for projects in practice.

19.9   Acknowledgements

We thank Taylor and Francis for allowing us to use our article 'Simulation modelling for food supply chain redesign; integrated decision making on product quality, sustainability and logistics', which was published in the *International Journal of Production Research* (van der Vorst, Tromp and van der Zee, 2009).

19.10   References

ALFIERI, A. and BRANDIMARTE, P. (1997). 'Object-oriented modelling and simulation of integrated production/distribution systems'. *Computer Integrated Manufacturing Systems*, **10**(4), 261–6.

AL-AOMAR, R., ULGEN, O. and WILLIAMS, E. (2009). *Process Simulation Using WITNESS: Including Lean and Six-Sigma Applications*, Wiley, New York.

BANKS, J., CARSON II, J.S. and NELSON, B.L. (1996). *Discrete-Event System Simulation*, 2nd edition, Prentice-Hall, New Jersey.

BARRATT, M. and OLIVEIRA, A. (2001). 'Exploring the experiences of collaborative planning initiatives'. *International Journal of Physical Distribution & Logistics Management*, **31**(4), 266–29.

BELL, P.C., ANDERSON, C.K., STAPLES, D.S. and ELDER, M. (1999). 'Decision-makers' perceptions of the value and impact of visual interactive modelling'. *Omega – The International Journal of Management Science*, **27**, 155–65.

BOURLAKIS, M.A. and WEIGHTMAN, P.W.H. (2004). *Food Supply Chain Management*, Blackwell Publishing, Oxford.

CHUNG, C.A. (2003). *Simulation Modeling Handbook: A Practical Approach*, CRC Press, Boca Raton, FL, USA.

DAVIS, T. (1993). 'Effective supply chain management', *Sloan Management Review*, Summer, 35–46.

EDWARDS-JONES, G., CANALS, L.M., MOUNSONE, N., TRUNINGER, M., KOERBER, G. *et al.* (2008). 'Testing the assertion that "local food is best": the challenges of an evidence-based approach'. *Trends in Food Science & Technology*, **19**, 265–74.

GIANNAKOUROU, M.C. and TAOUKIS, P.S. (2003). 'Application of a TTI-based distribution management system for quality optimisation of frozen vegetables at the consumer end'. *Journal of Food Science; Food Engineering and Physical Properties*, **68**(1), 201–9.

GUNASEKARAN, A. (2004). 'Supply chain management: Theory and applications'. *European Journal of Operational Research*, **159**(2), 265–8.

GUNASEKAREN, A., MACBETH, D.K. and LAMMING, R. (2000). 'Modelling and analysis of supply chain management systems'. *Journal of the Operational Research Society*, **51**, 1112–15.

HAGELAAR, J.L.F., VAN DER VORST, J.G.A.J. and MARCELIS, W.J. (2004). 'Organising life-cycles in supply chains: Linking environmental performance to managerial designs'. *Greener Management International*, **45**(Spring), 27–42.

HAUGE, J.W. and PAIGE, K.N. (2004). *Learning SIMUL8: The Complete Guide*, 2nd edition, Plain Vu Publishers, Bellingham, UK.

HUANG, G.Q., LAU, J.S.K. and MAK, K.L. (2003). 'The impacts of sharing production information on supply chain dynamics: a review of the literature'. *International Journal of Production Research*, **41**(7), 1483–517.

HURRION, R.D. (1991). 'Intelligent visual interactive model-ing'. *European Journal of Operational Research*, **54**(3), 349–56.

JONGEN, W.M.F. and MEULENBERG, M.T.G. (2005). *Innovation in Agri-Food Systems, Product Quality and Consumer Acceptance*, Wageningen Academic Publishers, Wageningen.

KARACAL, S.C. and MIZE, J.H. (1996). 'A formal structure for discrete event simulation. Part I: Modeling multiple level systems'. *IIE Transactions*, **28**(9), 753–60.

KETTENIS, D.L. and VAN DER VORST, J.G.A.J. (2007). 'Discrete-event simulation', in *Decision Sciences*, G.D.H. Claassen, T. Hendriks, E.M.T. Hendrix (eds), Chapter 13, Wageningen Academic Publishers, The Netherlands.

KLEIJNEN, J.P.C. (2005). 'Supply chain simulation tools and techniques: a survey'. *International Journal of Simulation & Process Modelling*, **1**(1/2), 82–9.

KLEIJNEN, J.P.C. and SMITS, M.T. (2003). 'Performance metrics in supply chain management'. *Journal of the Operational Research Society*, **3**, 1–8.

LAMBERT, D. and COOPER, M.C. (2000). 'Issues in supply chain management'. *Industrial Marketing Management*, **29**, 65–83.

LAW, A.M. (2007). *Simulation Modeling and Analysis*, 4th edition, McGraw-Hill, Boston, USA.

LAW, A.M. and KELTON, W.D. (1991). *Simulation Modelling and Analysis*, 2nd edition, McGraw-Hill, Boston, USA.

LAW, A.M. and KELTON, W.D. (2000). *Simulation Modeling and Analysis*, 3rd edition, McGraw-Hill, Boston, USA.

LAZZARINI, S.G., CHADDAD, F.R. and COOK, M.L. (2001). 'Integrating supply chain and network analyses, the study of netchains'. *Journal on Chain and Network Science*, **1**, 7–22.

LINTON, J.D., KLASSEN, R. and JAYARAMAN, A. (2007). 'Sustainable supply chains: an introduction'. *Journal of Operations Management*, **25**, 1075–82.

LUNING, P.A. and MARCELIS, W.J. (2006). 'A techno-managerial approach in food quality management research'. *Trends in Food Science and Technology*, **17**, 378–85.

MCHANEY, R. and CRONAN, T.P. (1998). 'Computer simulation success: on the use of the end-user computing satisfaction instrument: a comment'. *Decision Sciences*, **29**(2), 525–36.

MEIXELL, M.J. and GARGEYA, V.B. (2005). 'Global supply chain design: a literature review and critique'. *Transportation Research Part E*, **41**, 531–50.

MIN, H. and ZHOU, G. (2002). 'Supply chain modelling: past, present and future'. *Computers and Industrial Engineering*, **43**(1–2), 231–49.

MIZE, J.H., BHUSKUTE, H.C., PRATT, D.B. and KAMATH, M. (1992). 'Modeling of integrated manufacturing systems using an object-oriented approach'. *IIE Transactions*, **24**(3), 14–26.

PIDD, M. (1999). 'Just modelling through: A rough guide to modelling', *Interfaces*, **29**(2), 118–32.

QUARIGUASI FROTA NETO, J., BLOEMHOF-RUWAARD, J.M., VAN NUNEN, J.A.E.E. and VAN HECK, E. (2008). 'Designing and evaluating sustainable logistics networks'. *International Journal of Production Economics*, **111**, 195–208.

RIDALL, C.E., BENNET, S. and TIPI, N.S. (2000). 'Modeling the dynamics of supply chains'. *International Journal of Systems Science*, **31**(8), 969–76.

ROBINSON, S. (2002). 'General concepts of quality for discrete-event simulation'. *European Journal of Operational Research*, **138**(1), 103–17.

ROBINSON, S. (2004). *Simulation – The practice of model development and use*, Wiley, Chichester, UK.

ROSSETTI, M.D. (2010). *Simulation Modeling and Arena*, Wiley, Hoboken, USA.

SCHOUTEN, R.E., KESSLER, D., ORCARAY, L. and VAN KOOTEN, O. (2002a). 'Predictability of keeping quality of strawberry batches'. *Postharvest Biology and Technology*, **26**, 35–47.

SCHOUTEN, R.E., TIJSKENS, L.M.M. and VAN KOOTEN, O. (2002b). 'Predicting keeping quality of batches of cucumber fruit based on a physiological mechanism'. *Postharvest Biology and Technology*, **26**, 209–20.

SILVER, E.A., PYKE, D.F. and PETERSON, R. (1998). *Inventory Management and Production Planning and Scheduling*, 3rd edition, John Wiley & Sons, New York.

SRIVASTAVA, S.K. (2007). 'Green supply-chain management: A state-of-the-art literature review'. *International Journal of Management Reviews*, **9**(1), 53–80.

TAOUKIS, P.S. and LABUZA, T.P. (1999). 'Applicability of time temperature indicators as shelf life monitors of food products'. *Journal of Food Science*, **54**(4), 783–8.

TERZI, S. and CAVALIERI, S. (2004). 'Simulation in the supply chain context: a survey'. *Computers in Industry*, **53**, 3–16.

TIJSKENS, P. (2004). *Discovering the Future, Modelling Quality Matters*, Phd Thesis Wageningen University.

VAN DER VORST, J.G.A.J. and BEULENS, A.J.M. (2002). 'Identifying sources of uncertainty to generate supply chain redesign strategies'. *International Journal of Physical Distribution and Logistics Management*, **32**(6), 409–30.

VAN DER VORST, J.G.A.J., BEULENS, A.J.M. and VAN BEEK, P. (2005). 'Innovations in logistics and ICT in food supply chain networks', in *Innovation in Agri-Food Systems, Product Quality and Consumer Acceptance*, Jongen, W.M.F. and Meulenberg, M.T.G. (eds), Wageningen Academic Publishers, Wageningen, 245–92.

VAN DER VORST, J.G.A.J., VAN KOOTEN, O., MARCELIS, M. and LUNING, P. (2007). 'Quality controlled logistics in food supply chain networks: integrated decision-making on quality and logistics to meet advanced customer demands'. *Proceedings of the Euroma 2007 Conference*, Ankara, 18–20 June.

VAN DER VORST, J.G.A.J., TROMP, S. and VAN DER ZEE, D.J. (2009). 'Simulation modelling for food supply chain redesign – Integrated decision making on product quality, sustainability and logistics'. *International Journal of Production Research*, 47:23, 6611–6631.

VAN DER ZEE, D.J. (2006). 'Modelling decision making and control in manufacturing simulation'. *International Journal of Production Economics*, **100**(1), 155–67.

VAN DER ZEE, D.J. (2007). 'Developing participative simulation models – Framing decomposition principles for joint understanding'. *Journal of Simulation*, **1**(3), 187–202.

VAN DER ZEE, D.J. and VAN DER VORST, J.G.A.J. (2005). 'A modelling framework for supply chain simulation – opportunities for improved decision making'. *Decision Sciences*, **36**(1), 65–95.

VAN DER ZEE, D.J. and SLOMP, J. (2009). 'Simulation as a tool for gaming and training in operations management – a case study'. *Journal of Simulation*, **3**(1), 17–28.

VAN DONSELAAR, K., VAN WOENSEL, T., BROEKMEULEN, R. and FRANSOO, J. (2006). 'Inventory control of perishables in supermarkets'. *International Journal of Production Economics*, **104**(2), 462–72.

ZSIDISIN, G.A. and SIFERD, S.P. (2001). 'Environmental purchasing: A framework for theory development'. *European. Journal of Purchasing and Supply Management*, **7**, 1–73.

20

Adoption of e-business solutions in food supply chains

M. Vlachopoulou and A. Matopoulos, University of
Macedonia, Greece

Abstract: The aim of this chapter is to increase understanding of e-business adoption in the food supply chain. In the first part of the paper, an analysis of the concept and the context of e-business are presented along with a classification of the various e-business applications based on the dimensions of involvement and complexity. Next, emphasis is given to the issue of e-business adoption. In particular, an analysis of the factors that influence the adoption process is provided by exploring the specific characteristics of the food industry, as well as the potential impact of the use of e-business solutions. These factors are not only related to the company, but also to the supply chain of each company and the actual value of e-business applications. The chapter concludes that in many cases e-business is wrongly considered as a sole application and its adoption as mainly an intra-firm issue. Finally, future trends are considered and discussed.

Key words: e-business adoption, e-business research, food chain, ICT.

20.1   Introduction

The information and communication technology (ICT) revolution and the introduction of e-business applications in the mid-1990s brought to companies an excellent opportunity to facilitate, improve and in some cases to even transform their business processes and their way of doing business. In many business environments, ICT and e-business have been an established driver of change and a source of competitive advantage (Tatsis *et al.*, 2006). However, in the food industry, in contrast to other sectors, e-business adoption rates have been rather slow despite the potential benefits (E-business watch, 2006). What are the reasons for this? Which factors explain the low e-business adoption rates? What are the specific characteristics of the food industry that play a role in the adoption process? These are some of the issues that will be discussed in the chapter. In particular, the aim of this chapter is to increase understanding of e-business adoption in the food supply chain, by emphasizing analysis of the factors that influence the adoption process and also by exploring the specific characteristics of the food industry, as well as the potential impact of the use of e-business solutions. The chapter comprises four sections. In the first section, the concept and the context of e-business are presented. Next, the e-business adoption process is explored, along with the affecting factors. In the third section, the chapter explores the potential impact of e-business on the food supply chain by providing case studies and surveys that have been published in the literature. Finally, the chapter closes with a section referring to future trends for the sector.

20.2   Approaching e-business

20.2.1   Concept and context of e-business

More than a decade after the introduction of the Internet, what is included under the term e-business is still in debate, with the relevant literature offering a plethora of definitions and approaches. Some authors for example, do not consider the creation and the exchange of script messages as an e-business application (Watson *et al.*, 2000; Reedy *et al.*, 2000). In other surveys, e-commerce is perceived as an equivalent of e-business, ignoring the fact that e-business applications vary in complexity (Lockett and Brown, 2001). In Table 20.1, a short presentation of some of the definitions that have been proposed is provided. While definitions present some variations, most of them recognize the role of e-business in improving relationships between different business entities and in offering opportunities for the evolution and the improvement of the enterprise at many levels.

In this chapter e-business is approached, not as a single application, but rather as an all-encompassing term, which encapsulates a number of applications varying from the simple use of e-mail to more complicated collaborative platforms. The preferred definition for e-business in this chapter is the one by Brown and Lockett (2004). According to them: 'E-business includes a number of applications that vary in complexity and could be defined as the use of the Internet or any other electronic medium for the execution of transactions, the support of business processes and the improvement of collaboration opportunities among entities'. In Fig. 20.1, different e-business applications are depicted based on the dimensions of complexity and involvement. Complexity refers to the technical characteristics and the requirements of the applications, while involvement refers to organizational changes, capital investments and also to the focus of the application, whether it is intra-firm or extra-firm.

For example, applications of low complexity and low involvement require minimum investment in technology, minor organizational changes, take an

Table 20.1 Definitions of e-business

Author	E-business definition
Glover *et al.*, 2001	Use of information technology and electronic communication networks in business information exchanges for the completion of transactions
Sawhney and Zabin, 2001	Use of electronic networks and relative technologies for the facilitation, improvement and transformation of a business process or system in order to create superior value for current or future customers
Turban *et al.*, 2001	Use of telecommunication networks, not only for buying and selling, but also for customer service, business partners' cooperation and execution of transactions within an organisation
Kalakota and Robinson, 2001	Complex fusion of business processes, enterprise applications and organizational structure necessary to create a high-performance business model
Bocij *et al.*, 2003	All electronic exchanges of information, internally in the company or with external users, supporting a wide range of business processes
Brown and Lockett, 2004	Use of the Internet or other electronic means of communication for the execution of transactions, processes and cooperation between entities in markets
Koh and Maguire, 2004	E-business refers to information systems solutions packages used by enterprises for electronic business transaction purposes to meet customer requirements in business-to-business (B2B) and business-to-consumer (B2C) environments
Croom, 2005	Use of systems and open communication sources for information exchange, business transactions and knowledge-sharing between organizations

operations-related perspective, and their focus and major impact is intra-organizational. In contrast, applications of high complexity and involvement are more expensive, often require organizational changes, have a longer term perspective and the focus or impact expands the boundaries of the firm. The application of e-mail for example, is one which has minimum technical requirements, is a low capital investment application and requires minor or no changes to a company's *modus operandi*.

A more complex application is the one of electronic marketplaces. E-marketplaces are platforms that facilitate transactions, financial services

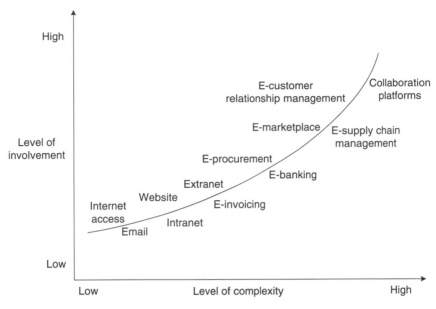

Fig. 20.1 Classifying e-business applications.

and order management. They bring together buyers and sellers or offers and requests, respectively. This type of application is more complex than the use of e-mail and often involves a certain level of investment and commitment. Usually, such e-marketplaces are imposed by business leaders or consortiums (who are responsible for organizing and controlling the e-marketplace).

Collaboration platforms, such as collaborative planning forecasting and replenishing (CPFR) are even more demanding applications, since they involve increased technical challenges (e.g. common standards for sharing information and security protocols for securing proprietary information of the parties involved). Therefore, these platforms require higher capital investment and often changes in the way the company operates so as to create shared processes among the supply chain parties. Increased trust between the supply chain parties is also needed. A very successful CPFR business paradigm is the one between Procter and Gamble (P&G) and Wal-Mart, where Wal-Mart's marketing information was integrated with P&G's manufacturing systems to streamline order and replenishment processes across their firm-level boundaries (Kim and Mahoney, 2006). This CPFR arrangement between Wal-Mart and P&G revealed that the successful implementation of these collaborative platforms depends not only on extensive information sharing but also on mutual understanding and commitment to the dedicated partners from repeated interactions (Holmstrom et al., 2002; Kim and Mahoney, 2006).

20.2.2   Critical issues in e-business research

Research into e-business covers a number of areas, such as e-readiness, e-activity and e-impact (E-business watch, 2007). E-readiness is the 'state of play' of a country's ICT infrastructure and the ability of its consumers, businesses and governments to use ICT for their benefit (Economic Intelligence Unit, 2007). Specifically, e-readiness is defined as a weighted function of the following criteria: connectivity and technology infrastructure, business environment, social and cultural environment, legal environment, government policy and vision, and consumer and business adoption. E-activities refer to the business processes that can be facilitated or altered by e-business applications, such as e-marketing, e-procurement and e-invoicing. Finally, e-impact refers to the implications for individual enterprises or for the value systems of the industry (E-business watch, 2007).

Most research conducted so far has focused mainly on the e-activity aspect of e-business, while less emphasis has been given to e-impact or to e-readiness. In addition, there are some other issues that are worth mentioning regarding e-business. The first issue is related to the fact that much of the existing research has focused on large firms (Fillis *et al.*, 2003), where successful examples have been identified in many cases. The second issue concerns the fact that often the benefits that have been reported in the various surveys are quite generic and are not linked to specific impact on specific business processes. Another basic problem in many surveys is the lack of a specific definition of e-business adoption and in many cases the adoption process has been approached in a very generic way.

In this chapter, in an effort to avoid generalizations regarding e-business adoption, it is proposed that the following dimensions should be taken into account when approaching the issue of adoption (Van der Veen, 2004):

- Dimension of activity: this is related to the way the company is supported by e-business. Companies may not fully adopt or fully reject e-business applications. The rationale is that there might be cases where companies adopt e-business applications for specific processes, while reject it for others.
- Dimension of application: this is related to the use of particular e-business applications, which present a specific level of complexity (e.g. e-business applications varying from very low complexity to very high complexity). This distinction goes beyond a theoretical and conceptual approach, but also has practical value in the sense that it affects the adoption of the applications. Considering e-business as a sole application would result in difficulties in understanding the specific impact. As a result, in this research the complexity levels proposed by Brown and Locket (2004) are taken into consideration.
- Dimension of value creation: refers to the value or the impact generated for the company as a result of the adoption of e-business.

- Dimension of the intensity of use: this is related to the frequency of e-business use. In other words, very intense use of a specific e-business application would mean that the company relies on this very application which will have significant implications if it is not used.

20.3   Factors affecting e-business adoption: taking a food chain-based approach

The issue of e-business adoption is not strictly an intra-firm issue owing to the interactive nature of e-business applications. The specific characteristics of the food sector, in addition to the structure of food supply chains, are expected to influence the adoption process greatly. Based on a literature review and specifically the suggestions of Kwon and Zmud (1987), Tornatzky and Fleischer (1990), Martin and Matlay (2001) and Patterson *et al.* (2003), factors affecting the adoption of e-business have been classified into three major categories (Fig. 20.2): factors related to e-business applications in the specific sector, intra-firm factors and factors related to the characteristics of food supply chains. An analysis of these factors follows.

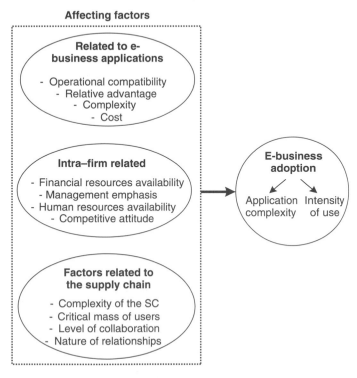

Fig. 20.2   Factors affecting e-business adoption (from Matopoulos *et al.*, 2007).

20.3.1   Factors related to e-business applications

Factors in this group deal with the appropriateness of e-business applications in specific business fields. The factors that have been recognized in the literature are the following: operational compatibility, relative advantage, complexity and the cost of e-business applications (Davis *et al.*, 1989; Rogers, 1995; Mehrtens *et al.*, 2001; Sadowski *et al.*, 2002).

Operational compatibility deals with how compatible the e-business applications are with the existing activities of companies (Rogers, 1995). Under the food prism, the element of operational compatibility is linked to the specific characteristics of the product and the sector. For example, a processing company has rather different e-business applications options to use in comparison to a company that sells fresh produce, owing to time constraints, the perishable nature of the product or even the need for physical interaction with the product. The very nature of food products is an important element, which needs to be taken into consideration when examining trust formation in the sector. Hofstede *et al.* (in press), for example, argue that trust between companies is not mediated appropriately by existing e-business solutions, because of the difficulty of examining food quality: no examination is possible in some electronic transaction environments and there is a perceived risk in performing a transaction via e-commerce.

The relative advantage refers to the expected benefits and the usefulness that will arise from e-business applications in comparison to other applications (Rogers, 1995). To a great extent, this is related to the existence or not of a killer application. These applications are perceived to offer a specific path of best practice, which companies find hard to ignore. Complexity refers to the difficulties that a company is expected to face in understanding and using the applications (Davis *et al.*, 1989; Rogers, 1995; Van der Veen, 2004). Finally, cost involves managers' perceptions regarding the capital needed for investment in order to use these e-business applications.

20.3.2   Intra-firm related factors

These factors are related to companies' characteristics. These characteristics are the availability of financial resources, the emphasis management puts on adoption by the administration of the company, the availability of human resources and the competitive attitude of the company. The availability of financial resources, although it is undoubtedly related to the cost of applications (analysed in previous section), in the current phase is correlated to the financial health of the company. Many authors believe that this factor is a critical one in the adoption process of e-business (Stokes, 2000; O'Gorman, 2000; Fillis, 2002). In the European food sector for example, low e-business adoption rates have been linked to the size of companies, since the majority of companies operating in the sectors are small and medium-sized enterprises (SMEs) characterized by limited availability of financial resources (European Commission, 2002).

Besides the size of the company, an important factor is the management emphasis and the commitment of top managers to the adoption process. It is also related to the business mentality of company's administration. In general, the role of top management in the adoption of ICT applications has been clearly recognized in the literature (Dewar and Dutton 1986; Damanpour, 1991; Henderson *et al.*, 2000). In many cases, particularly when it comes to SMEs, these intra-firm related factors become equally, if not more important than every other factor (Fillis *et al.*, 2003).

The availability of human resources is associated with the existence of employees who have the knowledge and experience to use e-business applications (Mehrtens *et al.*, 2001). Finally, the competitive attitude of a company is allied to its perception of the way in which it will improve its competitive position and whether or not this could be achieved by the adoption of e-business applications (Dos Santos and Peffers, 1998; Sadowski *et al.*, 2002; Waarts *et al.*, 2002). The food sector, in particular, is a very competitive one owing to a number of factors such as market globalization and deregulation, the abolition of constraints in the movement of products, the existence of powerful companies and the increased needs of customers for service (Nitchke and O'Keefe, 1997; Folkerts and Koehorst, 1998; Marion, 1998; Saxowvsky and Duncan, 1998). Companies are therefore seeking ways to respond to these changes and to add efficiency to their processes, by reducing costs and by shrinking response times in their effort to survive. E-business has been considered by many to be able to improve the performance of companies in many different business sectors, including the food sector (Tucker and Jones, 2000; Tan, 2001; Lee and Whang, 2002).

20.3.3 Factors related to the supply chain

These factors take into consideration the external environment of the company and its interaction with other companies. In particular, such factors are the complexity of the supply chain, the existence of critical mass of users, the level of collaboration and the nature of relationships. The complexity of the supply chain is defined by the number of entities that interact with a particular company, their proximity, and the number and the complexity of transactions. The more complex the structure of a supply chain the more imperative the adoption of e-business applications becomes, given that the company has increased coordination needs. The food supply chain is one containing many entities (Matopoulos *et al.*, 2005) and many transactions either as a result of a product's seasonality (implying changing suppliers) or as a result of the interaction of companies without proximity. The perishable nature of the products also affects order patterns, by necessitating small orders on a more frequent basis. E-business is considered to provide solutions to the above problems and as a result it is expected that companies with more complex supply chains will appear to have greater will to adopt e-business. Many global food retailers are using Internet-based platforms in order to manage their extensive supply base, to reduce lead times and to shrink the 'time to shelf' lifecycle for new products that have been developed by their suppliers.

The factor of critical mass of users is associated with the existence or not of companies in the supply chain of the firm that are using e-business applications (Markus, 1990). This factor has been recognized as a determinant of the adoption of e-business-based applications, particularly when it comes to applications which are related to communication, since it affects the expected benefits and the relative costs (Rogers, 1995; Kraut et al., 1998). Similarly, some industry organizations have agreed to adopt specific 'standards' or industry-accepted e-business solutions in an effort to increase the level of interaction between companies, for example, the RosettaNet® (in the ICT and electronics industries) or papiNet® (in the pulp and paper industry). Solutions for data exchange have been agreed between a limited number of companies operating in the same supply chain. In 2000, for example, 17 international retailers founded the WorldWide Retail Exchange (WWRE) to enable participating retailers and manufacturers to simplify, rationalize and automate supply chain processes, thereby eliminating inefficiencies in the supply chain. The WWRE was the premier Internet-based business-to-business exchange in the retail e-marketplace, enabling retailers and manufacturers to reduce costs substantially across product development, e-procurement and supply chain processes.

The level of collaboration is also an important factor. Based on the work of Winer and Ray (1994), Spekman et al. (1998) and Wang and Archer (2004), regarding the stages in the relationships between companies, as the interaction develops from cooperation to collaboration, the adoption of e-business applications increases. Long-term relationships between organizations which are characterized by trust have been proved that provide a basic motive for electronic integration (Konsynski and McFarlan, 1990). Lack of trust is a crucial barrier towards the uptake of e-business applications which needs further attention. Interorganizational trust has been recognized as central to the adaptation of e-business applications. A very important issue related to trust generation is the influence of culture. This is extremely important as the food sector is characterized by cross-country transactions and exchanges. Moreover, food SMEs have specific requirements concerning trust formation as they often do not trust the medium itself. This could be particularly the case in electronic marketplaces and electronic auctions for agricultural commodities, which are sold by farmers to other companies. Given that the aforementioned types of electronic transaction of commodities are not based on long-term relationships (in order to become more cost efficient and to control quality), but mostly on the price criterion, some kind of accreditation (third party institution-based trust) is needed for all participating entities in order to assure the quality of the products offered.

The nature of relationships includes characteristics such as power and dependence. Many researchers have shown that pressure from the environment of a company (e.g. suppliers and customers) affects the adoption of e-business (Norris, 1988; Mehrtens *et al.*, 2001; Daniel and Grimshaw, 2002). Pressure from supply chain partners could also affect the adoption of various ICT applications. Partner companies that are characterized by close collaboration present increased dependence and adopt easier applications such as e-business (Patterson *et al.*, 2003). For example, it is well known that retailers, such as Wal-Mart, put pressure on its suppliers to adopt electronic data exchange (EDI) (Premkumar *et al.*, 1997), and lately has done the same by putting pressure on its hundred biggest suppliers in an effort to adopt RFID technology (Wailgum, 2004).

20.4   View of the evolution and current state of e-business uptake in the food industry

The introduction of the Internet and the development of e-business more than a decade ago has changed, even transformed, much of the traditional way of doing business. In the food sector, similar to many others, the introduction of the Internet and the advent of e-business was initially expected to bring about radical advances in the business to consumer (B2C) relationship, rather than the business to business (B2B) relationship. However, in practice the opposite has occurred. Most of the benefits and opportunities brought by e-business were taken up by its B2B part. Only a very limited number of e-marketplaces proved sustainable and those that have particularly relate to B2B transactions. In contrast, B2B e-business has followed a growing pattern over the years, with companies gradually implementing e-business applications into their agendas. In Fig. 20.3, the relevance of ICT and e-business in ten sectors in Europe for the year 2006 is presented, according to a survey conducted by the organization E-business watch for the year 2006/07.

It is evident from Fig. 20.3, that the food sector in Europe has comparatively one of the lowest rates of involvement in terms of relevance and diffusion of the applications in the sector, particularly for e-sourcing and procurement, e-marketing and sales and ICT use for innovation. In contrast, higher rates seem to exist for more operational related applications such as e-logistics/SCM.

Analysing e-business deployment in the context of the food industry is a very complex task owing to the strong links of the industry with the different sectors in the food chain. In many cases food companies rely and have links with many non-food companies. For example, the industry has links with non-food sectors like chemicals, food technology, packaging and machinery, but the industry also forms the link between agriculture and food retailing. Moreover, the food sector comprises various subsectors with

Sector \ Application	e-source and procurement	e-logistics/ SCM	e-design and planning	e-marketing and sales	ICT use for innovation	Perceived ICT significance
Food and beverage	••	•••⊙	••	•⊙	••	••
Footwear	•	•	•	•⊙	••	•
Pulp and paper	•••	•••	••	•••⊙	••	••
ICT manufacturing	••••	•••	•••	•••⊙	••••	•••
Cons. electronics	••••	••⊙	•••	•••⊙	••••	•••
Shipbuilding	••	••	••••⊙	•	••⊙	••
Construction	••	•⊙	•••⊙	•	•••⊙	•⊙
Tourism	••⊙	•••	••⊙	••••⊙	••	•••⊙
Telecoms	••••	••	•••	••••⊙	••••	••••
Hospital activities	•••	••	••	•	•••	•••

• = below average relevance/diffusion; •• = average relevance/diffusion; ••• = above average relevance/diffusion;
•••• = high relevance/diffusion; ⊙ = applies only for some sub-sectors/types of firms.
SCM = supply chain management.

Fig. 20.3 Relevance of ICT and e-business in 10 sectors in 2006 (from E-business Watch, 2007).

distinct characteristics. Some of the major subsectors include fruit and vegetables, dairy products, beverages, snack foods, flour and bakery products, confectioneries, meat and poultry products, fish and marine products and fats and oils. Another, important issue that needs to be considered is the fact that for many companies their customers are the end consumers. From the results of the survey it is evident that the food industry has a good level of development of internal process integration and supply chain-related activities.

20.5 Conclusions and future trends

A review of e-business adoption research has revealed that in many cases e-business is wrongly considered as a sole application and its adoption as mainly an intra-firm issue. The initial goal of this chapter was, first of all, to provide a clear understanding of the context and concept of e-business and subsequently to propose and analyse factors that affect its adoption, particularly in the context of the food chain. In terms of future research, a review of the literature revealed several key areas where more research is needed. First, a significant weakness is that in many cases e-business studies have focused more on identifying the potential benefits of its adoption, rather than the actual benefits or the exact impact of the adoption. There is a need to measure the impact of e-business in specific processes, based on specific criteria. Up until now, the focus of research has been on the deployment of e-business and the differences between SMEs and large enterprises, whereas regional and sub-sector differences were rather neglected. Therefore, given the characteristics of the food sector, it would be very interesting to identify, if any, the differences across the different sub-sectors of the food industry and the regional differences in adoption across the world.

In terms of future trends, the B2B part of e-business will definitely follow a pattern of increasing uptake. It is very likely that pressure from large manufacturers and retailers on smaller players will continue to grow as the benefits of e-business become more obvious, more tangible and therefore more requisite. Particularly, more interest in its deployment will be expressed for those applications focusing on reducing the time for non-value adding activities, such as invoicing, order taking and processing and forecasting, with regard to the B2C part of e-business, it seems that currently we are experiencing the beginning of a new wave of B2C, with new companies entering the market. However, in comparison to the past, these new wave companies seem to have designed more sustainable business models and moreover their customer base consists now of 'new' more educated consumers than in the past. Thurow (2001), in an effort to explain B2C business failures, stated that: 'Sociology always beats technology, but eventually sociology changes ... What my generation finds strange and uncomfortable (e.g. buying a car on the Internet without a test ride) will seem completely normal to our grandchildren'. It seems that this change to which Thurow (2001) was referring, will be completed in less than a generation.

20.6   Sources of further information and advice

Key books
E Issues in Agribusiness: The What, Why and How by Kim Bryceson (2006) discusses the use of electronic-based applications by agribusiness. The book outlines what exactly 'electronically enabled' agribusiness is, why agribusiness wants to embrace the electronic era, and how it can go about doing it. It also discusses cutting edge innovations in agribusiness systems such as precision farming and livestock electronic identification, risk management, supply and value chain management, knowledge management and e-governance. Finally, it reviews the underlying technological challenges, e-enabled business models and e-strategies, management concepts and innovative education programmes.

Scientific associations
The European Scientific Association EFITA (European Federation for Information Technology in Agriculture, Food and the Environment) is a federation of European National Associations of scientists and professionals who deal with the design, development, use and management of modern information and communication systems in the agri-food sector. The focus of EFITA is on the utilization of modern information technology for the development and the economic and environmental sustainability of the

agri-food sector with its institutional infrastructure and the agricultural and commercial enterprises along the food production chain. http://www.efita. net/

Websites

http://www.etrustproject.eu/ A European Commission funded initiative which has the objective of facilitating the uptake of e-commerce technologies by food sector SMEs to support the exploitation and take up of the power of business-to-business (B2B) e-commerce technology for cost efficiency in food chains enhancing the competitiveness of the European food sector.

http://www.ebusiness-watch.org/ e-Business Watch is the European Observatory in the field of ICT and e-business policies.

http://www.oecd.org/sti/ict OECD Committee for Information, Computer and Communications Policy.

20.7   References

BOCIJ, P., CHAFFEY, D., GREASLEY, A. and HICKIE, S. (2003). *Business Information Systems, Technology, Development and Management for the e-business*, 2nd edition, Pearson Education, Harlow, Essex, UK.

BROWN, D.H. and LOCKETT, N. (2004). 'Potential of e-applications for engaging SMEs in e-business: a provider perspective', *European Journal of Information Systems*, **13**, 21–34.

BRYCESON, K.P. (2006). *E Issues in Agribusiness: The What, Why and How*, CABI Publishing, July 384 pp, Wallingford, UK.

CROOM, S.R. (2005). 'The impact of e-business on supply chain management: An empirical study of key developments', *International Journal of Operations and Production Management*, **25**(1), 55–73.

DAMANPOUR, F. (1991). 'Organizational innovation: A meta-analysis of effects of determinants and moderators', *Academy of Management Journal*, **34**(3), 555–90.

DANIEL, E.M. and GRIMSHAW, D.J. (2002). 'An exploratory comparison of electronic commerce adoption in large and small enterprises', *Journal of Information Technology*, **17**, 133–47.

DAVIS, F.D., BAGOZZI, R.P. and WARSHAW, P.R. (1989). 'User acceptance of computer technology: A comparison of two theoretical models', *Management Science*, **35**(8), 982–1003.

DEWAR, R.D. and DUTTON, E.J. (1986). 'The adoption of radical and incremental innovations: an incremental analysis', *Management Science*, **32**(November), 1422–33.

DOS SANTOS, B.L. and PEFFERS, K. (1998). 'Competitor and vendor influence on the adoption of innovative applications in electronic commerce', *Information and Management*, **34**(3), 175–84.

E-BUSINESS WATCH (2006). *ICT and e-Business in the Food and Beverages Industry*, Sector Report No 1/2006, European Commission, Enterprise & Industry Directorate General, 2006.

E-BUSINESS WATCH (2007). *The European e-Business Report, A portrait of e-business in 10 sectors of the EU economy 5th Synthesis Report of the e-Business Watch*, European Commission, Enterprise Directorate-General e-Business, ICT Industries and Services, January 2007.

ECONOMIC INTELLIGENCE UNIT (2007). *The 2007 e-readiness Rankings: Raising the Bar*, A white paper from the Economist Intelligence Unit, The Economist.

EUROPEAN COMMISSION (2002). *Benchmarking National and Regional E-business Policies for SMEs*, final report of the E-business Policy Group of the European Union, Brussels, 28 June.

FILLIS, I. (2002). 'Barriers to internationalisation: an investigation of the craft microenterprise', *European Journal of Marketing*, **36**(7/8), 912–27.

FILLIS, I., JOHANSSON, U. and WAGNER, B. (2003). 'E-business development in the smaller firm', *International Journal of Small Business and Enterprise Development*, **10**(3), 336–44.

FOLKERTS, H. and KOEHORST, H. (1998). 'Challenges in international food supply chains: vertical co-ordination in the European agribusiness and food industries', *British Food Journal*, **100**, 385–8.

GLOVER, S., LIDDLE, S. and PRAWITT, D. (2001). *E-business, Principles and Strategies for Accountants*, Prentice Hall, Upper Saddle Ricer, New Jersey.

HENDERSON, J., DOOLEY, F. and AKRIDGE, J. (2000). 'Adoption of e-commerce strategies for agribusiness firms', in *Proceedings of the American Agricultural Economics Association Annual Meeting*. Tampa, FL, USA.

HOFSTEDE, G.J., CANAVARI, M., FRITZ, M., OOSTERKAMP, E. and VAN SPRUNDEL, G.-J. (20xx). 'Towards a cross-cultural typology of trust in B2B food trade', *British Food Journal* (forthcoming).

HOLMSTRÖM, J., FRÄMLING, K., KAIPIA, R. and SARANEN, J. (2002). 'Collaborative planning forecasting and replenishment: new solutions needed for mass collaboration', *Supply Chain Management: An International Journal*, **7**(3), 136–45.

KALAKOTA, R. and ROBINSON, M. (2001). *E-business 2.0: Roadmap for Success*, 2nd edition, Addison Wesley, NJ.

KIM, S.M. and MAHONEY, J.T. (2006). *Collaborative Planning, Forecasting, and Replenishment (CPFR) as a Relational Contract: An Incomplete Contracting Perspective*, Working Paper Series 2006, University of Illinois at Urbana-Champaign.

KOH, S.C.L and MAGUIRE, S. (2004). 'Identifying the adoption of e-business and knowledge management within SME's', *Journal of Small Business and Enterprise Development*, **11**(3), 338–48.

KONSYNSKI, B.R. and MCFARLAN, W.F. (1990). 'Information partnerships – shared data, shared scale', *Harvard Business Review*, **68**(5), 114–20.

KRAUT, R.E., RICE, R.E., COOL, C. and FISH, R.S. (1998). 'Varieties of social influence: the role of utility and norms in the success of a new communication medium', *Organization Science*, **9**(4), 437–53.

KWON, T.H. and ZMUD, R.W. (1987). 'Unifying the fragmented models of information systems implementation', in *Critical Issues in Information Systems Research*, Boland, R.J. and Hirschheim, R.A. (eds), John Wiley, New York, 247–52.

LEE, H.L. and WHANG, S. (2002). 'Supply chain integration over the Internet', in *Supply Chain Management: Models, Applications and Research Directions*, Geunes, J., Pardalos, P. and Romeijn, E. (eds) Kluwer Academic Publishers, Dordrecht, The Netherlands, 3–17.

LOCKETT, N. and BROWN, D.H. (2001). 'A framework for the engagement of SMEs in E-business', in *Proceedings of the Americas Conference on Information Systems*, Boston, MA, Association of Information Systems, 656–62.

MARION, B.W. (1998). 'Changing power relationships in the US food industry: brokerage arrangement for private label product', *Agribusiness*, **14**, 85–93.

MARKUS, L. (1990). 'Toward a "critical mass" theory of interactive media', in *Organizations and Communication Technology*, Fulk, J. and Steinfield, C. (eds), Sage Publications, Newbury Park, CA, 194–218.

MARKUS, M.L., PETRIE, D. and AXLINE, S. (2002). 'Bucking the trends: what the future may hold for ERP packages', *Information System Frontiers*, **2**(2), 181–93.

MARTIN, L.M. and MATLAY, H. (2001). '"Blanket" approaches to promoting ICT in small firms: Some lessons from the DTI ladder adoption model in the UK', *Internet Research: Electronic Networking Applications and Policy*, **11**(5), 399–410.

MATOPOULOS, A., VLACHOPOULOU, M., MANTHOU, V. and MANOS, B. (2005). 'A conceptual framework for e-business adoption and development for enterprises in the agri-food industry', in *Proceedings of the: EFITA & WCCA 2005 joint Conference*, Boaventura Cuhna, J. and Valente, A. (eds), 25–28 July, Villa Real, Portugal, 370–6.

MATOPOULOS, A., VLACHOPOULOU, M. and MANTHOU, V. (2007). 'Factors affecting e-business adoption and impact in the supply chain: empirical research on the Greek agri-food industry', in *Proceedings of the 5th International Workshop on Supply Chain Management and Information Systems*, Beaumont, N. and Sohal, A. (eds), 9–12 December, Melbourne, Australia, 1–15.

MEHRTENS, J., CRAGG, P.B. and MILLS, A.M. (2001). 'A model of Internet adoption by SMEs', *Information and Management*, **38**, 165–76.

NITCHKE, T. and O'KEEFFE, M. (1997). 'Managing the linkage with primary producers: experiences in the Australian grain industry', *Supply Chain Management: An International Journal*, **2**(1), 4–6.

NORRIS, R.C. (1988). *The ADL Grocery Report Revisited*, EDI Forum, 44–48.

O'GORMAN, C. (2000). 'Strategy and the small firm', in *Enterprise and Small Business: Principles, Practice and Policy*, Carter, S. and Jones Evens, D. (eds), Pearson Education, London, 283–99.

PATTERSON, K.A., GRIMM, C.M. and CORSI, T.M. (2003). 'Adopting new technologies for supply chain management', *Transportation Research Part E*, **39**, 95–121.

PREMKUMAR, G., RAMAMURTHY, K. and CRUM, M.R. (1997). 'Determinants of EDI adoption in the transportation industry', *European Journal of Information Systems*, **6**(2), 107–21.

REEDY, J., SCHULLO, S. and ZIMMERMAN, K. (2000). *Electronic Marketing*, The Dryden Press, New York.

ROGERS, E.M. (1995). *Diffusion of Innovations*, 4th edition, The Free Press, New York.

SADOWSKI, B.M., MAITLAND, C. and VAN DONGEN, J. (2002). 'Strategic use of the Internet by small- and medium-sized enterprises: an exploratory study', *Information Economics and Policy*, **14**, 75–93.

SAWHNEY, M. and ZABIN, J. (2001). *The Seven Steps to Nirvana*, Mc Graw-Hill, New York.

SAXOWVSKY, D.M. and DUNCAN, M.R. (1998). *Understanding Agriculture's Transition into the 21st Century: Challenges, Opportunities, Consequences and Alternatives*, Report No. 181. Department of Agricultural Economics, North Dakota State University.

SPEKMAN, R.E., KAMAUFF, J.W. and MYHR, N. (1998). 'An empirical investigation into supply chain management', *International Journal of Physical Distribution and Logistics Management*, **28**(8), 630–50.

STOKES, D. (2000). 'Marketing and the small firm', in *Enterprise and Small Business: Principles, Practice and Policy*, Carter, S. and Jones Evens, D. (eds), Pearson Education, London, 354–83.

TAN, K.C. (2001). 'A framework of supply chain management literature', *European Journal of Purchasing and Supply Management*, **7**(1), 39–48.

TATSIS, V., MENA, C., VAN WASSENHOVE, L.N. and WHICKER, L. (2006). 'E-Procurement in the Greek Food and Drink Industry: Drivers and Impediments', *Journal of Purchasing and Supply Management*, **12**, 63–74.

THUROW, L.C. (2001). 'Does the "E" in E-Business Stand for "Exit"?', *MIT Sloan Management Review*, **42**(2), 112.

TORNATZKY, L.G. and FLEISCHER, M. (1990). *The Processes of Technological Innovation*, Lexington Books, Lexington, MA.

TUCKER, D. and JONES, L. (2000). 'Leveraging the power of the Internet for optimal supplier sourcing', *International Journal of Physical Distribution and Logistics Management*, **30**(3–4), 255–67.

TURBAN, E., MCLEAN, E. and WETHERBE, J. (2001). *Information Technology for management: transforming business in the digital economy*, 3rd edition, John Willey and Sons, New York.

VAN DER VEEN, M. (2004). *Explaining e-business adoption: innovation and entrepreneurship in Dutch SMEs*, PhD Thesis, University of Twente, The Netherlands.

WAARTS, E., VAN EVERDINGEN, Y.M. and VAN HILLEGERSBERG, J. (2002). 'The dynamics of factors affecting the adoption of innovations', *The Journal of Product Innovation Management*, **19**(6), 412–23.

WAILGUM, T. (2004). 'Tag, you're late', *CIO magazine*, November Issue, Available at: http://www.cio.com/article/143701/Tag_You_re_Late.

WATSON, R.T., BERTHON, P., PITT, L.F. and ZINKHAN, G.M. (2000). *Electronic Commerce: The Strategic Perspective*, Dryden Press, Orlando, FL.

WINER, M. and RAY, K. (1994). *Collaboration Handbook: Creating, Sustaining and Enjoying the Journey*, Amherst H. Wilder Foundation, Saint Paul, MN.

21

Radio frequency identification (RFID) as a catalyst for improvements in food supply chain operations

K. Pramatari, A. Karagiannaki and C. Bardaki, Athens University of Economics and Business, Greece

Abstract: The objective of effective supply chain management is the coordination of information, materials and financial flows between organisations. The recent trends of globalisation, consumer pressure for responsiveness and reliability, and intense competition in the global trading community have made effective supply chain management a very challenging issue. New information technologies are promising for optimising supply chain operations and solving many related issues. Indeed, supply chain management information systems have greatly benefited companies that use them, minimising information processing costs and raising great potentials like information sharing and fast communication that were not feasible before. RFID is an emerging technology that can further contribute to supply chain optimisation. RFID enables accurate real time product location information provision in high volumes and at very low (or even zero) labour costs. This chapter looks closely into the technology of RFID and the way it is employed in supply chain management and, particularly, in the food supply chain by describing two applications. The first describes the requirements' analysis, development and pilot implementation of a RFID-enabled traceability system for a company that deals with frozen food. The second describes a distributed, service-oriented architecture that supports RFID-integrated decision support and collaboration practices in a networked business environment. In the context of retail industry, a RFID-integrated 'dynamic pricing' service is described regarding its functionality and implementation. Several considerations from the cases are presented which could provide valuable feedback to other organisations interested in moving to a RFID-based scheme.

Key words: food supply chain, RFID, supply chain optimisation.

21.1 Introduction

With the competitive differentiators of cost reduction, service enhancement and operations' velocity, the deployment of new information initiatives has become a market mandate for every firm that struggles to streamline its supply chain. This implies that the introduction of new information technologies should be perceived and positioned as a catalyst for better business practices and not as a cost to a business or as a voluntary responsibility.

Nowadays, the emerging radio frequency identification (RFID) technology is expected to meet the above requirements and thus revolutionise many supply chain operations. RFID is a technology that uses radio waves to identify objects automatically. The identification is done by storing a serial number, and perhaps other information, on a microchip that is attached to an antenna. This bundle is called an RFID tag. The antenna enables the chip to transmit the identification information to a reader. The reader converts the radio waves reflected back from the RFID tag into digital information that can be passed on to an enterprise information system (Kelepouris et al., 2007). The advanced data capture capabilities of RFID technology coupled with unique product identification and real-time information coming from different data sources, such as environmental sensors, define a new and rich information environment that opens up new horizons for efficient management of supply chain processes and decision support.

RFID offers a wide range of applications across several industries, such as healthcare, transport and textiles. The value of RFID may diverge and its effect or change can be greater in specific industries. A predominant industry that seems to benefit largely through RFID is the food industry. All in all, the food industry represents a supply chain that is increasingly being challenged by legal compliance, safety and quality assurance, risk prevention, efficient recalls/withdrawals and the consumers' right to know. These characteristics make the piloting and implementation of a RFID system a particularly appealing investment. As a result, RFID applications in the food supply chain range from upstream warehouse and distribution management down to retail-outlet operations, including shelf management, promotions management and innovative consumer services, as well as applications spanning the whole supply chain, such as product traceability (Pramatari et al., 2005).

However, despite all the areas of opportunity and the fact that many companies (Metro, Tesco, Delhaize, Ahold, Rewe) have pilot tested the technology or have already started roll-out, the level of RFID implementation can be considered as pre-mature. Moreover, several white papers and reports published recently either focus on related technical aspects or are mainly qualitative studies of business cases for RFID deployment (Angeles, 2005; Jones et al., 2005; Curtin et al., 2007; Attaran, 2007; Reyes and Jaska, 2007). In addition, there is a small, but growing body of literature trying to give a quantitative assessment of the deployment of RFID (Lee et al., 2004; Fleisch and Tellkamp, 2005; Atali et al., 2005; Gaukler et al., 2006; Wang et al., 2008). In view of the pre-mature level of RFID research and implementation, it is obvious that as with all novel technologies, there is a credibility gap: 'To make robust investment decisions we need a much more credible assessment of the true value of RFID ... based on the operating characteristics of the underlying supply chain processes' (Lee, 2007). Evidently, the value of investment in RFID constitutes a matter of considerable concern and debate for both practitioners and academics alike.

This chapter looks closely into the technology of RFID and the way it is employed in supply chain management and, particularly, in the food supply chain. Section 21.2 presents a general description of the RFID technology within supply chain management. Section 21.3 demonstrates the value of RFID based on empirical evidence by describing two applications. The first describes work undertaken for a company that deals with frozen food regarding the requirements' analysis, development and pilot implementation of a RFID-enabled traceability system. The second describes a distributed, service-oriented architecture that supports RFID-integrated decision support and collaboration practices in a networked business environment. In the context of retail industry, the functionality and implementation of a RFID-integrated 'dynamic pricing' service are described. Finally, Section 21.4 provides several considerations from cases that could provide valuable feedback to other organisations interested in moving to a RFID-based scheme. While offering an immense learning value for academics and researchers, it is hoped that this chapter will help professionals and executives to understand the far-reaching applications of RFID better.

21.2   Radio frequency identification technology within supply chain management

RFID is a generic technology concept that refers to the use of radio waves to identify objects (Auto-ID Center, 2002). RFID tags have both a microchip and an antenna. The microchip is used to store object information such as a unique serial number. The antenna enables the microchip to transmit object information to a reader, which transforms the information on the RFID tag to a format understandable by computers (Angeles, 2005).

There are two types of RFID tags, with regard to the source of electric energy they use:

- Passive tags: these are the most common tags used for identification. The tags receive energy from the reflected radio waves transmitted by the reader and transmit the digital information stored in the microchip. The reading range of passive tags is limited to a few metres, owing to the way the energy is received. However, passive tags are far cheaper than active tags. The cost of a simple passive tag starts from US $0.05 and is expected to fall by 10% annually in the future.

- Active tags: these tags carry an embedded battery and transmit radio waves by themselves. The range of active tags can reach a few hundred metres but their cost is much higher than that of passive tags.

RFID technology has been extensively used for a diversity of applications ranging from access control systems to airport baggage handling, livestock management systems, automated toll collection systems, theft-prevention systems, electronic payment systems and automated production systems (Agarwal 2001; Hou and Huang, 2006; Kelly and Erickson, 2005; Smith and Konsynski, 2003). Nevertheless, what has made this technology extremely popular nowadays is the application of RFID for the identification of consumer products and the management of supply chain processes.

The application of RFID technology in the supply chain refers to the attachment of an RFID tag in every single product in the supply chain that will uniquely identify it globally. The number of tags to be deployed is immense and, for cost reasons, passive tags are the most appropriate for deployment in the supply chain. The deployment of RFID tags in the supply chain will start from pallet level, then at case level and as soon as the tag's cost decreases sufficiently, tags will be applied to all products in the supply chain. However, it is not believed that this will happen in the next 7–10 years (Shutzberg, 2004).

The deployment of RFID in the supply chain was initially supported by the AUTO-ID research centre (http://www.autoidlabs.org/), which was founded in 1999. The AUTO-ID Centre is a not-for-profit federation of seven research universities (including MIT, Cambridge and St. Gallen) and its objective is to develop an open standard architecture for creating a seamless global network of physical objects. In 2003, the AUTO-ID Centre was substituted by AUTO-ID Labs. Since October 2003, the research and standardisation of RFID deployment in the supply chain has been coordinated by EPCglobal Inc (http://www.epcglobalinc.org). EPCglobal is a joint venture between EAN International and the Uniform Code Council (UCC). It is a not-for-profit organisation entrusted by industry to establish and support the Electronic Product Code (EPC) Network as the global standard for immediate, automatic and accurate identification of any item in the supply chain of any company, in any industry, anywhere in the world. The objective of EPCglobal is to drive global adoption of the EPCglobal Network.

According to Agarwal (2001), RFID technology is believed to benefit organisations in the following crucial supply chain issues:

- Improving product availability: RFID tagging can tame the phenomena of out-of-stock and out-of-shelf. The ability to provide the necessary solutions lies in the ability to capture the behaviour of each stock keeping unit (SKU) at each location through tagging. Moreover, readers can provide accurate data on shelf and backroom availability, as well as the time of delivery of incoming products, triggering the necessary alarms to the store personnel about a reordering or a replenishment that needs to take place.

- Mass customisation: Use of RFID technology, will provide each product with a unique product identification from very early in its life. This carries details of how a standard machine cell would have to be configured to make the product as well as the required specification of the final product. In this instance such standard cells would become highly flexible with the ability to reconfigure on the demand of product arriving, enabling mass customisation.
- Automatic proof of delivery: Using RFID tagging, the process of obtaining proof of delivery can be automated, eliminating manual errors. Products can simply be passed through tag readers at the retailers rather than have to be checked and counted manually. In addition, manufacturers will not just save on the cost of products unpaid for, but also on the man hours spent processing and negotiating the claims of the retailers.
- Security: RFID technology can increase security in the supply chain and combat retail theft. The level of theft can be reduced, as readers could warn personnel of any attempted theft in the vicinity. In addition, stolen products can be identified and restored to their rightful owners.
- Eliminating stock verification: Using RFID technology products can be passed through tag readers and the stock verification can be automated, speeding up the process and eliminating manual effort. Automated operation also eliminates errors in scanning and labelling. RFID can provide continuous, accurate and real-time information on the type of product, the amount and its location in a company's site.
- Reducing inventory levels: RFID technology can reduce inventory across a supply chain. This can be done by:
  - reducing order cycle times through better information flow
  - enabling the improvement of forecast accuracy though better and quicker information on current demand patterns
  - understanding promotions better through visibility of consumer behaviour
  - reducing variability in ordering
  - increasing speed and accuracy of planning
  - increasing flexibility in responding to unexpected demand.

In the food supply chain, RFID can potentially empower a broad spectrum of applications, ranging from upstream warehouse and distribution management down to retail-outlet operations, including shelf management, promotions management and innovative consumer services, as well as applications spanning the whole supply chain, such as product traceability (Pramatari *et al.*, 2005). Despite the broad spectrum of applications, RFID implementation currently takes place internally within a company, mainly with the

objective of automating warehouse management processes or store operations when making the first move. In the future, an industry report (GCI,
2005) identifies certain application areas (specifically store operations, distribution operations, direct store delivery, promotion/event execution, total
inventory management and shrink management) as major opportunities for
the deployment of RFID technology in the short- and mid-term. These
application areas have been selected based on their performance in comparison to the ratio of expected benefits over associated costs, including
process transformation difficulties. The same report identifies further opportunities in several 'track and trace' activities (such as anti-counterfeiting,
product diversion, recalls/reverse logistics, fresh/code-dated product management, cold chain monitoring and legal compliance), although it is noted
that 'more work is required to understand its potential applications and
benefits in these areas' (GCI, 2005). The 'more work' refers to the need to
connect supply chain partners and streamline the flow of information for
the applications to operate.

The contribution of RFID can be sought in the following applications:

- the automation of existing processes, leading to time/cost savings and
  more efficient operations;
- the enablement of new or transformed business processes and innovative consumer services, such as monitoring product shelf availability or
  consumer self checkout;
- the improvement achieved in different dimensions of information
  quality, such as accuracy, timeliness, and so on (Ballou *et al.*, 1998);
- the formation of new types of information, leading to a more precise
  representation of the physical environment, e.g. a product's exact position in the store, a specific product's production, distribution and sales
  history, etc.

The last two applications in particular, need new decision support algo-
rithms and tools for the associated benefits to be exploited, opening up a
whole new research area for decision support systems. Furthermore, for the
full benefits to be reaped, the information needs not to be exploited locally
but shared with supply chain partners in a complex network of relationships
and decision making.

Leading companies in the global market have already made moves
towards the application of RFID technology to monitor product flow in
their supply chain. Wal-Mart, the biggest retail chain in the USA, has mandated its suppliers to apply RFID tags to each pallet arriving in its central
warehouse. Pallet-level tagging is expected to be rolled out chainwide in
2010, while the deadline for tagging sellable units is 'under review'. Metro,
a big retail chain in Europe, has implemented a store (called 'the future
store') that operates using RFID tags applied to each product. Metro has
optimised many internal processes in the store utilising RFID technology
and provides its customers with innovative services such as semi-automatic

checkout and a smart trolley, which carries a TFT display and provides the customer with information about the products on the shelves and the trolley. Furthermore, future store gives Metro the opportunity to asses the benefits of RFID in a real case, measuring the impact of RFID deployment on stock reduction, increased availability and other issues of supply chain management (Hamner, 2005). RFID technology has already been adopted by some suppliers at the product level. Gillette is the most striking example, having already applied RFID tags to some razor products.

The adoption of RFID technology in supply chain management has initiated intense research interest in the following research streams:

- Tag–RF technology: during the last decade extended research has been conducted on the technology of tag implementation and radio wave communication with readers. Since 1999, the AUTO-ID Centre has made some important steps forward in improving tag reliability and readability. Universities and companies have also contributed to this field. The research objective in this area is to make RFID tags as reliable as possible and to reduce production costs to minimum. Research into radio wave communication involves the adoption of appropriate frequencies and protocol implementation for tag–reader communication. The objective of this research area is the improvement of read speed and read reliability. In June 2005, the University of Arkansas announced the opening of the RFID Research Center, funded by a number of companies including Deloitte and ACNielsen. The objective of this research centre is to study the read rates and other deployment issues at each point in the retail supply chain. Many companies (e.g. Wal-Mart) own their own RFID research centres in which they study similar issues (rfidjournal.com).
- Information systems: there has also been significant research activity in the design and implementation of information systems that will store and use the valuable information flows that derive from the adoption of RFID in the supply chain. This research area not only includes technological issues regarding the integration of information systems with the readers and data acquisition, but also ways that this information will be used to optimise crucial supply chain operations. Recently, the biggest enterprise resource planning (ERP) providers (e.g SAP) have updated their software products to support integration with RFID infrastructure. Moreover, extended research in developing middleware that will make it easier for companies to integrate data from RFID systems with their existing enterprise applications. Hong Kong University (www.hku.hk) in collaboration with IBM (www.ibm.com) has already begun a research project in this field.
- Privacy: an important issue that has to be solved before applying RFID on the broad scale in the supply chain (i.e. at the product level) is consumer privacy and personal data protection. Consumers have already

reacted in the application of RFID in products, creating boycotting campaigns against the companies that apply RFID to their products. The campaign against Gillette is a good example of this (http://www. boycottgillette.com/). EPCglobal Inc, through the Gen 2 tags specifications has tried to solve this problem, adding a bit into the electronic product code that disables the tag. Therefore, each tag can be disabled as soon as the product leaves the store (EPCglobal, 2005).

• Operations management: extended research has taken place for the exploitation of RFID for the optimisation of internal and inter-organisational operations. Apart from the analytical approaches of researchers, companies all over the world have already begun to gain the benefits of RFID technology in optimising supply chain operations. RFID has the potential to revolutionise the efficiency and accuracy of manufacturing and other service operations. Indicatively, RFID has significantly contributed to the optimisation of just in time production and an increase of factory throughput, in which delivery docks are usually a serious bottleneck.

• Marketing: the marketing research stream with regard to RFID deployment is still in its infancy. However, when RFID deployment comes to the product level, consumers must be convinced that the new technology can act also for their own benefit and not to spy on their consumer preferences. Companies must take advantage of RFID to provide consumers with smart high-tech services (smart shelves, smart trolleys, etc) that will encourage consumers to adopt the new technology. These new features should be promoted in a way that will dispel any misconception the public hold that RFID will invade their private life.

21.3   RFID applications in food supply chain

21.3.1   Application A: a proposed RFID-enabled traceability system for the food industry

The application concerns a leading food company in Greece (more than 30% of market share) that is also one of the largest in Europe. Its brands are recognised by millions, reaching consumers in 30 countries whilst expanding across the world map. Its success is based first and foremost on its respect for the consumer and its tireless daily efforts to supply the best possible value in the form of healthy, quality products. The company now comprises four divisions: dairy and drinks, bakery and confectionery, food-services and entertainment and frozen foods.

The frozen foods division is involved in the production and processing of frozen vegetables and foods in Greece and abroad. The range of the division's products is constantly developing. It is active in the production of frozen vegetables, pre-cooked meals, mixtures of frozen vegetables and, more recently, fresh salads. Over its 35 years in the market, it has always been innovative and generated new products. Realising the potential of RFID to improve different aspects of the warehouse, the company decided to participate in a project partly funded by the General Secretariat for Research and Technology, Ministry of Development of the Hellenic Republic, investigating the requirements' analysis, development and pilot implementation of a RFID-enabled traceability system within the central warehouse.

The warehouse and its operations

The company has a central warehouse that stocks frozen vegetables and includes a production unit where vegetables are packaged in bags. This section includes the description of the as-is operations within the central warehouse and aims to understand the relationships between various activities and identifying operations that are troublesome and can be improved by the deployment of RFID. This is accomplished by interviewing and visually examining operations including queues, bottlenecks, and human errors and, as a result, gain insight into the problems that are expected to be improved by RFID deployment.

The raw materials constitute domestic fresh vegetables (e.g. green beans, peas) or imported frozen vegetables. The incoming fresh product is frozen immediately and packaged in large containers. The imported frozen vegetables that arrive in the factory from approved suppliers from abroad are packaged directly in large containers. The semi-finished product is either from freezing and packaging domestic fresh vegetables in large containers or the packaging imported frozen vegetables in large containers. The large containers are then stored in a chamber for semi-finished products until there is a need to put them into consumer packaging. Packaging follows a rolling and controlled programme based on the sales target. Then the semi-finished product is conveyed to a machine which bags it and puts it into a sachet/pouch. Workers stack the sachets in cases and, finally, palletise the cases. Consequently, the finished product is derived by packaging the semi-finished product in sachets, cases and pallets (see Fig 21.1).

In the receiving process, a container carrying fresh vegetable arrives at the assigned docks outside the warehouse. The products on the bed of the container are unloaded onto a conveyor, which triggers the production line.

In the manufacturing process, freezing is seasonal, based on a harvest that starts in May. By the end of December the last vegetables have been picked, processed and stored as semi-finished products. The actual process of freezing a food item varies depending on what is to be frozen. Peas are the most common frozen vegetable. The pea process is typical of many vegetables. A typical process for a frozen product involves the following steps:

- cultivating the peas
- picking and washing

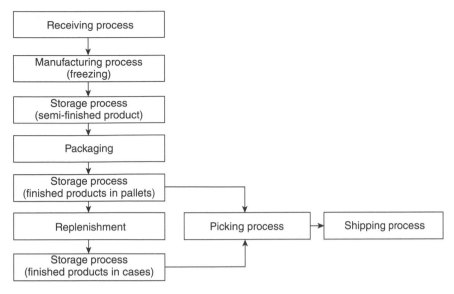

Fig. 21.1 The most important processes that take place within the central warehouse.

- blanching
- sorting
- inspecting
- transferring from the oven to the freezer

In quality control, frozen foods must be carefully inspected both before and after freezing to ensure quality. When vegetables arrive at the processing plant, they are given a quick overall inspection for general quality. The peas are inspected visually to make sure that only the appropriate quality peas go on to the packaging and freezing step. Laboratory workers also test the peas for bacteria and foreign matter, pulling random samples from the production line at various points.

In the packaging process, the frozen vegetable passes on a belt to mechanical equipment that bags it and puts it in a case. Then, workers dressed in cold-weather gear for protection palletise the cases. The pallets are stored in a warehouse cooled to between −17.8 and −28.9 °C. They remain there until demanded by the customer.

In the storage process, although a warehouse management system (WMS) is in place, the process of storage of semi-finished products depends heavily on quality variation and first-in-first-out (FIFO). It is impractical to have predetermined fixed positions because of the characteristics of the particular product. As a result, the assignment of the semi-finished product to storage locations is haphazard, indicating that products are not stored in designated fixed locations (random storage scheme). The system considers only the production and the expiration date of the products. Product enters the warehouse to a location designated by quality control and they are released automatically by the system in a certain number of days after the production date. The finished product is stored in two chambers: one that consists of pallets of finished products and one that consists of cases of finished products.

In the picking process, whenever an order is requested, a picking list is generated. The operator of an indoor forklift truck picks up the corresponding products from the designated locations using the WMS and their own perception. This policy of marshalling products for delivery is chosen because it is easily employed and order integrity can be maintained. Multiple orders are picked consecutively and are accumulated applying a first-come-first-served (FCFS) logic that combines orders as they arrive until the maximum cubic and weight capacity of a container has been reached.

In the shipping process, before being transported, products destined for one truck are held together at a provisional position. During this time, an operator gives them a compliance check. After that, a truck arrives at the designated docks and all products are loaded on the bed of the truck. Then, the truck departs for its destination.

Alternative radio frequency identification implementation

Alternative RFID implementation scenarios result from the recognition of inefficiencies within the warehouse. The implemented RFID scenario should satisfy and improve most of these inefficiencies. Through the interviews that were conducted, the inefficiencies that occur within the warehouse were found to be the following:

- need for automatic tracing of quality controls – requires electronic data entry of the files of agronomists and food technologists
- need for a more efficient picking of the semi-finished products' pallets – they are stored without precise recording of their location
- need for automation of the replenishment process between the chamber where the finished products are stored in pallets and the chamber where the finished products are stored in cases
- need for precise tracking at the case level within the dispatch locations.

The proposed RFID implementations are:

1. electronic data entry of information regarding the quality controls (tracing): lot number relates a particular lot with information regarding the manufacturing process (quality tests of agronomists and food technologists)
2. tracking for order picking (internal traceability): efficient picking of pallets of semi-finished products using FIFO.

3. automation of storage and replenishment (in-bound, out-bound) (internal traceability): automation of the replenishment process between the chamber where the finished products are stored in pallets and the chamber where the finished products are stored in cases

4. tracking at case level:
   - shipments to central warehouses and stores
   - lot number relates cases with dispatch locations.

Alternative RFID implementations based on improvement opportunities are depicted in Fig 21.2.

Selected radio frequency identification scenarios
Among the RFID alternative scenarios, it was decided to implement the first and fourth scenario based mainly on cost considerations. A detailed description of the final scenario that incorporates the two alternatives follows.

When the freezing process is finished, the technologist performs sample tests (microbiological, natural and chemical) and according to their results, the product quality is defined. The results of these tests will henceforth be recorded electronically, in order to be able automatically to relate the lot number of the semi-finished product with the files of the technologists. The semi-finished product is then stored in the chamber for semi-finished products until it is needed for consumer packaging. Packaging follows a rolling and controlled programme based on the sales target. Then the semi-finished product is conveyed to a machine which bags it and puts it in a sachet/

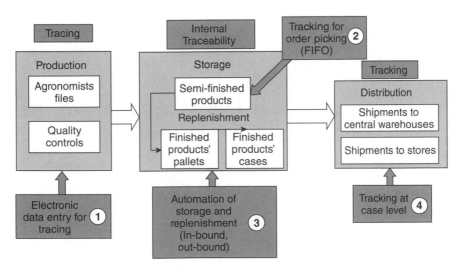

Fig. 21.2 Alternative RFID implementations based on improvement opportunities.

pouch. When, the product is packed in sachets, a change of state from semi-finished to finished product is recorded. This means that the lot number of the semi-finished product will be related to the lot number of the finished product. At this point an RFID will be used at case level. Afterwards, RFID readers will be placed in the shipping areas. During loading the overall bill of goods, the RFID tag of boxes and shipping gate will be recorded. Through the interconnection of the RFID system with the WMS which maintains all the data related to the orders and routes, the lot number of the finished product can be related to the dispatch locations. Finally, the system will provide the possibility of two types of report: (a) tracing, which will allow the data from qualitative control to be retrieved by providing a specific product lot number and (b) tracking, which will allow the current position of products with a specific lot number to be recovered. Implementation of this scenario provides full traceability, as indicated in Fig. 21.3.

Figure 21.4 summarises the implementation scenario.

The functionality of the proposed traceability system
The RFID-enabled traceability system functionality can be summarised as:

- Electronic data entry of the information from quality tests: cross-correlation of lot number of semi-finished product with the files of food technologists at all test stages (microbiological, natural, chemical) of the semi-finished product
- Record of the change from semi-finished to finished product: cross-correlation of lot number of semi-finished product with lot number of finished product
- Record of product shipments: cross-correlation of lot number of finished product with dispatch locations
- Generation of reports: tracing and tracking.

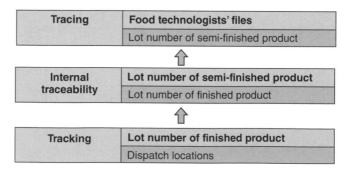

Fig. 21.3 The RFID scenario offers full traceability.

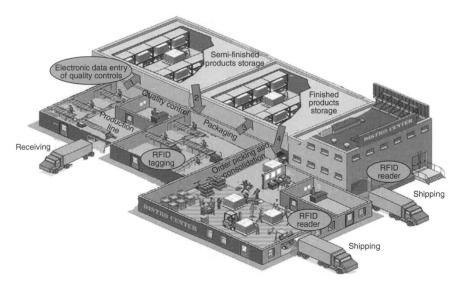

Fig. 21.4   The RFID scenario.

21.3.2   Application B: a distributed service-oriented architecture for radio frequency identification-integrated supply chain collaboration services: the case of dynamic pricing in the retail supply chain

In this section, we describe a service-oriented architecture that utilises the automatic, unique identification capabilities of RFID technology, data stream management systems and web services to support RFID-integrated supply chain decision support and collaboration services in a networked business environment. As a business context, the grocery retail sector has been selected as it is characterised by an intense supply chain environment on one hand, handling thousands of products and supply-chain relationships on a daily basis, and increased competition and consumer demands on the other. In this context, a RFID-integrated dynamic pricing service has been designed, developed, pilot-deployed and evaluated (SMART, 2007b; SMART, 2008a) in a 'SMART' project (IST-20005, FP6) with European grocery retailers and suppliers from the fast-moving consumer goods sector as participating user companies.

In this context, the key requirements include the following:

- An immense amount of data needs to be processed in real time: today when products are identified at product-type level through barcodes, the handling of information in real time for decision-support purposes is quite a technical challenge.
- Synchronised product information between supply chain partners must be ensured (Roland-Berger, 2003): although barcode technology has

been adopted as a standard to identify products, the information is maintained at different levels in either the retailers' or the manufacturers' systems causing serious integrity issues when data exchange and synchronisation are required.

- Many different business relationships need to be supported: each retailer may collaborate with hundreds of suppliers and vice versa.
- Different collaboration scenarios, perhaps applied in each supply chain relationship, need to be supported: a retailer may collaborate with one supplier on efficient warehouse replenishment following continuous replenishment product (CRP)/vendor managed inventory (VMI) or on category management with another supplier etc. (Pramatari, 2006).
- Seamless information sharing and collaborative decision-support, through automated and secure inter-organisational system links, should be ensured.

In order to cope with the above requirements, the architecture employs the following technologies in an integrated network infrastructure:

- Data stream management systems (DSMS), supporting real-time analytics and decision support based on continuous queries of transient data streams.

  Many applications involve data items that arrive on-line from multiple sources in a continuous, rapid and time-varying fashion. Under these circumstances, it may not be possible to process queries within a database management system. For this reason, data are not modelled as being persistent, but rather as transient data streams where 'continuous', not 'one-time', queries are produced over time, reflecting the data stream seen so far. A good example of such an application would be one that constantly receives data from electronic product code observations (e.g. RFID tagged product observations) across a supply chain. Computing real-time analytics (potentially complex) on top of data streams is an essential component of modern organisations (Chatziantoniou and Johnson, 2005).

  This architecture employs a data stream management system to perform complex real time analysis efficiently on top of streams of RFID observations. In the context of the retail supply chain, the main purpose of the DSMS it to translate the data streams coming from the RFID infrastructure (tags and readers) into meaningful product-related data, enabling aggregation at various levels and to provide them with business meaning for the retail supply chain partners through the RFID-integrated services.

- Web services orchestration, enabling secure and seamless information sharing and collaboration in a distributed environment (Muehlen et al., 2005).

  A web service, as defined by the W3C Web Services Architecture Working Group, is 'a software application identified by a URI, whose interfaces and bindings are capable of being defined, described, and discovered as XML artifacts' (W3C, 2002). In general, a web service is an application that provides a web API, supporting application-to-application communication using XML and the web.

The term 'orchestration' has been employed to describe the collaboration of web services. Orchestration describes how web services can interact at the message level, including the business logic and execution order of the interactions. These interactions may span applications and/or organisations, and result in a long-lived, transactional process. With orchestration, the process is always controlled from the perspective of one of the business parties.

Figure 21.5 illustrates a high-level logical view of the architecture. It is a distributed architecture, where the application layer runs on the system of

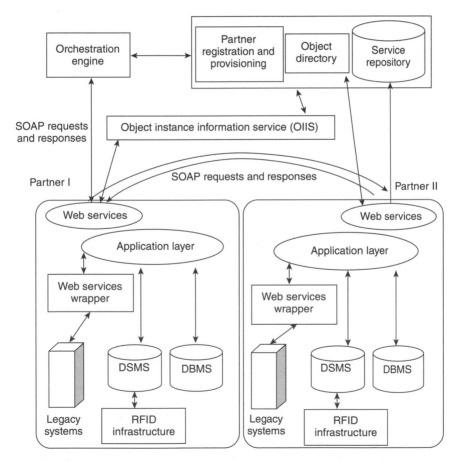

Fig. 21.5   Logical view of the distributed service-oriented architecture.

each collaborating partner and web services implement the interface between the different partners' systems using simple object access protocol (SOAP) requests and responses. The data layer is implemented by both a relational database management system (DBMS) and a DSMS providing the application layer with continuous real time reports after processing unique product identification data streams generated from the RFID infrastructure. The orchestration engine coordinates the exchange of messages between the partners' web services following the logic of the specific collaborative services. The service repository component provides an interface for the orchestration engine to execute queries and discover the exposed services from the partners. The object directory component stores partners' identifiers that can provide information for unique object instances. It accepts queries about unique object instances (electronic product codes-EPCs) and replies with the partner's identifier which can provide the required object instance information. The partner registration and provisioning component holds all partner information and relationships among partners. It also holds application related information that affects partner relationships. Since the RFID infrastructure recognises unique object instances and not object descriptions, when higher level architectural components need to get information for a specific object, they use the appropriate Object Instance Information Service (OIIS) of the partner which can provide this information (for more information on the architecture please refer to SMART, 2008b).

Ultimately, this architecture is a generic distributed service-oriented architecture that can potentially support various supply chain collaboration and decision support services, whether these are integrated with RFID technology or not. However, in the course of the SMART project, RFID-integrated services were deployed. Each RFID-integrated collaboration service carries its own value proposition and can be characterised, on a high-level, by the information shared between supply chain partners, the level of RFID tagging (pallet/case/item level) and the location of the RFID readers (Bardaki *et al.*, 2007). The following subsection presents the RFID-integrated 'dynamic pricing' service that was supported by the architecture in the course of SMART project.

Radio frequency identification-integrated dynamic pricing in retail supply chain

The generic distributed service-oriented architecture can potentially support various RFID-integrated supply chain services in retail industry. However, the research focused on eight RFID-integrated services, such as back-room and shelf visibility, smart product recall, promotion management and so on (Bardaki *et al.*, 2007). To evaluate the business relevance of the alternative services, an industry survey was conducted, addressed to top executives representing retailers and suppliers/manufacturers of European fast-moving consumer goods (SMART, 2007a; Lekakos, 2007).

In addition, a consumer survey provided useful input regarding the evaluation of innovative retail consumer services. The findings of the two surveys (SMART, 2007a) prioritised, among others, the development and implementation of the RFID-integrated dynamic pricing service.

Purpose of dynamic pricing

Dynamic pricing is suitable for products that require frequent price adjustments and it supports different product instances to be sold at different prices (a strategy that is widely used for example in airline tickets). It can be used in the food industry to generate demand for products with a short lifecycle, such as fresh or frozen products that are approaching their expiration date and are soon to become out-of-date gathered stock. Generally, this service can be applied to any product where both the retailer and the supplier have decided to adjust its price according to its expiration date, sales volume, the available stock, and so on. Dynamic pricing requires manual monitoring of the products' expiration date and available stock with the purpose of pricing mark-down of products approaching their 'sell-by' date.

Ultimately, the main concept inspiring this service is to reduce the price of products in order to motivate consumers to buy the nearly out-of-date product items. The goal of the service is to increase the customers' satisfaction and loyalty by providing them with price mark downs; simultaneously, the freshness of the products will be better ensured and the products' waste and the labour cost are expected to be reduced.

Functionality of the RFID-integrated dynamic pricing

The placement of RFID tags on product items enables the monitoring of individual stock items' availability and expiration date from a distance, without requiring line-of-sight; thus, products approaching their sell-by-date can be individually priced or discounted.

Motivated by the aforementioned RFID-offered product monitoring capability, the RFID-integrated service automates the process of dynamic pricing. Specifically, the RFID-integrated dynamic pricing service provides retailers, suppliers and consumers with the following functionalities, via a friendly, graphical user interface:

- Monitor product availability: A number of charts show the availability of the selected product(s) for each expiration date. Information on product availability on the shelf and in the backroom/cold room can be provided.
- Monitor product sales: A number of charts show the performance of the selected product(s) in terms of sales volume. The users (retailers and suppliers) can be informed of the progress of the sales by studying daily or total sales. They are also given the opportunity to monitor the daily shelf replenishment of the products. It should be noted that users can select the period (in days) they are interested in.

- Set new product price: The service proposes a dynamic pricing algorithm, which uses the availability of the products, their expiration date, and so on, in order to provide retailers and suppliers with price mark-down suggestions per product expiration date and stock availability. It is worth mentioning that users can select the criteria based on which the algorithm provides the price suggestions. Then, the retailer and the supplier collaborate to confirm price mark-downs, after considering the algorithm's price suggestions.
- Electronic shelf display for the consumers: A small electronic display is placed on the shelves to inform consumers in real time about the price mark-down and the shelf availability of the products per expiration date. In this way, the information provided to the back-office via the 'promotion availability' functionality also reaches consumers.

Implementation of the RFID-integrated dynamic pricing
A step-by-step approach was adopted to decide on 'realistic' RFID-integrated implementation of the dynamic pricing service by assessing its technical feasibility and to what extent the potential benefits gained by RFID outweigh the value of investment in such an initiative. This approach is based on a detailed business process analysis, technical laboratory experiments and a cost–benefit assessment (Bardaki *et al.*, 2008).

Based on the results of the technical tests, the cost versus benefit and the business needs of the users, the dynamic pricing service was implemented as follows in the course of the SMART project:

- Packaged fresh minced meat, kept in the fridge, has been selected to test the service.
- Both case and product item tagging have been selected to satisfy the needs of the service.
- An RFID reader with antennas will be placed at the conveyor belt of the supplier's packaging facilities to capture the electronic product code (EPC) of the products and cases tagged.
- An RFID reader with antennas will be placed at the exit–entrance of the cold room to capture the delivery of cases and the replenishment of the sales floor (case-level reading).
- An RFID reader will be placed in the regular shelf in the fridge to monitor the product availability (item-level reading).

Figure 21.6 depicts the topology of the RFID readers and the level of tag reading at each location in the dynamic pricing service.

21.4   Conclusions

This chapter looks closely into the technology of RFID and the way it is employed in the supply chain and describes two applications. The first

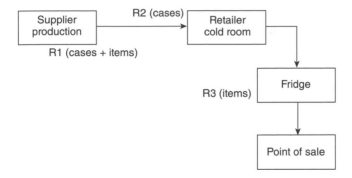

Fig. 21.6   Topology of RFID readers in a dynamic pricing service.

concerns the requirements' analysis, development and pilot implementation of a RFID-enabled traceability system for a frozen food company. The main implication is that the introduction of an RFID-enabled traceability system would alleviate the ill effects of manual processes caused by automation, eliminate transaction errors, labour intervention and the required cycle time while increasing the throughput speed of product and inventory accuracy. In addition, it would increase the visibility across the supply chain by providing better tracing (electronic data entry of information regarding the quality controls) and precise tracking at case level within the dispatch locations.

Following developments in the RFID field and in supply chain collaboration, the second application describes a distributed, service-oriented architecture that supports RFID-integrated decision support and collaboration practices in a networked business environment. In the context of retail industry, a RFID-integrated 'dynamic pricing' service is described regarding its functionality and implementation. The main findings can be categorised into two foci: the proposed architecture and the dynamic pricing service implications. The architecture can potentially support various supply chain collaboration and decision support services, whether these are integrated with RFID technology or not. RFID can help in the introduction of a service that allows individual stock items to be separately identified, as they move off the shelves, and be individually priced or discounted, by monitoring product's availability on the shelf and in the store and guaranteeing freshness of products by relying on the constant monitoring of the expiration date and reduce waste, labour time and cost.

21.5   Sources of further information and advice

The following is a collection of online materials that anyone can draw upon. This collection is partial since there are thousands of sources in RFID. No endorsement is intended for any source listed, nor any slight intended for any source not listed.

Overviews

- RFID Journal: A guide to understanding RFID (http://www.rfidjournal. com/article/gettingstarted)
- Wikipedia: RFID (http://en.wikipedia.org/wiki/RFID)
- HowStuffWorks: How RFIDs Work (http://electronics.howstuffworks. com/rfid.htm)
- Other online RFID Resources:
  - RFID News (http://www.rfidnews.org/)
  - RFID Gazette (http://www.rfidgazette.org/)
  - RFID Weblog (http://www.rfid-weblog.com/)
  - RFID Times (http://rfidtimes.org/).

Organizations

- EPCglobal Inc. (http://www.epcglobalinc.org/home)
  EPCglobal leads the development of industry driven standards for the electronic product code (EPC) to support the use of radio frequency identification in today's fast-moving, information rich, trading networks.
- Auto-ID Labs (USA) (http://www.autoidlabs.org/)
  The Auto-ID Labs are the leading global network of academic research laboratories in the field of networked RFID. The labs comprise seven of the world's most renowned research universities located on four different continents. These institutions were chosen by the Auto-ID Center to architect the Internet of Things together with EPCglobal.
- RACE networkRFID (http://www.race-networkrfid.eu/)
  RACE networkRFID is designed to become a federating platform for the benefit of all European stakeholders in the development, adoption and usage of RFID. The network considers its mission is to create opportunities and increase the competitiveness of European Member States in the area of RFID through leadership, development and implementation. At the same time it will position RFID technology within the mainstream of information and communications technology (ICT).
- Ubiquitous ID Center (Japan) (http://www.uidcenter.org/index-en.html)
  The Ubiquitous ID Center was set up within the T-Engine Forum to establish and popularise the core technology for automatically identifying physical objects and locations and to work toward the ultimate objective of realising a ubiquitous computing environment.
- RFID Alliance (http://www.rfidalliancelab.org/index.shtml)
  The RFID Alliance Lab, housed at the University of Kansas, is a not-for-profit testing facility that provides objective benchmarking reports on RFID equipment.

Companies

- Large companies with a large volume of RFID production
  - Texas Instruments (http://www.ti.com/rfid/)
  - Philips Electronics (http://www.nxp.com/products/identification/)
- Smaller companies with a primary focus on RFID solutions
  - Alien Technology (http://www.alientechnology.com/)
  - Intermec (http://www.intermec.com/)
  - Impinj (http://www.impinj.com/)
  - Ascendent ID (http://www.ascendentid.com/)
- Large companies with an RFID solution capability
  - Hewlett Packard (http://welcome.hp.com/country/us/en/welcome. html#Product)
  - IBM (http://www-03.ibm.com/solutions/sensors/us/index.html)
  - Microsoft (http://www.microsoft.com/dynamics/businessneeds/scm_ chip.mspx)
  - Sun Microsystems (http://java.sun.com/developer/technicalArticles/ Ecommerce/rfid/)

21.6 Acknowledgement

Application A has been partly funded by the General Secretariat for Research & Technology, Ministry of Development of the Hellenic Republic. Application B is partly funded by the SMART research project (IST-2005), Information Societies Technology Programme, 6th Framework, Commission of the European Union (www.smart-rfid.eu).

21.7 References

AGARWAL, V. (2001). *Assessing the Benefits of Auto-ID Technology in the Consumer Goods Industry*, Cambridge University Auto-ID Centre, Cambridge.

ANGELES, R. (2005). 'RFID technologies: supply-chain applications and implementation issues', *Information Systems Management*, **22**(1), 51–65.

ATALI A., LEE H. and OZER O. (2005). 'If the inventory manager knew: value of RFID under imperfect inventory information', *MSOM Conference, the Annual Meeting of the INFORMS Society on Manufacturing and Service Operations Management*, June 27–28, Northwestern University, Evanston, Illinois.

ATTARAN, M. (2007). 'RFID: an enabler of supply chain operations', *Supply Chain Management: An International Journal*, **12**(4), 249–57.

AUTO-ID CENTER (2002). *Technology Guide*, Auto-ID Center, www.autoidcenter.org.

BALLOU, D., WANG, R., PAZER, H. and TAYI, G. (1998). 'Modelling information manufacturing systems to determine information product quality', *Management Science*, **44**(4), 462.

BARDAKI, C., PRAMATARI, K. and DOUKIDIS, G. J. (2007). 'RFID-enabled supply chain collaboration services in a networked retail business environment', *Proceedings of the 20th Bled eConference*, June 3–6 2007, Bled, Slovenia.

BARDAKI, C., KARAGIANNAKI, A. and PRAMATARI, K. (2008). 'A systematic approach for the design of RFID implementations in the supply chain', *Proceedings of the Panhellenic Conference on Informatics (PCI 2008)*, August 28–30, Samos, Greece.

CHATZIANTONIOU, D. and JOHNSON, T. (2005). 'Decision support queries on a taperesident data warehouse', *Information Systems Journal*, **30**(2), 133–149.

CURTIN, J., KAUFFMAN, R. J. and RIGGINS, F. J. (2007). 'Making the "MOST" out of RFID technology: a research agenda for the study of the adoption, usage and impact of RFID', *Information Technology and Management*, **8**, 87–110.

EPCGLOBAL (2005). *The EPCglobal Network and The Global Data Synchronisation Network (GDSN): Understanding the Information and the Information Networks*. Retrieved from http://www.epcglobalinc.org/about/media_centre/ EPCglobal_and_GDSN_v4_0_Final.pdf

FLEISCH E. and TELLKAMP C. (2005). 'Inventory inaccuracy and supply chain performance: a simulation study of a retail supply chain', *International Journal of Production Economics*, **95**(3), 373–85.

GAUKLER, G., SEIFERT, R. and HAUSMAN, W. (2006). 'Item-level RFID in the retail supply Chain', *Production and Operations Management*, **16**(1), 65–76.

GLOBAL COMMERCE INITIATIVE-GCI (2005). *EPC: A Shared Vision for Transforming Business Processes*. Retrieved from: http://www.fmi.org/technology/GCI_IBM_ EPC_report.pdf

HAMNER, S. (2005). *The Grocery Store of the Future*, Business 2.0 Magazine. Retrieved from http://money.cnn.com/magazines/business2/business2_archive/2005/03/01/ 8253112/index.htm

HOU, J.L. and HUANG, C.H. (2006). 'Quantitative performance evaluation of RFID applications in the supply chain of the printing industry', *Industrial Management and Data Systems*, **106**(1), 96–120.

JONES P., CLARKE-HILL C., HILLER D. and COMFORT D. (2005). 'The benefits, challenges and impacts of radio frequency identification technology (RFID) for retailers in the UK', *Marketing Intelligence & Planning*, **23**(4), 395–402.

KELEPOURIS, T., PRAMATARI, K. and DOUKIDIS, G. (2007). 'RFID-enabled traceability in the food supply chain'. *Industrial Management and Data Systems*, **107**(2), 183–200.

KELLY, E.P. and ERICKSON, G.S. (2005). 'RFID tags: commercial applications v. privacy rights', *Industrial Management and Data Systems*, **105**(6), 703–13.

LEE, H.L. (2007). 'Peering through a glass darkly', *International Commerce Review*, **7**(1), 60–8.

LEE, Y. M., CHENG, F. and LEUNG, Y. T. (2004). 'Exploring the impact of RFID on supply chain dynamics', in *Proceedings of the 2004 Winter Simulation Conference*, Washington, USA.

LEKAKOS, G. (2007). 'Exploiting RFID digital information in enterprise collaboration', *Industrial Management and Data Systems*, **107**(8), 1110–22.

MUEHLEN, Z.M., NICKERSON, J.V. and SWENSON, K.D. (2005). 'Developing web services choreography standards – the case of REST vs. SOAP', *Decision Support Systems*, **40**, 9–29.

PRAMATARI, K. (2006). 'Collaborative supply chain practices and evolving technological approaches', *Supply Chain Management: an International Journal*, **12**(3), 210–20.

PRAMATARI, K.C., DOUKIDIS, G.I. and KOUROUTHANASSIS, P. (2005). *Towards 'Smarter' Supply and Demand-chain Collaboration Practices Enabled by RFID Technology*, in *Smart Business Networks*, Vervest, P., Van Heck, E., Preiss, K. and Pau, L.F. (eds) Springer Verlag, New York, NY.

REYES, P.M. and JASKA, P. (2007). 'Is RFID right for your organisation or application?' *Management Research News*, **30**(8), 570–80.

ROLAND-BERGER (2003). *ECR-optimal Shelf Availability*, ECR Europe, available at: www.ecr-net.org

SHUTZBERG, L. (2004). *RFID in the Consumer Goods Supply Chain: Mandated Compliance of Remarkable Innovation?* Rock-Tenn Company, Norcross, GA.

SMART (2007a). *Deliverable 1.2 Requirements Analysis Report*, EU Project ST-5-034957-STP, ELTRUN, Athens University of Economics and Business, Athens, Greece (www.smart-rfid.eu).

SMART (2007b). *Deliverable 1.4 Final Specifications of the SMART Scenarios*, EU Project ST-5–034957-STP, ELTRUN, Athens University of Economics and Business, Athens, Greece (www.smart-rfid.eu).

SMART (2008a). *Deliverable 4.4 Business Validation Impact and Consumer Survey results*, EU Project ST-5-034957-STP, ELTRUN, Athens University of Economics and Business, Athens, Greece (www.smart-rfid.eu).

SMART (2008b). *Deliverable 2.1 Overall System Architecture*, EU Project ST-5-034957-STP, ELTRUN, Athens University of Economics and Business, Athens, Greece (www.smart-rfid.eu).

SMITH, H. and KONSYNSKI, B. (2003). 'Developments in practice X: radio frequency identification (RFID) – an internet for physical objects', *Communications of the AIS*, **12**, 301–11.

WANG, S., LIU, S. and WANG, W. (2008). 'The simulated impact of RFID-enabled supply chain on pull-based inventory replenishment in TFT-LCD industry', *International Journal of Production Economics*, **112**(2), 570–86.

W3C WEB SERVICES ARCHITECTURE WORKING GROUP (2002). *Web Services Architecture Requirements*. Retrieved from http://www.w3.org/2002/ws/arch/2/wd-wsawg-reqs-03262002.html

Part VI

Delivering food sustainably and responsibly

22

Reducing the external costs of food distribution in the UK

D. Fisher, Aecom, UK, A. McKinnon, Heriot-Watt University, UK, and A. Palmer, Cranfield University, UK

Abstract: This chapter summarises the results of a study undertaken for the UK government which examined the opportunities for reducing the external costs of food distribution. These costs comprised environmental, infrastructural and congestion elements. In consultation with an industry advisory group, the researchers identified a 'long list' of over 60 measures, 15 of which were subjected to external cost modelling. The chapter focuses on a short list of six of these measures which were considered to be practical and offer significant benefit. These included transport collaboration, increasing maximum truck weight and size, redesigning logistics systems and increasing the use of vehicle routing and telematics systems. It is estimated that the application of the six measures could cut the external cost of UK food distribution in 2012 by 17%.

Key words: food distribution, external costs, environmental impact, UK.

22.1 Introduction

Campaigning by environmental groups has given the impression that sourcing food over long distances is unsustainable. They have highlighted the so-called 'food miles' issue and advocated a return to more localised sourcing (Sustain, 1999). 'Food miles' have been shown, however, to be only a partial, and in some cases misleading, measure of the environmental impact of the food supply chain (AEA Technology, 2005). This impact is affected not just by the distance travelled but also the nature of the transport. The long distance movement of food in full loads in a modern articulated vehicle along a motorway is less environmentally damaging than its distribution by van over short distances on urban roads. Lifecycle analysis also reveals that it can be environmentally beneficial to source food products from distant locations where the production operation uses less energy and emits less pollution (Mason *et al.*, 2002).

Cutting the external costs of food distribution, therefore, involves more than simply returning to more localised supply systems. This is only one of a range of measures that can be applied to 'green' the movement of food products. In this chapter we summarise the results of a study commissioned by the British government to review these measures, quantify their environmental benefits and thus rank them in order of importance. The work was undertaken for the Food Industry Sustainable Strategy (FISS) Champions Group on Food Transport, which comprised representatives of the food industry and three government departments.

As the measures varied in their physical effects on particular externalities, such as air pollutants, noise and accidents, a common metric had to be found to compare them on a consistent basis. This metric was money, in other words the monetary valuation of the environmental and social impacts. Numerous attempts have been made in the UK to attach monetary values to the external costs of transport (e.g. Sansom, 1998; NERA *et al.*, 2000; INFRAS, 2004; Piecyk and McKinnon, 2007). They have recently been reviewed by a study undertaken for the European Commission as part of a wider initiative to internalise these costs in higher taxes (CE Delft, 2007). The study reported in this chapter employed official valuations obtained from UK government departments. These related to six externalities: accidents, air quality, climate change, noise, traffic congestion and infrastructural wear.

The scope of the work was bounded by a 'farm-to-shelf' principle which excluded the movement of agricultural inputs upstream of the farm, consumer shopping trips and home deliveries of food. For the purposes of this research the food supply chain extended from the farm gate to the shop, passing through varying numbers of factories and warehouses along the way. Only food movements[1] within the UK were considered and all the data analysed related to 2005.

22.2   Estimate of the total external costs of UK food distribution

Food transport externalities were calculated on a vehicle-km basis (or tonne-km for non-road modes). Data on the movement of food by road was obtained from the Continuing Survey of Road Goods Transport (Department for Transport, 2006). Figures for air pollution and CO_2 emissions came from the National Atmospheric Emissions Inventory (NAEI) adjusted using information from NERA (2005) on the effects of vehicle weights. Monetary values for externalities came from various sources

[1] Although reference is made to food, the analysis covered both food and drink products.

Table 22.1 Estimated external costs of transporting food within the UK (2005)

	Accidents	Air quality	Congestion	CO$_2$	Infrastructure	Noise	Total
Costs*	£356 m (19%)	£143 m (8%)	£783 m (42%)	£175 m (9%)	£381 m (20%)	£39 m (2%)	£1877 m (100%)

* 2005 values.

including DEFRA (2006), Department for Transport (2007a) and personal communication with government officials. Values for the cost of peak and off-peak congestion were obtained from the Strategic Rail Authority, SRA (2003), while a case study on urban road noise carried out in Stuttgart distinguished the valuation of night-time noise from that of daytime noise (Schmid *et al.*, 2001). Table 22.1 presents estimates of the total cost of each of the six externalities in 2005. Congestion costs were by far the largest element in the calculation, followed infrastructure wear and accidents. Other environmental costs, associated with air pollution, global warming and noise, together accounted for less than 20% of the total.

The external costs varied substantially by vehicle type reflecting differences in the nature of their deliveries and the sensitivity of the environments in which they tend to operate (Fig. 22.1). For example, light rigid HGVs had the highest congestion cost per vehicle-km because much of their mileage is run during peak hours in urban environments. In contrast, articulated HGVs had a higher proportion of their external costs in the infrastructure and climate change categories owing to their heavier axle weights and higher fuel consumption per km.

The original objective of the research was to find a combination of measures that would reduce the total external cost of food transport in the UK by 20%. It was recognised that the choice of measures would be strongly influenced by the distribution of external costs across the six externalities, different transport modes and the various types of vehicle. The study also examined the barriers likely to inhibit the implementation of these measures.

22.3 Identifying and modelling options for reducing external costs

An initial 'long list' of over 60 options was compiled on the basis of the study team's expert knowledge and desktop research. This preliminary review classified options into four categories:

- wider implementation of existing best practice in the management of food logistics

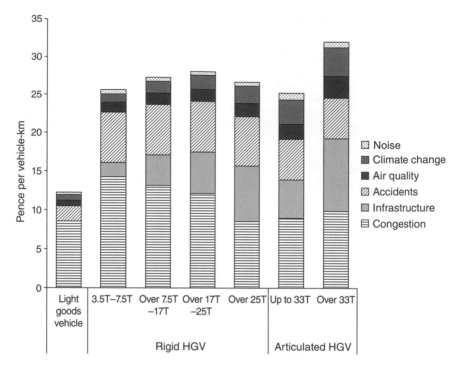

Fig. 22.1 External costs associated with different classes of freight vehicle transporting food products.

- redesign of local supply and distribution networks for food
- step changes in transport methods and / or infrastructure
- regulatory changes affecting logistics.

Following consultation with various stakeholder groups, the long list of options was reduced to a short list of 15 measures whose impact on external costs was quantitatively modelled. The choice of options was mainly determined by the scale of potential environmental benefits and likelihood of implementation, but it was also influenced by the availability of data. Some options, which could yield significant environmental benefit, were excluded because no means could be found of quantifying their impact. There had to be sufficient information on each option to give the FISS Group confidence in the accuracy and realism of the results. The nature of the modelling, nevertheless, varied across the 15 short-listed options depending on the amount, type and quality of data available. Broadly speaking, the modelling was conducted at four levels of realism:

1. complete set of operational data available permitting accurate estimates based on real-world experience

2. extensive operational data available from which industry-wide estimates can be extrapolated
3. limited case study data available to calibrate the models
4. no operational data available, but opportunities exist to simulate impact using hypothetical data.

The option screening process also excluded measures which would have an indirect effect on food transport and/or have an adverse environmental effect on other parts of the food supply system. Examples of such measures were:

• Relocation of food shops to suburban or out-of-town locations: While this would reduce the external costs of freight deliveries it might substantially increase the frequency and length of car-borne shopping trips, transferring the external costs from freight to personal transport.
• Encouraging consumers to switch their demand to less transport-intensive products: This would raise wider issues about the sustainability of consumption patterns and, as discussed earlier, need not yield a net environmental benefit.
• Reducing the volume of packaging used on finished products to improve vehicle utilisation: This would have wider implications for production operations, product wastage and the recycling/disposal of packaging material

The short-listed options were considered and approved by the FISS Group before being tested against the 'base case' external cost values for 2005. Reductions in external costs were estimated for 2012, though the modelling assumed that there would be no change in traffic levels during the intervening period. As a result, the estimated reductions in external costs could be conservative. Options were, nevertheless, tested against low, central and high implementation scenarios, representing differences in the applicability of the measures to different sectors of the food industry and the potential for improving the take-up rate for measures within each sector.

Table 22.2 shows the three levels of estimated savings in external costs accruing from the 15 measures, expressed both as monetary values and as a percentage of the total external costs of food transport.

Following a review of the modelling results, the FISS Group selected six options which it believed would offer the best combination of external cost saving and practicality. These were:

1. upgrading vehicles to higher emission standards
2. transport collaboration
3. expanding vehicle capacity
4. logistics system redesign
5. out-of-hours deliveries
6. computerised vehicle routing and scheduling (CVRS)/telematics.

Table 22.2   Options subject to external cost modelling

Options	Low case	Central case	High case
Transport collaboration	0.5%–£10 m	3.2%–£60 m	8.1%–£153 m
Out-of-hours deliveries	1%–£18 m	2%–£37 m	3.9%–£74 m
Logistics system redesign (Strategy 1)	1%–£18 m	2.3%–£44 m	3.2%– £61 m
Logistics system redesign (Strategy 2)	1.7%–£34 m	2.5%–£51 m	3%–£60 m
Logistics system redesign (Strategy 3)	0.9%–£17 m	4.4%–£82 m	5.8%–£108 m
Expanding vehicle capacity	1.4%–£26 m	3%–£57 m	5.3%–£99 m
Increased backloading	0.4%–£8 m	1.1%–£20 m	1.7%–£32 m
Switch to bio-diesel mix	0.01%–£0.1 m	0.05%–£1 m	0.26%–£5 m
Switch to compressed natural gas	0.12%–£2 m	0.29%–£5 m	0.58%–£11 m
Upgrading to vehicles with higher emission standards	2.7%–£50 m	3.5%–£66 m	4.1%–£77 m
Fuel management and driver training	0.09%–£2 m	0.28%–£5 m	0.73%–£14 m
CVRS and telematics	0.5%–£9 m	1.5%–£28 m	3.0%–£57 m
Adoption of vendor managed inventory	0.9%–£16 m	1.3%–£24 m	1.7%–£32 m
Local sourcing		0.2%–£4 m	
Modal shift – road to rail	0.9%–£17 m	2.3%–£42.5 m	4.5%–£85.1 m

It was recognised that these six options were not mutually exclusive and, therefore, that their combined impact on total external costs would not necessarily be cumulative. Further modelling revealed that in combination these measures would reduce the external cost of domestic food transport by 17.3% (or £326 million per annum).

The remainder of the chapter examines each of these six options, assessing their potential and factors inhibiting their implementation.

22.4   The key options

These options will be reviewed in descending order of the potential reduction in external costs. None of the options are exclusive to the food supply chain. Indeed some have already been more effectively applied in other sectors where their beneficial effect on the environment has been demonstrated.

22.4.1 Upgrading vehicles to higher emission standards

Since 1991 new trucks and vans have had to meet tightening emission standards. Table 22.3 shows how limits on the amount of nitrogen oxides (NO_x) and particulate matter (PM10) have been drastically reduced over the intervening period. The gradual renewal of companies' fleets results in the upgrading of freight vehicles to higher emission standards. The NAEI contains forecasts of future vehicle emissions based on projected upgrades to these higher standards. For example, it is predicted that by 2012, 73% of vans will meet Euro IV standard, 48% of rigid HGVs will meet Euro V and 57% of articulated HGVs will conform to the Euro V standard. Allowance is also made for increases in the proportion of vans run on diesel fuel from 88% to 90%, offering fuel efficiency benefits.

Overall it can be seen that the gradual adoption of these higher emissions standards significantly reduces the external cost of air pollution (Table 22.4). This measure has a very marginal effect on climate change, however, CO_2 is not one of the exhaust gases controlled by Euro emission standards. On the contrary, in an effort to cut emissions of other pollutants, particularly NO_x, vehicle manufacturers have had to sacrifice potential fuel efficiency gains and CO_2 reductions. The climate change benefits quoted in Table 22.4 resulted mainly from the marginal switch from petrol to diesel fuel in the van fleet but also from improvements in the average fuel efficiency of new freight vehicles.

The high and low estimates for external cost savings from this option reflect the rate at which higher emission standards are adopted across the

Table 22.3 Euro emission standards for heavy goods vehicles (g $kW^{-1}h^{-1}$)

Standard	Date	NO_x	PM	CO	HC
Euro I	1992	8.0	0.36	4.5	1.1
Euro II	1998	7.0	0.15	4.0	1.1
Euro III	2000	5.0	0.10	2.1	0.66
Euro IV	2005	3.5	0.02	1.5	0.46
Euro V	2008	2.0	0.02	1.5	0.46

Source: www.nao.org.uk

Table 22.4 External cost savings from upgrading vehicles to higher emission standards

Scenario	Air quality saving	CO_2 saving	Overall saving
Low	£49.8 m (36%)	£0.6 m (0.3%)	£50.4 m (2.7%)
Central	£65.3 m (47%)	£0.6 m (0.4%)	£66 m (3.5%)
High	£76.3 m (55%)	£0.7 m (0.4%)	£76.9 m (4.1%)

vehicle fleet transporting food products. The lower values assume that it will take two years longer to achieve the forecast emission level, while the high values assume that they will be reached two years earlier. The introduction of the low emissions zone in London, which penalises companies whose vehicles do not meet Euro III NO$_x$ standards, is promoting more rapid adoption of cleaner engines. On the other hand, during periods of economic recession, hauliers have less capital to invest in new trucks and often extend the vehicle replacement cycle.

22.4.2   Transport collaboration

There is a limit to the extent that any individual company can improve the efficiency of its transport operation. This limit can be relaxed when it is prepared to collaborate with other companies to consolidate loads, improve backloading and reschedule deliveries. In recent years, businesses in the food sector have shown increased willingness to explore opportunities for collaboration. The European Logistics Users Providers and Enablers Group (ELUPEG) and Efficient Consumer Response (ECR) (Institute of Grocery Distribution, 2003 and 2007) have been actively promoting transport collaboration among food businesses. The English Farming and Food Partnerships (EFFP) have also been working with companies such as ASDA, Tesco and McDonalds to help create more transparent and efficient supply chains (Anon, 2005).

Collaboration can take various forms. It can involve cooperation between manufacturers distributing their products to similar outlets. This 'horizontal' type of collaboration is well illustrated in the food sector by a collaborative initiative established by Douwe Egberts, Unipro and Masterfoods within the Dutch catering sector (Cruijssens, 2006). All three companies supply frozen products to catering outlets and had a 68% customer overlap. As distribution costs constituted a relatively high proportion of sales revenue it was agreed that joint distribution presented a commercially attractive opportunity. Overall, joint route planning and sharing of vehicles reduced total distance travelled by 31%, increased load factors by over 95% and permitted a 50% reduction in fleet size. Although the collaboration was financially motivated, it yielded major environmental benefits.

Also in the Netherlands, Unilever and Kimberley Clark have set up a joint distribution centre (DC) at Raamsdonksveer where they consolidate loads. This has achieved substantial savings in logistics costs, improved service quality and reduced environmental impact. In this case the distribution operation is outsourced to a logistics service provider (LSP) (Kuehne and Nagel). LSPs can play a critical role in collaborative initiatives. In the USA, collaborative transportation management (CTM) has been strongly advocated as a holistic process that brings together supply chain trading partners and logistics providers to drive inefficiencies out of the transport planning and execution process (VICS Logistics Committee, 2004). The objective of CTM is to improve vehicle utilisation through the sharing of information and forecasts and collaborative planning of freight movements. Carriers participating in CTM pilot projects reported reductions in empty mileage of 15% and fleet utilisation improvements of 33%.

Organisations belonging to ECR UK have trialled 'collaborative multi-partner trunking' which involves a retailer and a group of their suppliers working together to reduce empty running. It was estimated that across 20 organisations participating in the trial, a total of around 3.2 million vehicle-km could be saved. More detailed modelling of the potential savings in inner London from collaboration between Boots, Musgraves-Budgens-Londis and Sainsburys suggested that 2.5% of vehicle-km could be eliminated. This potential saving was constrained by restrictions on delivery times imposed by local authorities both at individual shops and at a zonal level. The case for easing these restrictions is discussed in Section 22.4.5.

A more radical collaborative option was tested during the study. This involved the integration of two large supermarket chains' distribution systems and sharing of DCs and vehicles fleets. This hypothetical analysis indicated that, if the existing DC locations and capacities were retained, lorry traffic could be cut by approximately 6%. Optimising the DC locations and capacities to supply the combined retail chains would raise this percentage to 10%. These reductions in lorry-km may seem rather low given the scale of restructuring required. This can be explained partly by the fact that the analysis only included outbound distribution to shops and was confined to two of the largest UK supermarket chains which generated a large proportion of full loads. A similar analysis conducted on smaller retail chains or food manufacturers would probably have revealed greater opportunity for combining part loads and cutting lorry-km.

On the basis of more realistic estimates of the degree of transport collaboration that might be achieved by 2012, potential savings in external costs might average around £60 million per annum (Table 22.5).

The main barriers to collaboration between suppliers and distributors in the food supply chain are often management culture and lack of trust. Equitable sharing of the costs and benefits and effective inter-organisational exchange of information play a pivotal role in any collaborative scheme. Although the potential benefits of transport collaboration

Table 22.5 External cost savings from transport collaboration

Scenario	Congestion saving (million £)	Infrastructure saving (million £)	Accident saving (million £)	Air quality saving (million £)	CO_2 saving (million £)	Noise saving (million £)	Overall saving (million £)
Low	3.1	2.7	2.4	0.7	1.1	0.2	10.3
Central	17	16.6	14.4	4.5	6.5	1.3	60.4
High	41.8	42.4	36.6	11.7	16.9	3.3	152.8

have been widely publicised for many years, it is only recently that many large companies in the food sector have begun to make a serious commitment to rationalising their transport operations on a collective basis (Institute of Grocery Distribution, 2009).

22.4.3   Expanding vehicle capacity

On roughly 6.5 of every 10 km that a truck travels carrying food products, the load is constrained by the cubic capacity of the vehicle (Table 22.6). Twice as many of these 'laden-km' are subject to a volume constraint as to a weight constraint. Increasing the size of vehicles can therefore create greater opportunities for load consolidation in the food sector than raising their maximum weight. Capacity can be expanded either vertically in double-deck vehicles or horizontally in longer vehicles. The study focused on the former option as this would not require any relaxation of current regulations. Two separate studies, sponsored by the UK government (Knight et al., 2008) and European Commission (Transport and Mobility Leuven, 2008) have examined the costs and benefits of increasing the length and weight of trucks.

Unlike in much of mainland Europe where truck height is restricted to 4 or 4.2 m, there is no legal limit on the maximum height of lorries in the UK. By 'custom and practice' 5 m is considered to be the height limit across the trunk road network. This greater height clearance in the UK has allowed companies moving lower density products, such as packaged food, to double-deck their trailers and carry two layers of pallets or roll cages. Double-deck lorries are now extensively used in the UK food distribution system. In 2006, approximately 6% of the trailers operated by large British supermarket retailers were double decked (Institute of Grocery Distribution, 2007). These vehicles are well suited to fixed-route interdepot trunking, factory gate pickups and deliveries to larger supermarkets. Depending on the trailer configuration, double decks can carry 60–90% more product than a single deck vehicle. Tesco estimates that its double-deck trailers have 67% more carrying capacity than single decks. As they are heavier and have

Table 22.6   Proportion of HGV laden-km limited by weight and/or volume (2005)

	Total laden vehicle-km	Constrained by weight (%)	Constrained by volume (%)
Agricultural products	11 650	54	70
Beverages	6 031	36	71
Other foodstuffs	22 895	22	60
Food, drink and tobacco	40 577	33	65

Source: Department for Transport, 2006.

Table 22.7   External cost savings from greater use of double-deck vehicles

Scenario	Congestion saving (million £)	Infrastructure saving (million £)	Accident saving (million £)	Air quality saving (million £)	CO₂ saving (million £)	Noise saving (million £)	Overall saving (million £)
Low	8.1	7.5	4.3	2.2	2.9	0.6	25.6
Central	18	16.8	9.6	4.8	6.4	1.3	56.8
High	31.4	29.3	16.7	8.4	11.2	2.3	99.4

a poorer aerodynamic profile, double-deck vehicles consume more fuel per vehicle-km, but their overall energy efficiency (expressed as tonne-km or pallet-kms per litre of fuel) is much higher. One case study of a British retailer's store delivery operations recorded fuel savings of 48% (Department for Transport, 2005a). Efforts have been made recently to improve the aerodynamic styling of double decks, although this reduces available space. One streamlined double-deck trailer sacrifices 9% of internal cubic capacity to achieve a 15% reduction in fuel consumption.

The main factors inhibiting the use of double-deck trailers are inadequate reception facilities at factories, DCs and shops, their higher capital costs and the limited proportion of loads that are large enough to exploit the extra vehicle capacity. Double decks account for around 5% of the articulated vehicle-km generated by major grocery retailers and wholesalers (Institute of Grocery Distribution, 2007). The modelling of potential external cost savings assumed that by 2012 this share would increase to 25% of food laden-km subject to a volume constraint (Table 22.7).

22.4.4 Logistics system redesign

The expansion of retail chains, changing spatial distribution of demand, the reconfiguration of the upstream supply chains (e.g. partly as a result of the growth of imports) and company mergers can leave DCs in sub-optimal locations and/or with inadequate capacity. This impairs the efficiency of distribution operations, causing vehicle-km and the associated externalities to be higher than necessary. Relocating and resizing DCs, and reallocating stores between them, can significantly improve the efficiency of the system and reduce its environmental footprint. This 'reoptimising' of the system also needs to take account of seasonal fluctuations in throughput. Lack of capacity at peak times can result in the additional shuttling of product to/from 'overspill' warehouses. One biscuit manufacturer, for example, had a single DC which was unable to hold its ambient Christmas stock. It used an external ambient storage facility about 16 km from the DC. This generated an extra 104 000 vehicle-km miles per annum, representing about 5.2% of the total. Another problem is that some companies do not revise their delivery boundaries frequently enough to take account of changes in the patterns of demand and road accessibility. Indeed, many companies have now abandoned fixed boundaries around DC service areas and plan deliveries across the system as a whole. This can achieve much higher levels of vehicle utilisation although, in the case of more complex systems, presents formidable analytical problems.

To assess the possible contribution of system redesign to external cost reduction, transport flows for four supermarket chains were modelled using data on DC and store locations, the volume of supplies moved, delivery frequencies, transport costs and other operating parameters, split by product categories. Several strategic options were tested for each of the supermarket chains:

- Retaining the existing DC store allocation and DC size, but relocating DCs to optimal locations.
- Retaining the existing DC locations, but optimising the allocation of supermarkets to DCs. In this strategy, the size of the DC was variable, depending on the store volumes supplied
- Optimising current DC locations with flexibility in the allocation of shops to DCs. In this strategy, the capacity of the DC could again be varied in line with the required throughput.

Each of these strategic options would offer significant savings in vehicle-km, vehicle numbers and operating costs, with the second being the most beneficial to the environment (Table 22.8). The modelling revealed that simply by optimising DC size and store allocation, a reduction in vehicle-km of around 10.5% could be achieved. It should be noted too that this modelling only took account of outbound deliveries to shops. As generally happens in the planning of DC locations, no allowance was made for the geography of inbound flows into the DCs.

The results from this strategic modelling were factored into the externalities model on the assumption that distance savings would apply equally across all vehicle classes, types of route and time periods (Table 22.9). The logistical systems of food manufacturers and food service companies could be subjected to similar redesign to augment these external cost savings.

Table 22.8 Effects of three logistics system redesign strategies

Strategy	Min saving Veh-km (%)	Average saving Veh-km (%)	Max saving Veh-km (%)	Min saving vehicles (%)	Max saving vehicles (%)	Min saving vehicle operation costs (%)	Max saving vehicle operation costs (%)
1	3.2	7.6	10.5	1.8	5.8	2.4	6.2
2	5.9	8.9	10.5	1	7	2.9	8.3
3	2.9	14.3	18.9	0.9	13.2	1.7	15.4

Table 22.9 External cost savings from logistics system redesign (Strategy 2)

Scenario	Congestion saving (million £)	Infrastructure saving (million £)	Accident saving (million £)	Air quality saving (million £)	CO_2 saving (million £)	Noise costs (million £)	Overall saving (million £)
Low	6.2	4.9	3.3	1.5	2	0.4	18.4
Central	14.6	11.7	7.7	3.7	4.8	1	43.6
High	20.2	16.2	10.7	5.1	6.7	1.4	60.2

The main constraints on the redesign of logistics systems are the fixed nature of the assets, the additional capital investment required and the scale of the associated replanning and potential disruption. The re-sizing and relocation of DCs is clearly a longer term and more expensive option than the other five.

22.4.5 Out-of-hours deliveries

The rescheduling of road deliveries to the evening and night reduces their exposure to traffic congestion during the working day. Vehicles can then travel at more fuel efficient speeds, releasing less pollutants per kilometre travelled at times of day when there are few people on the street to inhale them. Accident risk is also reduced as there is less traffic on the roads and fewer pedestrians crossing them. The only externality which tends to increase with greater evening and night-time running is noise irritation, although in recent years this has reduced on a vehicle-km basis as a result of several trends:

- tightening noise limits on new vehicles
- technological advances permitting the quietening of vehicles below legal limits: for example, one vehicle manufacturer has managed to reduce running noise to around 60 dB(A), well below the 78 dB(A) that a standard vehicle of comparable size creates when accelerating from stationary
- investment in on-vehicle noise mitigation systems such as air brake silencers, load restraint systems, quieter refrigeration units and the use of alternative fuels such as compressed natural gas (CNG)
- adoption of a new generation of quieter tyres
- resurfacing of roads with low-noise pavements.

As many factories, warehouses and even supermarkets are now open for 24 hours a day, the opportunity exists to send and receive deliveries around the clock. Even when premises are closed, unattended delivery systems can be used to give delivery staff secure access out-of-hours. The main constraints on evening/night-time delivery appear to be the rigidity of existing working practices and curfews imposed by local authorities. Many

Table 22.10   External cost savings from increased out-of-hours deliveries

Scenario	Congestion saving (million £)	Infrastructure saving (million £)	Accident saving (million £)	Air quality saving (million £)	CO_2 saving (million £)	Noise saving (million £)	Overall saving (million £)
Low	29.5	0	−10.5	0	0	−0.5	18.4
Central	59	0	−21	0	0	−1.1	36.9
High	117.9	0	−42	0	0	−2.1	73.8

supermarkets in urban areas are close to noise-sensitive residential areas and have night-time delivery restrictions written into their planning consents. A survey by the British Retail Consortium (2004) found that 32% of retail outlets were affected by delivery curfews while the Freight Transport Association (2005a) estimated that over 40% of supermarkets throughout the UK were subject to some form of restriction on night delivery, usually between 10pm and 7am. For instance, 104 of the 360 supermarkets operated by the Morrisons supermarket chain were affected by night curfews.

In addition to site-specific restrictions, there are zonal bans on night-time lorry movement. The largest area covered by such a ban in the UK is in London. Sainsbury's has estimated that the London lorry ban results in its delivery vehicles travelling an additional 160 000 km per week because they must circumnavigate the restricted area to position vehicles for early morning store delivery, in many cases running this extra distance through environmentally sensitive areas.

Relaxing some of these regulations and other operational constraints would permit greater rescheduling of deliveries. This measure would yield both economic and environmental benefits (Cabinet Office, 2005). The modelling undertaken for this study suggested that the increase in noise costs would be marginal and far exceeded by congestion and accident savings (Table 22.10). It should be noted, however, that the noise costs related solely to moving vehicles and excluded the noise associated with loading and unloading operations at collection/delivery points. On the other hand, the analysis did not make allowance for reductions in emission costs accruing from vehicles running more fuel efficiently on congestion-free roads.

22.4.6   Computerised vehicle routing and scheduling/telematics

CVRS is now widely adopted by companies across the UK food supply chain. It allows them to reduce journey planning time, cut transit times, minimise vehicle-km, improve vehicle loading and save fuel (Department for Transport, 2005b). The software is usually calibrated to minimise transport costs and these costs are a function of both time and distance travelled. This does not necessarily minimise the environmental costs of the delivery. The available case study evidence suggests, however, that the application of CVRS generally reduces both vehicle-km and fuel consumption. Reductions in transport costs and distance travelled can be in the order of 5–10% (Eibl, 1996). The magnitude of the resulting saving partly depends on the quality of the previous system of manual vehicle routing and the nature of the transport operation. The potential benefits tend to increase with the number of drops and collections per delivery round, the range and complexity of delivery restrictions, the variability of collection and delivery times and the diversity of vehicle types and capacities. It should be noted, too, that some of the more popular CVRS packages now include modules which can plot routes that minimise carbon emissions.

An e-mail survey by the Freight Transport Association (2004) of a sample of member companies found that 22% were using some form of CVRS with a further 7% actively considering adopting it. This relatively low level of CVRS penetration is perhaps surprising since over 90% of CVRS users in the survey indicated that they were either 'satisfied' or 'very satisfied' with their choice of system and supplier. A large proportion of the users reported CVRS benefits which had obvious environmental spin-offs: improved efficiency (75%), reduced fuel cost (38%), better matching of demand and resource (38%) and reduced mileage (29%).

The level of CVRS adoption in the food sector may be higher than the Freight Transport Association average. Of 13 large retailers of grocery products profiled by the Institute of Grocery Distribution (2007), 11, accounting for 85% of total lorry-km, used CVRS systems. From industry consultation, however, it was concluded that across the food sector as a whole, CVRS could be more fully exploited to reduce the environmental impact of distribution. Evidence to support this claim came from a retrospective analysis of the routing of a large sample of food deliveries surveyed over a 48-hour period in May 2002 (as part of a so-called transport KPI audit) (Department for Transport, 2003; McKinnon and Ge, 2004). The actual routing was compared with the optimal routes constructed within seven delivery scenarios in which the length of driver's shifts and delivery time windows at shops and warehouses varied (McKinnon et al., 2004). The researchers concluded that:

> the actual pattern of delivery observed during the transport KPI survey for the sample fleets was sub-optimal. The degree of sub-optimality is difficult to determine as the KPI survey did not collect any information about opening times, delivery time windows or driver shifts. The analysis has shown, however, that if deliveries had been optimised within a realistic range of scheduling constraints, substantial reductions in vehicle-km, empty running, transit time and vehicle numbers could have been achieved. This would have translated into significant economic cost savings and environmental benefits (p.36).

Telematics can also enhance the efficiency of delivery operations by tracking vehicles, establishing communication with drivers and remotely monitoring various aspects of vehicle performance. By giving companies visibility of their vehicle fleets, it permits more effective planning and management of the transport operation (Department for Transport, 2007b). Routes can be replanned at short notice to take account of real-time information on traffic conditions and changing customer requirements. When used in combination with CVRS, a higher level of transport efficiency can be achieved with consequent benefits for the environment.

Surveys by the Freight Transport Association (2005b) of telematics usage among member companies found that it almost doubled between 2002 and 2004, rising from 17% to 33%. An analysis of the logistical profiles of 13 large grocery distributors indicated that eight (62%) employed vehicle tracking, a similar number had installed in-cab communication and five (38%) were using real-time traffic information in their delivery planning. Research by McKinnon et al. (2004), however, found that some of the companies that had installed telematics systems in their vehicles were using only a small part of the available functionality. On the other hand, some larger businesses were deploying them in a more sophisticated way to improve transport collaboration. In 2007, for example, Coca Cola, Masterfoods and Procter & Gamble jointly launched the Skylark Project which used vehicle tracking to identify and exploit opportunities for shared use of vehicles.

On the basis of the available evidence and industry consultation, the study concluded that there was still considerable scope for increasing both the diffusion of telematics technology and the average environmental benefit obtained from individual applications. This is reflected in the estimated external cost savings for this 'green logistics' option (Table 22.11).

Barriers to the uptake of CVRS and telematics include a lack of technical knowledge, particularly of the potential benefits, scepticism about new technology, the fragmentation of the road haulage industry and cost. Steady reductions in the real cost of both the hardware and software are, nevertheless, making these systems more affordable, while their functionality and user-friendiless is improving.

Table 22.11 External cost savings from increased use of CVRS and telematics

Scenario	Congestion saving (million £)	Infrastructure saving (million £)	Accident saving (million £)	Air quality saving (million £)	CO$_2$ saving (million £)	Noise saving (million £)	Overall saving (million £)
Low	4.3	1.3	1.6	0.5	0.6	0.2	8.5
Central	14.1	4.3	5.2	1.8	1.9	0.5	27.7
High	28.7	8.7	10.6	3.6	3.9	1	56.5

22.5 Conclusions

The research summarised in this chapter identified a series of sustainable logistics measures and attempted to quantify their impact on the external costs of food distribution in the UK. It was undertaken for the UK government in close cooperation with an industry group which provided both advice and data. The original goal was to find a set of measures which collectively would reduce the external costs by a fifth. The short list of six measures that was finally selected would, in combination, cut these costs by around 17% by 2012 against a 2005 baseline (DEFRA, 2007). The choice of measures was based mainly on their potential contribution to the reduction in external costs, but also took account of the likelihood of them being implemented and the availability of data to model their impact. No attempt was made to estimate the financial cost of applying these measures. In most cases, however, they would yield both economic and environmental benefits and be likely to prove self-financing within a relatively short payback period.

The FISS Group excluded from the final short list two other measures which have generated much discussion in recent years: localised sourcing of food and modal shift to rail and waterborne transport. The former option was largely excluded because a full-life analysis would be required to determine the net environmental impact, not only of changes to transport but also to related production, storage and handling operations. This was beyond the scope of the research. The exclusion of local sourcing also helped to deflect attention from the 'food miles' issue which has tended to dominate much of the debate on the sustainability of food supplies in recent years.

The modal shift option was not short listed for more detailed analysis because at the time of the study very little food was moved by rail and most food retailers and manufacturers did not consider it to be a viable logistical option in the UK. Over the past three years, however, two major retailers, Tesco and ASDA, have begun to make significant use of rail for long haul deliveries, while Tesco has also started moving wine and spirits by canal. This highlights the need to keep the range of sustainability measures under review and be prepared to expand the list of priorities.

Finally, the results of the modelling are very sensitive to the monetary values attached to the various externalities. The values will vary through time as the relative severity of the various environmental problems changes. As congestion worsens, the marginal cost of adding extra lorry journeys to the road network rises, while the cost of CO_2 emissions is projected to increase as efforts intensify to decarbonise the economy. The analysis will therefore require regular updating to provide policymakers and logistics managers with the advice they require to maintain the 'greening' of food transport operations.

22.6   Acknowledgement

The authors are grateful to the Department for the Environment, Food and Rural Affairs (DEFRA) for funding the study and granting permission for the main results to summarised in this chapter. They also wish to acknowledge the support of Aecom, the main contractor on the project, and the members of the FISS Champions' Group on Food Transport who provided valuable advice and data. Responsibility for the accuracy of this summary, nevertheless, rests with the authors.

22.7   References

AEA TECHNOLOGY (2005), *The Validity of Food Miles as an Indicator of Sustainable Development*, Defra, London.

ANON (2005), 'FCBs link to retailers and food service', *Farmers Weekly*, 10/11/2005, Vol. 143 Issue 19.

BRITISH RETAIL CONSORTIUM (2004), *Counting the Cost of Delivery Curfews*, British Retail Consortium, London.

CABINET OFFICE (2005), *Best Practice Toolkit: Delivery Restrictions*, Cabinet Office, London.

CE DELFT (2007), *Handbook on Estimation of External Cost in the Transport Sector*, CE Delft, Delft.

CRUIJSSENS, F. (2006), *Horizontal Collaboration in Transport and Logistics*, Tilburg University.

DEFRA (2006), *Air Quality Damage Cost Guidance: Interdepartment Group on Costs and Benefits Report to Defra*, Defra, London.

DEFRA (2007), *Report of the Food Industry Sustainability Strategy Champions Group on Food Transport*, Defra, London.

DEPARTMENT FOR TRANSPORT (2003), *Analysis of Transport Efficiency in the UK Food Supply Chain*, Freight Best Practice Programme, London.

DEPARTMENT FOR TRANSPORT (2005a), *Focus on Double Decks*, Freight Best Practice Programme, London.

DEPARTMENT FOR TRANSPORT (2005b), *Computerised Vehicle Routing and Scheduling for Efficient Logistics*, Freight Best Practice Programme, London.

DEPARTMENT FOR TRANSPORT (2006), *Road Freight Statistics*, Department for Transport, London.

DEPARTMENT FOR TRANSPORT (2007a), *Guidance on Noise and Greenhouse Gas Evaluation*, Department for Transport, London.

DEPARTMENT FOR TRANSPORT (2007b), *Telematics for Efficient Road Freight Operations*, Freight Best Practice Programme, London.

EIBL, P. (1996), *Computerised Vehicle Routing and Scheduling in Road Transport*, Aldershot, Brookfield USA.

FREIGHT TRANSPORT ASSOCIATION (2004), *CVRS Survey*, Freight Transport Association, Tunbridge Wells.

FREIGHT TRANSPORT ASSOCIATION (2005a), *Delivering the Goods: Guidance on delivery restrictions*, Freight Transport Association, Tunbridge Wells.

FREIGHT TRANSPORT ASSOCIATION (2005b), *Telematics Guide*, Freight Transport Association, Tunbridge Wells.

INFRAS (2004), *External Costs of Transport, Update Study*, INFRAS, Zurich.

INSTITUTE OF GROCERY DISTRIBUTION (2003), *Transport Optimisation: Sharing Best Practice in Distribution Management, ECR*, Institute of Grocery Distribution, Watford.

INSTITUTE OF GROCERY DISTRIBUTION (2007a), *ECR UK Collaborative Green Distribution*, Institute of Grocery Distribution, Watford.

INSTITUTE OF GROCERY DISTRIBUTION (2007b), *Retail Logistics*, Institute of Grocery Distribution, Watford.

INSTITUTE OF GROCERY DISTRIBUTION (2009), *On the Road to Greener Distribution*, Letchmore Heath. http://www.igd.com/index.asp?id=1&fid=1&sid=5&tid=47&cid=564 (accessed 31 August 2009).

KNIGHT, I., NEWTON, W., MCKINNON, A., PALMER, A., BARLOW, T., MCCRAE, I., DODD, M., COUPER, G., DAVIES, H., DALY, A., MCMAHON., COOK, E., RAMDAS, V. and TYLOR, N. (2008), *Longer and/or Longer and Heavier Vehicles (LHVs): A Study of the Likely Effects if Permitted in the UK – Final Report*, TRL Ltd, Crowthorne.

MASON, R., SIMONS, D., PECKHAM, C. and WAKEMAN, T. (2002), *Life Cycle Modelling CO₂ Emissions for Lettuce, Apples and Cherries*, Department for Transport, London.

MCKINNON, A.C. and GE, Y. (2004), 'Use of a synchronised vehicle audit to determine the opportunities for improving transport efficiency in a supply chain', *International Journal of Logistics: Research and Applications*, **7**(3), 219–38.

MCKINNON, A.C., GE, Y. and MCCLELLAND, D. (2004), *Assessing the Potential for Rationalising Road Freight Operations*, Logistics Research Centre, Heriot-Watt University, Edinburgh.

NERA (2005), *Costs Imposed By Foreign Registered Lorries on Britain's Roads*, NERA, London.

NERA, AEA TECHNOLOGY, TRL (2000), *Lorry Track and Environmental Costs*, DETR, London.

PIECYK, M. and MCKINNON, A.C. (2007), *Internalising the External Cost of Road Freight Transport in the UK*, Logistics Research Centre, Heriot-Watt University, Edinburgh.

SANSOM, T., NASH, C., MACKIE, P., SHIRES, J. and WATKISS, P. (1998), *Surface Transport Costs and Charges, Great Britain*, Institute for Transport Studies, University of Leeds in association with AEA Technology Environment, Leeds.

SCHMID, S.A., BICKEL, P. and FRIEDRICH, R. (2001), *RECORDIT project – Deliverable 4: External Cost Calculation for Selected Corridors*, European Commission, Brussels.

STRATEGIC RAIL AUTHORITY (2003), *Sensitive Lorry Miles*, Strategic Rail Authority, London.

SUSTAIN (1999), *Food Miles – Still on the Road to Ruin?* Sustain, London.

TRANSPORT and MOBILITY LEUVEN (2008), *Effects of Adapting the Rules on Weights and Dimensions of Heavy Commercial Vehicles as Established within Directive 96/53/EC, TREN/G3/318/2007*, European Commission, Brussels.

VICS LOGISTICS COMMITTEE (2004), *Collaborative Transportation Management White Paper Version 1.0*, prepared by the CTM Sub-Committee, VICS, New Jersey.

23

Fair trade and beyond: voluntary standards and sustainable supply chains

G. Alvarez, Cranfield University, UK

Abstract: Over the last decade, a plethora of voluntary initiatives and standards have emerged as a response to increasing awareness of the social and environmental implications of modern production and trading systems. Within these initiatives, sustainability issues have been particularly relevant to manufacturers and retailers who trade on their brand image as providers of safe, environmental and ethically produced goods. Initiatives like Fairtrade, GLOBAL GAP and Rainforest Alliance are examples of a broad range of voluntary standards covering a wide scope of requirements that include quality and hygiene, working conditions, terms of trade and soil management practices. These standards open new opportunities for retailers, brand owners and producers but also pose new risks to food and beverage supply chains. This chapter provides an introduction and classification of the range of voluntary standards in the industry today, the opportunities and challenges that emerging trends in this area represent for supply chain managers and an overview of alternative strategic options for corporations to become engaged in this field.

Key words: sustainability, voluntary standards, ethical trade, food safety standards, food and beverage.

23.1 Introduction: rise of voluntary standards

The history of social and environmental concerns about business is probably as old as trade itself. In Ancient Mesopotamia, around 1700 BC, King Hammurabi introduced a code in which builders, innkeepers or farmers were put to death if their negligence caused the deaths of others or major inconvenience to local citizens (Centre for Business Relationships, Accountability, Sustainability and Society, 2007). Even in the absence of royal or government regulation, companies have, for a long time, voluntarily engaged in activities that go well beyond strictly business. Over a century ago, Quakers such as Barclays and Cadbury, as well as socialists like Engels and Morris, experimented with socially responsible and values-based forms of business (Henriques, 2003).

Historically, consumer activism has also played a role in pressuring companies to take action independent of existing regulations. One of the first large-scale consumer boycotts took place in England in the 1790s over slave-harvested sugar, succeeding in forcing the importer to switch to free-labour sources (Micklethwait and Wooldridge, 2003).

Sustainability concerns, including food safety, environment and social issues, have been around for a long time, but the debate has intensified with recent food safety scares and an increased interest by consumers and civil society in the sustainability of production and consumption patterns. A number of food safety scares like *salmonella* and *E.coli* outbreaks, avian flu and toxic milk in China (Traub, 2008; Jaffee and Henson, 2004) shook consumer confidence in the industry and raised questions on the public mechanisms of food safety control (Henson, 2006).

An increasing awareness about the rate of depletion of natural resources and wide disparities in well being around the world also led to a new interest across all sectors of society towards sustainability issues, that is 'the ability to meet current needs without compromising the ability of future generations to meet their needs' (World Commission on Environment and Development, 1987).

Consumer interest, powered by media attention and non-governmental organisation activism, created a powerful mixture that pressured corporations to rethink some of their sourcing practices and these pressures became a major force behind the launch of many sustainability initiatives by large corporations (Christmann and Taylor, 2002; Feltmate, 1997). NGOs have emerged as new powerful forces in shaping consumer attitudes over the last 20 years. Using media and the more egalitarian and pervasive channel for distribution of information created by the Internet, NGOs have become powerful actors in threatening corporate reputations and market positions (Argenti, 2004).

As one example of NGO action, Oxfam's campaign in the coffee industry illustrates the potential force that NGOs can have to influence corporate action. In 2002, amidst a severe crisis that was affecting the coffee industry, Oxfam drew attention to the plight of millions of coffee farmers and pressured large coffee roasters to react to the situation or risk damaging their brands. A report published by the NGO entitled *Mugged: Poverty in your coffee cup* analysed the origins and effects of the coffee crisis and urged American consumers to join Oxfam in bringing relief to farmers and in calling for the major players in the coffee industry to support a Coffee Rescue Plan (Oxfam America, 2002).

Over the following five years, most coffee roasting corporations and large retailers got involved, at least to some extent, in sustainable coffee schemes such as Fairtrade, Rainforest Alliance or Utz Kapeh (now called Utz certified), thus making the coffee industry one of the first areas in which social and environmental voluntary standards flourished and became a key element for producing, trading and marketing coffee (Agritrade, 2008).

In the food safety area, government regulation has also increasingly played an important role in promoting new supply chain coordination mechanisms. Regulations like the UK's Food Safety Act of 1990 marked a shift of responsibility from the seller of food products to the buyer who, in the case of a charge, now needed to prove that he had taken 'all reasonable precautions and exercised all due diligence to avoid the commission of the offence by himself or by a person under his control' (Hobbs and Kerr, 1992). These regulations led large food retailers and importers increasingly to scrutinise their supply chains, specifying their own set of private safety and quality standards and enforcing these through systems of second-party certification (Henson, 2006).

At the European level, CEC regulation No 178/2002 established a European Food Safety Authority (CEC, 2002) and shifted the focus from product inspection to ensuring food safety across the food value chain as a whole. It established food safety based on the principle of containing risk and increasing traceability, relying on HACCP (hazard analysis and critical control point) methodology as a central component of food safety mechanisms. It also placed primary legal responsibility for ensuring food safety on the private sector stating that: 'A food business operator is best placed to devise a safe system for supplying food and ensuring that the food it supplies is safe; thus it should have primary legal responsibility for ensuring food safety' (CEC, 2002).

But sustainability initiatives have also gone beyond being just risk management mechanisms. They have become an important variable on which suppliers compete to differentiate food products to end consumers. In high-income economies, consumers increasingly demand full information on a product so that they can make individual choices in relation to taste preferences as well as in relation to personal beliefs (e.g. food safety, labour rights) (Gibbon and Ponte, 2005). In this context, sustainability includes notions such as 'environment conservation', 'environment friendly', 'corporate social responsibility', 'economic viability for farmers', 'fairtrade' and 'ethical trade', among others. The common denominator of these products is that they include a statement that the product offered contains attributes referring to the environmental or social conditions surrounding the production of these food products.

Overall, consumer interest, NGO activism and regulatory changes have led to significant changes in supply chain management by leaders in the food and beverage industry. As a risk management strategy to avoid legal exposure or brand damage, or as an opportunity to differentiate brands and products in an increasingly ethical and environmentally aware marketplace, corporations are in the process of reviewing their supply chains and defining new ways of doing business. An important way of doing this has been through the development of voluntary standards and increased engagement of stakeholders along the entire value chain.

23.2 What are voluntary standards?

Standards are agreed upon criteria whereby performance of a product or service, its technical and physical characteristics and/or the process and conditions under which it has been produced or delivered can be assessed (Navdi and Wältring, 2004). Standards communicate information about the characteristics of a product or the process by which this product was produced (Ponte, 2004) and they can be based on quality, search, experience or credence attributes (Gibbon and Ponte, 2005; Ponte, 2004). Quality standards communicate information about a product's degree of excellence, as embodied in specific attributes. Search attributes are those that can be verified at the time of the transaction (e.g. colour, size). Experience attributes can only be assessed after the transaction has taken place (e.g. flavour). Finally, credence attributes cannot be objectively verified and are based on trust (e.g. if a product is organic). Labels, codes and certifications can be a visible means of signalling that a certain product or process has met the requirements of a given standard, but not all standards have an associated label.

While mandatory standards are set by governments and take the form of regulation, voluntary standards are non-mandatory. They are established as a formal coordinated process in which key participants in a market or sector seek consensus (Ponte and Gibbon, 2005) and which can be monitored internally by individual enterprises or through third parties.

23.3 Typologies of voluntary standards

Voluntary standards vary along multiple dimensions, with a number of classifications having been proposed in the literature. The most frequently used typologies distinguish standards according to four variables: (1) the organisation(s) leading and influencing the setting up of the standard, (2) the motivation in the organisation(s) to set a new voluntary standard, (3) the scope of the initiative and (4) the enforcement mechanisms.

23.3.1 Leading organisation(s)
Probably the most common dimension used to distinguish between voluntary standards is based on the parties involved in determining the requirements (Acutt, 2002; Mazurkiewicz, 2005). This distinguishes three

types of approach: unilateral initiatives, multi-stakeholder agreements and third-party initiatives.

Unilateral initiatives can be led by a single company or by companies jointly establishing a group or industry code of conduct. Company-specific programmes include initiatives such as UK retailer Tesco's Nature Choice, Marks & Spencer's field-to-fork initiative and Nestlé Nespresso's AAA Sustainable Quality Programme. In this case, the programme is directly linked with the operations of the company and, although it can be determined in consultation with multiple stakeholders, the company ultimately defines the programme's elements and standard requirements.

Companies can also work in cooperation with others in precompetitive, voluntary standards initiatives or in multistakeholder agreements. The latter are developed between different sectors and generally include representation from retailers, manufacturers, NGOs, academia and, increasingly so, representatives of the public sector. An example of this type of agreement is the Common Code for the Coffee Community (4C), established in 2006 as part of a collaboration between retailers, industry, coffee producers, civil society, government and development organisations. While many of the initiatives at the time were geared to small-scale niche products, 4C's objective is to foster sustainability in the 'mainstream' green coffee chain and to increase the quantity of coffee meeting basic social, economic and environmental sustainability criteria (Common Code for the Coffee Community Association, 2008).

Third party initiatives are established by other stakeholders, particularly NGOs, as part of an advocacy and campaigning initiative to have an impact on corporations and producers. Corporations can adopt these standards on a voluntary basis and thus be associated with the claims made by the voluntary standard which are often communicated to the consumer through a label. An example of this is the emergence of the Fairtrade initiative, aimed at offering better trading conditions to small-scale producers and which has been adopted by large corporations such as Tesco's own-label Fairtrade line on a range of groceries or Nestlé's Partners Blend Fairtrade certified coffee brand.

23.3.2 Motivation: risk management versus differentiation

A second dimension of differentiation between standards is the rationale for the creation or adoption of a particular voluntary standard. Jaffee *et al.* (2005) distinguish between a standard and initiative geared to managing risk and one where a competitive differentiation is sought.

The mitigation of reputational and/or commercial risks associated with the safety of food products has been a driver behind the establishment of some of the most popular voluntary standards such as the British Retail Consortium (BRC) and the Euro-Retailer Produce Group Good Agricultural Practice (EUREPGAP) Standards. Ethical or environmental concerns have also increasingly become an area of risk mitigation. As mentioned in the introduction, the risk of reputational loss owing to mismanagement of sustainability concerns, or becoming the target of activists and media campaigns, has also been recognised as a driver behind the adoption of certain ethical sourcing standards.

On the other hand, some companies have adopted food safety or sustainability voluntary standards as a means to differentiate their products and compete in quality-defined markets (Henson, 2006). Quality in this case can be expressed not only in physical attributes but also in production and process methods (Ponte, 2004; Reardon and Farina, 2002). These standards are mainly targeted at consumers and so are generally associated with a brand or label that distinguishes the products. Examples of differentiation of process or product quality include UK retailer Tesco's Nature Choice, Marks & Spencer's field-to-fork initiative and Nestlé Nespresso's AAA Sustainable Quality initiative, where the retailers communicate directly to the consumer the claim that their products and processes are better than other products and processes (Humphrey, 2008).

23.3.3   Scope: emphasis, geography and product

A third difference between standards is the scope of the initiative. Although many standards tend to incorporate multiple dimensions as they evolve and expand, most voluntary initiatives were launched with an emphasis either on food safety or on environmental, social or economic sustainability. For example, the EUREPGAP standard, created partly as a response to food safety legislation, was dominated by such concerns, with social or indirect environmental elements being secondary in the standard. Organic standards, on the other hand, have a focus on environmental criteria, while the Fairtrade movement focused on modifying trade relations along the conventional commodity chain.

Standards also differ in the products covered. Safety-based standards tended to focus on the products exposed to a higher risk, such as fruits, vegetables and meat products. Social or economic sustainability driven standards, on the other hand, started on products considered the most subject to poor social or economic conditions such as coffee, bananas and cocoa.

23.3.4   Enforcement: 1st, 2nd, 3rd party monitoring

Enforcement of a voluntary standard refers to the process whereby the actors in the chain (producers-traders-consumers) coordinate with each other to ensure the objectives of the standard are being met. Navdi and Wältring (2004 cited by Ellis and Keane, 2008) differentiate between company voluntary standards that are self-monitored (1st party) and a monitoring process where a sector-based standard is monitored by associated sector members (2nd party). In the case of third-party monitoring, an independent organisation verifies that the process and/or product satisfies all the standard's criteria. A certification body can also give a written guarantee (and the right to use a label if available) that a product, process or service conforms to the specifications of the standard. In this case, an accreditation organisation is an authoritative body for a specific standard that grants formal recognition to a certifying body or person to certify to the standard (Trade Standards Practitioners Network, 2008).

Although the costs of implementation can be significantly higher in the third-party certification process, this enforcement mechanism has become widely used among voluntary standards trying to establish their legitimacy among consumers. Unilateral standards tend to attract a higher level of scepticism, especially if they 'lack independent third-party verification, show a lower level of transparency and lower participation of affected stakeholders' (Ponte, 2004). As a result, most standards have, over time, become more transparent and many have incorporated third-party verification or certification to increase their credibility in the marketplace.

23.4   Overview of the main voluntary standards in the food industry

Using the dimensions outlined in the previous section, Table 23.1 provides examples of the major types of voluntary standards that exist today in the food and beverage industry, followed by short descriptions of some of the major standards.

23.4.1   Fairtrade

Fairtrade is probably the most recognised label among consumers and it is sometimes used as a generic term encompassing all sustainable trade initiatives. Fairtrade's focus has been on smallholder producers and is driven by a threefold stated strategic intent: to work with marginalised producers and workers in order to help them move from a position of vulnerability to one of security and economic self-sufficiency; to empower producers and workers as stakeholders in their own organisations; and actively to play a wider role in the global arena to achieve greater equity in international trade (FairTrade Labelling Organisation, 2008).

Fairtrade has its origins in the 1950s and 1960s in alternative trade organisations (ATO) like Traidcraft in the UK and SOS Wereldhandel in The Netherlands. These were later followed by labelling initiatives, such as Max Havelaar in The Netherlands and the Fairtrade Foundation in the UK, both of which were later involved in the establishment of the Fairtrade Labelling Organisation (FLO) in 1997 to coordinate the expanding number of labelling organisations. The Fairtrade label now applies to products

Table 23.1 Examples of voluntary standards in the food and beverage industry

Primary emphasis	Name – short description	Motivation	Promoters, participation	Enforcement	More information
Food safety	BRC – British Retail Consortium (food standard) Covers food safety and environmental issues such as: • HACCP system • Quality management • Factory environment standard • Product and process control	Risk management: designed to meet the requirements of the EU General Product Safety Directive and the UK Food Safety Act.	Retailers belonging to the UK retailer association.	Independent third-party certification by accredited organisation.	http://www. brcglobalstandards. com
	GLOBALGAP (based on previous EUREPGAP): Food safety standard for agricultural products sourced by participating European suppliers and processors. The standard is primarily designed to reassure consumers about how food is produced on the farm by minimising detrimental environmental impacts of farming operations, reducing the use of chemical inputs and ensuring a responsible approach to workers' health and safety as well as animal welfare.	Risk management: developed as a response by European retailers to food scares in Europe and a public demand to know the source and safety of food purchased in Europe.	Retailers belonging to the EUREP group. Associated producers and other related organisations can participate in the definition of the standards.	Independent third party certification by accredited organisation.	http://www.globalgap. org

Table 23.1 *Cont'd*

Primary emphasis	Name – short description	Motivation	Promoters, participation	Enforcement	More information
	SQF 1000: based on CODEX Alimentarius HACCP Guidelines to manage food safety and quality.	Risk management: enabling tool for producers and manufacturers to demonstrate 'due diligence' and compliance with regulatory and product traceability requirements.	Retailers grouped in the US-based Food Marketing Institute.	Independent third-party certification by accredited organisation.	http://www.sqfi.com/sqf_program.htm
	ISO 22000 Food Safety Management System: A generic food safety management system standard. It defines a set of general food safety requirements that apply to all organisations in the food chain.	Risk management: to demonstrate ability to control food safety hazards and ensure food safety at time of human consumption.	International Organisation for Standardisation.	Self-assessment or certification.	http://www.22000-tools.com
	Tesco's Nature Choice: All fresh product suppliers to Tesco have to comply with the standard, aimed at chemical usage control and environmentally sustainable production standards. Growers can be certified at different levels, including a 'Gold Standard'.	Differentiation: the standard is positioned as being the highest standard in the industry.	Corporation – Tesco.	Independent third-party certification by accredited organisation.	http://www.tescofarming.com/tnc.asp

Environmental	MSC – Marine Stewardship Council Encourages sustainable fishing practices to ensure healthy species population and ecosystem integrity. Addresses ecosystem integrity and sustainable exploitation through socially and economically sound methods.	Differentiation: MSC principles and criteria for sustainable fishing use product label to reward environmentally responsible fishery management and practices.	NGO driven – Marine Stewardship Council.	Independent third-party certification by accredited organisation.	http://www.msc.org/
	ISO 14001 – Environmental Management System Standard to develop an EMS so that businesses can track and improve environmental performance.	Differentiation: label can be used to communicate achievement of standard (mainly for B2B relationships).	International Organisation for Standardisation.	Self-assessment or certification.	http://www.22000-tools. com
	Rainforest Alliance: Standard designed to conserve ecosystems and promote dignified living conditions for workers and communities. Ten areas covered, including, among others: • Social and environmental management system • Ecosystem conservation • Wildlife protection • Fair conditions and good treatment for workers • Integrated waste management	Differentiation: label can be used to communicate achievement of standard (consumer oriented).	NGO driven – Rainforest Alliance and the Sustainable Agriculture Network.	Independent third-party certification of farms by accredited organisation.	http://www.ra.org/ programs/agriculture/ index.html

Table 23.1 Cont'd

Primary emphasis	Name – short description	Motivation	Promoters, participation	Enforcement	More information
	Utz Certified: The standard provides guidelines for food safety, environmentally and socially appropriate growing practices, and quality processing. Requirements include traceability and documentation; varieties and rootstocks; soil and substrate management; fertiliser use; crop protection; product handling; waste and pollution management/ recycling and reuse; workers' health, safety, and welfare. Initially a coffee standard, expanding to cocoa, tea and palm oil.	Differentiation: label can be used to communicate achievement of standard (consumer oriented).	Codeveloped by retailer Ahold and Guatemalan grower–exporters.	Independent third-party certification of farms by accredited organisation.	http://www.utzcertified.org/
Organic	IFOAM The International Federation of Organic Agriculture Movements Basic Standards – The IFOAM Basic Standards are largely considered the benchmark for most national programmes for organic standards. There is a general set of standards for agricultural products, with separate standards for animal husbandry, bee keeping, aquaculture, processing and handling, textiles and sustainable forest management. Requirements include ecosystem management (water, soil, biodiversity), genetically modified materials, pesticides, fertilisers, social justice, transport and processing.	Base standard to be used as benchmark for other organic standards.	NGO driven.	Independent third-party certification by accredited organisation.	http://www.ifoam.org/

Labour rights and conditions	Fairtrade: Main objective is to improve the position of poor, small-scale and marginalised producers in the developing world by influencing the conditions of trade in their favour. FLO International is the umbrella association of 20 labelling initiatives including Max Havelaar, TransFair and Fairtrade Foundation. The standard includes agreed minimum prices, payment of a social premium, pre-financing if required and promotes long-term buyer–seller relationships.	Differentiation: label can be used to communicate achievement of standard (consumer oriented).	NGO driven.	Certification done by an independent international certification company, FLO-CERT.	http://www.fairtrade.net
	Ethical Trade Initiative: the main objective is to promote and improve implementation of corporate codes of practice which cover supply chain working conditions.	Baseline standard on working conditions based on International Labour Organisation conventions.	Multi-stakeholder initiative including companies, NGOs, trade unions in the UK.	Principle of continuous improvement. Implementation assessed through monitoring and independent verification.	http://www.ethicaltrade.org/
	SA 8000: Social accountability system standard oriented at maintaining just and decent working conditions throughout the supply chain. Requirements address child labour, forced labour, health and safety, working hours and remuneration among others.	Risk management: to demonstrate compliance with labour standards across the supply chain. No label is associated with the initiative.	NGO driven, multi-stakeholder initiative including companies, NGOs and trade unions.	Certification done by third-party, SAI -accredited certifying bodies.	http://www.sa-intl.org

Table 23.1 *Cont'd*

Primary emphasis	Name – short description	Motivation	Promoters, participation	Enforcement	More information
Round tables and commodity-based initiatives	Roundtable of Sustainable Palm Oil (RSPO): A joint initiative of producers, distributors, conservationists and other stakeholders to support the production and trade in sustainable palm oil. Projects include plantation management practices, development of new plantations, responsible investment in oil palm and chain of custody.	Baseline standard involving environmental, social and economic sustainability.	NGO (WWF) as a key driver in establishing and facilitating the round table discussions.	Certification by third party, RSPO accredited certifying bodies.	http://www.rspo.org
	Common Code for the Coffee Community (4C): Multi-stakeholder initiative fostering sustainability in the 'mainstream' green coffee chain and increase the quantity of coffee meeting basic social, economic and environmental sustainability criteria. Main pillars of 4C are a code of conduct, participation rules for trade & industry, support mechanisms for coffee farmers, and participatory governance structure.	Baseline standard.	Initially driven by a public/private collaboration between the German Coffee Association and the German Ministry of Cooperation and Development.	Verification system.	http://www.4c-coffeeassociation.org/index.html

Source: Trade Standards Practitioners Network and standard setting organisation websites.

promoted by 20 labelling initiatives, grouped under the FLO and sharing a Fairtrade Certification process.

Although it is among the best-known voluntary standards, Fairtrade products represent today €2.38 billion (FairTrade Labelling Organization, 2008), accounting for a very small percentage of market share in the product categories where it participates. Albeit from a low base, Fairtrade labelled products, have been growing at an average rate of 40% per year over the last five years (Fig. 23.1).

Overall, the main requirements associated with this standard from a buyer's perspective are:

- Agreed minimum prices, usually set ahead of market minimums: producer organisations guarantee a minimum price that covers the cost of production and provision for family members and farm improvements
- Payment of an agreed social premium: often set at 10% or more of the cost price of goods and intended to support community projects or cooperative business investments
- Direct purchasing from producers
- Promotion of long-term trading partnerships, including long-term contracts
- Provision of credit when requested: Importers are required to prefinance up to 60% of the total purchase of seasonal crops
- No labour abuses occurring during the production process and workers must be allowed to unionise.

(The above is based on material from the Fairtrade Labelling Organisations International FLO.)

One distinguishing feature of this programme, compared to other sustainability standards, is the establishment of agreed minimum prices. These are usually above or independent of world market prices and allow for a living wage for producers as well as the provision of a social premium.

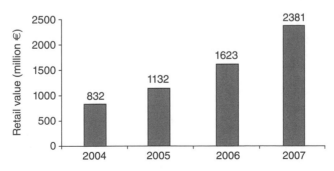

Fig. 23.1 Estimated retail value of Fairtrade certified products (from Fairtrade Labelling Organisation, 2008).

Although this represents an additional cost for the buyer, the cost has typically been passed on to the consumer in the form of higher prices for most labelled goods.

On the producer side, in order to be certified, small-scale producers have to be democratically organised (often in a cooperative), ensure there are no abuses and must be allowed to unionise. The producers must also invest the Fairtrade price premium in projects that aid their social, economic and environmental development. The standard further differentiates between small producers grouped in cooperatives, which is the case of coffee (one of the highest traded products under this scheme), and larger operations with hired labour, as in the case of large tea or banana plantations.

Fairtrade today covers nine food and beverage product groups. The origins of alternative trade schemes can be found in handcrafts and textiles, but since 1991 food has taken over as the most important category in terms of sales volumes and growth (Nicholls and Opal, 2005), with bananas and coffee as the two leading products. In core Fairtrade markets such as UK, the Fairtrade label represented 18% and 7%, respectively of the total imports of these two products in 2006 (Ellis and Keane, 2008). Still, despite these specific markets and a high growth rate, total global volumes traded are marginal with respect to the total volume traded for these commodities, the highest category being bananas with only 1.36% of the total global market (Table 23.2).

Geographically, the main markets for Fairtrade products today are the USA and the United Kingdom, which together account for 60% of total sales (Fig. 23.2). Even within Europe there is high divergence in Fairtrade spending. Switzerland shows the highest consumption of Fairtrade products, spending an average of €20.80 per capita in 2007, while German

Table 23.2   Fairtrade certified products as a percentage of global trade

Product	Fairtrade certified estimated sales volume 2006/7 (metric tonnes)	Total volumes traded worldwide (2007) (metric tonnes)	Estimated trade covered by Fairtrade (%)[a]
Bananas	233 791	17 198 800	1.36
Cocoa	7 306	3 010 391	0.24
Coffee	62 209	5 676 749	1.10
Honey	1 683	419 915	0.40
Juices	24 919	12 499 040	0.20
Rice	4 208	2 306 983	0.18
Sugar	15 074	18 203 968	0.08
Tea	5 421	1 490 012	0.36
Wine	5 740	8 537 996	0.07

[a] Volume of Fairtrade certified products is relative to global trade.
Source: Fairtrade Labelling Organisation (2008) and International Trade Center Trademap.

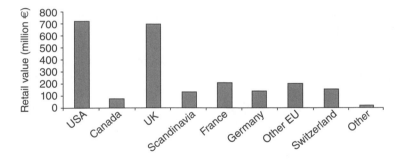

Fig. 23.2 Fairtrade certified products estimated retail value (millions of euros). Scandinavia: Denmark, Finland, Sweden and Norway. Other: Australia, New Zealand and Japan (from Fairtrade Labelling Organisation, 2008).

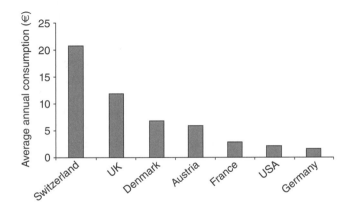

Fig. 23.3 Average annual consumption of Fairtrade products (euros per capita) 2007 (from Max Havelaar, 2007).

consumers spent on average less than €3 per capita per year on Fairtrade products (Fig. 23.3) during the same year.

One of the important advantages of the Fairtrade label is the very high consumer recognition it enjoys in several markets. A recent market survey in the UK showed that 70% of the population recognise the Fairtrade trademark and that 64% of the UK population linked the label to a better deal for producers in the developing world (Fairtrade Foundation, 2008). This high level of consumer recognition is an attractive feature of the Fairtrade label as it makes it easier for branded products or retailers to communicate clearly – and less expensively – with end consumers on sustainability issues.

From the producer's perspective, there is a clear benefit in profiting from minimum prices, credit and longer-term relationships. But less tangible benefits have also been identified such as capacity building and improved management practices in cooperatives (Sidwell, 2008).

Fairtrade also has its critics. A report published by the Adam Smith Institute (Sidwell, 2008) argues that an economy works best when prices are the direct reflection of supply and demand, not when they are politically motivated and artificially determined. As an arbitrary subsidy on a particular crop, Fairtrade pricing both distorts the comparative advantage of that economy and hurts farmers elsewhere. The report also estimates that only about 10% of the premium paid by consumers actually makes it to the producer, leading to the conclusion that there are other, more efficient ways of helping to reduce poverty.

23.4.2 Ethical trade initiative

Like Fairtrade, the Ethical Trading Initiative (ETI) is an 'ethical sourcing' initiative. But rather than focusing on the trading conditions of small producers, the ETI emphasises the conditions of workers in the producer countries. Thus, the main objective of the alliance is to 'promote and improve the implementation of corporate codes of practice which cover supply chain working conditions' (Ethical Trading Initiative, 2008).

The initiative is centred on the purchasing practices of companies importing products into the UK market and has a membership of 54 companies, including leading retailers such as Tesco, Sainsbury and Marks and Spencer as well as branded manufacturers like Chiquita Brands International and Associated British Foods. When companies join the initiative they commit to implement a base code and the accompanying principles of implementation in all or part of their supply chain. This code contains nine clauses reflecting the most relevant international standards with respect to labour practices, based on International Labour Organisation conventions.

The code requirements include the right to choose employment freely, freedom of association and collective bargaining, restrictions on child labour as well as ensuring fair wages and working conditions. The implementation principles set out general rules that govern execution of the base code. These are based on a commitment expressed by the company and communicated to the suppliers, followed by implementation of the code on the basis of continuous improvement. Companies accept the principle that the implementation of codes will be assessed through 'monitoring and independent verification; and that performance with regard to monitoring, practice and implementation of codes will be reported annually' (Ethical Trading Initiative, 2008).

Supporters of the initiative point to the increased efforts to address international labour standards on a larger scale through the market power of large retailers. A study commissioned by the Ethical Trade Initiative reported widespread adoption of the initiative, with ETI member companies registering more than 20,000 supplier sites in over 100 countries

(Barrientos and Smith, 2006). Producers can also initiate the process, taking responsibility for meeting the standards and assuming auditing costs. After assessing their compliance with the requirements, suppliers can submit this information to the Suppliers Ethical Data Exchange (SEDEX), which serves as a repository of information available to buyers (www.sedex.org.uk).

Among the criticisms, Ellis and Keane (2008) observe that though compliance costs can be potentially high, given that membership is about progress towards the goal of compliance and not about fulfilling the full base code, actual costs of compliance can be much lower and are sometimes passed on to producers. It has also been criticised for reaching mostly regular and permanent workers and being less effective in reaching more insecure and marginal workers (Barrientos and Dolan, 2006).

23.4.3   Organic

Born with the early regional groups of organic farmers developing organic standards in the 1940s, the organic movement has evolved to include hundreds of private standards worldwide alongside technical regulations of more than 60 governments (IFOAM, 2009). Organic standards aim at 'the worldwide adoption of ecologically, socially and economically sound systems that are based on the principles of Organic Agriculture' (IFOAM, 2009).

Organic products represent a large percentage of the products that are certified under voluntary standards. The world market of organic food and beverage products was estimated at US$52 billion (€40.5 billion) in 2008 (Fig. 23.4). Sales have grown almost 80% since 2004 and are expected to reach US$85 billion (€66 billion) by 2013 (Datamonitor, 2008).

Consumer demand is heavily concentrated in North America, representing 49.1% of total consumption, followed by Europe at 47.4% and Asia-Pacific at 3.5% (Datamonitor, 2008; Willer *et al.*, 2008).

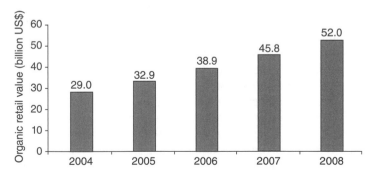

Fig. 23.4   Estimated retail value of organic certified products in billion US dollars (from Datamonitor, 2008).

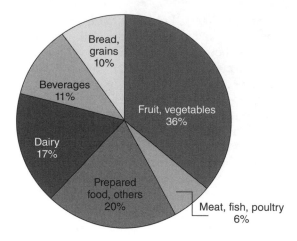

Fig. 23.5   Global organic food market segmentation percentage share, by value, 2008 (from Datamonitor, 2008).

Fruits and vegetables represent the largest category in organic products, accounting for 36% of the total market. Prepared food and dairy products are the next two categories with an additional 20% and 17%, respectively (Fig. 23.5).

As the number of organic codes and voluntary standards proliferated in the 1970s and 1980s, two main voluntary 'standards of standards' emerged. The International Federation of Organic Agriculture Movements was created in 1972 and published a first set of basic standards in 1980 to facilitate the development of organic standards and third-party certification worldwide and to provide an international guarantee of these standards and organic certification. On the inter-governmental side, a Committee on Food Labelling of the FAO/WHO Codex Alimentarius Commission adopted in 1999 the 'Guidelines for the Production, Processing, Labelling and Marketing of Organically Produced Food'.

The IFOAM Basic Standards and the Accreditation Criteria are two of the main components of the Organic Guarantee System (OGS). Based on the decision of the IFOAM General Assembly in September 2005, IFOAM has been revising the OGS with the aim of creating more access to the system by increasing its flexibility.

Organic markets are more regulated than other sustainable sourcing activities. EU countries have endorsed a common organic standard, initially described in 1991 by Regulation EEC 2092/91 and revised between 2005 and 2008. Canada, the United States, Japan and over 60 other countries have also implemented regulations on organic farming. Organic agriculture is an area where voluntary and mandatory standards are interrelated. In an effort to harmonise private and government standards, an International

Task Force established jointly by FAO, IFOAM and UNCTAD was launched in 2003.

Advocates of organic agriculture highlight its positive effect on soil structures, water conservation, mitigation of climate change and sustained biodiversity (IFOAM, 2009). Challengers of the organic certification cite lower yields and difficulties in expanding supply at the scale required for mainstream consumption. There is also still an open debate concerning harmonisation of certification requirements, which have continued growing in number and complexity. While considerable consistencies exist across different specific standards, differences still remain on issues like conversion period, permitted substances for fertilisation, disease control and food processing (UNCTAD, 2007), with producers and buyers often following multiple certification programmes and bearing the costs and complexity associated with managing multiple certifications.

23.4.4 GLOBALGAP

GLOBALGAP (Global Good Agricultural Practice Standard) was created in 2007; built on the basis of a previous standard, the EUREPGAP that was established in 1997 by a European group of 13 retailers.

The standard traces its roots to the response of European retailers to changes in regulation such as the UK Food Safety Act of 1990, which marked a shift of responsibility from the seller of food products to the buyer. The standard focused initially on fruits and vegetables and aimed at 'making a first step towards European-wide harmonisation of minimum standards for integrated production' (van der Grijp *et al.*, 2004).

GLOBALGAP is a business-to-business standard that is not made evident to consumers via a logo or any other identification. It is based on third-party certification and its main focus is to 'achieve its goals of safe food under reasonable labour conditions and without damage to the environment by identifying risk factors and establishing practices to counter these risks' (Humphrey, 2008).

The standard consists of a set of rules organised in three areas: (1) 'major musts' that must be met in their entirety, (2) 'minor musts' for which 95% of all control points is compulsory and (3) 'recommendations' for which no minimum compliance level is set (GLOBALGAP, 2007). Traceability is an essential part of the standard. As shown in Fig. 23.6, information is collected and audited along the process flow using a common farm base and multiple scope and sub-scope requirements. Producers can be certified on an individual or group basis, or by having the industry's own quality and safety management system deemed equivalent to that of GLOBALGAP. Certification requirements include limits on pesticide residue as well as a ban on non-essential animals being around packinghouses. It also integrates social considerations such as workers' health and safety requirements.

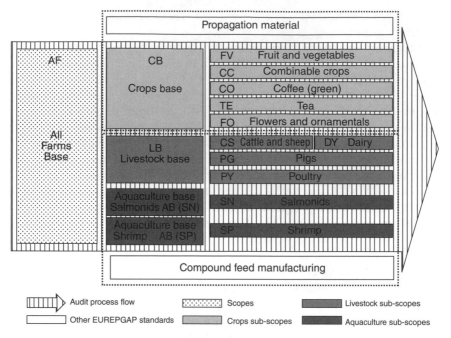

Fig. 23.6 Stages of production covered by GLOBALGAP (from GLOBALGAP, 2008).

Although the standard is still heavily focused on supplying the European Union market, it has become an important element of the food safety practices of most major European retailers and a '*de facto*' requirement for exporters trying to sell fruits and vegetables to this market as well as, and increasingly so, other products covered by the GLOBALGAP standard. Over time, membership of the EUREP and later of the GLOBALGAP grew to include over 40 retailers (mostly European with the exception of one Japanese retailer) and an increasing range of food products including tea, salmon and wheat, involving over 92,000 producers from 88 countries certified as EUREPGAP or GLOBALGAP producers and 134 certifying organisations.

Already the most widely implemented farm certification scheme for fruits and vegetables, the standard has also expanded its reach driven by the expansion of European retailers to North America and the work of the Global Food Safety Initiative (GFSI) to benchmark the five leading food safety standards: United States' Safe Quality Food (SQF) Institute standard, UK's British Retail Consortium (BRC), International Food Standard (IFS), US Safe Quality Food (SQF) standard and the Dutch HACCP initiative.

An important strength of GLOBALGAP rests on its widespread acceptance among retailers, especially in Europe. The process towards the harmonisation of food safety standards has also been an important development, avoiding multiple competing audits of standards required by buyers. Criticism of the standard is centred on the high compliance costs and the difficulties faced by many producers, particularly smallholders, to comply (Henson, 2006; Ellis and Keane, 2008; UNCTAD, 2008).

23.5   Trends

Voluntary standards are gaining importance in the food and beverage industry. New standards are constantly being introduced, multiple stakeholders are progressively getting involved in the development and implementation of these standards and their relationship with regulatory standards is being closely observed. As voluntary standards increase their role in the industry, five important trends can be identified: (1) expansion of scope (emphasis, geography, product) among leading voluntary initiatives, (2) increasing awareness and need to clarify compliance costs associated with voluntary initiatives and standards, (3) emergence of a new set of indicators and measurement methodologies to assess the impact of sustainability initiatives, (4) closer supply chain relationships resulting from participation and compliance with voluntary standards, (5) expanding role of multiple stakeholders in the supply chains. These trends are reviewed below.

23.5.1   Scope expansion and harmonisation efforts

Most voluntary standards were introduced as a response to a specific issue, one area of the supply chain, and covered a limited range of products. EUREPGAP, for example, initially covered only fruits and vegetables exported to Europe. Its expansion in membership and the change in name to GLOBALGAP signalled its intended evolution in geographic scope. A fruit and vegetables protocol, the initial focus of the standard, later became just one module of an expanded initiative that included new modules for livestock production and aquaculture (GLOBALGAP, 2007). In addition to increasing product and geographic scope, there has been a trend to expand the emphasis of the standards to integrate safety and quality with social and environmental sustainability concerns under the same umbrella initiative. Social and environmental sustainability standards such as Fairtrade, Rainforest Alliance and Utz Certified also expanded the range of products and geographies covered by their schemes.

As a consequence of the expanding scope for many initiatives, both buyers and producers are faced with multiple overlapping but non-aligned standards, resulting in increased management costs and complexity. Safety standards might be the leaders in progressing towards increased harmonisation. GLOBALGAP, for example, sets out a 'benchmarking process' that qualifies and harmonises different standards around the globe (GLOBAL-GAP, 2008). By the end of 2008 this process had resulted in 13 standards from 11 different countries being approved, and a further 10 being considered. Further, as mentioned in Section 23.4.4, the Global Safety Initiative was set up to define a benchmark across the five leading food safety standards: United States' Safe Quality Food (SQF) Institute standard, UK's BRC, International Food Standard (IFS), US Safe Quality Food (SQF) standard and the Dutch HACCP initiative (not included in the above 13 standards).

A process of harmonisation and mutual recognition has only recently started in response to growing dominance of certain safety standards like GLOBALGAP (Henson, 2006). However, the progress is still limited in many voluntary standard areas such as ethical or environmental sustainability, and the harmonisation process has made very little progress, limited to certain farms opting for multiple simultaneous certifications (for example organic and Fairtrade) (Ponte, 2004).

23.5.2   Compliance costs

Voluntary initiatives are not cheap to implement. Complying with the requirements specified in the standard, setting up management systems, infrastructure and certification mechanisms all add up to making adoption of voluntary initiatives an expensive undertaking. Two questions frequently asked in the debate on sustainability and compliance are: (1) how much does it cost? and (2) who pays for what?

With exceptions, sustainability is a new field for all players involved. Estimating the cost to set up, promote and enforce a new voluntary standard is a difficult undertaking and significant uncertainty still exists about how to assess the levels of investment required to 'green' a supply chain or to run a 'sustainable' operation. Most companies enter this field without an accurate idea of what kind of investments will be required, frequently miscalculating the costs involved, including the management time that sustainability initiatives can take up. As one trader involved in a sustainable coffee sourcing programme said: 'We didn't know what it would take... maybe that was better' (Alvarez and Wilding, 2008).

It is also unclear what an appropriate division of costs among the different players is. In most schemes, producers are required to cover the costs of certification and/or transformation (in the case of organic agriculture) in the expectation that there will be a market (and an associated premium or long-term relationship) with a buyer. Indeed, the cost of compliance has been identified as one of the major limitations resulting in reduced participation among producers in developing countries, particularly smallholders (Jaffee *et al.*, 2005; Humphrey, 2008). Ellis and Keane (2008) propose creating a 'good for development label' that would indicate to consumers a retailer's or brand's efforts towards providing technical or financial assistance to small producers, contributing to local infrastructure development, bearing costs of compliance themselves, and so on.

Balancing this view, promoters of incorporating sustainability considerations in corporate strategy point out to the even larger costs of ignoring the issue. Adapting the sentence by Derek Bok, former president of Harvard University, 'If you think education is expensive, try ignorance . . .', a participant in a recent food industry conference offered 'If you think food safety and sustainability initiatives are expensive, try the alternative . . .'. Even though correctly assessing the costs of increased involvement in sustainability initiatives can be difficult, the costs of not participating can be even more burdensome.

The debate on costs is, of course, only one side of the argument and can only be appropriately assessed by incorporating the benefit side of the equation. Again, it can be argued that it is not only about the benefits that are generated, but where the benefits are accrued and what the resulting impact is.

23.5.3   Impact measurement

As sustainability initiatives start to be espoused by large corporations, a natural consequence has been the need to define a set of measurements and to assess the impact of the initiative from a corporate as well as from a broader perspective. The so-called triple bottom line (Elkington, 1999) requires an enterprise's accounting system to incorporate not only the traditional financial measurements but also social and environmental outcomes. Measurement of 'direct' (within the direct control of the enterprise) water and energy savings is probably the area where most progress has been made so far.

New methodologies have also been introduced aiming at one or more of the areas of impact. Examples are social accounting methodology, sustainability balanced scored, social return on investment (SROI), the Committee on Sustainability Assessment (COSA) Project, and several carbon footprint measurement methodologies like the one endorsed by the UK Carbon Trust (Carbon Trust, 2008).

In terms of communication and reporting, the most comprehensive framework at the present time is probably The Global Reporting Initiative (GRI), a large multi-stakeholder network of thousands of experts. The framework proposed by the organisation sets out the principles and indicators that organisations can use to measure and report their economic, environmental and social performance. These Sustainability Reporting Guidelines are known as the GRI or G3 (third version) and are free public goods. Other framework components include sector supplements (unique indicators for industry sectors), protocols (detailed reporting guidance) and national annexes (unique country-level information) (Global Reporting Initiative, 2008).

Although frameworks like the one proposed by the GRI are increasingly being adopted, agreeing on indicators, methodologies and boundaries for accounting for environmental and social impact throughout the entire supply chain, the indirect impacts on producer communities and among consumers is still an important outstanding challenge and will continue to be an important issue for voluntary sustainability initiatives.

23.5.4   Closer buyer–seller relationships

Participating in and complying with sustainability voluntary standards requires deeper knowledge about what happens beyond the traditional boundaries of the organisation. Safety requirements such as traceability, environmental or impact assessment imply establishing a higher level of cooperation and exchange of information beyond the immediate trade partners and across the entire supply chain. Although not originally pursued in most cases, the high level of coordination required has also disposed buyers and sellers to establish closer relationships. It has also meant an additional role for intermediaries. In pure commodity markets, arms-length relationships have traditionally been the norm, with trading companies expected to act as intermediaries between producers and buyers. In a recent research on the coffee industry, large trading companies were found to play an important role in sustainability initiatives, going far beyond pure finan-cial and product intermediation and into coordination of a wide range of activities including training, micro-lending and environmental projects (Alvarez and Wilding, 2008). This research also identified the need to acquire a new set of capabilities and resources as an important enabler for supporting the new requirements in the now expanded buyer–seller relationships.

23.5.5   Multi-stakeholder dialogue and consultation

In addition to closer buyer–seller relationships, working together with mul-tiple stakeholders is becoming common practice for most voluntary stand-ards in the food and beverage industry. Many voluntary standards, particularly those focused on food safety, originated as a list of rules and requirements determined by the expectations of retailers or manufacturers. Jaffee *et al.* (2005) argues that exercising an effective voice in the definition of standards is usually quite difficult for smaller countries and individual suppliers, although opportunities often arise for dialogue about how the standards are to be applied.

Even standards 'with the best intentions' have been often criticised for lacking input from the producers (specially smallholders) in relation to the appropriateness and the costs/benefits of the standards (Ponte, 2004) or for lacking a participatory approach to auditing impact (Auret and Barrientos, 2006). As standards become a larger portion of total trade, the organisations behind each standard are becoming more participatory and are making efforts to incorporate the views of multiple stakeholders through more or less structured mechanisms. Initial dialogues between corporations and NGOs are giving way to increasing involvement of producers, development institutions and governments.

Besides participating in the crafting of the standards, the role of NGOs and development organisations has been crucial in activities that go beyond traditional production and trading. The implementation of sustainability initiatives that involve activities such as technical training, transfer of management skills, access to finance and infrastructure investments requires the involvement of and coordination with other 'non-traditional' players such as NGOs, development organisations and education institutions. This multi-stakeholder participatory management is relatively new for many corporations which need to work at a strategy level, but also for supply chain managers who need to make this cooperation work on the ground. Again, as in the case of shifting of the buyer–seller relationship, a new set of skills is now required from managers working on the ground and who now need increasingly to work with multiple stakeholders.

23.6   Corporate strategies

While there exists significant awareness of the increasing role played by voluntary safety and sustainability standards among companies, for most of them it is much less clear how to react to them, which standards to incorporate, their impact on supply chain operations and the role they can play in the overall corporate strategy. Defining a sustainability strategy requires analysing the risks and opportunities the company faces, not only across its operations but also across the extended supply chain limits and within the broader community. It also requires a comprehensive view of the corporation's values and overriding corporate strategy. A sustainability strategy cannot operate in a vacuum, it needs to be aligned with the rest of the company and what it stands for. An analysis of risks, opportunities and the company's overall strategy involves identifying the key issues that the corporation needs to assess in determining an appropriate sustainability strategy.

23.6.1   Risks

Sustainability initiatives can be an important tool to deal with risks in the supply chain. The three main categories of risk to assess are:

- Safety risks: Which are the areas within my direct or the extended supply chain that present potential food safety, environmental or social risks?

- Reputational risks: What are the potential events that could have a negative reputational impact on the industry? On the firm? On a particular business unit?
- Supply chain continuity: Securing the long-term supply of critical or specialised inputs can be an important driver to engage in sustainability initiatives.

23.6.2   Opportunities

Sustainability initiatives can also represent a strategy to profit from existing opportunities. The three main areas to identify opportunities are:

- Market driven: With double digit growth rates for many sustainability labelled products, the category can be an attractive niche for branded products and for retailers. Natural resource saving innovations can also find new markets that were not available only a few years ago.
- Brand image: Opportunities may exist associated with enhanced brand reputation through a differentiation of products with a sustainability claim. Retailers communicating a higher level of commitment to safety or to environmental or social sustainability have the opportunity to increase their brand reputation among sustainability-aware consumers and environmentally minded investors. But it can also be a double-edged sword and become a risk as companies making claims on the sustainability of their brands face increased scrutiny by consumers and NGOs.
- Employee engagement: According to a study conducted in 2007 by management consultancy McKinsey, 48% of surveyed CEOs believed that employees have the greatest impact on the way that a company manages societal expectations (McKinsey & Company, 2007). Employees can be an important driver for engaging in sustainability initiatives and can also be an important resource for execution of these initiatives.

23.6.3   Corporate strategy

In an article on business and society, Porter and Kramer observed 'The fact is, the prevailing approaches to corporate social responsibility are so fragmented and so disconnected from business and strategy as to obscure many of the greatest opportunities for companies to benefit society' (Porter and Kramer, 2006). The set of values espoused by the organisation, its mission, vision and objectives should serve as guidance as to the role that sustainability initiatives should play in the corporation and what should be the role of the various stakeholders in supporting the chosen strategy.

23.6.4   Strategy options

Although each company's strategy is different and should devise a unique strategy, Fig. 23.7 illustrates three archetypal choices in response to alternative configurations of risks and opportunities.

Strategy option 1: cooperate to establish common baseline standards
In a situation with high safety or industry reputational risk but with reduced opportunities for differentiation, a common agreement on baseline standards can provide a higher level of protection with a lower implementation cost. It can also act to pre-empt or shape regulation. Examples of this type of initiative include the Roundtable for Sustainable Palm Oil (RSPO) and the Common Code for the Coffee Community (4C). There are, however, significant challenges in the coordination of such broad initiatives involving multiple stakeholders with different agendas. It can take a significant time to coordinate such a large group and some initiatives have been criticised as being a 'race to the bottom', agreeing only on the lowest common denominator and therefore compromising the relevance of the initiative.

Strategy option 2: selectively engage for differentiation
In situations where there is a lower level of risk but where sustainability offers an opportunity for differentiation from competitors, corporations can choose to engage selectively in the areas that offer the highest opportunity to increase revenues or to generate efficiencies across the entire supply chain. Many corporations in the industry have recently incorporated lines of products or purchased brands that address the small but growing 'sustainability' market. Others have set up focused initiatives that result in

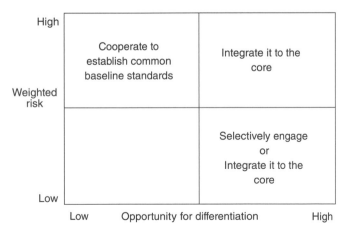

Fig. 23.7 Corporate responses to voluntary initiatives.

water, energy or waste savings. Venturing into a focused initiative can also serve as a first experiment for the company, without involving the entire set of operations. The risk, on the other hand, is that once the company communicates on these selected initiatives, the rest of the operation can come under scrutiny and backfire on the corporate reputation.

Strategy option 3: integrate it to the core
In certain situations, sustainability issues simultaneously represent a significant risk and important opportunities for the company. For example, a situation with a risk of scarcity of raw materials but where closer and broader relationships with stakeholders can result in sustainable value creation can represent an important opportunity. Sustainability can thus become an integral part of how a business unit or an entire corporation does business, linking it to the company's core values, its strategy and its business model, even if the level of risk is low. UK retailer Marks and Spencer has been recognised as successfully integrating its sustainability strategy into the overall corporate strategy, with sustainability-related activities that span across all major activities of the company and integrated with the traditional core operations (Marks and Spencer, 2009). In the extreme version of this type of strategy some companies' identity is not only integrated but is driven by sustainability goals. For example, USA retailer Whole Foods Market's value proposition is to sell organic, natural, healthy food products. Social issues are also fundamental to what makes Whole Foods unique and the commitment to sustainability extends beyond sourcing to the store construction material, wind energy credits to offset its electricity consumption, composting residues, and so on (Whole Foods Market, 2009).

Again, as in the previous option, engaging in sustainability activities and making them visible can raise the standards by which the company is judged and therefore companies that espouse responsibility as a core value may be more exposed to NGO campaigns 'holding a company to a higher standard because they feel the company is using a socially responsible positioning to enhance its reputation rather than honestly trying to do the right thing' (Argenti, 2004).

23.7   Conclusion

Voluntary standards, while not historically new, increasingly play an important role in shaping the supply chains for manufacturers and retailers across both product and geographic markets. These standards, whether based on consumer, producer, NGO or governmental interests, have become important elements for manufacturers differentiating products and managing potential supply risks, and consumers making purchase choices. Whether associated with consumer interests, NGO activism, manufacturer differentiation or regulatory protection, standards play a growing role in the supply chain strategies of a growing number of companies across product and geographic markets.

Voluntary standards can focus on a product's technical performance or characteristics, convey information to consumers and place demands on suppliers. These standards traditionally vary dramatically across many dimensions, including the lead organisation controlling the standard, the motivation of the party employing the standard, the scope and focus of the standard and the enforcement mechanism.

Despite traditional differences, a number of common trends are having an impact on voluntary standards today. Most voluntary standards, while initially focused on specific products or markets, are expanding their scope in terms of the range of products or markets covered. This expansion in scope is resulting in increasingly overlapping but not necessarily aligned standards, increasing the cost and complexity of compliance. A second trend is growing costs associated with compliance, as well as debate over both understanding the true total cost of compliance and over who should bear these rising costs. A third trend is growing awareness of the need to measure the true impact of standards, whether it is on producers or suppliers or the environment itself. The complexity of measurement across the entire value chain, whether through a so-called triple bottom line or another approach, is still an area to be explored and developed. Further trends are associated with the impact of standards on relationships across buyers, sellers and other stakeholders and the need for dialogue and consultation across the entire range of supply chain activities.

In general, voluntary standards are playing a more central role in overall corporate strategies. The impact on corporate strategies is associated with risk (safety, reputation and continuity) identification and management, and opportunities for strategic differentiation. Important strategic opportunities for firms today vary based on balancing the opportunities for differentiation against management of risks. Strategic options can include focusing on cooperation to establish common benchmarks, expanded engagement to increase differentiation, or full scale integration into the core strategy of a firm. In any case, corporations are increasingly expected to participate in areas not traditionally associated with business and to go beyond the boundaries of their operations to a much broader playing field. The opportunity to achieve important benefits for the corporation and for the broader society exist. But there are also significant costs involved in adopting sustainable sourcing strategies. A higher cost, however, might well be the cost of ignoring the issue.

23.8   Acronyms and abbreviations

- 4C: Common Code for the Coffee Community
- ATO: Alternative Trade Organisation

- BRC: British Retail Consortium
- CEO: Chief Executive Officer
- COSA: Committee on Sustainability Assessment
- ETI: Ethical Trade Initiative
- EU: European Union
- EUREPGAP: Good Agricultural Practice standard developed by EUREP group of firms
- FAO: Food and Agriculture Organisation of the United Nations
- FLO: Fairtrade Labelling Organisation International
- GLOBALGAP: The new brand name of EUREPGAP
- GRI: Global Reporting Initiative
- HACCP: Hazard Analysis and Critical Control Point
- IFOAM: International Federation of Organic Agriculture Movements
- ILO: International Labour Organisation
- ISO: International Organisation for Standardisation
- MSC: Marine Stewardship Council
- NGO: Non-Governmental Organisation
- OGS: Organic Guarantee System
- RSPO: Roundtable on Sustainable Palm Oil
- SA 8000: Social Accountability 8000
- SAI: Social Accountability International
- SAN: Sustainable Agriculture Network
- SEDEX: Supplier Ethical Data Exchange
- SROI: Social Return on Investment
- UK: United Kingdom
- UNCTAD: United Nations Conference on Trade and Development
- US: United States
- WHO: World Health Organisation

23.9   References

ACUTT, N. (2002), 'The South African experience with voluntary environmental initiatives', *Proceedings of the 3rd International Conference on Environmental Management*, 29 August 2002, Johannesburg, South Africa.

AGRITRADE (2008), *Coffee: Executive brief*, Agritrade.

ALVAREZ, G. and WILDING, R. (2008), 'Governance mechanisms dynamics in a multi-stakeholder network: The case of Nespresso AAA Sustainable Quality Program', *BAM 2008: The Academy Goes Relevant*, 9–11 September 2008, Harrogate, British Academy of Management, London.

ARGENTI, P.A. (2004), 'Collaborating with activists: how Starbucks works with NGOs', *California Management Review*, **47**(1), 91.

AURET, D. and BARRIENTOS, S. (2006), 'Participatory social auditing: developing a worker-focused approach', in *Ethical Sourcing in the Global Food System*, Barrientos, S. and Dolan, C. (eds), Earthscan, London, 129–48.

BARRIENTOS, S. and DOLAN, C. (2006), 'Reflections on the future of ethical sourcing', in *Ethical Sourcing in the Global Food System*, Barrientos, S. and Dolan, C. (eds), Earthscan, London, 179–85.

BARRIENTOS, S. and SMITH, S. (2006), *ETI Impact Assessment 2006*, Institute of Development Studies, Sussex.

CARBON TRUST (2008), *How to calculate a full carbon footprint*, available at: http://www.carbontrust.co.uk/solutions/CarbonFootprinting/how_to_calculate_a_full_carbon_footprint.htm (accessed January 21, 2009).

CEC (2002), *Regulation (EC) No 178/2002 Laying Down the General Principles and Requirements of Food Law, Establishing the European Food Safety Authority and Laying Down Procedures in Matters of Food Safety*, available at: http://ec.europa.eu/food/food/foodlaw/principles/index_en.htm (accessed January 18, 2009).

CENTRE FOR BUSINESS RELATIONSHIPS, ACCOUNTABILITY, SUSTAINABILITY AND SOCIETY (2007), *History of Corporate Social Responsibility and Sustainability*, available at: http://www.brass.cf.ac.uk/uploads/History_L3.pdf.

CHRISTMANN, P. and TAYLOR, G. (2002), 'Globalization and the environment: Strategies for international voluntary environmental initiatives', *The Academy of Management Executive*, **16**(3), 121.

COMMON CODE FOR THE COFFEE COMMUNITY ASSOCIATION (2008), *4 C Association*, available at: http://www.4c-coffeeassociation.org/ (accessed January 10, 2009).

DATAMONITOR (2008), *Global Organic Food*, 0199–0853, Datamonitor, New York.

ELKINGTON, J. (1999), *Cannibals with Forks: The Triple Bottom Line of 21st Century Business*, Capstone, Oxford.

ELLIS, K. and KEANE, J. (2008), *A review of ethical standards and labels: Is there a gap in the market for a new 'Good for Development' label?* Working Paper 297, Overseas Development Institute, London.

ETHICAL TRADING INITIATIVE (2008), *The base code principles of implementation*, available at: http://www.ethicaltrade.org/resources (accessed January 10, 2009).

FAIRTRADE FOUNDATION (2008), *Facts and Figures on Fairtrade*, available at: http://www.fairtrade.org.uk/what_is_fairtrade/facts_and_figures.aspx (accessed January 21, 2009).

FAIRTRADE LABELLING ORGANISATION (2008), *FairTrade Labelling Organisation*, available at: http://www.fairtrade.net/.

FELTMATE, B.W. (1997), 'Making sustainable development a corporate reality', *CMA*, **71**(2), 9.

GIBBON, P. and PONTE, S. (2005), *Trading Down: Africa, Value Chains and the Global Economy*, 1st edition, Temple University Press, Philadelphia, PA.

GLOBAL REPORTING INITIATIVE (2008), *What we do: The framework and its benefits*, available at: http://www.globalreporting.org/Home (accessed January 21, 2009).

GLOBALGAP (2007), *General Regulations Integrated Farm Assurance*, GLOBALGAP, Köln, Germany.

GLOBALGAP (2008), *Harmonisation via Benchmarking*, available at: http://www.globalgap.org/cms/front_content.php?idcat=29 (accessed January 21, 2009).

HAVELAAR, M. (2007), Max Havelaar Fair Trade Switzerland, *Jahresbericht 2007*.

HENRIQUES, A. (2003), *Ten Things You Always Wanted to Know About CSR (but were afraid to ask)*, Ethical Corporation Institute, London, United Kingdom.

HENSON, S. (2006), 'The role of public and private standards in regulating international food markets', *IATRC Summer Symposium*, IATRC, Bonn, Germany.

HOBBS, J.E. and KERR, W.A. (1992), 'Costs of monitoring food safety and vertical coordination in agribusiness: what can be learned from the British Food Safety Act 1990?', *Agribusiness (1986–1998)*, **8**(6), 575.

HUMPHREY, J. (2008), *Private Standards, Small Farmers and Donor Policy: EurepGap in Kenya*, 308, Institute of Development Studies, Sussex.

IFOAM (2009), *About IFOAM*, available at: www.ifoam.org/about_ifoam/ (accessed January 10, 2009).

JAFFEE, S. and HENSON, S.J. (2004), *Standards and Agri-food Exports from Developing Countries: Rebalancing the Debate,* 3348, World Bank Publications, Washington, DC.

JAFFEE, S., VAN DER MEER, K., HENSON, S., DE HAAN, C., SEWADEH, M., IGNACIO, L., LAMB, J. and BERGOVOY LISAZO, M. (2005), *Food Safety and Agricultural Health Standards: Challenges and Opportunities for Developing Country Exports,* 31207, Poverty Reduction and Economic Management Trade Unit and Agriculture and Rural Development Department, World Bank, Washington, DC.

MARKS and SPENCER (2009), *How we do business,* available at: http://corporate.marksandspencer.com/howwedobusiness (accessed January 21, 2009).

MAZURKIEWICZ, A. (2005), 'Corporate selfregulation and multistakeholder dialogue', in *The Handbook of Environmental Voluntary Agreements: Design, Implementation and Evaluation Issues,* Croci, E. (ed.), first edition, Springer, 31–48.

MCKINSEY & COMPANY (2007), *Shaping the New Rules of Competition: UN Global Compact Participant Mirror,* McKinsey & Company, February.

MICKLETHWAIT, J. and WOOLDRIDGE, A. (2003), *The Company: A Short History of a Revolutionary Idea,* Modern Library, New York, NY.

NAVDI, K. and WÄLTRING, F. (2004), 'Making sense of global standards', in *Local Enterprises in the Global Economy,* Schmitz, H. (ed.), Edward Elgar, Cheltenham.

NICHOLLS, A. and OPAL, C. (2005), *Fair Trade: Market-driven ethical consumption,* Sage Publications, London, UK.

OXFAM AMERICA (2002), *Mugged: Poverty in your coffee cup,* Oxfam International, Washington, DC.

PONTE, S. (2004), *Standards and Sustainability in the Coffee Sector,* International Institute for Sustainable Development, Winnipeg, Manitoba, Canada.

PONTE, S. and GIBBON, P. (2005), 'Quality standards, conventions and the governance of global value chains', *Economy and Society,* **34**(1), 1.

PORTER, M.E. and KRAMER, M.R. (2006), 'Strategy & society: the link between competitive advantage and corporate social responsibility', *Harvard Business Review,* **84**(12), 78.

REARDON, T. and FARINA, E. (2002), 'The rise of private good quality and safety standards: illustrations from Brazil', *International Food and Agribusiness Management Review,* **4**(4), 413–421.

SIDWELL, M. (2008), *Unfair Trade,* Adam Smith Institute, London.

TRADE STANDARDS PRACTITIONERS NETWORK (2008), *TradeStandards.org,* available at: http://tradestandards.org/en/Index.aspx (accessed January 12, 2009).

TRAUB, D. (2008), 'China milk scandal: Families of sick children fight to find out true scale of the problem', *The Telegraph,* Kent, United Kingdom, 3rd December.

UNCTAD (2007), *Sector background note – Organic Agriculture,* UNCTAD, Geneva.

UNCTAD (2008), *Private-Sector Standards and National Schemes for Good Agricultural Practices: Implications for Exports of Fresh Fruit and Vegetables from sub-Saharan Africa,* UNCTAD/DITC/TED/2007/13, UNCTAD, New York, Geneva.

VAN DER GRIJP, N., MARSDEN, T. and CAVALCANTI, J. (2004), 'Retailers as agents of change towards pesticide reduction', *88th Seminar of the EAAE, Retailing and Producer-Retailer Relationships in Food Chains,* May 2004, Paris.

WHOLE FOODS MARKET (2009), *Values and Actions Overview,* available at: http://www.wholefoodsmarket.com/values/index.php (accessed January 21, 2009).

WILLER, H., YUSSEFI, M. and SORENSEN, N. (2008), *The World of Organic Agriculture: Statistics and Emerging Trends,* International Federation of Organic Agriculture Movements (IFOAM), Bonn, Germany.

WORLD COMMISSION ON ENVIRONMENT AND DEVELOPMENT (1987), *Our Common Future,* Oxford University Press, Oxford.

24

Trends in food supply chain management

M. Bourlakis, Brunel University, UK and A. Matopoulos,
University of Macedonia, Greece

Abstract: This chapter analyses specific trends in food supply chain management. We focus our analysis on information technology and on sustainability. Relevant applications, implementation practices and challenges are examined in the context of contemporary food supply chains. This analysis will benefit supply chain managers, researchers and other stakeholders involved with the food sector.

Key words: food supply chain management, information technology, RFID, sustainability, trends.

24.1 Introduction

Food supply chain management is developing as a research discipline spanning local, regional, national and international arenas and it has progressed from a series of shorter, independent transactions to more collaborative relationships between producers, processors, manufacturers and retailers (Bourlakis and Weightman, 2004). This chapter discusses the key trends in food supply chain management building on and extending previous work published on this topic (see for example, Bourlakis, 2001; Bourlakis and Bourlakis, 2004). In the next section, we illustrate the various phases of food chain evolution analysing the key forces which have been most influential. It is followed by a section on information technology and its key applications. One of these is radio frequency identification (RFID) which is gaining importance in contemporary food chains. A key section of this chapter is devoted to sustainability in food supply chains where we discuss relevant sustainability-related challenges and specific implementation practices faced by food chain members.

Many key trends are evident in food supply chains. However, our aim is to focus on two major trends (information technology and sustainability) that have already had a large influence on the development of contemporary food supply chains and will continue to be very influential in the future.

24.2   Evolution of the food chain

There are a range of forces that have an impact on the food chain and its evolution. For example, Strak and Morgan (1995) have suggested five key forces including globalisation, market structure and power, consumer tastes and lifestyles, technological change and regulation. These forces, which are highly interrelated, affect the chain structure, its dynamics and its evolution. This interrelationship is evident when, for example, food retailers nowadays source various products (e.g. fresh fruit and vegetables) from all over the world, an issue that emanated from consumer demand (Sparks *et al.*, 2006). This demand is met by the use of various sophisticated logistics processes and information technologies in the food chain (Bourlakis and Bourlakis, 2006). In turn, this sourcing and large selection of fresh produce from all over the world provides these multiple food retailers with a competitive advantage against smaller retailers. This strengthens their market power which could also attract extra enquiries by regulators as is seen in the UK with the work of the Competition Commission (Blythman, 2005).

In terms of the food chain evolution, Ramsey (2000) noted four phases. In the first phase, the era of early competition (from 1900 to 1920), regional wholesalers dominated the food supply chain. In the second phase, the consolidation period (from 1920 to 1945), national manufacturers played the most prominent role in the chain. In the third phase, the internationalisation period (from 1945 to 1980), many food manufacturers expanded outside their domestic base and, although they were key players in that chain, they faced many challenges from food retailers. It is not surprising then that in the fourth phase (from 1980 to 2000), the globalisation phase, retailers took control over the food supply chain and continuously challenged the manufacturers (see Table 24.1).

From 2000 onwards, the food retailers become the 'chain captains' and one key initiative here is factory gate pricing (Finegan, 2002). Under this initiative, retailers collect the suppliers'/manufacturers' products from the manufacturing plants (factory gate) and as retailers become responsible for that product collection (and its logistics cost); they pay a smaller product price to manufacturers. As well as that initiative, other key food retail logistics developments include centralisation, composite distribution and the increasing use of third party logistics services by retailers (Bourlakis and Bourlakis, 2005). Under centralisation, food retailers make use of large regional warehouses to manage successfully their complex supply chain comprising thousands of products and hundreds of stores (Bourlakis and

Table 24.1 Evolution of the food system (from Cotterill, 2001; Ramsey, 2000)

Phase	Structure	Scope	Competitive environment	Control of supply chain	Degree of marketing sophistication	Consumer demand
I The era of early competition 1900–1920	Small to medium sized firms	Local/regional within one country except for commodity movement	Best example of 'perfect' competition with entrepreneurial farming	Run by regional wholesalers	Limited branding, mainly commodities	Food a major part of disposable income – up to 50% in some countries
II National consolidation 1920–1945	Rise of large manufacturers via publicly owned corporations with many small processors and retailers	Move to national/major regional level in one country Limited export	'Imperfect' competition and start of acquisition activity; Food manufacturing is an important factor in national economies	Run by national manufacturers with wholesaler and retail-chain support	Rise of national branding, sales, advertising and R&D. Private label appears	Rise in per capita income and demand for wide range of branded convenience foods
III Internationalisation 1945–1980	Mix of publicly owned manufacturing oligopolies, retail chain concentration and many smaller entrepreneurs	Multinational expansion of major manufacturers with significant increase in turnover	Golden age of manufacturer branding and mass marketing	National manufacturers dominate but some retail challenge resulting from concentration	Brand management at national and international level; Increased demand for market data/information	Food expenditure declines as percentage of disposable income; Move to larger retail outlets; Growth of eating out
IV Globalisation 1980–2000	Polarisation of manufacturing and retailer structures via concentration, acquisition and divestment	Manufacturers extend globally and retailers go multinational	Retailer branding increases level of penetration and begins to challenge manufacturer branding. Both now 'oligopolistic'	Supply chain in Europe run by retailers and challenging for dominance in North America	Major manufacturers identify core categories. Superbrands, retailer brands, address vertical coordination issues in a concentrated channel, Internet grocers	Turnover in food service now challenging for leadership as slowdown in food sales at retail

Bourlakis, 2005). In the food chain, the composite distribution principle (for both warehouses and transportation) is also evident where a multi-temperature lorry or warehouse can manage products under various temperatures (ambient, chilled, frozen) all together (Bourlakis and Weightman, 2004; Smith and Sparks, 2004).

To be able to accommodate all these changes in their chain, food retailers make increasing use of third party logistics firms and it is anticipated that these logistics firms will play a more important role in the years to come (Bourlakis and Bourlakis, 2005). This anticipation is verified by the fact that a considerable variation of logistics externalisation exists between various European Union members. For example, although logistics externalisation is an established practice in many countries like the UK, it is less prominent in other European environments such as Greece (Marketline International, 1997). The above discussion illustrates the increasing role of the downstream members of the chain including retailers and third party logistics firms. Other downstream members include the food service firms (catering firms) which are part of a separate and unique food chain in its own right and the online grocery chain that can be also considered a separate and distinctive chain.

Aiming to discuss issues that have a universal appeal across the whole food chain system, in the next section we provide an analysis of specific trends. We start this analysis by examining the increasing role of information technology and relevant applications in contemporary food supply chains.

24.3   Use of information and communication technology (ICT)-based applications

24.3.1   Role of information technology (IT) and technology for food supply chains

Successful companies seem to be those that have carefully linked their internal processes to external suppliers and customers in unique supply chains (Frohlich and Westbrook, 2001; Boyer et al., 2004). In this effort towards linking internal processes to external suppliers and customers, the sharing of information among enterprises is absolutely critical. Electronic data interchange (EDI) in the past and the Internet in the last decade have enabled supply chain partners to share information and to act upon the same data (Christopher, 2005). Internet-based applications in general, can change the way companies conduct business mainly in two ways: (1) by facilitating and improving already existing business activities and processes and (2) by changing completely the traditional business model of the company (Manthou et al., 2005).

In the food context, Internet developments have mostly facilitated some of the activities taking place, without transforming companies into virtual enterprises. Only a limited number of food-related companies have managed to change completely their business model or to become successful e-business start-ups. Particularly, the last decade has witnessed more e-business start-up failures than success stories. In contrast, Internet-based applications seem to have influenced the food chain, mainly in terms of process improvement and facilitation.

The food industry is one of the most important industries and presents some interesting characteristics. Food is perhaps the most universal commodity; thus competition, along with varying customers' tastes and the perishable nature of products, often spurs retailers to go to great lengths to develop new technologies and methods of streamlining their supply chain efforts (Boyer and Hult, 2005).

Another characteristic of the sector is its diversity, since it comprises companies ranging from very small manufacturers to global retailers. A number of changes have occurred in the last decade in this sector. The entrance of global retailers, the industry's consolidation, changing consumer consumption attitudes and the existence of stricter regulations and laws regarding food production as a result of the recent food crises (Hughes, 1994; Kaufman, 1999; Fearne et al., 2001) have altered the business environment for most companies operating in the sector, encouraging a positive attitude towards integration, improvement and therefore uptake of information and communication technology (ICT)-based applications.

Despite the specific characteristics of the sector, still the food sector is one where pioneering work in electronic integration has been conducted (Harrison and van Hoek, 2008) and where even competitors decided to invest in electronic integration. In 2000, for example, 17 international retailers founded the Worldwide Retail Exchange (WWRE) to enable participating retailers and manufacturers to simplify, rationalise and automate supply chain processes, thereby eliminating inefficiencies in the supply chain. The WWRE was the premier Internet-based business-to-business exchange in the retail e-marketplace, which enabled retailers and manufacturers to reduce costs substantially across product development, e-procurement, and supply chain processes.

Regarding the impact of the Internet, it has been important but probably not equally so across food chain entities (Matopoulos et al., 2007). For example, Internet-based applications have played an important role in integrating food manufacturers and retailers. Particularly, forward integration has been more intense at the retailer level, given the size of these companies, but not at the consumer level, which essentially remains at a distance and isolated from food production (Kaufman, 1999; Fearne et al., 2001; GMA, 2000). Backward integration however, has been weak, as most of retailers' suppliers are small to medium-sized companies, such as farmers, producer groups and cooperatives, which have not adopted Internet-based applications (E-business Watch, 2006).

However, some big manufacturers have established links with the retailers and have moved towards an increased level of electronic integration. For example, in many cases manufacturers are scheduling in the short- or mid-term, planning forecasts and promotion plans (for example, the use of the CPFR platform), in collaboration with the retailers. Sainsbury, one of the biggest UK retailers, developed a package to reduce new product development time by up to a third of the traditional time by linking over its website a number of entities involved in the new product development process (e.g chefs, concept developers, manufacturers, nutritionists, marketers, design studios, artwork and reproductive houses, and product safety and legal experts). For example, a technologist on a supplier visit to South America could log on and approve packaging details online, keeping the project on time. The system enabled all parties to access the data (same versions) online as needed. For example, if an ingredient was changed by the chef, an e-mail would be sent to the packaging designer alerting them to alter the packaging too, which results in reductions in mistakes.

24.3.2   Radio frequency identification (RFID) in the food chain
View on RFID characteristics and benefits
In this subsection, a brief introduction to radio frequency identification (RFID) technology is provided followed by a discussion of the advantages and disadvantages of this technology in the food chain setting. Similar to barcode technology, RFID technology has been introduced as an improved technology permitting contactless object identification. RFID is by no means a new and radical technological initiative, but it is rather a new view of an existing technology. The RFID system is not a complex one, neither in terms of architecture nor in terms of function. It consists of the RFID tag (consisting of a chip and antenna), an RFID reader device and a backend IT system. The main function of the system is the communication between the tag and the reader by sending and receiving radio frequency waves. Several frequency bands, tags and readers are available, each linked to specific benefits and drawbacks.

The literature reveals a number of important advantages of the RFID technology, many of which derive from its ability to identify objects in an automatic and contactless way. Gaukler and Seifert (2007), for example, suggested from an application standpoint, that there are four major advantages of RFID over barcoding:

- No line of sight required.
- Multiple parallel reads are possible.
- Individual items instead of an item class can be identified.
- It has a read/write capability.

In addition, McFarlane and Sheffi (2003) pointed out two more advantages: data can be obtained continuously (providing more updated information) and no human interaction is needed (implying more error-free, cost reduction). Overall, the advantages of RFID technology have been approached from different angles, and one of the major distinctions is between 'closed loop'-based versus 'open-loop'-based benefits. Closed loop benefits refer to those that can be achieved internally in a company, while, open loop benefits refer to those that span organisational boundaries.

RFID implementation in the food chain
Owing to the fact that the technology is relatively new, there is no critical mass of research or reports on actual implementations of the technology. Most of the literature has focused so far on presenting the potential benefits, advantages or disadvantages of the technology as well as a presentation of implementation and adoption constraints (Prater *et al.*, 2005). In addition, most of the implementations seem to be in the grocery sector, where lead retailer organisations are trialling RFID systems with partner suppliers. For example, Kärkkäinen (2003) discussed the potential benefits of RFID for short shelf-life products retailers and analysed the potential impact of RFID for other supply chain participants. Jones *et al.* (2004), presented the opportunities and the challenges of RFID implementation by UK retailers, and also reported on the efforts and trials undertaken in order to implement the technology.

Michael and McCathie (2005), presented the advantages and disadvantages of using RFID technology in supply chain management and in comparison to the barcode. It is also evident that RFID implementation changes the dynamics and competitive structures of the food chain. For example, Boyle (2003) provides the case of Wal-Mart, the leading USA grocery retailer, which expected its major suppliers to implement RFID on each product box and pallet sent to that retailer by 2005.

Apart, from the process-specific benefits, RFID is expected to have very significant benefits for traceability systems. Traceability systems have been adopted on a significant scale only in the last years within the agri-food industry. In many cases, food manufacturers still lack an efficient traceability system, but nowadays new opportunities for the food traceability come from RFID technology. For example, a new RFID logger combines tracking and tracing with temperature readings, giving processors a way of identifying when food safety may have been compromised (Food Quality, 2007).

24.4   Sustainable food supply chains

24.4.1   Why food chains are 'unsustainable'?
The food sector is one where significant questions have been raised concerning sustainability practices particularly, as the local production–local consumption model, in most cases, is no longer a reality. The deregulation

of the markets, the emergence of global companies and changing consumption and shopping patterns are just some of the factors that have caused 'unsustainability' in the food industry (Fritz and Matopoulos, 2008). For example, globalisation of the sector has resulted in increased imports and exports and ever wider sourcing of products from all over the world.

The changes in consumer requirements and consumption patterns have also been very influential, with consumers nowadays demanding a wider choice of food products, often out of season. In this context, locally processed products are complemented by products that are processed at longer distances. Consumers generate a direct environmental impact through the way they transport, store and prepare food, how much waste they generate, and how they dispose of it. Recent figures show that up to 20–30% of food is wasted in households, losing all resource inputs used in its production (CIAA, 2007).

Most of the abovementioned changes are directly linked to decreased environmental performance, as a result of the increased need for transportation. There is no doubt that the agri-food industry is a logistics–transport intensive industry. Transport impact is overwhelmingly in the areas of road congestion, damage to infrastructure and road accidents. There is also an impact on greenhouse gas emissions, air and noise pollution. According to a report by the Department for Environment, Food and Rural Affairs (DEFRA) (2005), the direct environmental, social and economic costs of food transport in the UK are over £9 billion each year, and are dominated by congestion. In the Organisation for Economic Cooperation and Development (OECD) area, food and drink manufacturing accounts for about 8% of industrial energy use, ranking fifth behind primary metals, chemicals, paper and pulp and non-metallic minerals (CIAA, 2007).

Another dimension in the sustainability agenda is the stark contrast between the needs of developed and developing countries and the subsequent challenges emanated for the food distribution system coping with those. Kohl (2001, p 1) elaborating on these issues notes: 'The global food market will be comprised of two major segments. The developed countries' marketplace is large and very important but slow growing. The marketplace of developing countries is smaller and faster growing, with enormous potential for food imports but limited income'. Kohl (2001, p 1) stresses that developing countries aim to provide 'basic nutrition in a low-cost efficient manner' whilst developed countries aim to provide social and self-actualisation aspects including 'convenience, entertainment value and bio-security'.

24.4.2   Implementing sustainable practices: driving forces and current status

Sustainability expectations in recent years are increasing; governments and policy makers develop strategies for sustainable development and establish frameworks for sustainable consumption and production and consumers also emphasise nowadays the moral and ethical and environmental values of the products they consume (Svensson, 2007). For example, the Food Industry Sustainable Strategy (FISS) which was presented by the UK Department for Environment, Food and Rural Affairs (DEFRA, 2006) aims to improve the food industry's environmental, social and economic performance by encouraging the widespread adoption of best sustainable practises by the industry. The goal in the case of food transportation is to reduce both social and environmental costs of domestic food transport by 20% by 2012.

In addition, there is no doubt that in the years to come, companies in the food industry will have to incorporate social and environmental objectives, in addition to economic ones in order to meet the growing sustainability expectations. The role of food retailers is expected to be critical in the successful implementation of sustainability practices, as they play a pivotal role by linking primary production and manufacturing to consumers (Fritz and Schiefer, 2008). Retailers are only one link in the food supply chain. However, they are now more than ever before, in the position to control and influence some of the companies, if not all of them in their supply chain (O'Keeffe and Fearne, 2002; Van der Grijp *et al.*, 2005). The latter situation is described by Spence and Bourlakis (2009). They examine Waitrose, a leading British grocery retailer, and the good practices followed by that retailer in dealing with suppliers. These practices include a plethora of issues including responsible sourcing from suppliers, risk assessment, supplier accreditation schemes and quality control, supplier ethical audit and managing the supplier–retailer partnership. Spence and Bourlakis (2009) also present an evolutionary framework arguing that food retailers could move from just managing and dealing with corporate social responsibility to implementing supply chain-wide responsibility, illustrating the increasingly influential role of retailers in the food chain.

All these issues in many cases force manufacturers to comply, or fall out of the supply chain. Wal-Mart for example, audited nearly 9000 of its suppliers' factories in 2006, in an effort to portray itself as an ethical and environmentally conscious company (Food Navigator, 2007a, 2007b). Often, food retailers attempt internal company sustainability improvements. Sainsbury's, the UK retailer, has established a new warehouse designed to reduce carbon emissions through devices such as air-tight freezers, energy-efficient lighting, wall-mounted photovoltaic panels, solar walls and a power plant that reuses heat generated by air conditioning.

In Table 24.2, an indicative and selective presentation of some of the very latest sustainability actions undertaken by different level supply chain actors is provided, based on sustainability objectives and dimensions. The practices mentioned have been published and advertised in various newsletters, company reports or corporate websites. Despite the existence of an increased number of companies following similar actions, in many

Table 24.2 Overview of uptake of sustainability practices (from Fritz and Matopoulos, 2008)

Company	Type	Action	Sustainability objective	Sustainability dimension
Ahold	Retailer	Jobs and schooling for local people in Ghana	Community	Social
Marks & Spencer	Retailer	Increased UK sourcing for several food products	Community and food transportation	Social–environmental
Tesco	Retailer	Change transport mode from truck to rail	Food transportation	Environmental
Wal-Mart	Retailer	Support free trade and cut tariffs initiatives	Ethical trading schemes	Social
		Measure energy used to create its products	Energy	Environmental
		Apply packaging scorecard to 800 suppliers	Waste	Environmental
Sara Lee	Manufacturer	Double procurement of sustainable coffee	Community–water–waste	Social–environmental
Fonterra	Manufacturer (Dairy)	Measure CO_2 emissions across the entire milk supply chain	Energy–food transportation	Environmental
Dole	Grower-trader	Brought electricity and drinking water to a Peruvian village	Community	Social
		Trained local farmers	Workplace improvements	Social

cases it is still not clear whether or not and the extent to which these actions are measurable, longitudinal and have a significant impact or whether they are related with gains related solely to the establishment and preservation of the good image of these companies. Another issue requiring further investigation is the extent to which the economic dimension of sustainability (e.g. productivity, training, diversity of the industry) is seriously taken into consideration. Most company releases seem to focus heavily primarily on the environmental and social perspectives of sustainability.

Supporting the above discussion, a report commissioned by the Chatham House (see Ambler-Edwards *et al.*, 2009) examined the future of the UK food supply system and pinpointed four characteristics that will be critical in a future food supply system: sustainability, resilience, competitiveness and managing consumer expectations. The definitions of these four characteristics are given below (Ambler-Edwards *et al.*, 2009, p 26):

- *resilience* – a system able to assure longer term availability in the light of increasing global uncertainties;
- *sustainability* – a system that can supply safe, healthy food with positive social benefits and low environmental impacts;
- *competitiveness* – a system capable of delivering affordable food around a potentially higher baseline of costs;
- *managing consumer expectations* – a system which shapes and responds to consumer preferences in line with societal needs.

In that report, the three pillars of sustainability were acknowledged (economic prosperity, social development and environmental protection). It was also stressed that the issue of sustainability should not be confined to agriculture only but it should be examined as a system-wide problem; we expect that the supply chain can be perceived as that system. Finally, in that work, new and emerging food supply chain requirements were illustrated (see Table 24.3). These were based on the requirements of the four aforementioned characteristics (resilience, sustainability, competitiveness, managing consumer expectations) and have an impact on a range of system-wide/ supply chain issues.

24.5   Conclusions

This chapter has introduced two key trends (information technology and sustainability) that have an influential role on current and modern food supply chains. We envisage that this chapter will be useful reading for, inter alia, academic colleagues, researchers, policy makers, managers and practitioners dealing with the food supply chain system and its management.

Table 24.3 New and emerging supply chain requirements (from Ambler-Edwards *et al.*, 2009)

	Old models	Emerging models
Farming system	• High input systems • Established practices	• Low input/high output optimised systems • High levels of experimentation • Waste reuse
	• Subsidies without condition, price support	• Subsidies dependent on sustainable production practices • Support structures for investment, knowledge transfer and technology access (rather than direct support)
	• High levels of inefficiencies pre-2005 • Mid-term reform of CAP begins process of consolidation of farming businesses. Ownership and production start to separate	• Competitiveness through horizontal collaborative models • Increase in farm scale along with separation of ownership and production
	• Focus on volume production pre-2005. Mid-term review begins process of aligning production to market requirement	• Optimisation of resources to align with sustainable goals • Minimisation of losses in whole system through horizontal production networks and vertical supply chain efficiency
Supply processes	• Organisational risk management	• System-based risk management • Crisis management for whole chain
	• Individual measures based on quality, cost and delivery performance	• Shared measurement systems based on cost-competitiveness. Compliance with public requirements for resilience and sustainability
	• Process efficiency • Management of product flows	• Resource efficiency • Integration and management of waste streams with product flows

Products	• Wide product choice	• Product rationalisation and editing of choice, based on higher standards • Use of substitutes and alternative ingredients
Assets and structures	• Large, capital-intensive assets • Investment decisions based on production cost per unit • Efficient distribution systems	• More flexible use of assets • Increased investment in smaller-scale assets • New assets related to waste reuse lead to more horizontal collaboration, particularly in producer networks • Investment decisions based on total cost of ownership (inc. environmental costs) • National models developed alongside regional sourcing overlaid on efficient distribution models. Inefficient local models replaced by local solutions integrated with existing efficient distribution models
Relationships	• Vertical transactional relationships. Limited horizontal collaboration • Linear relationships • Sector-based engagement	• Greater horizontal collaborative relationships • Longer term supply contracts where power is equilibrium • Partnerships with other sectors/industries • Interlinkage of whole chain, from farm through to consumer • Engagement with all stakeholders
Strategies	• Price vs product differentiation	• Additional competition based on low environmental impacts due to high public incentives and costs of non-sustainable supply

24.6   Sources of further information and advice

Key books
Bourlakis, M. and Weightman, P. (eds) (2004). *Food Supply Chain Management*, Blackwell Publications, Oxford.
Eastham J.F., Ball S.D. and Sharples A.E. (eds) *Food and Drink Supply Chain Management for the Hospitality and Retail Sectors*, Butterworth-Heinemann, UK.
Gustafsson, K., Jonson, G., Smith, D. and Sparks, L. (2006). *Retailing Logistics and Fresh Food Packaging: Managing Change in the Supply Chain*, Kogan Page, London.

Research networks
The Auto-ID Labs are the leading global network of academic research laboratories in the field of networked RFID. The labs comprise seven of the world's most renowned research universities located on four different continents. These institutions were chosen by the Auto-ID Center to architect the Internet of Things together with EPC global. http://www.autoidlabs.org

Websites
http://www.sustainweb.org/ An alliance for better food and farming that advocates food and agriculture policies and practices that enhance the health and welfare of people and animals, improve the working and living environment, enrich society and culture and promote equity. They represent around 100 UK public interest organisations working at international, national, regional and local level.

http://www.greenlogisticsforum.com/index.shtml The Green Logistics Forum is dedicated to turning environmental business goals into real collaborative achievements by providing a forum in which all areas of the transportation and logistics industry can come together to discuss the issues and make firm plans to green their logistics in a informed and strategic way.

http://www.wrap.org.uk/retail/ The Retail Innovation Programme was established as a result of research undertaken by WRAP on behalf of the Strategy Unit, which found that as much as 50% of household waste, which ultimately ends up in landfill, has originated from a purchase from the top five retail supermarket chains. These supermarkets link massive supply chains with households' behaviour and are therefore well placed to influence change.

24.7   References

AMBLER-EDWARDS, S., BAILEY, K., KIFF, A., LANG, T., LEE, R., MARSDEN, T., SIMONS, D. and TIBBS, H. (2009). *Food Futures: Rethinking UK Strategy*, A Chatham House Report, Royal Institute of International Affairs, London.

BLYTHMAN, J. (2005). *Shopped: The Shocking Power of British Supermarkets*, Harper Perennial, London.

BOURLAKIS, M. (2001). 'Future issues in supply chain management', in Eastham, J.F., Ball, S.D. and Sharples, A.E. (eds) *Food and Drink Supply Chain Management for the Hospitality and Retail Sectors*, Butterworth-Heinemann, Oxford, 297–303.

BOURLAKIS, C. and BOURLAKIS, M. (2004). 'The future of supply chain management', in *Food Supply Chain Management*, Bourlakis, M. and Weightman, P. (eds) Blackwell, Oxford, 221–31.

BOURLAKIS, C. and BOURLAKIS, M. (2005). 'Information technology safeguards, logistics asset specificity and 4th party logistics network creation in the food retail chain', *Journal of Business and Industrial Marketing*, **20**(2/3), 88–98.

BOURLAKIS, M. and BOURLAKIS, C. (2006). 'Integrating logistics and information technology strategies for sustainable competitive advantage', *Journal of Enterprise Information Management*, **19**(2), 389–402.

BOURLAKIS, M. and WEIGHTMAN, P. (eds) (2004). *Food Supply Chain Management*, Blackwell Publications, Oxford.

BOYER, K.K. and HULT, T.G. (2005). 'Extending the supply chain: Integrating operations and marketing in the on-line grocery industry', *Journal of Operations Management*, **23**(6), 642–61.

BOYER, K.K., FROHLICH, M.T. and HULT, T.G. (2004). *Extending the Supply Chain: How Cutting Edge Companies Bridge the Critical Last Mile into Customers' Homes*, American Management Association, New York.

BOYLE, M. (2003). 'Wal-Mart keeps the change', *Fortune*, November 10, p. 46.

CHRISTOPHER, M. (2005). *Logistics and Supply Chain Management: Creating Value-Adding Networks*, 3rd edition, Pearson Education, London.

CIAA, CONFEDERATION OF FOOD AND DRINK INDUSTRIES OF THE EU (2007). *Managing Environmental Responsibilities in European Food and Drink Industries*, Report, October 2007.

COTTERILL, R.W. (2001). 'Dynamic explanations of industry structure and performance', *British Food Journal*, **103**(10), 679–714.

DEFRA, (2005). *The Validity of Food Miles as an Indicator of Sustainable Development*, Final report, by Paul Watkiss, July 2005.

DEFRA, (2006). *Food Industry Sustainable Strategy Report* http://www.defra.gov.uk/farm/policy/sustain/fiss/pdf/fiss2006.pdf (last access 20/11/08).

E-BUSINESS WATCH (2006). *ICT and e-business in the Food, Beverage and Tobacco Industry*, Sector report.

FEARNE, A., HUGHES, D. and DUFFY, R. (2001). 'Concepts of collaboration-supply chain management in a global food industry', in *Food and Drink Supply Chain Management Issues for the Hospitality and Retail Sectors*, Eastham, J.F., Sharples, L. and Ball, S.D. (eds), Butterworth-Heinemann, Oxford, 55–89.

FINEGAN, N. (2002). *Backhauling and Factory Gate Pricing*, Institute of Grocery Distribution, UK.

FOOD NAVIGATOR (2007a). *Wal-Mart beefs up audits of supplier factories*, by Ahmed El Amin, Available at: http://www.foodproductiondaily.com/Processing/Wal-Mart-beefs-up-audits-of-supplier-factories (last accessed 10/9/09).

FOOD NAVIGATOR (2007b). *Wal-Mart Canada unveils supply chain scorecard*, by George Reynolds, Available at: http://www.foodproductiondaily.com/Supply-Chain/Wal-Mart-Canada-unveils-supply-chain-scorecard (last accessed 10/9/09).

FOOD QUALITY (2007). *RFID tag a cold chain management tool*, By Ahmed ElAmin, Available at: http://www.foodqualitynews.com/Innovation/RFID-tag-a-cold-chain-management-tool (08/11/07).

FRITZ, M. and MATOPOULOS, A. (2008). 'Sustainability in the agri-food industry: a literature review and overview of current trends', in *Proceedings of the 8th*

International Conference on Chain and Network Management in Agribusiness and the Food Industry, 26–28 May 2008, Ede, The Netherlands.

FRITZ, M. and SCHIEFER, G. (2008). 'Food chain management for sustainable food system development: a European research agenda', *Agribusiness*, **24**(4), 440–52.

FROHLICH, M.T. and WESTBROOK, R. (2001). 'Arcs of integration: an international study of supply chain strategies', *Journal of Operations Management*, **19**, 185–200.

GAUKLER, G.M. AND SEIFERT, R.W. (2007). 'Applications of RFID in supply chains', in *Trends in Supply Chain Design and Management: Technologies and Methodologies*, Jung, H. Chen, F. and Jeong, B. (eds), Springer, London.

GMA, (2000). *Food manufacturers take first step toward real B2B e-commerce for grocery industry*, accessed 5 July 2005, http://www.gmabrands.com/news/docs/NewsRelease.cfm?DocID=615

HARRISON, A. and VAN HOEK, R. (2008). *Logistics Management and Strategy*, 3rd edition, Pearson Education, Harlow, 238.

HUGHES, D. (1994). *Breaking with Tradition: Building Partnerships and Alliances in the European Food Industry*, Wye College Press, Wye.

JONES, P., CLARKE-HILL, C., SHEARS, P., COMFORT, D. and HILLIER, D. (2004). 'Radio frequency identification in the UK: opportunities and challenges', *International Journal of Retail and Distribution Management*, **32**, 164–71.

KÄRKKÄINEN, M. (2003). Increasing efficiency in the supply chain for short shelf life goods using RFID tagging, *International Journal of Retail and Distribution Management*, **31**(10), 529–36.

KAUFMAN, P. (1999). Food retailing consolidation: Implications for supply chain management practices, *Journal of Food Distribution Research*, **30**(1), 6–11.

KOHL, D.M. (2001). Mega trends in agriculture: Implications for the food distribution system, *Journal of Food Distribution Research*, March, 1–4.

MATOPOULOS, A., MANTHOU, V. A. and VLACHOPOULOU, M. (2007). 'Integrating supply chain operations in the internet era', *International Journal of Logistics Systems and Management*, **3**(3), 305–14.

MANTHOU, V., MATOPOULOS, A. and VLACHOPOULOU, M. (2005). 'Internet-based applications in Agri-Food logistics: a survey on the Greek canning sector', *Journal of Food Engineering: Special Issue on Operational Research and Food Logistics*, **70**(3), 447–54.

MARKETLINE INTERNATIONAL, (1997). *EU Logistics*, Market International, London.

MCFARLANE, D. and SHEFFI, Y. (2003). 'The impact of automatic identification on supply chain operations'. *International Journal of Logistics Management*, **14**(1), 1–17.

MICHAEL, K., and MCCATHIE, L. (2005). 'The pros and cons of RFID in supply chain management', *Proceedings of the International Conference on Mobile Business*, 11–13 July 2005, IEEE, 623–29.

O'KEEFFE, M. and FEARNE, A. (2002). 'From commodity marketing to category management: insights from the Waitrose category leadership program in fresh produce', *Supply Chain Management: An International Journal*, **7**(5), 296–301.

PRATER, E., FRAZIER, G.V. and REYES, P.M. (2005). 'Future impacts of RFID on e-supply chains in grocery retailing', *Supply Chain Management: An International Journal*, **10**(2), 134–42.

RAMSEY, B. (ED.) (2000). *The Global Food Industry: Strategic Directions*, Financial Times Retail and Consumer Publishing, London.

SMITH, D. and SPARKS, L. (2004). 'Temperature-controlled supply chains', in *Food Supply Chain Management*, Bourlakis, M. and Weightman, P. (eds), Blackwell, Oxford, 179–98.

SPARKS, L., GUSTAFSSON, K., JONSON, G. and SMITH, D. (2006). *Retailing Logistics and Fresh Food Packaging: Managing Change in the Supply Chain*, Kogan Page.

SPENCE, L. and BOURLAKIS, M. (2009). 'The evolution from corporate social responsibility to supply chain responsibility: The case of waitrose', *Supply Chain Management: An International Journal*, **14**(4), 291–302.

STRAK, J. and MORGAN, W. (1995). *The UK Food and Drink Industry*, Euro PA & Associates, UK.

SVENSSON, G. (2007). 'Aspects of sustainable supply chain management', *Supply Chain Management: An International Journal*, **12**(4), 262–6.

VAN DER GRIJP, N.M., MARSDEN, T. and CAVALCANTI, J.S.B. (2005). European retailers as agents of change towards sustainability: The case of fruit production in Brazil, *Environmental Sciences*, **2**(4), 445–60.

Index